AF777291

SAFETY AND RISK MANAGEMENT TOOLS AND TECHNIQUES IN THE CPI

Edited by
G. Sam Samdani, Ph.D.
and
The Staff of Chemical Engineering

McGraw-Hill
New York San Francisco Washington, D.C. Auckland Bogota
Caracas Lisbon London Madrid Mexico City Milan
Montreal New Delhi San Juan Singapore
Sydney Tokyo Toronto

Copyright © 1996 by Chemical Engineering McGraw-Hill Pub. Co.
1221 Avenue of the Americas, New York, New York 10020

All rights reserved. No parts of this work may be reproduced or utilized in any form or by any means, electronic or mechanical, including photocopying, microfilm and recording, or by any information storage and retrieval system without permission in writing from the publisher.

Printed in the United States of America.

Library of Congress Cataloging-in-Publication Data

Safety and risk management tools and techniques in the CPI / edited by
G. Sam Samdani and the staff of Chemical engineering.
p. cm.
Includes index.
ISBN 0-07-057768-4
1. Chemical plants--Safety measures. I. Samdani. G. Sam.
II. Chemical engineering (Albany, N.Y. : 1946)
TP 155.5.S193 1996
660'.2804--dc20 92-11101
CIP

CONTENTS

Introduction

Safety and risk management are two sides of the same coin. While safe practices can lower risks, they cannot eliminate the hazards altogether. Hence the need for the management of risks through reliability engineering, for example. To this end, government regulations often provide guidelines as to what must be done to ensure safety in the workplace and minimize the impact of potential mishaps resulting from "human error" or equipment failure.

The articles presented in this volume have been selected from the recent issues of *Chemical Engineering* for their timeliness. Section I sets the regulatory scene for the reader. The subsequent sections offer the tools and techniques for implementing safety and risk management in various segments of the chemical process industries (CPI). For example, Section II on the do's and don'ts in safety is followed by another section on modern methods for quantifying risks, and for estimating the reliability of safety systems.

Safe handling of toxic materials in the plant is the focus of Section IV. The "best practices" as well as regulatory requirements provide the reader with essential information for dealing with hazardous fluids.

Tips on accident prevention and safe operation should arm the reader with the basic knowledge and expertise for coping with the inherent hazards in an operating process plant. High temperatures and pressures are often the "triggers" for mishaps. The articles offer techniques on how to guard against such mishaps.

Section VII is a primer on how to protect the plant and personnel against crises and liabilities. The ideas should help practicing engineers ask the right questions when they discuss the available options with a lawyer or an insurance agent.

The final chapter is devoted to the various testing methods and diagnostic tools for detecting faults or deficiencies in process equipment. These tools are becoming highly sophisticated and require the skills of trained professionals for their proper use and interpretation of the results. The articles are meant to provide the reader with enough background to seek the right expert for a given problem.

We hope the articles will provide the reader with the perspective and knowledge to lead the efforts in safety and risk management during the design of a new facility or the operation of an existing plant.

Best of luck.

G. Sam Samdani, Ph.D.

Section I

The Regulatory Scene

RISK & the CPI

Bhopal and Flixborough and accidents like them have left an indelible imprint on the public, making risk management an imperative for the chemical process industries (CPI). It was just such disasters that prompted CPI leaders to set up their own risk management programs in the 1970s and 1980s. Today, all agree that risk assessment is a valuable way to improve safety and efficiency. However, it's a new science that's still evolving, and techniques still need refining.

One could argue that the CPI are already managing risk well. Although deaths have occurred within plants, no one has been killed outside a U.S. plant during an accident in the past 50 years, says Isadore (Irv) Rosenthal, senior research fellow at the Wharton School's Risk Management and Decision Processes Center (Philadelphia).

Nevertheless, risk minimization is a moving target. Adding urgency is a lack of understanding of risk by the public, the media and government regulators, who frequently lump together plant safety and toxicology when the subject of risk comes up. The result, CPI firms complain, is regulations that inflate potential risk and alarm the public.

For example, a Risk Management Program (Box, p. 4) proposed by the U.S. Dept of Environmental Protection (EPA; Washington, D.C.) mandated "worst-case scenarios" to formulate strategies. The worst case was defined as instantaneous loss of containment of all hazardous chemicals within a facility—a situation that engineers agree is impossible in most cases

Further, lack of consensus in the U.S. is reflected in ambivalence toward risk management, particularly among smaller CPI companies. Compliance with U.S. regulations such as the Occupational Safety and Health Admin.'s (OSHA; Washington, D.C.) Process Safety Management rule and the amended Clean Air Act ensures a level of accident prevention, but this is a "lowest-common denominator" approach, says Scott Stricoff, vice-presi-

Risk analysis aims to reduce the likelihood of accidents like this polymerization reactor vessel failure, which killed four. Opinion is divided on the best way to do this. Recent U.S. state regulations call for repeated risk analysis at hazardous plants (table, below). To date, only the Netherlands actually requires use of quantitative techniques (bottom table)

STATES GET TOUGH ON RISK ASSESSMENT
Sample penalties from New Jersey's toxic catastrophe prevention act
(Figures in U.S. dollars)

Offense	First	Second	Third
Starting up a new extraordinarily hazardous substance (EHS) facility without an approved risk management program (RMP)	10,000	20,000	50,000
Failure to complete requirements of an established RMP and perform a safety review and hazard analysis before modifying an existing EHS facility (per requirement)	2,500	5,000	12,500
Failure of new owner or operator of existing EHS facility to adopt or obtain an approved RMP before operating EHS equipment	5,000	10,000	25,000
Failure of owner or operator to submit summary RMP statement at least 90 days before building a newEHS facility	5,000	10,000	25,000

Source: N.J. Dept. of Environmental Protection

SOCIETAL RISK* ASSESSMENT METHODS USED TO OBTAIN OPERATING LICENSES FOR LAND BASED NON-NUCLEAR PLANTS

Country	Classification of installation	Hazard identification	Partial quantification	Probabilistic assessment	Quantitative objectives
Belgium	Yes	Mandatory	Sometimes	No	No
Canada (Nat'l level)	No	No	Rare	Some	No
Denmark	Yes	Mandatory	Rare	Some	In some cases
Finland	Yes	Mandatory	Sometimes	Some	In some cases
France	Yes	Mandatory	Frequent	Rare	Occasionally
Germany	Yes	Mandatory	Sometimes	No	Occasionally
Greece	Yes	Mandatory	-	No	No
Italy	Yes	Mandatory	-	No	No
Japan	Yes	Mandatory	Rare	Rare	No
Luxembourg	Yes	Mandatory	-	No	No
Netherlands	Yes	Mandatory	Frequent	Yes (by law)	Yes (by law)
New Zealand	Yes	-	-	Some	No
Norway	Yes	Mandatory	Sometimes	Some	In some cases
Sweden	Yes	Mandatory	Sometimes	Some	In some cases
U.K.	Yes	Mandatory	Frequent	Some	In some cases
U.S. (Federal level)	No	Yes	Yes	Some	No

*Societal risk, as defined by the Institution of Chemical Engineers (U.K.), is the relationship between frequency and the number of people suffering from a specified level of harm in a given population from the realization of specified hazards.

Source: U.K. Health & Safety Executive

Originally published February 1995

MANAGING OR IMAGINING RISKS?

EPA's proposed regulation, Risk Management Programs for Chemical Accidental Release Prevention (40CFR Part 68), will require facilities to address potential chemical hazards. Its content is not yet set, as the final rule is not due out until March 29, 1996. However, companies will be required to assess potentially hazardous situations, provide emergency response for potential accidents, and keep citizens informed of both.

According to EPA's Lyse Helsing, the rule, as originally proposed by EPA, received "tons" of comments, and the Agency is now formulating a supplemental proposal due on February 28. One thorny issue was the definition of the "worst case scenario." The original definition was seen by many as physically impossible and unnecessarily alarming. "It's as if a liquefied-petroleum-gas sphere were to become unzipped all at once," explains Raymond French, Engineering associate at Exxon Chemical Co. (Baytown, Tex.). "There's never been such a case,"

And the very definition of "worst cases" is in question. "These scenarios are limited only by the imagination of the person doing them," says Raymond Freeman, senior fellow at Monsanto Co. (St. Louis, Mo.). Compromises are being worked out, among them, "reasonable worst cases." However, definitions have not been settled. "EPA wants a simple definition for reasonable worst cases, and it's bound to be conservative," says Paul Baybutt, president of Primatech, Inc. (Columbus, Ohio).

Current discussions also address companies' compliance efforts with related regulations, such as OSHA's Process Safety Management rule. "If a source is already meeting these requirements, some commenters feel it should not have to duplicate its effort under RMP," says Helsing. Funding is yet another issue. So far, EPA is planning to raise funds for the program locally. The "Unfunded Mandates" bill, however, could have an impact in the future.

After the rule becomes final, companies will have three years to meet requirements, which include:

- Estimating the potential offsite consequences of accidental releases to public health and the environment
- Compiling a five-year accident history

The relationship of worst-case scenarios to hazard assessments is also addressed in the supplemental notice. Although the proposed rule steers clear of dictating any absolute measures, such as quantitative risk assessment, EPA is working with businesses to develop generic models for non-CPI businesses, e.g., ammonia refrigeration plants, that use quantities of hazardous chemicals above the EPA's threshold level. ❑

dent for safety and environmental health at Arthur D. Little, Inc. (ADL: Cambridge). Beyond compliance, he says, risk-assessment activities are voluntary, and implemented unevenly across the board.

There is also general confusion about different types of risk assessment. Several years ago, legislators of one state, in drafting an accident-prevention law, counted the toxicologists available to conduct risk assessment, rather than the number of safety engineers, relates Raymond (Randy) Freeman, senior fellow at Monsanto Co. (St. Louis, Mo.). As it turned out, toxicologists vastly outnumbered engineers, and the regulators had to make plenty of revisions.

Europe has concrete standards for risk acceptability. In the U.K. and the ity. The technique is long and complex, but has dictated controls less stringent than New Jersey's. Nevertheless, says Rosenthal, "U.S. industry would scream if EPA used these thresholds."

As the CPI and regulators struggle to optimize definitions of risk, controversy is growing over the need for QRA. Required in the Netherlands, and widely used throughout Europe, QRA is slowly gaining ground in the U.S. and Japan, to allow companies to do "self checks."

Engineers in the U.S. use QRA only as the last resort, and German regulators are wary of the technique. "QRA is no panacea," points out Dennis Hendershot, a technical fellow with Rohm and Haas Co.'s (Bristol, Pa.) engineering division and head of the American Institute for Chemical Engineers

Risk assessment is a science that's still evolving, but it's critical for the CPI

Netherlands, says Rosenthal, plants cannot be built if their potential safety and toxicological risk exceeds 1 death out of 100,000 exposures. In the U.S., however, the concept of an "acceptable number of deaths" remains anathema.

This made for some challenges in drafting New Jersey's Toxic Catastrophe Prevention Act program, says Reginald Baldini, acting chief of the Bureau of Release Prevention in the Dept. of Environmental Protection (Trenton). In writing the law, which requires hazardous chemical plants to conduct periodic risk assessment (Table, p. 26), New Jersey regulators settled on a 10^{-4} frequency cutoff point for toxic hazards. If a toxic cloud, which they defined as the concentration in air of a chemical that will kill one in 20 people, has a 1-in-10,000 chance of reaching plant property limits, controls are required.

In contrast, the Netherlands' acceptable-risk numbers are reached using quantitative risk assessment (QRA) techniques to extrapolate the toxic cloud's risk of actually reaching a populated area. QRA requires factoring in the likelihood of wind carrying the toxic vapor that distance, which may be 2-3% of the time, says Baldini, and depends on wind and atmospheric stabil-

(AIChE; New York, N.Y.) Center for Chemical Process Safety (CCPS). He and others stress that quantitative techniques should be used only when qualitative methods fail to determine whether or not a risk is hazardous.

The crucial parts of a quantitative assessment come before and after the actual analysis — i.e., making the correct initial assumptions and then interpreting results. An assumption for one case may not be appropriate for another, says Jim Wilson, regulatory issues director at Monsanto. In one 1988 study, for example, 11 teams used QRA on a small ammonia plant. Different teams' results for one hazard varied from 1 in 400 to 1 in 10-million.

Another problem is that the models drastically simplify what happens in nature. Last June, eight CPI firms participated in a worst-case-scenario exercise in West Virginia's Kanawha Valley. Based on the same set of data, companies came up with different toxic-cloud footprints, depending on the program they used, the weather and temperature, says Lester Wittenberg manager of the CCPS. Modelling a chlorine release posed problems, as Monsanto's Freeman points out. "There are plenty of models for chlorine, which forms a

This 1993 oleum spill released a plume, causing thousands to seek medical treatment. General Chemical paid $1.18 million in damages, while the incident helped spur EPA to include oleum in the final list of chemicals subject to accidental release prevention requirements (59 FR 4478)

Primatech, Inc

SAME THEORY, DIFFERENT PERSPECTIVE

Though related to process-safety risk assessment, health and environmental risk assessments have a different emphasis, and primarily involve toxicologists and epidemiologists. "Chronic health assessment predicts the effects of some chemical on people in the environment," explains Monsanto's Freeman. "You assume that everyone is affected, and exposed for 70 years."

Unlike safety analysis, which focuses on probability, environmental and health risk assessments focus on the consequence of risk. Thus, threshold emission limits for U.S. permits are based on the effects of exposure on someone living on the plant's fence line for 70 years, says Colin Park, a research scientist with Dow Chemical Co. (Midland, Mich.).

Industry groups such as the Chemical Manufacturers' Assn. (CMA; Washington, D.C.) are lobbying for more realistic approaches to health and environmental risks, to make toxicological limits more reasonable. For example, says Park, "Instead of assuming that someone lives on the fence line, you could assume he or she lives on the closest block." Other refinements could also be made, he says, since "not everyone weighs 70 kg and breathes exactly a certain amount of air per day." One improvement he mentions is Monte Carlo analysis, which considers variable, rather than fixed, exposures.

As far as health and environmental risk assessment is concerned, "We've got a lot of learning yet to do," says Jim Wilson, regulatory issues director at Monsanto Co. (St. Louis, Mo.). For example, how does one set an exposure threshold for a substance that isn't clearly hazardous? "Consider the Ames test," says Daniel Crowl, professor of chemical engineering at Michigan Technological University (Houghton). "You expose cells to a substance; and if the cells mutate, you sometimes consider the substance a mutagen." Rosenthal points out that possible carcinogens' risks are often listed as "zero to x," indicating that the substance may not even be carcinogenic. Another major problem, says Wilson, is understanding when animal responses are predictive of human responses. ❐

dense cloud, but they're all for flat terrain. How do you model a dense cloud in a valley flanked by 1,500-ft cliffs?"

Another drawback to QRA is the need for accident and equipment failure data, which become scarcer, the safer plants become. Nevertheless, trends can be seen. One common cause of failure is "correlated failure," in which backing up one piece of equipment is assumed to increase safety, says Joe Fragola, vice president of advanced technology for Science Applications Inc. (New York, N.Y.). In an example of "external" correlated failure, an explosion would disable two generators located next to each other. An example of internal correlated failure would be when environmental factors damage the Teflon seals in two pumps of the same type, and a pressure surge takes them both out. Human error is also becoming a more-prominent factor in failures, says Fragola, as the trend to automated equipment continues.

Despite the drawbacks of QRA, when competently performed, it generally provides accuracy to within a factor of ten, according to Michael Hogh, head of safety, fire and risk management for BP International Ltd. (Sunbury-on-Thames, U.K.). In the U.K. and in Europe outside of Germany, both the CPI and regulators find QRA "a good starting point for discussions," says Hogh.

In Japan, most large CPI companies use statistical techniques to analyze risk, says Satoru Sayama, a professor of engineering at Okayama University (Okayama). Idemitsu Kosan (Tokyo), he says, is now using a method developed by ICI plc (London), while 75% of the 85 companies recently surveyed by Japan's High Pressure Gas Safety Institute use QRA, says Takeo Kaneko, managing director of the safety consultancy, Technostaff (Yokohama).

Successful risk management requires prioritizing. "You don't want to spend time on something that contributes 1% of the total risk, when you could have eliminated something that contributes 50 or 60%," explains Rohm and Haas' Hendershot.

Early spadework is critical, says ADL's Stricoff. Success requires a thorough understanding of chemical plants, processes and design, adds Hendershot. Not surprisingly, cross functional teams are the rule. Nippon Petrochemical Co., for example, teams up operators, maintenance staff, safety experts, design engineers and retired engineers.

Frequently, findings are surprising. Hendershot recalls one installation, in which chlorine was to have been piped in from shipping containers. Engineers initially thought the risk centered around the chlorine in the trucks, but it turned out that the transfer pipeline was the biggest factor.

In general, the more frequently a plant's risk is analyzed, the better. "The analysis must be a living document, not something you do once and put away," says Hendershot. OSHA's PSM rules already call for periodic assessments, as do rules in New Jersey, Nevada, California and Delaware.

Japan, Europe, and the U.S. differ in the degree to which risk is communicated to the public. Recently, for example, Japan's Rokkasho Municipal Assembly rejected a proposal that residents be allowed to vote on whether to allow a radioactive waste storage facility in their village. This would be unthinkable in the U.S. or Europe.

Yet, despite differences in attitudes toward risk communication and quantitative methods, risk analysis is becoming more important globally. "There's going to be lot more use of risk analysis, because people recognize it as a very powerful tool that can help them spend their money more intelligently," says ADL's Stricoff. Engineers always see room for improvement. "The only acceptable risk is really zero," says Rohm and Haas' Hendershot. ■

Irene Kim with Stephen Moore and Gerald Ondrey

PIERRE FORTIN

Today, environmental auditing is both a technical exercise and a legal art. To perform a comprehensive and effective environmental audit requires a team of individuals who combined have broad engineering skills, a knowledge of chemistry, and specialized legal experience with all the relevant environmental regulations.

Because the audit involves such highly qualified individuals, it is typically quite expensive. Nonetheless, the cost is well worth the investment, because of the high stakes involved in environmental legislation and litigation. Simply put, the cost of non-compliance can far outweigh the costs associated with a comprehensive audit.

PERFORMING ENVI-

An Engineer's Guide

Janis A. Morelli
Cohen, Shapiro, Polisher, Shiekman and Cohen

A carefully designed audit can pinpoint and eliminate potential environmental violations

For example, the laws enacted over the past two decades to control the handling, storage and disposal of hazardous waste have today turned into a complex maze. Any failure to correctly negotiate through this regulatory maze can result not only in expensive civil and criminal penalties for a corporation, but can lead to imprisonment of company personnel.

Before assembling an audit team or beginning an audit itself, it is imperative that the project have the unconditional support of corporate management. Without such support, an environmental audit is an effort in futility, since there may be no viable commitment to approve capital expenditures to correct violations

Further, if violations are not corrected, the audit report can create more technical and legal problems than it solves. For example, the report can be used against a plant's managers as evidence of knowledge of potential problems in courts of law.

Therefore, it is unconscionable to operate a plant today without paying careful attention to the laws and making sure that they are complied with. And, the only way to do so — that is, to establish environmental compliance and avoid undesirable legal consequences — is by auditing the facility

Originally published February 1994

RONMENTAL AUDITS

Starting the audit

Basically, an environmental audit consists of three steps: data gathering, document review and physical inspection of a site. At first, gathering the data may appear difficult because it is a typical chicken-and-egg story. For example, novice auditors may not know what regulations a plant is subject to if they do not know which materials are being processed; at the same time, the team may not be aware of which materials are regulated unless they also know which regulations to apply.

Next, the information is compared with federal, state and local regulations, to ensure that all the requisite standards are being met. The results of the comparison enable company management to determine whether their plants are being run in a compliant manner. It also isolates non-compliance issues that need to be addressed, and outstanding violations that have to be immediately corrected.

Depending on the overall goal, an environmental audit can take on a variety of forms. For example, an *internal compliance audit* is carried out if the objective is to ensure that a particular plant meets all the environmental regulations that it is subject to. Or if the goal is to purchase a plant or property, a pre-acquisition "due diligence" audit is performed. In addition, treatment, storage and disposal (TSD) facilities, the ultimate destination points of hazardous wastes, require very specific and intensive audits.

In fact, an internal-compliance audit of a plant is incomplete without also auditing the TSD facility that receives the plant's hazardous wastes. This is because the U.S. Resource Conservation and Recovery Act (RCRA) establishes "cradle to grave" regulations for the storage, handling, transportation and disposal of hazardous waste.

As a result, a plant's responsibilities for hazardous wastes do not end when the materials leave the site. Therefore, a plant's own internal-compliance audit must also encompass that of the TSD facility to ensure that it is meeting its environmental obligations.

Besides RCRA, the Comprehensive Environmental Response, Compensation and Liability Act (Cercla), also known as Superfund, imposes liability upon those who improperly dispose of hazardous materials. Superfund liability is extremely encompassing, being shared not only among the owners and operators of a disposal site, but also by the generators and transporters of the wastes. Further, Superfund liability is extremely costly, the tab typically ranging into several million dollars for each responsible party.

What this means is that hazardous materials processed at a particular facility and improperly disposed of away from the site can impose cleanup obligations upon the facility's corporate owner, as well as the individuals involved. Therefore, an internal-compliance audit of a plant addresses only half the requirements and a careful review is needed of both the waste transporter and the TSD facility.

Also, if sampling of process streams is required, some technical and legal issues must be considered. The taking and testing of samples are strictly describes by federal and state regulations, and failure to comply with these procedures can invalidate the analyses. If the sampling is required by the government, then "chain of custody" paperwork must provide a trail of documents from the sampling location to the final analysis. This is needed to prove that the reported results are indeed those indentified by the company.

The ideal team

Ideally, the audit team should include individuals from within a corporation and some who are independent of the corporation. This is critical because most company employees may not be objective, especially when they have to criticize the actions of coworkers. At the same time, inhouse expertise is essential to fully comprehend the technical processes involved.

An audit team should have at least three members: an environmental attorney, an environmental consultant, and an environmental, safety and health expert. The attorney and consultant are usually recruited from outside the company. In general, it is advisable to have members with multidisciplinary backgrounds.

It should be emphasized that the team members are not the only individuals involved in the auditing process. While visiting each department within a facility, the audit team should be joined by the supervisor (and other necessary personnel) of each particular area. The supervisor should review the process and the materials involved, and explain the various process operations. Additionally, the team should interview a plant's emergency coordinators, since they are familiar with a facility's safety and environmental aspects.

Team members should have the following background: legal knowledge of the applicable regulations, engineering knowledge of the processes, chemical knowledge of the materials involved, and experience in occupational health and safety. In fact, an environmental audit provides the perfect opportunity to review occupational health, safety and industrial hygiene issues facing a plant. That's because federal and state environmental regulators have of late begun to work more closely with their counterparts in the occupational, safety and and health departments.

Thus, if a facility was recently inspected by a federal or state environmental agency, and if the inspector had questions related to safety and health, the inspector would probably alert safety regulators of potential violations. The reverse also occurs, as when safety inspectors review a facility and observe environmental hazards.

A standard way

In May, 1993, the American Soc. for Testing and Materials (ASTM) published guidelines to define good commercial practices for conducting an environmental site assessment. The guidelines are the *Standard Practice for Environmental Site Assessments: Phase I Environmental Site Assessment Process E 1527-93* and *Standard Practice for Environmental Site Assessments: Transaction Screen Process E 1528-93*.

Ostensibly, the ASTM standard defines the appropriate level of inquiry, or "due diligence," necessary to support an "innocent landowner defense" under Cercla. The standard determines the range of contaminants under Cercla. However, the ASTM standard does not cover materials not regulated by Cercla, such as asbestos, radon and lead, which, as major environmental problems by themselves, should be part of the audit.

Nonetheless, the ASTM standards provide a good starting position for a novice auditor. However, more experienced auditors are aware that additional items should be addressed to make the audit comprehensive.

The ASTM standards include Appendices that explain the legal background to federal law and the practices of environmental assessments in commercial real-estate transactions. They explain the elements of liability under Cercla, and list the various defenses to Superfund liability.

The key issue regarding site assessment in real-estate transactions is deciding the appropriate level of inquiry regarding the previous ownership and uses of the property. The ASTM appendices refer to a number of cases that discuss such due diligence inquiries that are deemed adequate for an innocent-purchaser defense. As such, the ASTM standards provide a due diligence benchmark for pre-acquisition site investigations. This benchmark has yet to be tested in court.

Although the ASTM standards are not devised for the internal-compliance audit, they may provide some useful background information. For example, ASTM's Transaction Screen Process E 1528-93 includes a questionnaire and guide for owner-occupant inquiry that essentially amounts to an environmental checklist. Although this checklist is designed for pre-acquisition site reviews, it provides some background for internal-compliance audits.

A plant's environmental permits, violations, emission sources, waste manifests, training manuals and safety logs must be reviewed to ensure that regulations are being met

The definitions that precede both ASTM standards are extremely useful. Not only do the standards define the key terms used in environmental auditing process, but many definitions are annotated to government documents that explain the terms in greater detail.

Another significant aspect of the standards is the attention they focus on the potential for offsite contamination. The standards recommend that the potential for hazardous-waste sites adjoining the property or situated in such a way that they could affect the subject property be fully investigated. Also, the standards seek to determine if any adjoining land may be on specific environmental lists, such as the National Priority List (NPL).

Doing a compliance audit

Depending upon the size of the facility, a team can usually spend between one and five days performing an audit. Facility management should be forewarned that the audit team will be visiting. It is strongly recommended that all relevant documents pertaining to environmental regulations be assembled for the audit team, since much time can be wasted if they are not ready.

Team members should not split up, but should stay together through all facets of the audit. It is recommended that the document review precede the onsite physical inspection. There is a very good reason for this: It is through review of the documents that the audit team will be apprised of the types of processes, operations and hazardous materials used at the facility.

In a way, the document review provides a narrative overview of facility operations. This solves two problems: The audit team will become familiar with the operations and, when doing the inspection, can confirm where the regulations apply and if they are being met. In the document-review process the team should:

- Review all environmental permits and related licenses, ensure that they are current, and ascertain that there is a system in place to re-apply for expiring permits and licenses
- Review any past or current violations, and ensure that there is a mechanism to correct violations and document the corrective action
- Review all sources of emissions including air and wastewater, carefully analyze any monitoring data for contaminants, and confirm that the contaminants do not exceed applicable environmental standards
- Review manifests regarding the disposal of hazardous wastes off-site, and ensure that they are completed correctly and returned to the facility within the regulatory time period
- Review all reporting and recordkeeping requirements under the environmental regulations, and confirm that they are not outdated

• Review all necessary training manuals and testing documentation
• Review occupational safety and health logs, and review repeat injuries or illnesses that indicate a potential problem

A review of all legally required documentation is critical. Failure to maintain legally required documents is one of the prime areas for environmental and safety violations. It is relatively easy for an inspector to charge a facility with violation of a missing manifest, permit, report or document that is required under an environmental law. In fact, the absence of a legally required environmental document is easily detected by inspectors, more so than other types of violations.

The physical review of a site may be the more interesting part of the audit. Each department of a facility must be physically inspected unless there is a certainty that no hazardous materials are present. However, be aware of office areas that adjoin or are part of a processing operation.

The site inspection should include a walk of the periphery of the property. This is an opportunity for auditors to look for signs of stressed vegetation, which indicates contamination in the area, to see if there are any unpermitted outfalls. It also enables auditors to determine the proximity of the plant to environmentally sensitive areas, such as wetlands and residential areas.

At this point, auditors can determine whether there is any potential offsite contamination coming onto the property from adjoining neighbors. Vice versa, auditors should look for any fugitive air emissions that are discharged from an audited facility that may impact neighboring residents. Any process noise or odors that may be objectionable should be investigated.

While doing the audit, the audit team is gathering information on the processes and operations, and materials used at the site. Experienced auditors will be able to relate the applicable regulation with what they are witnessing during the audit itself. However, auditors should also analyze their observations back in their offices to relate them to the appropriate environmental regulations.

The table summarizes some of the main federal regulations that may affect a facility. Remember, too, that a site review does not end with federal requirements. Many states have their own regulations that impose stricter requirements upon industry. In these cases, the stricter state regulations would apply.

The audit team should thus not only be apprised of the federal regulations, but should know when the state, and possibly local, regulations take precedence in a particular case. Thus, one needs to ensure that knowledgeable attorneys and consultants are either on the team or accessible to it.

For pre-acquisition site assessments, many states have disclosure laws that require the seller of a subject property to disclose the environmental conditions on that site. Since some of these laws are very strictly interpreted, the state could possibly invalidate a sale or transaction if the disclosure application is not properly submitted prior to the sale. Thus, it is essential to review laws where the facility is located, to ensure proper compliance with all legal obligations.

After the audit

Once an audit is complete and the audit team is satisfied with the information gathered, it could take anywhere from a couple of days to a couple of weeks to complete a report. The report must be carefully worded by incorporating the advice of both environmental counsel and in-house counsel. It is wise to represent any outstanding commitments as "needs" and not as omissions. This a subtle distinction that could become critical in the event of a lawsuit.

The report should be addressed to the appropriate corporate management, and copied to only those individuals at a facility who have a need to know the results and put into effect its findings. No other copies should be made available. Indeed, the original and all copies should be kept in strictly confidential files.

These steps are essential in maintaining the attorney-client privilege that

Any failure to correctly negotiate the maze of environmental and safety laws can lead to costly civil and criminal penalties, and to imprisonment of plant personnel

may be associated with this document if it was initiated by an attorney. As such, environmental counsel or inhouse counsel should generate the audit report, and the report along with accompanying documents should be stamped "attorney-client privilege."

If the audit report is seen by parties not copied on the report and do not have a need to know the findings, this privilege can be dissolved. Another essential aspect of the report is that it should contain a schedule of recommendations attached to the findings. This schedule should prioritize those "needs" observed during the audit that are to be addressed in order of the highest level of importance first. The schedule should include a blank space for management to insert the person responsible to address a need, and the due date by which the need should be resolved.

This is where management support becomes vital. Management must supervise the efforts to achieve compliance and provide the money to achieve these ends.

All compliance efforts and achievements must be documented. This document must be attached to or included with the audit report. This way, if for any reason the audit report is disclosed to the government, there is also accompanying documentation to prove that the company addressed the audit findings. An audit report without documented corrective actions attached is a ticking bomb.

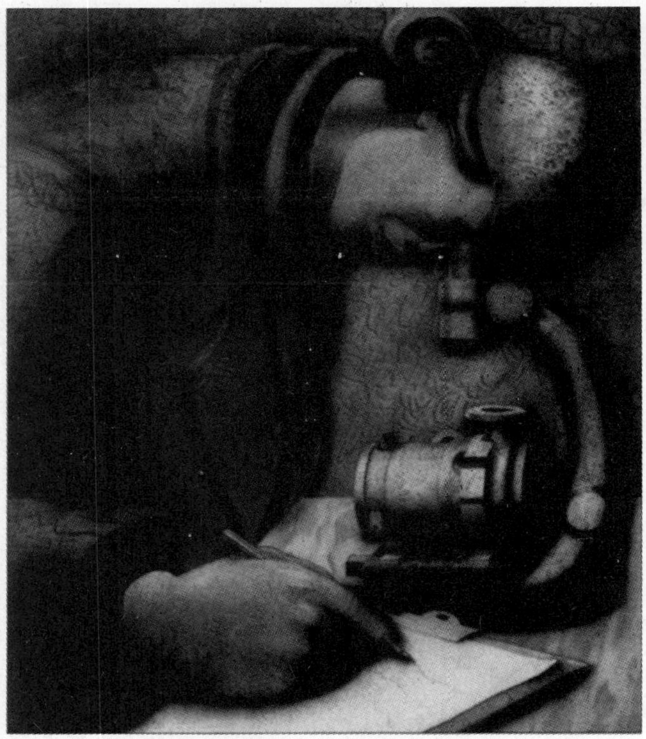

All compliance efforts resulting from an environmental audit must be documented. An audit report without documented corrective actions is a ticking bomb

Pros and cons

The primary benefit of an internal compliance audit is to determine that the facility is working within the environmental regulations. If this is not the case, then the audit highlights those areas that need to be addressed.

Another important benefit of an audit is to prepare the facility in the event of an inspection. This could be either a federal or state environmental or safety agency. In the event of a surprise audit, the facility would already be apprised of its strengths and weaknesses, and be able to explain to the inspector what steps it is currently undertaking. This could be vital in avoiding costly fines and penalties, and in negotiating settlements. It may even avert the issuance of an environmental violation.

The audit report is also important in the sale of business assets. If the report and accompanying corrective action documents support the conclusion that the facility is in environmental compliance, the sale can proceed more quickly. On the other hand, documents reporting significant environmental violations can discourage, or perhaps even prevent, a sale.

Finally, the most serious consequence of an environmental audit is when the report goes unheeded and no corrective actions are initiated. If the report finds its way into the hands of the authorities, serious consequences may follow. Federal and state government can impose fines up to $25,000 and even higher per day per violation. More seriously, corporate management and even employees can be individually fined and imprisoned.

Policies on auditing

In 1986, the U.S. Environmental Protection Agency (EPA; Washington, D.C.) issued a final policy statement on environmental auditing. In it, EPA endorsed self-auditing by companies, to avoid environmental penalties. However, the agency did not guarantee that audit findings would not be used by the government against the facility and its management, but merely stated that EPA would not "routinely" request audit reports.

The agency has recognized that routine requests for audit reports restrict the auditing process in the long run, and decrease both the quantity and quality of audits conducted. Essentially, EPA is saying that frequent requests for audit reports have a chilling effect upon industry, because industry would then be reluctant to perform audits if the agency routinely asks for the reports.

However, EPA has reserved its authority to request relevant portions or the entire report on a case by case, basis. There are four circumstances where EPA will request the audit report:

- To accomplish its statutory mission
- When the audit is performed pursuant to a consent decree
- When the audit material is essential to a criminal investigation
- And, when a company asserts management practices as an affirmative defense in an enforcement action.

Another important policy document on environmental audits was issued by the U.S. Department of Justice (DOJ)

Regulations	Subject
40 CFR 1-49	General and procedural rules governing the U.S. EPA, and federal grants
Clean Air Act (CAA)	
40 CFR 50	National ambient air quality standards (NAAQS)
51-52	State implementation plans (SIPs)
53	Ambient air monitoring — test methods
60	New-source performance standards (NSPS)
61	National emission standards for hazardous air pollutants (Neshaps)
62	State plans for designated facilities and pollutants
65	Delayed compliance orders
66	Assessment of penalties
Clean Water Act (CWA)	
33 CFR 153-157	U.S. Coast Guard oil spill regulations
40 CFR 110-117	Oil spills and hazardous substances (SPCC) reporting, removal and penalties
121-129	National pollutant discharge elimination system (NPDES) program
131	Water-quality planning and standards
136	Test procedures for pollutant analyses
220–229	Ocean dumping
401–463	Effluent standards for specific industries:
403	Pretreatment standards (general)
411	Cement
413	Electroplating
414	Organic chemicals
415	Inorganic chemicals
416	Plastics and synthetics
417	Soaps and detergents
418	Fertilizers
419	Petroleum refining
420	Iron and steel
421	Nonferrous metals
422	Phosphate
423	Steam electric-power generating
424	Ferroalloys
425	Leather tanning
426	Glass
427	Asbestos
428	Rubber
430	Pulp and paper
433	Metal finishing
439	Pharmaceutical
443	Paving and roofing materials (tars and asphalt)
446	Paint formulating
447	Ink formulating
454	Gum and wood chemicals

Regulations	Subject
455	Pesticides
457	Explosives
458	Carbon black
Safe Drinking Water Act (SDWA)	
40 CFR 141-143	Primary and secondary drinking-water standards
144-147	Underground injection
Resource Conservation and Recovery Act (RCRA)	
40 CFR 240-257	Solid-waste processing, recycle and disposal
260	Hazardous wastes — general rules
261	Definition of a hazardous waste
262	Hazardous-waste generator standards
263	Hazardous-waste transporter standards
264	Standards for existing treatment, storage and disposal facilities (TSDs)
265	Interim status regulations for TSDs
267	Interim standards for new hazardous-waste disposal facilities
270	Hazardous-waste permit program
280	Standards for USTs
Toxic Substances Control Act (TSCA)	
40 CFR 702	General practices and policies
704	Reporting and recordkeeping
707	Import and export
710	Inventory reporting
712	Chemical Information rules
716–717	Health and safety data reporting
720	Premanufacture notification (PMN)
761	Polychlorinated biphenyls (PCBs)
762	Fully halogenated chlorofluoroalkanes
763	Asbestos
775	Storage and disposal of waste material
792	Good laboratory-practice standards
Comprehensive Environmental Response, Compensation and Liability Act (Cercla), or Superfund	
40 CFR 300	National contingency plan
40 CFR 302	Hazardous Substances
Occupational Safety and Health Administration (OSHA)	
29 CFR 1900–1999	
29 CFR 1910	OSHA Standards
Subpart H	Hazardous materials
Subpart I	Personal protective equipment
29 CFR 1910.119	Process Safety Management Standard
29 CFR 1910.1000	Limits for Air Contaminants (Z-Table)
29 CFR 1910.1200	Hazard Communication
29 CFR 1910.1450	Laboratory Hygiene

TABLE. Some of the laws and standards that may impact an environmental audit

on July 1, 1991. Like EPA, DOJ stated that its policy is to encourage self-auditing, self-policing and voluntary disclosure of environmental violations by the regulated community. These activities are viewed by the DOJ as mitigating factors in its exercise of criminal environmental enforcement.

The DOJ document describes the four factors that it considers in deciding whether to bring criminal prosecution for violation of an environmental statute. DOJ further claims that it is unlikely that any one factor will be dispositive in any particular case.

Factor 1: Voluntary disclosure. Consideration is given to any person who voluntarily makes a timely and complete disclosure of the matter under investigation. Such disclosure would not be considered voluntary if it is already specifically required by law, regulation or a permit. (Most environmental regulations require disclosure.)

Factor 2: Cooperation. The government will review the violator's degree of cooperation and timeliness. Full and prompt cooperation is essential for DOJ consideration. The violator's willingness to make all relevant information available to the government, including the results of any internal or external investigation and the names of all potential witnesses, would be looked upon favorably by the DOJ.

Factor 3: Preventive measures and

compliance program. The government will investigate the extent and scope of any scheduled intensive and comprehensive environmental-compliance program, such as the performance of annual environmental audits. Corporate policy to eliminate environmental violations and promote sound environmental policy would be perceived as good faith.

Factor 4: "Pervasiveness of non-compliance" may indicate continual repeated participation in environmental criminal behavior. It may also appear to the government that such behavior is the result of a lack of any meaningful environmental compliance program.

"Internal disciplinary action" should be implemented where consistent environmental violations are found. A lack of disciplinary response indicates acquiescence. "Subsequent compliance efforts" strongly indicate plans to remedy any non-compliance. The timeliness and completeness of corrective actions is viewed favorably by the government.

An environmental audit is required when the government imposes a consent decree, or if there is a history of non-compliance. An audit, although not "legally" required, is essential for pre-acquisition assessment.

In the U.S., the Occupational Safety and Health Administration's Process Safety Management Standard (CFR, 1910.119) stipulates the use of compliance audits. For those chemical process industries that are subject to this standard, a compliance audit must be done at least every three years to verify that the procedures and practices developed under the standard are adequate and are being followed. ■

Edited by Jayadev Chowdhury

The author

Janis A. Morelli is an environmental attorney with Cohen, Shapiro, Polisher, Shiekman and Cohen (Lawrenceville, N.J.; Tel: 609-895-1600). She received a B.S. from Ramapo College in 1978, an M.S. in Chemistry from Stevens Institute of Technology in 1982, and her Juris Doctor with a Certificate in Environmental Law from Pace School of Law in 1989. Prior to practicing law, she was a research and development chemist for Fortune 500 companies. Ms. Morelli has audited industrial facilities in the U.S., as well as in Taiwan and Mexico, inspected major treatment, storage and disposal facilities, and counselled clients in environmental compliance. She lectures and publishes articles on environmental compliance, auditing, and health and safety regulations. Ms. Morelli is a member of the New Jersey and New York bars.

STEPHEN WALSH

TSCA TRAPS FOR THE UNWARY PLANT MANAGER

Seemingly simple process changes might add up to a costly violation

Matthew Kuryla
Jones, Day, Reavis & Pogue

Plant managers in the CPI should proceed with caution in making some changes in chemical processes or waste disposal. A number of seemingly routine decisions, often made at the level of the plant manager and even below, can turn into costly compliance "traps" under the Toxic Substances Control Act (TSCA). TSCA is a federal statute designed to give the U.S Environmental Protection Agency (EPA) information and control over commercial chemicals. Originally passed in 1976, the regulations it supports have been reworked at various times over the years.

EPA enforcement blitz

EPA is on the hunt for TSCA violators. In a December 1992 blitz, EPA assessed combined TSCA penalties of $9 million against 22 major chemical companies, including several of the top 20 international producers. According to a recent EPA press release, these actions are only the first in a series of planned enforcement initiatives.

Among the various environmental statutes, TSCA is one of the most costly to violate. TSCA authorizes $23,000-per-day penalties for violations of its rules. EPA's internal penalty policies apply rigid formulas, often resulting in million-dollar assessments for essentially "paperwork" violations.

About half of the recent citations allege violations of Sec. 5 of TSCA, which requires the submission of a premanufacture notification (PMN) to EPA before manufacturing or importing a "new" chemical substance. A new chemical is one that does not appear on an EPA list known as the TSCA Inventory. The TSCA Inventory is a constantly changing list, accessible through many private on-line data retrieval services. A fully comprehensive search of the TSCA Inventory, including the confidential portion, requires a written request to EPA certifying a bona fide intent to manufacture a chemical.

Unless an exemption applies, a manufacturer must file a PMN with EPA before commencing production or import of a new chemical. (In certain instances, a PMN must also be filed for existing chemical production that falls under a regulation known as the "significant new use rule.") The PMN must include the identity of the chemical, along with information about its proposed use and quantity, its byproducts, and all available data concerning potential worker exposure and environmental or public-health effects.

Ninety days after filing a PMN, a company may commence manufacture or import of the chemical as long as a Notice of Commencement (NOC) is filed. When EPA receives an NOC, it places the chemical on the TSCA Inventory. Then, the chemical is no longer considered new, but an "existing" chemical subject to other TSCA rules.

Exemptions

A number of chemicals and processes are exempt from the PMN requirements of TSCA, including:

- foods, drugs and cosmetics (including their intermediates)
- pesticides (however, not their intermediates)
- chemicals used solely for research

Originally published September 1993

and development purposes, in small quantities
• chemicals manufactured solely for export
• impurities unintentionally present in another chemical
• byproducts whose only commercial purpose is for burning as fuel, disposal as waste or reclamation
• non-isolated intermediates (i.e., those mixed with other products and reactants) or incidental reaction byproducts

In addition to these exemptions, TSCA provides a specialized PMN process for certain limited uses of a new chemical. These rules are known as the "test market," "low volume" and "polymer" exemptions. These exemptions are subject to detailed rules of their own. They do not apply across the board, and frequently do not have specific, quantitative limits.

TSCA and the plant manager

At first glance, it would appear that the decision to manufacture or import a new chemical would be an issue for company management alone. Not so: The plant manager also must factor TSCA into ordinary operating decisions. In particular, plant managers should proceed with caution in taking the following ordinary actions:
• substituting a raw material imported from outside the U.S. for a similar domestic raw material
• changing a production process
• isolating intermediates not previously isolated
• changing raw materials or suppliers
• employing different catalysts
• changing monomer content, cross-linking agents, initiators or other reactants in a polymer
• changing disposal practices
• entering into tolling arrangements with outside contractors

These actions, in and of themselves, are not PMN violations. They do, however, bring increased risk of an inadvertent failure to abide by PMN rules. Here are a few illustrations of the ways in which these actions could have major TSCA impact. They are drawn from actual incidents that have occurred in recent years.*

*The examples were provided by Dr. William C. Kuryla, a frequent lecturer on TSCA compliance strategies for industrial corporations.

Example 1: polymer production

Most polymers are listed on the TSCA Inventory in terms of their constituent monomers, and other reactants used in concentrations of 2 wt.% or greater in the reactor mix. However, some polymer listings include monomers and other reactants charged at 2% *or less.* (Minor structural changes are permitted without the need to file a PMN.)

If a new reactant is charged to the reactor at more than 2% of the dry weight of the polymer produced, the polymer is considered a different chemical, and a PMN must be filed if it is not on the Inventory in the new form. This is known as the "2% rule."

For example, assume that a plant's production manager proposes what is believed to be a minor substitution of a reactant in a polymerization process, such as the substitution for monomer A (charged at 1.5% in the polymer synthesis reaction) by monomer B (also charged at 1.5%). Under the 2% rule, no PMN would be required for this substitution unless monomer A is part of the name of the chemical as it appears on the TSCA inventory. When monomer A is part of the TSCA Inventory name, however, the 2% rule applies, and a PMN would be required.

In another polymer example, consider monomer C, which is not listed as part of a product's TSCA Inventory name. A production manager proposes to increase the level of monomer C from 1.5% to 3.0%. Because of the 2% rule, this increase would require a PMN to be filed.

Example 2: catalyst changes

Changes in a catalyst may also have PMN implications. A production manager wishes to use a new catalyst in the manufacture of solvent Z. From a marketing standpoint, this new product works even better than the old one in certain outdoor applications. R&D determines that this new product contains an increased amount of a chemical that was formerly considered an unintentional byproduct or impurity in the original product. This increased level of byproduct is responsible for the improved performance in these applications. The byproduct is not on the TSCA Inventory. If the former byproduct is responsible for the primary properties of the new product's use, then a PMN must be filed.

Example 3: changed waste stream

Waste streams containing "new" chemicals are exempted from the PMN requirement, but only under certain circumstances. In this example, the waste stream from Unit N had previously been sent to the power house to be burned for its fuel value. Recently, R&D tests showed this stream to be nearly as good a neutralizer in a textile process as a commercial chemical.

A major customer has indicated a desire to purchase this waste stream for this application, at 15¢/lb over its fuel value. Unfortunately, several major components of the stream are not on the TSCA Inventory. The sale of this stream without a PMN would be a serious TSCA violation.

Example 4: raw material change

In this example, the purchasing department has been developing cheaper sources for Unit Y's raw materials. XYZ Chemicals, a British firm, has agreed to export to the U.S. a process intermediate for a price below domestically available material. The R&D staff confirms that the XYZ chemical appears to be identical in all respects to the domestic intermediate with regard to the end-use application.

The XYZ chemical has the same

trade name and product code as the domestic material. However, analysis shows the XYZ material to be mainly a 2-methyl-3-ethyl derivative, whereas the domestic material is the 2-methyl-4-ethyl derivative. The domestic material is listed on the TSCA Inventory, while the XYZ material is not. Under these circumstances, a PMN must be filed for the imported XYZ material.

Example 5: isolated intermediate

Unit X makes a catalyst mixture by mixing a solvent and three intermediates in a closed reaction vessel, and then feeding the resultant mixture to continuous closed process to make the final product. In order to increase efficiency, production management proposes first to mix two of the three intermediates in the solvent, then to store this mixture in drums until it is needed. At that point, the mixture would be transferred to a reaction vessel with the third intermediate, and the final mixture would be fed into the continuous process.

However, while the two intermediates are in the drums, they react to form a new substance, not isolated in the earlier process. If the new intermediate is not listed in the TSCA Inventory, a PMN must be filed. It is the physical separation from the process that "isolates" the intermediate for PMN purposes. Even if the new intermediate is isolated for only a very short period of time, without a PMN, EPA would likely view the process change as a violation of TSCA.

Example 6: toll manufacturing

Such arrangements give rise to special risks. Company A has offered its services to perform a manufacturing process on a contract basis, at a cost below that which Company B's facility can deliver. After the contract is signed and the arrangement is underway, Company A decides, on its own initiative, to substitute a different intermediate. The result is a product that is not listed in the TSCA Inventory. No PMN is filed for the product, and the violation is soon caught. Both companies are subject to penalties for violating TSCA, regardless of whether Company A has breached its contract.

The foregoing illustrations are real-world examples of situations where rational decisions have led to inadvertent TSCA violations. They represent only a few of the ways well-meaning plant and process unit managers can get their companies into.

Under TSCA, PMN violations by manufacturers or importers can be based on strict liability. That is, a knowing or intentional violation is not required in order for a civil penalty to be imposed. Knowledge or intent will be relevant in a criminal prosecution — and this is an enforcement tool that EPA has begun to employ more frequently against companies and individuals that EPA considers to be culpable.

CPI plant and process unit managers should gain a close understanding of the TSCA rules governing their routine decisions. Corporate TSCA compliance handbooks and training seminars should be given more than passing attention by all decisionmakers at the plant level. Also, plant managers in larger corporations should participate actively in coordinated TSCA compliance networks within their corporations. This network would entail regular meetings and communications among managers for whom TSCA compliance is important. Participants can share experiences and provide each other guidance.

Finally, internal TSCA audits are a safeguard well worth the expense. Although TSCA compliance can be an unwanted headache for plant managers geared toward maximizing the bottom line, a far worse headache could occur after EPA's next inspection. ■

Edited by Nicholas Basta

The author

Matthew L. Kuryla is an associate with the Cleveland office of Jones, Day, Reavis & Pogue, specializing in environmental matters. He is a graduate of the University of Virginia, and was admitted to the bar in Ohio in 1989.

NEW SAFETY RULES ADD TO PLANT MANAGERS' WORRIES

OSHA's process-safety standards expand management responsibilities

Matthew L. Kuryla
and **Stephen C. Yohay**
Jones, Day, Reavis & Pogue

Most manufacturers throughout the chemical process industries (CPI) have a new federal regulation to comply with: *Process Safety Management of Highly Hazardous Chemicals, Explosives and Blasting.* The rule will expand the safety planning procedures that CPI facilities must undertake and will involve federal inspectors more intimately with manufacturing processes. Although the rule came out of specific incidents in the chemical and refining industries, it applies to a long list of chemical users, and to plants using many standard unit operations.

The new rule, designated as 29 CFR 1910.119 in the Code of Federal Regulations, originated as a response to a series of catastrophic accidents in the chemical and petrochemical industries over the past three years. Workers were killed in each of these accidents, triggering an intense investigation by federal, state and local agencies particularly the federal Occupational Safety and Health Administration (OSHA). Recordbreaking fines were assessed. Despite OSHA's activities. Congressional committees held hearings to admonish OSHA to do more to prevent such disasters.

Now, OSHA has issued a new standard that mandates the use of state-of-the-art techniques for assuring chemical process safety. Consequently, those with responsibility for process safety need to be aware of the requirements of the new standard. Also, those with the credentials and technical experience are likely to find themselves increasingly in demand.

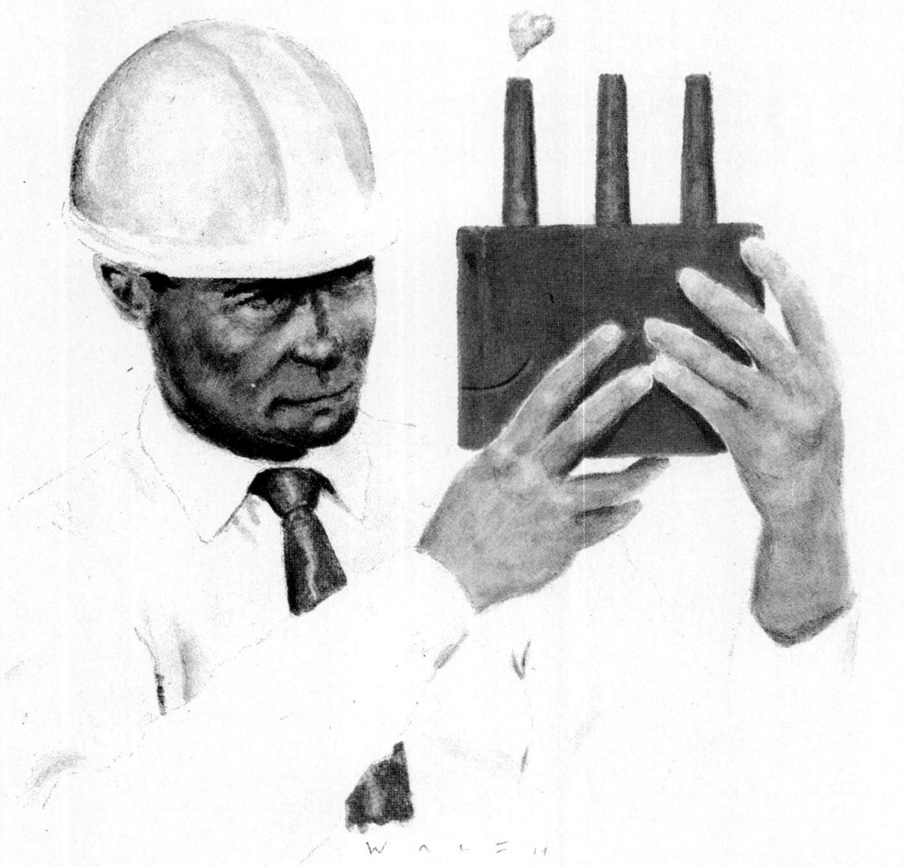

STEPHEN WALSH

OSHA initiatives

The new standard ensures that the chemical and petrochemical industries will experience much closer interaction with OSHA than they typically experienced during the first 15 years of OSHA enforcement. The object of OSHA's initiative is to require those industries to analyze and address the safety of their most-hazardous chemical processes.

When OSHA began its investigation of the 1989 explosion at the Phillips 66 refinery in Houston, it found that it could not identify any clear violations of specific safety standards in the process that caused the accident. Instead, OSHA was forced to rely on the "general duty clause" of the Occupational Safety and Health Act. That provision essentially requires employers to provide workplaces that are free of recognized hazards that are likely to cause death or serious physical harm to employees.

In April 1990, OSHA cited Phillips for violating the general duty clause. OSHA charged that the explosion and fire at the refinery occurred in part because Phillips did not conduct internal audits that included items such as a process hazard review, a job-safety analysis and a human-factor analysis of the process involved in the accident. The citation suggested that Phillips 66 could correct this assumed hazard by conducting a general hazard and operability study (Hazop), and a review of conditions that influence human performance of operators, maintenance staff, technicians and other personnel.

The Phillips case provided little guidance as to what employers must do to assure chemical process safety. It did, however, give significant impetus to OSHA's development of a specific process safety standard.

New safety rules

On Feb. 24, 1992, OSHA made its mandate on process safety clear when it published *Process Safety Management of Highly Hazardous Chemicals, Explosives and Blasting.* OSHA stated that the purpose of the standard is to establish procedures that "would protect employees by preventing or minimizing the consequences of chemical accidents involving highly hazardous chemicals."

Generally, the standard applies to:

1. Processes involving one of a list of highly hazardous chemicals (Appendix A of the standard) in amounts exceeding the listed threshold levels — in some cases, as low as 100 lb

2. Processes involving flammable liquids or gases in quantities exceeding 10,000 lb in one location, except for hydrocarbon fuels to be consumed at the workplace

Requirements

OSHA's process safety rule will impose legal requirements, for all covered processes, in seven areas:

1. Process Hazard Analysis

Perhaps the most significant aspect of the standard is the requirement that employers perform hazard analysis for identifying and controlling process hazards. Employers must use either one of the six specified methodologies (box), such as a Hazard and Operability Study (HAZOP) or a Failure Mode and Effects Analysis (FMEA), or a method that is technically equivalent.

The hazard analysis must address process hazards, controls, consequences of control failures, and effects on employees. A system for addressing and implementing the results of the hazard analysis must be established. Hazard reviews must be updated every five years to assure consistency with the current process.

The standard allows a phased-in implementation schedule for completing process hazard analyses:

- At least 25% of the initial analyses must be completed by May 26, 1994
- 50% complete by May 26, 1995
- 75% complete by May 26, 1996
- 100% complete by May 26, 1997

2. Process safety information

Under the standard, employers are required to compile, maintain and communicate written safety information to employees, identifying the hazards posed by its processes involving highly hazardous chemicals. The information must include:

A) Hazards of the chemicals themselves, such as toxicity, permissible exposure levels (PELs), physical data, reactivity, corrosivity, thermal and chemical stability, and the effects of inadvertent mixing

B) Information on the process technology, such as block flow diagrams, process chemistry, and safe limits for temperature, pressures, flows, compositions and deviations

C) Information about the equipment in the process, such as construction materials, piping and instrumentation diagrams (P&IDs), safety systems, and design codes

3. Operating procedures

Records and written documentation of operating procedures must now be formalized. The documentation should include the steps for each operating phase, including startup, shutdown, emergency procedures, operating lim-

PREVENTION IS THE BEST CURE

The Occupational Health and Safety Administration explicitly identifies six methods of analyzing the risks of chemical manufacturing processes when performing a process safety analysis. They are:

What-if
Intended for relatively uncomplicated processes, this method involves the review of a manufacturing process from raw materials to product. At each step, "What if?" questions are posed and answered, to evaluate the effects of component failures or human errors

Checklist
For more complex processes, a standardized checklist is developed and used for each aspect of the process. Once developed, it is used regularly to ensure compliance with a set of standard procedures

What-if and Checklist
A broad-based hazard assessment technique that involves a brainstorming "what-if" session, followed by a methodical checklist development procedure. An answer is developed for each question, and a series of recommendations is prepared in a comprehensive report

Hazard and operability study (Hazop)
This systematic study of each process element requires an examination of diagrams, followed by an analysis of deviations in a series of parameters, such as flow, temperature, pressure and time. A formal report identifies and assesses all sources of potential failure

Failure mode and effects analysis (FMEA)
A methodical study of component failures, this method starts with a listing of all components of a system on a matrix identifying potential failure mode, consequences, probability, hazard class, detection methods and compensating provisions. The data is then analyzed and recommendations developed

Fault tree analysis
This method starts with a graphic diagram of all sequences of events that could result in an incident. From this analysis, recommendations can be made for avoiding unsafe upsets □

its and safety and health considerations relevant to the process. Copies of the operating procedures must be readily accessible to employees, and operating procedures must be reviewed and updated as necessary.

4. Employees and contractors

The standard contains requirements for training new employees and refresher training for current employees. It also requires host employers to evaluate the safety and health records and performance of contractors selected to perform process work, and to inform contractors of the hazards to which their employees may be exposed. The host employer must explain the facility's emergency action plan directly to the contractor's employees.

5. Prestartup safety reviews

The standard requires prestartup safety reviews to be performed for all new and modified facilities. These reviews must confirm a detailed list of design and construction specifications, safety, maintenance and emergency procedures, process hazard analysis recommendations, and related issues.

6. Compliance audits and management of change

Employers must now implement written procedures to "manage changes" in process chemicals, technology, equipment and facilities. Employees must be informed of, and trained for, the changes. "Compliance safety audits," to evaluate compliance with the standard, are required every three years. A report of findings must be made and deficiencies must be addressed.

7. Incident investigation and reporting

Employers must investigate within 48 hours every incident that results in — or could have resulted in — a major accident. A report of the incident must be prepared, discussed with employees and implemented.

Industry initiatives

Even before it was promulgated, many of the essential tasks under the standard were becoming recognized in the CPI as sound operating, safety and health practices. For example, in January 1990, the American Petroleum Institute (Washington, D.C.) published its Recommended Practice 750, "Management of Process Hazards." The Chemical Manufacturers Assn. (Washington, D.C.) has adopted similar procedures as part of its Responsible Care program. The provisions of RP 750 are similar to the OSHA standard in many respects.

In coming years, workers and others who are injured as a result of chemical process failures will likely assert in private lawsuits that procedures such as those outlined in RP 750 or the Responsible Care program should have been followed. Similar claims, using the OSHA standards, are also likely.

While the value of process hazard analysis as a tool for preventing catastrophic accidents is becoming increasingly accepted, so too is the realization that there is a national shortage of professionals qualified and experienced in performing this work. Some CPI representatives argued that a phased implementation was necessary because of the manpower shortage, which OSHA recognized in its schedule.

Regardless of the implementation schedule, when accidents occur injured workers and their representatives are likely to argue that safety analysis should already have been performed. Given the emotion surrounding any serious accident, it may be difficult for an employer to argue before a jury that it had planned to undertake the required safety analysis, but had not gotten to it yet, or could not find qualified professionals to do the job.

The pressure is building in the CPI to develop additional qualified personnel to perform these analyses. These industries must now intensify and systematize their efforts to assess and improve the safety of their operations.

Given the significant liability involved with catastrophic accidents, many employers now conduct investigations and fault analyses of accidents under the cloak of the attorney-client or work-product privileges. The purpose of preserving these privileges is to allow the employer to develop a candid assessment of the facts and to obtain appropriate legal advice without fear that the information generated will be disclosed in a subsequent lawsuit.

The OSHA standard could make the preservation of these privileges more difficult. Information that has been voluntarily disclosed to a third party cannot be protected. In the portion of the standard that requires employers to investigate each incident that caused or could reasonably have resulted in a catastrophic release of hazardous chemicals, OSHA requires the report to be reviewed with all affected personnel, including contract employees. Moreover, where an incident involves a contractor, its employees must be included on the investigation team.

There may be ways, however, to satisfy the legitimate purposes of the standard while preserving the privilege. Therefore, employers should immediately review with their attorneys how to structure their investigations. ■

Edited by Nicholas Basta

The authors

Matthew L. Kuryla is an associate with the Cleveland office of Jones, Day, Reavis & Pogue, specializing in environmental matters. He is a graduate of the University of Virginia, and was admitted to the bar in Ohio in 1990.

Stephen C. Yohay is of counsel in the Washington, D.C., office of Jones, Day, as part of that firm's Environmental Health and Safety Section. A member of the bar in New York and Washington, he attended Union College, and has graduate law degrees from Georgetown University and George Washington University.

Section II

The Do's and Don'ts in Safety

MANAGING SAFETY

Do's and Don'ts to 'OSHA-proof' your process hazard analyses

David A. Walker, Thomas R. Williams, Dale F. Nawrocki, William G. Bridges, and **J. Steven Arendt**
JBF Associates, Inc.

The chemical process industries (CPI) are extremely diverse, producing hundreds of chemicals that are used in the manufacturing of consumer goods. The materials created and used in chemical process plants range from the relatively innocuous to those that pose significant risks to people and the environment. Industry is constantly being challenged to manage the risks associated with the production and use of these materials, while remaining competitive in the global marketplace.

From the CPI's early days, it has been apparent that facilities need to be reliable and safe. Indeed, strategies to increase workplace safety have become a normal part of plant design, operating and maintenance philosophies.

Nonetheless, because of the presence of extremely hazardous chemicals, the potential for catastrophes exists. But, since such major accidents have been infrequent, plant designers have usually addressed the potential for such major accidents in indirect, and perhaps inconsistent, ways.

Investigations of major accidents show that failures are associated with four key areas: technology, management systems, human factors and external events. For years, companies have concentrated on making their equipment more reliable, and on correcting perceived problems with technology.

But while use of better technology has usually resulted in tremendous increases in productivity, there may not have been commensurate improvements in the prevention of major accidents. In fact, recent reviews show that human

Originally published March 1993

error and the lack of adequate management systems are the most significant contributors to major plant accidents.

It's the law

While no one can guarantee a completely accident-free facility, companies should have "zero accidents" as a goal. Furthermore, the public expects industry to manage the risks of their operations in a responsible way.

While a variety of safety laws and regulations have been on the books for many years, those dealing specifically with catastrophic chemical accidents have come about only recently. Among the first attempts to deal with chemical disasters were those in Europe, following the Flixborough, U.K., vapor-cloud explosion in 1974 and the release of toxic materials in Seveso, Italy, two years later.

The bulk of U.S. efforts for prevention of major chemical plant accidents began in the aftermath of the Bhopal tragedy in 1984 (table, p. 92). First, legislators passed the framework for emergency planning efforts through Title III of the Superfund Amendments and Reauthorization Act (SARA). Subsequently, the U.S. Environmental Protection Agency (EPA) released its Emergency Preparedness and Community Right-to-Know regulations.

These initiatives have generated an avalanche of hitherto unavailable technical information about the identity, use and release of toxic chemicals. This, in turn, has created a heightened awareness on the part of regulators and the public over the potential for major chemical accidents.

A more recent U.S. regulation that addresses accidental releases stems from the passing of the Clean Air Act Amendments (CAAA) in November 1990. This legislation mandates that EPA create regulations to require facilities possessing listed chemicals above a threshold amount to develop and implement extensive risk-management plans (RMPs).

These RMPs must contain three specific components: a hazard assessment of potential worst-case accidents; an accident-prevention program; and an emergency-response program. The focus of EPA's regulations — which are lagging behind the statutory schedule established in the CAAA and have not yet been published — will be to protect the public and the environment from the effects of catastrophic accidents.

Meanwhile, in Bhopal's aftermath, the U.S. Dept. of Labor's Occupational Safety and Health Administration (OSHA) began performing wall-to-wall inspections of chemical plants. One conclusion that OSHA drew from these inspections was that it could not consistently address the safety concerns associated with catastrophic accidents without a new regulatory standard.

A performance orientation

OSHA's efforts culminated in a new rule, *Process Safety Management of Highly Hazardous Chemicals (29 CFR 1910.119)*, which became effective last May. The rule requires that companies possessing a hazardous substance in excess of a specified threshold amount must develop and implement a process safety management (PSM) system (*CE*, May, 1992, p. 37).

OSHA's 14-element system must be maintained over the life of the facility, and is primarily directed at protecting workers within the facility. The cornerstone of the PSM regulation is the requirement to perform process hazard analyses (PHAs) to anticipate the causes of potential accidents, and to ensure that the plant has sufficient safeguards.

As the latest — and perhaps most comprehensive — of recent regulatory efforts, OSHA's "performance based" regulation is a significant departure from previous safety legislation that prescribes specific equipment requirements or actions to be taken. OSHA's PSM rule establishes goals, describes the basic elements of a program designed to fulfill these goals, and outlines important activities associated with each element.

While the performance-based approach does not define specific methods and approaches, it gives companies the flexibility for customizing the methods to best meet their needs and organizational structures. However, the price of such flexibility is that it leaves managers and engineers speculating about what will actually be required to achieve compliance. (The rest of this article centers on compliance with OSHA's PSM rule, but the underlying principles and strategies can be applied to all process-hazard-analysis and risk-assessment activities, whether these are required by law or not.)

OSHA plans to routinely audit CPI facilities for compliance with its PSM rule, with special audits being precipitated by plant employee complaints or accidents. However, OSHA does not spell out what it will find acceptable in its process safety audits. This is especially true in the area of performing and documenting PHAs, which — as paragraph (e) of the OSHA standard shows — covers a host of technical issues (table, p. 94):

- Prioritizing and scheduling initial PHAs
- Choosing appropriate hazard-evaluation techniques
- Addressing specific topics in a PHA
- Selecting an analysis team
- Documenting and communicating the team's findings and recommendations, as well as the company's resolution of them
- Updating and revalidating PHAs
- Retaining PHA records

Specifically, OSHA personnel will base their inspections on OSHA's Instruction CPL-2-2.45A, which is the compliance directive for the PSM regulation. The directive is the primary guidance for OSHA safety and health inspectors, and includes the following activities relating to PHAs:

Audits of documentation associated with the overall PHA approach at a facility, such as prioritized schedules for completing initial PHAs of covered processes. The results of specific PHA studies, including the resolutions of recommendations from selected PHAs, are also covered. The audits review the trail of documentation verifying that: formal PHA activities exist and have the appropriate parts; PHAs are being implemented according to the procedures defined by the company; specific PHA studies are being conducted in an appropriate and thorough manner; and results of PHAs are being acted upon in a timely fashion.

Employee interviews with those who participate in PHA activities, and with those whose jobs are affected by the implementation of PHA recommendations. The interviews help ensure that: employees are involved in the process of conducting PHAs and resolving PHA recommendations; employees believe PHAs are being conducted in an appropriate manner; employees believe no important hazards have been overlooked; and PHA results have been communicated to affected employees.

Inspections of process equipment are intended to help ensure that PHAs effectively address the obvious hazards of the process and that no significant equipment deficiencies exist. Key areas of interest in these inspections may include hydrocarbon- or toxic-gas monitors and alarms, impact protection for piping, pressure-relief devices, flares and other destruction systems, electrical classifications, and control-room siting issues.

To be "OSHA-proof," PHAs must convincingly fulfill the goals and specific requirements in paragraph (e) of the PSM regulation. Three factors primarily affect the likelihood that PHA activities can withstand OSHA's sharpest attention:

- ***Comprehensiveness.*** Clearly, PHA activities should address each and every PHA issue described in the PSM regulation, including such specific requirements as completion deadlines. A PHA should also document *how* the requirements of the PHA are met
- ***Technical thoroughness.*** Not only must PHAs address all the items mentioned in the regulation, they should also be technically thorough. Technical expertise will be judged against recommended practices and approaches from other regulations, professional organizations, consensus standards and industry practices. For example, a hazard-evaluation technique should be performed according to the recommended guidelines for that technique, such as by the American Institute of Chemical Engineers' Center for Chemical Process Safety
- ***Credibility.*** Addressing the required features for PHAs and having technically thorough PHAs will help establish

COMMON ELEMENTS OF SOME PROCESS SAFETY MANAGEMENT PROGRAMS

PROGRAM ELEMENT	OSHA[1] Paragraph	CAAA[2] C items	API[3] Section	CMA[4] Practices
Process safety information	d	1	2	7,8
Process hazards analysis	e	2,3,4	3	9
Operating procedures	f	6	5	7,18
Training	g	7	7	14,17,19,20
Prestartup safety review	i	12	9	13
Mechanical integrity	j	10,11	8	2,12,14,18
Management of change	l	14	4	10,11
Incident investigation	m	13	11	4,5
Audits	o	5	12	3
Safe work practices	*k	*	6	18
Emergency planning, control and response	n	9	10	5,6,11,16
Contractors	h	8	—	12,22
Employee fitness for duty	—	—	—	21
Multiple safeguards	*	*	*	15
Employee participation	c	*	—	1,5
Trade secrets	p	—	—	—
Number of toxic chemicals	135	100 (min.)	95	—

*Either stated or inherent in other elements
Credit: W. F. Early, II

[1]The U.S. Occupational Safety and Health Administration's *Process Safety Management of Highly Hazardous Chemicals* (29CFR 1910.119).
[2]The U.S. Clean Air Act Amendments of 1990, Sections 301 and 304
[3]American Petroleum Institute's *Guidelines for Management of Process Hazards* (Recommended Practice 750).
[4]The Chemical Manufacturers Assn.'s *Process Safety Code of Management Practices*

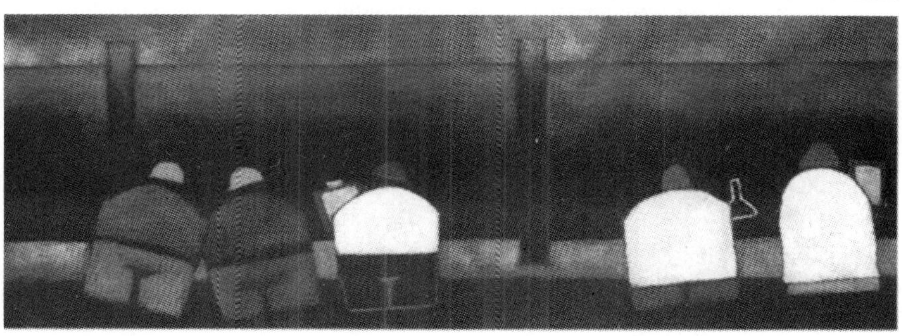

credibility for the individual PHAs and the overall PHA approach. However, unless the PHA activities create auditable documentation, even a very competently performed PHA could suffer criticism. Therefore, key PHA activities, such as documentation of PHA results and communication of PHA results to employees, deserve special attention to help ensure that auditors readily recognize the quality of the PHA efforts

Setting up a PHA

The first step in carrying out a PHA is to determine which processes are covered by the OSHA PSM regulation. The OSHA regulation requires that PHAs be performed for all processes involving highly hazardous chemicals at or above threshold quantities (the specific amounts are listed in Appendix A of the regulation), or if the quantities of flammable liquids and gases onsite in one location are in excess of five tons.

This regulation excludes hydrocarbon fuels that are not part of a processing unit, as well as flammable liquids stored in atmospheric tanks or transferred (without chilling) below their boiling point. The issue of covered processes is continually evolving (with almost weekly clarifications of the rule from OSHA), but here we will focus on conducting PHAs under the assumption that a process is covered by the regulation.

The PSM regulation mandates that PHAs be completed as soon as possible, and comply, at the least, with a five-year, phase-in schedule. PHAs must be performed in priority order, with the rationale for the priority documented by the employer. The prioritization rationale must include consideration of the process hazards, the number of employees who may be potentially affected, the age of the process, and its operating history.

Some companies have gone so far as to develop quantitative approaches with various inputs to prioritize processes for evaluation. This level of detail is probably not necessary — and may in fact be counterproductive — because this type of ranking scheme takes substantial effort to develop, document and defend.

A more straightforward qualitative prioritization based on the judgments of knowledgeable individuals who are familiar with the processes should be adequate. Involving employees familiar with the process operations during the prioritization activity is strongly recommended.

In individual facilities, OSHA is suggesting a commonsense approach for addressing the highest-risk processes first. For example, a process that has experienced numerous incidents over the past few years would probably merit higher priority than one that has operated safely for 20 years. Likewise, a small unit having a low inventory of hazardous chemicals and few operators would probably merit a lower priority than a major unit having a large inventory of hazardous chemicals.

The goal is to establish priorities that will address the units with the highest risks first, and to have a defendable basis for determining the priorities. The key to having a defendable basis is to document the rationale used in determining the priorities.

Once the priorities are established, a schedule for performing PHAs must be developed. OSHA requires 25% of the initial PHAs be completed by May 1994, 50% by May 1995, 75% by May 1996, and 100% by May 1997. A schedule must therefore not exceed this timetable.

However, OSHA also requires that the initial PHAs be performed "as soon as possible." The schedule should, therefore, reflect a reasonable effort to perform the PHAs quickly. For example, it would be difficult to defend a schedule that calls for completion of 50% of the PHAs in 1993, then shows no further activity until 1995.

Likewise, a facility with only one or two covered processes will find it extremely difficult to defend a prolonged schedule, for example, taking five years to perform an analysis of a water treatment system if there are no other covered processes at a facility. The schedule should be documented and revised as necessary to reflect the best estimate of when the PHAs will be completed.

Keep in mind that the regulation targets individual facilities rather than entire companies and corporations. Therefore, the prioritization of processes should be made in relation to other processes at the same facility. The schedule deadlines from the regulation also apply to an individual facility, and not to the entire corporation.

Although very large processing operations may be divided into distinct operating units for analysis, OSHA has indicated that large interconnected processes should not be artificially divided — or "gerrymandered" — into subsections so that the smaller studies can be scheduled over a longer period of time. PHA teams should divide large units on technically defendable bases.

Such divisions of large units may be based on whether adequate distance and isolation exist between parts of a large unit, or if operating responsibilities are clearly divided between parts

LAWS ARE GETTING TIGHTER ALL OVER

The continuing threat of a chemical catastrophe is providing the impetus, worldwide, for authorities to develop legislation for eliminating or minimizing the potential for such events. International efforts include the Seveso Directive covering members in the European Community (Brussels). Many other nations also have regulations, such as Mexico's Sedesol guidelines for carrying out process risk audits, and India's post-Bhopal accident-prevention law. The World Bank (Washington, D.C.) has developed guidelines for identifying and controlling hazards, and The International Labor Organization (Geneva, Switzerland) has developed a code of practice for preventing major accidents.

In the U.S., federal agencies are not the only government regulators active in the chemical accident-prevention arena. Several states have empowered their health, safety and environmental organizations to create regulations requiring companies to establish specific programs for managing safety. These include New Jersey, Delaware, California, Texas and Nevada. Many of these regulations require that facilities must register with the state, and prepare some sort of risk-management plan that contains provisions very similar to OSHA's process safety management regulation.

One notable exception is Texas, which does not actually have a chemical-accident regulation. However, through its environmental permitting program, state inspectors perform a "disaster potential" review, and negotiate facility design and operating changes to satisfy the state's major chemical-accident-prevention concerns. New Jersey and Nevada require facilities to do extensive quantitative risk-assessment studies that address the likelihood of potentially catastrophic accidents. □

of a large unit. For example, a large organic-chemical plant could be divided into the major distinct processing units at the facility, such as the feed-material processing units, intermediate-product units, and storage and handling units.

However, dividing an intermediate product unit into smaller sections with separate analyses for each section might be more difficult to defend. Regardless of how you decide to divide your system for analysis, be sure to document the criteria used to guide the decisions as well as how the criteria are applied in making the decisions.

CHOOSING THE TECHNIQUE

The first step in performing a PHA is choosing appropriate hazard-evaluation techniques, as specified in paragraph (e)(2) of the PSM regulation. The methods approved by OSHA include, but are not limited to: What-If Analysis; Checklist Analysis; What-If/Checklist Analysis; Hazard and Operability (Hazop) Analysis; Failure Modes and Effects Analysis (FMEA); and FTA, or Fault Tree Analysis (*1,2*).

What-If Analysis. This technique systematically examines the responses of process systems to equipment failures, human errors and abnormal process conditions. It requires participation by team members who know and understand the basic hazards associated with the process and its operation. The team leader helps a team develop "what if" questions about the process, such as "What if the pressure regulator failed to close?" By answering these questions, the team identifies potential hazards and suggests ways to improve safety.

The results of a What-If Analysis are generally documented by listing the specific questions, responses, and recommendations generated in the meetings. One of the strengths of this method is that it can be applied to any system at any stage of its design or development. However, the results of the analysis are highly dependent upon the experience of the leader and the team, and the completeness of the list of questions.

Because it is not as systematic as other techniques, What-If Analysis is not generally used as a stand-alone PHA technique. However, it is an excel-

HOW OSHA-PROOF IS YOUR PHA PROGRAM?

PRIORITIZING AND SCHEDULING PHAs	YES	NO	DOCUMENTATION
• Have you prioritized PHAs for all of your covered processes?	❑	❑	How is your prioritization rationale documented?
• Have you scheduled your PHAs to be completed "as soon as possible" ?	❑	❑	
• Does your schedule meet at least the minimum completion requirements?	❑	❑	
CHOOSING THE PHA TECHNIQUE			
• Do you have an appropriate basis for selecting a hazard-evaluation technique for a PHA?	❑	❑	How is that basis documented?
PERFORMING A PHA			
• Have you identified and documented the process hazards to be analyzed?	❑	❑	
• Have you evaluated and documented the range of health effects of these hazards to employees?	❑	❑	
• Have you reviewed previous incidents, both documented and undocumented, for contributions to process hazards?	❑	❑	How are the results of this review documented?
• Have you identified and reviewed the existing engineering and administrative controls applicable to the process?	❑	❑	How is this review documented?
• Have you addressed the consequences of failure of these controls?	❑	❑	
• Have you documented the safeguards identified for all events that could result in a hazardous incident?	❑	❑	How have you documented this?
• Have you reviewed hazards related to the relative location of equipment within a process, as well as the relative location of that process to other processes?	❑	❑	How is the review documented?
• Have you evaluated the contribution of human errors to process hazards?	❑	❑	
• Is the review documented?	❑	❑	
• Have you reviewed operating procedures?	❑	❑	
SELECTING A PHA TEAM			
• Have you selected a qualified team to perform the analysis?	❑	❑	How are the qualifications of each team member documented?
MANAGING PHA RESULTS			
• Do you have a written program for addressing recommendations arising from a PHA?	❑	❑	
• Does your program ensure that recommendations are resolved quickly, actions taken are completed "as soon as possible," and resolution of the recommendations is completely documented?	❑	❑	
• For existing studies, have you resolved all recommendations in a reasonably timely manner?	❑	❑	
• Do your employees know how your PHA program works?	❑	❑	
• Are they made aware of any actions resulting from a PHA that might affect your employees?	❑	❑	
RETAINING PHA RECORDS			
• Do you have a means to store all the documentation of your PHAs for the life of the process?	❑	❑	
OTHER PROCESS SAFETY PROGRAMS			
• Are your other PHA-related process safety management programs adequate to support PHAs?	❑	❑	
• Was the process safety information used for the PHA accurate?	❑	❑	

lent supplement to more-structured techniques, such as Hazop, Checklist and FMEA.

Checklist Analysis. Checklists are commonly used for identifying known types of hazards associated with a process and for helping ensure compliance with standard industry practices. This technique compares aspects of a system against a list of items to discover and document possible deficiencies.

The items included in a checklist should be based on features of similar systems, on company standards, and on common industry practice or experience. The quality of a Checklist Analysis is highly dependent upon the relevance of the checklist items, and how well the person completing the checklist understands the items. Therefore, the analysis should be conducted by a team with industry experience, knowledge of similar processes, and understanding of the particular system being analyzed.

Checklist Analysis is an experience-based technique that strengthens any PHA effort; however, the method alone does not generate the brainstorming of potential process problems that more-creative hazard evaluation techniques, such as What-If, Hazop and FMEA, offer. A checklist review of one specific piece of equipment for which an appropriate checklist exists, such as a safety checklist that was developed by experts for evaluating new-compressor installations, may be adequate. However, conducting a PHA using only a checklist is not generally recommended.

What-If/Checklist Analysis. As the name suggests, this technique combines the brainstorming of What-If Analysis with the systematic nature of Checklist Analysis. It is an improvement over either of the two. The checklists used in these analyses usually include more-general items intended to inspire creative thinking about known types of hazards and potential incidents. Because such checklists are inherently based on experience, and because applicable checklist items may differ from site to site, a team may choose to "What-If" potential events that are beyond the experience base of a checklist.

For instance, docking, mooring and unloading of barges involves specific procedures that, although common throughout the industry, can include subtle differences for each facility. A checklist of marine-related items could be used to analyze common docking equipment and procedures, but What-If Analysis might be needed to identify hazards specific to the subject facility. What-If Analysis also uses the team members' individual creativity and experience to allow extensive brainstorming of general issues introduced by the checklist.

Hazop Analysis. The Hazard and Operability Analysis technique provides an efficient, detailed and auditable study of process variables. It systematically identifies ways the process equipment can malfunction or be improperly operated, thus leading to undesirable conditions.

OSHA's performance-based safety rule provides flexibility but raises questions on how to achieve compliance

To apply Hazop Analysis, the team analyzes each process section (vessels, pumps, interconnecting piping, instrumentation, safety controls, and so on) or procedural step to identify consequences of deviations from the design or procedural intention. For each deviation, the team:

- Decides whether any consequences of interest would result from credible causes of the deviation
- Identifies the engineering and administrative control features that guard against the deviation
- Recommends ways to reduce the likelihood of the deviations or severity of the consequences

Results of a Hazop Analysis are usually recorded in a table listing item numbers, descriptions of the deviations, possible causes identified by the team, potential consequences of the deviations, existing safeguards (control features), and recommendations of the team. Because it is thorough and easy to use, Hazop Analysis is the most popular technique for PHAs.

In fact, many people still refer to PHAs as "Hazops," although OSHA's requirements for PHAs exceed the features generally associated with a classic Hazop Analysis alone. In general, supplementary techniques, such as small-scale What-If or Checklist Analyses, help ensure coverage of special requirements for OSHA compliance.

Failure Mode and Effects Analysis. The FMEA technique involves the methodical study of component failures. Here, all the failure modes for each component in a system are identified, and the effect of each component failure on the system is evaluated. As in Hazop Analysis, existing safeguards are identified, and recommendations for further improvements are documented.

Because the FMEA technique focuses on component failures, problems related to process chemistry, such as autocatalytic decomposition, may be overlooked. For this reason, FMEA is most appropriately used for processes that do not involve reactive chemistry. These include crushing, blending, mixing, conveying, transferring, distilling, coating and packaging. Complex control systems, such as emergency shutdown or interlock systems, are also often analyzed by FMEA.

Fault Tree Analysis. FTA is a deductive reasoning method for determining the logical relationships of component failures and human errors that result in a specific failure of a system. The fault tree portrays the combination of failure events that cause a specific undesired system failure, commonly referred to as the "Top" event.

The technique can be used to quantify the frequency or likelihood of a system failure; however, quantification is often not needed to identify the most important contributors to system failure. This method is generally too detailed and too specific for frequent use in PHAs. It is best used when the PHA team needs to evaluate specific system failures that result from multiple equipment failures or human errors, or both.

The documentation of an FTA usually

includes the graphical fault-tree model, the combinations of failures that cause the Top event, and numerical evaluations of the Top event. This technique is most often used to complement a Hazop, FMEA or What-If/Checklist Analysis to model specific system failures, such as a furnace explosion.

Which to choose?

Rarely is there a single best hazard-evaluation technique for a PHA. In fact, the selection often involves choosing not just one technique, but rather several complementary ones for analyzing different parts of a process or different types of hazards associated with a process. When selecting hazard evaluation techniques, one must consider the severity of the hazards of the process, such as the inherent dangers of the chemicals and the process conditions, the complexity of the process, such as the nature of the control systems, and the efficiency of the techniques for addressing the hazards of the process.

Most often, a detailed, well-structured hazard evaluation technique, such as Hazop Analysis or FMEA, is used as the basic technique for the PHA of a process. In sections of the process where the hazards are less severe or the process is less complex, a less-detailed — but more efficient — technique, such as What-If/Checklist Analysis, is used.

Conversely, sections of the process with more-severe hazards or more-complex control logic may warrant the use of a more-detailed technique, such as Fault Tree Analysis. Additionally, analyses using specific checklists are almost always useful for supplementing the other hazard-evaluation techniques by addressing special concerns, such as facility siting, human factors, or critical-equipment design issues. The key is to select techniques that best meet the needs of the analysis and complement each other to produce an efficient study that does not overwork the problem.

Moving to specifics

Regardless of the analysis technique chosen, the PHA must first address a number of technical issues (box, p. 97). Then, the key to establishing the credibility of each PHA is to describe in the PHA report how the PHA team addresses each of these issues. Most importantly, missing any piece of the PHA puzzle could have severe consequences.

Process hazards. The team should evaluate process hazards based on the nature of the chemicals involved (toxicity, flammability, and so on), the process conditions (flow, temperature, pressure and composition), the team's experience, and information about previous incidents in the facility and at other sites. Material safety data sheets (MSDSs) and similar information can be helpful in understanding the potential effects that specific process materials have on personnel. However MSDSs alone do not always identify all of the pertinent hazards, such as the properties of potential contaminants. For the purposes of the OSHA PSM regulation, the hazards of concern are those that could lead to toxic material releases, fires or explosions that could result in catastrophic consequences — worker fatalities or serious injuries — within the facility.

Previous incidents. A discussion of prior incidents is a valuable part of any hazard evaluation. One way to do this is to collect relevant incident reports filed within the past five years (or since the most recent PHA) and review them as a team during the course of the analysis. Because many companies have not maintained thorough incident records in the past (especially for near-misses), this type of information will be more difficult to obtain for the initial PHAs.

In the absence of thorough incident reports, the PHA team may find that employee injury reports, environmental release reports, and interviews of employees with extensive knowledge of the process' history help identify many of the potentially catastrophic incidents and near-misses at the facility. (Be aware that these reports often contain additional information that will

INDUSTRY LEADS WITH STRONG INITIATIVES

Led by various industry and trade groups, the CPI have taken significant strides in developing and applying process safety and accident prevention programs. In 1985, the American Institute of Chemical Engineers (AIChE; New York City) created the Center for Chemical Process Safety (CCPS). In 1988, the Chemical Manufacturers Assn. (CMA; Washington, D.C.) began its Responsible Care initiative to declare industry's commitment to responsibly manage the chemicals it uses.

CMA's initiative essentially establishes a "safety value" system that its 190 or so member companies must embrace as an obligation of membership. In doing so, the members adopt CMA's Process Safety Code of Management Practices, a 22-point program that covers the process-safety-related aspects of the design, construction, operation and maintenance of a plant throughout its operating lifetime.

In 1990, the American Petroleum Institute developed its Recommended Practice 750, called *Guidelines for Management of Process Hazards.* This document establishes practices and guidelines for petroleum and gas facilities to follow in developing programs for managing process hazards. □

not be particularly useful during a PHA, such as reports on employee slips, trips, and falls.)

The team should evaluate the results of the incident investigations, as well as the initially suggested remedies, and, if necessary, make recommendations for improvements. Reviewing incident information will help focus the analysis on known problems and help reduce the potential for recurrence. This exercise may also uncover weaknesses in the existing incident reporting and investigation system. For example, a PHA team may identify that an instrumentation change made for one vessel as a result of a near-miss should also be made to a similar vessel in the process.

One way to give credibility to discussions of previous incidents is to create a table summarizing the previous incidents and any suggestions the team develops pertaining to the previous incidents. Alternatively, copies of incident reports could be included with the PHA report as an appendix, or the PHA report could refer readers to the location where the incident reports are filed.

Engineering and administrative controls. The PHA team should consider the engineering and administrative controls applicable to the hazards of the process. These controls may include detection devices that provide early warnings of releases, inventory reduction policies, fire protection systems, criteria for equipment spacing, pressure relief devices, and process instrumentation and interlocks. All the hazard evaluation methods described above (including What-If and Checklist Analyses), if used creatively, can include consideration of engineered and administrative controls (often referred to as existing safeguards), and how they play a role in detecting and mitigating releases of hazardous materials.

Consequences of failure of controls. An analysis team should document the reasonable worst-case consequences for a particular deviation, component failure or human error without taking into account existing safeguards. This approach is analogous to examining the consequences associated with failure of engineering and administrative controls. Once the consequences are documented, the team should focus on identifying existing safeguards that protect against the deviation. The more severe the consequences, the more the team should focus on the need for specific, reliable safeguards (both engineering and administrative control features).

To add credibility to the PHA team's efforts, a well-documented PHA should describe the types of events that may result in catastrophic consequences, the specific causes and consequences of those events, the safeguards that are in place to protect against those events, and any recommendations suggested by the team. This information should be documented for not only those events for which the team makes recommendations, but also for the other events for which the team believes existing safeguards are adequate.

The information helps establish credibility. First, describing all of the events that the team has reviewed demonstrates technical thoroughness. Secondly, documenting the potential events for which no additional safeguards have been suggested helps defend the PHA team's rationale for not suggesting additional protections. And, finally, documenting potential events for which additional safeguards are suggested helps explain the PHA team's bases for proposing additional protections.

This raises a dilemma. On the one hand, more documentation of the PHA team's discussions can provide more opportunities for second-guessing the team's opinions — especially if an incident occurs. On the other, being able to completely explain the PHA team's approach, considerations and rationale is more defendable than having no record of what the team considered, or even that the team considered a particular issue at all.

TECHNICAL ISSUES A PHA MUST ADDRESS

- **Process hazards**
- **Previous incidents**
- **Engineering and administrative controls**
- **Failure of controls**
- **Facility siting**
- **Human factors**
- **Range of effects**

Facility siting. Siting reviews identify hazards caused by the proximity of process equipment containing hazardous materials to other equipment, buildings and personnel locations. They also recommend ways to reduce the hazards.

Facility siting can be addressed in a number of ways during a PHA. The team can estimate (qualitatively) consequences associated with an event based on the location of the process equipment with respect to the control rooms, other population centers (maintenance shops, engineering offices, administration buildings, and so on), and other process equipment (such as motors that could provide ignition sources). Valuable resources for facility siting reviews include design specifications for control rooms, separation distances for key areas from the process equipment, locations of drains and fresh-air intakes, proximity to fire protection equipment, and the location of potential ignition sources for flammable materials.

If a team includes current, experienced operators and engineers (as required by the regulation), the team will be very familiar with the layout of the process. This knowledge can be supplemented with plot plans and tours of the facility for team members unfamiliar with the physical layout.

Individual team members should take additional tours of the facility to answer specific questions posed during the team meetings. These questions can include ones regarding distances between equipment, location of controls and mitigation systems (and whether these systems might be damaged or be inaccessible in a catastrophe), labeling, area lighting, and so on. A What-If/Checklist Analysis can be used to ensure thorough coverage of facility siting issues, and to document that these issues are covered during the PHA.

Human factors. Because human errors are one of the most significant contributors to process accidents, team members should consider human factors throughout the course of a PHA. Potential human errors should be considered as causes of process upsets during the analysis. The team should estimate whether operators have adequate time, information and equipment to respond to deviations, and to contribute to incident prevention and mitigation.

Although not specifically required in the PSM regulation, formally reviewing routine operating procedures and other nonroutine procedures (startup, shutdown and maintenance procedures) by means of a hazard evaluation is an excellent way to uncover error-likely situations. In fact, reviewing these procedures is the best way to help identify hazards during off-normal operating modes.

In their enforcement actions after major process accidents, OSHA has repeatedly required companies to perform hazard-evaluation studies that focus on operating procedures. Deciding to review these procedures may exceed the minimal requirements for complying with the current PSM regulation, but the reviews can also significantly reduce the likelihood of process accidents related to human errors.

One of the best hazard-evaluation approaches for reviewing procedures is a modified form of the Hazop Analysis (*3,4*). Checklists can be used to evaluate the engineering considerations related to such human-factors issues as accessibility, labeling, workloads, the use of automatic versus manual controls, simplicity of instrument and control displays, and the clarity of signs and alarms. In addition, a team leader can use prepared questionnaires to solicit input from operators, supervisors, and managers regarding perceived facility hazards. The results of these interviews should then be reviewed with the rest of the team.

A well-documented Hazop Analysis table includes human factors in the discussion of causes and safeguards related to process deviations. Similarly, including a copy of What-If questions or Checklist Analysis items will help to show that your FHA team has done a thorough analysis of human factors.

Effects on employees. An analysis team should document consequences of interest in terms of potential releases, fires, and explosions, realizing that employees could suffer a range of health effects if such accidents occurred. The documented consequences should include reasonable worst-case effects assuming the safeguards fail. To assure coverage of a range of effects, the team should also document any consequences that are less severe than the worst-case effects, but still of interest, such as those that would have potentially adverse effects on employees.

For example, even if the ultimate consequence of high pressure in a vessel could be a large release of hazardous material from a rupture, the team should also document the potentially less-severe consequence of relieving hazardous material through a relief device. The team should then consider whether the vessel is adequately protected against high pressure and whether the relief device discharges to a safe location. This examination of a range of consequences of interest for a given deviation provides a more complete analysis than one that focuses only on worst-case events.

Selecting the team

The process safety management regulation requires that the PHA be performed by a team with expertise in engineering and process operations. The team must include at least one employee who has experience and knowledge specific to the process being evaluated (especially the day-to-day operations of the process), and one person who is knowledgeable in the specific PHA method being used.

OSHA has indicated that PHAs conducted by a single individual — even if that person satisfies the criteria above — are likely to miss important hazards. OSHA believes that involving multiple people in the PHA process, even for techniques such as FTA and FMEA, which are often performed by single individuals, reduces the likelihood that important hazards will be overlooked. For this reason, PHA teams must be composed of more than one person, but not so many that the size of the team becomes unmanageable. Typically, no more than six to

FOR THE FUTURE

The thoughtful and rational consideration of how to structure a technological and regulatory environment for ensuring that tomorrow's industrial use of chemicals is "safe enough" requires acknowledgement of several basic facts and premises:

- The chemical process industries (CPI) are extremely diverse in technology, hazards, and corporate composition. No single accident-prevention approach, particularly one that is rigidly prescriptive, can hope to efficiently manage the risks while helping companies remain competitive
- The CPI generally continue to move toward increased levels of safety in their operations, although the rate of movement in different sectors is not consistent
- Governments are moving even faster in prescribing regulations to ensure that industrial plants are "safe enough"
- The resources available for improving safety are finite, and spending for unnecessary risk-control systems imposes an economic penalty on both industry and consumers, as well as diverting funds from more-effective risk-management activities

With this as a basis, all stakeholders must forge new alliances and working relationships that will enable them to create effective partnerships for chemical-accident-prevention efforts. Industry has the internal motivation and technology to ensure chemical process safety, but the continuing occurrence of major accidents has eroded the public's trust. Government has neither the resources nor the track record to indicate that, acting alone, it can do the job. The problem of ensuring process safety and preventing accidents demands synergistic action by industry, government, labor and the public. □

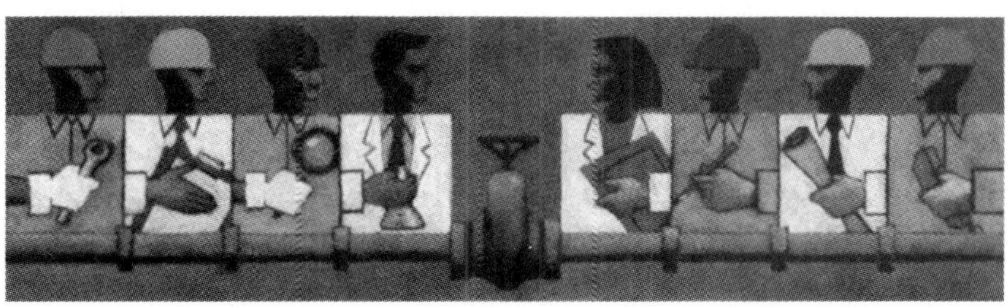

eight people are involved in a single PHA session.

Personnel responsible for designing and operating the specific process possess the bulk of the knowledge and experience needed for a detailed hazard evaluation. Thus, the team should include a process or design engineer — or both — who is knowledgeable about the process and the facility's design, and a senior operator or supervisor who has significant experience in operating the various processes to be reviewed. Other personnel familiar with the maintenance, instrumentation, and inspection and testing of equipment in the process and facility should be part of the team; at the very least, employees with such specialized knowledge should be "on call" to answer specific questions during the course of the review.

Whether an in-house employee, corporate specialist or outside consultant, the PHA methodology specialist should have experience in using the hazard evaluation tools chosen for the PHA. Ideally, the PHA leader should also have the breadth of experience in the various PHA techniques. This allows the leader to choose appropriate techniques for each aspect of a study, thus ensuring the quality of the study without overworking the problem and wasting valuable resources. Listing the team members and their job titles or functions in the PHA report will help verify compliance with the team composition requirement of the regulation.

Managing the results

A critical element of a PHA program is the resolution of team findings and recommendations. Whether it is a one-page letter or a formal policy manual, there should be a written description of the system for addressing team recommendations.

The resolution system should ensure that the recommendations are resolved in a timely manner, that the management response to each recommendation is documented, and that any actions to be taken are completed as soon as possible. A written schedule for completion of any actions should also be developed, and documentation of completion should be included in the resolution files.

Also, it is important to document the rationale for the completion schedule, particularly for any actions that may not appear to an OSHA inspector to be scheduled "as soon as possible." Having all of this documentation readily available will lend a great deal of credibility to the thoroughness of the PHA recommendation resolution program.

Communication of PHA results to affected employees is an area that can be easily overlooked in a PHA management system, but it is probably as important as complete documentation for developing a comprehensive PHA. OSHA plans to use both interviews of employees and reviews of PHA documentation to verify compliance, and it is crucial that both sources of information agree. Therefore, it is necessary to familiarize all PHA team members with: the goals of the PHA program; the methods used in the study; how the study addresses each of the requirements of the PSM regulation; and how the team's findings and recommendations have been resolved.

Performance improvements resulting from PHAs can more than pay for the cost of the studies themselves

In fact, nothing could be more damaging to the credibility of a PHA than to have team members tell an OSHA inspector that facility siting or human factors have not been covered (when in fact the employees just did not understand what the study did), or that they had no idea what became of the team's recommendations. It is also important that all employees affected by actions resulting from PHA team recommendations be informed of the actions and trained appropriately. For example, if an emergency isolation valve is added to a line, the operators need to know when and how to use the valve.

Updating and revalidating

OSHA requires that PHAs be updated and revalidated at least every five years. Although not an immediate issue in establishing the credibility of initial PHAs, a well-defined updating plan can help a firm allocate its PHA resources most efficiently over the life of a process (for ways of updating operating procedures, see *CE*, January, p. 108). Plants having good documentation from PHAs and other PSM programs, such as management of change and incident investigation, will reap benefits in this area.

Updating and revalidating can be accomplished on a routine schedule or may be initiated by significant changes in the process or process operations — for example, prior to the pre-startup safety review for a significantly modified process. Depending upon the differences between the process at the time the PHA is performed and the process at the time of updating or revalidating, a PHA leader may choose to simply modify the existing PHA or to completely redo the PHA, using the previous PHA as a guide to expedite the process.

Processes that have had few incidents, few changes in equipment, and few changes in operating procedures are the best candidates for simple modifications to the existing PHA. If many incidents have been occurring, or if many changes have been made to the process, redoing the PHA is likely to be more efficient and more prudent.

Retaining records. The records of all PHAs, including updates and revalidations, as well as the documented resolutions of PHA recommendations, must be retained for the life of the process. The information must be stored in a manner that allows retrieval within a reasonable period of time and provide access to the information for all employees (as required by the employee participation requirements of the PSM regulation).

The retention system must ensure that permanent records are safely stored for future reference. Storage of documents on electronic media is not prohibited by OSHA.

Interfacing PSM programs. Every element of a PSM program for a facility is closely related to the other PSM elements. This interdependence of elements is an important consideration.

PHAs are based on the information created from other PSM elements (including process safety information, operating procedures, management-of-change records, and documentation on mechanical integrity). If there are deficiencies in PSM activities that support PHA activities, the PHAs could be deficient, even if the PHA approach is consistent with the requirements in paragraph (e) of the PSM regulation. OSHA recognizes the interdependence of PSM elements, especially that between process safety information and PHAs.

For example, conducting a PHA without having adequate process safety information available would not satisfy the regulation. OSHA intends to specifically review the interactions of individual PSM elements during inspections. Therefore, when carrying out PHAs, one must ensure that the information intended to support PHA activities will meet OSHA's expectations. Compliance audits of directly related PSM elements can help identify weaknesses that could adversely affect PHAs.

To be confident that your company's PHAs are OSHA-proof, you must not only do everything that OSHA expects, but also be able to show how you comply with the requirements. For some organizations, simply meeting the basic PHA requirements will be a significant endeavor. Other companies may choose to go well beyond minimal compliance, in an effort to gain significant economic and operational benefits through the efficient use of PHA results in future PSM activities.

As a company's PHA efforts increase, its confidence in withstanding OSHA's scrutiny should also increase. However, the costs of performing the PHAs will likewise increase. Nevertheless, a prudent expansion of PHA efforts can yield many benefits — such as improvements in plant performance — that will begin to offset the direct costs of the analyses themselves. In fact, in many facilities, PHAs can actually produce a net profit, while satisfying regulatory requirements and significantly improving process safety. ■

Edited by Jayadev Chowdhury

References

1. JBF Associates, Inc., *Guidelines for Hazard Evaluation Procedures, Second Edition with Worked Examples*, Center for Chemical Process Safety (CCPS) of the American Institute of Chemical Engineers (AIChE), New York, NY, 1992.

2. Montague, David F., "Process Risk Evaluation: What Method to Use?" *Reliability Engineering and System Safety 29* Elsevier Science Publishers Ltd, England, 1990, pp. 27–53.

3. Bridges, W. G., Kirkman, J. Q., and Lorenzo, D. K., *Strategies for Integrating Human Reliability Analysis into Process Hazard Evaluations*, International Conference on Hazard Identification and Risk Analysis, Human Factors and Human Reliability in Process Safety, CCPS/AIChE, January 1992.

4. Lorenzo, D. K., *A Manager's Guide to Reducing Human Errors*, Chemical Manufacturers Assn., Washington, D.C., 1990.

The authors

J. Steven Arendt is president of JBF Associates, Inc. (Technology Drive, 1000 Technology Park Center, Knoxville, TN 37932-3341; tel: (615) 966-5232). He is a senior project manager involved with developing and implementing process safety management (PSM) programs in the chemical process industries. Steve has extensive experience with hazard-evaluation methodologies, applications and computer programs, and has directed more than 50 risk assessments and accident investigations in the course of his 16-year tenure in the process safety field. In that time, he has also published more than 40 articles, and written several major process-safety publications, including *A Manager's Guide to Quantitative Risk Assessment* and *A Resource Guide for the Process Safety Code of Management Practices* for the Chemical Manufacturers Assn. He is also the lead author of the *Guidelines for Hazard Evaluation Procedures, Second Edition with Worked Examples,* published by the American Institute of Chemical Engineers's (AIChE) Center for Chemical Process Safety. Steve holds an M.E. from the University of Tennessee, and he is a registered professional engineer.

William G. Bridges is a senior engineer at JBF Associates, Inc. He has 13 years of experience in process engineering, process and product development, engineering and manufacturing management, and safety evaluation. He has been an instructor in process hazard analysis (PHA) techniques since 1987, and has led or participated in more than 65 PHAs, using various analysis techniques. Bill has experience in all the elements of process safety management, including developing an overall program. To help understand, evaluate and reduce the human factor in industrial accident scenarios, he is continuing to refine methods for analyzing and preventing human erors. He has helped develop and currently teaches a Process Safety Institute course entitled *How to Implement 29 CFR 1910.119.* He received a B.S. and M.S. in chemical engineering from the University of Akron and is a member of AIChE and Tau Beta Pi.

Dale F. Nawrocki is a process safety engineer at JBF Associates, Inc. Dale has nine years of experience in research and process engineering, supporting petroleum refinery and chemical plant operations. He has led and participated in PHAs for the petroleum, chemical and paper industries. He has also audited PSM systems and written PSM program plans for clients. His experience further includes participating in and documenting a PHA of a large petroleum refinery that addressed all issues of paragraph (e) of the OSHA PSM regulation (*29 CFR 1910.119*). He has also been involved in several large petroleum-refinery-integration feasibility studies, including hydrotreating, reforming, catalytic and thermal cracking, and H-Oil units. In addition, Dale has an extensive background in computer modeling, and personal computer support (hardware and software) and training. He received a B.S. in chemical engineering from the University of Wisconsin-Madison.

David A. Walker is a process safety engineer at JBF Associates, Inc. He has led or participated in risk management activities for industry and government clients, including many PHAs (using the various PHA techniques) and accident investigations. Furthermore, he has audited PSM systems and written elements of PSM programs for clients. David has taught portions of courses on process safety analysis techniques and methods, and he has participated in the development of computer software packages for PHAs. In addition to working on his Master's thesis *Creating Reliability in the Design Process: An Evaluation Approach, Design Tool and Example Application,* he also coauthored the AIChE Center for Chemical Process Safety's *Guidelines for Hazard Evaluation Procedures, Second Edition with Worked Examples.* David holds a B.S. and M.S. in mechanical engineering from the University of Tennessee, and is a member of the American Soc. of Mechanical Engineers, Tau Beta Pi, and Pi Tau Sigma.

Thomas R. Williams is a process safety engineer at JBF Associates, Inc. He has led and participated in extensive PHAs, Hazop analyses, facility siting studies, human factors evaluations, and PSM program evaluations for the chemical and petroleum industries. He is an expert in using the Leader software during analysis meetings, to create clear, concise Hazop tables for clients. In addition to performing quality-assurance reviews of PHA documentation, Tom also assists in teaching hazard evaluation techniques, including detailed Hazop Analysis of operating procedures and the use of the Leader computer software. He was involved in the production of *A Manager's Guide to Reducing Human Errors,* published by the Chemical Manufacturers Assn. Tom holds a B.S. in mechanical engineering from the University of Tennessee and is a member of the American Soc. of Mechanical Engineers.

FORESTALLING

Charles E. Witherell, P.E.
Consulting Engineer

The chemical process industries (CPI) are particularly vulnerable to plant, process and product failures. Among the main reasons are the complexity of the technology involved, the broad spectrum of products, and the often-subtle nature of compositional changes, process shifts and hidden product flaws [*1-3*].

There is much incentive to minimize failures. They erode profits by triggering lawsuits and product recalls and stimulating higher insurance rates and tighter environmental regulations. And they make it harder to maintain a favorable public image and investor confidence.

"Failure" should be defined more broadly than the average engineer might realize. It includes any incident or condition wherein a plant, process, product, material, or service becomes unable to perform its intended purpose safely, reliably and cost effectively. In fact, any operation or product application that leads to dissatisfaction, unattained goals or undesirable side-effects should be considered a failure.

Failures rarely arise from complex, obscure, or unknown technological phenomena. Instead, they are usually the result of common, well-understood causes and foreseeable circumstances. They can be minimized only via increased reliability: not only in product and equipment quality but also in process control and plant operation. This must include activities such as materials handling, transportation, storage, and waste disposal.

Two key strategies for increasing reliability are the systematic and careful analysis of whatever failures do occur, and failure avoidance via predictive modeling or system-life-cycle management or both. Failure avoidance is discussed in the sections immediately following, and failure analysis on pp. 128 and 129.

Failure avoidance calls for developing a thorough understanding of everything that has failure potential, devising a program to analyze each situation, and implementing practices that will minimize the number of damaging failures and recurrences. Each of the aforementioned two major avoidance approaches is based on that strategy.

Predictive modeling

This approach begins with extensive information gathering. The engineer must assemble and organize detailed information on all known and relevant failures, hazards, injuries, damages, loss incidents, and even near-misses.

Especially pertinent are accounts of past events from the same plant, or from other corporate divisions that use the same processes and materials. But information derived from other sources for similar operations and equipment can also be helpful. Ideally, the information gathering should cover not only the firm's own experiences but also to the extent feasible, those of other firms engaged in similar activities.

Details on minor equipment and apparatus, controls and instrumentation, and fabricated items such as piping, fittings, tanks, vessels, combustors,

REBECCA PERRY

Originally published October 1991

FAILURE IN THE PLANT

Analysis of past incidents, modeling of potential ones and life-cycle management of key process equipment can minimize shutdowns and serious problems

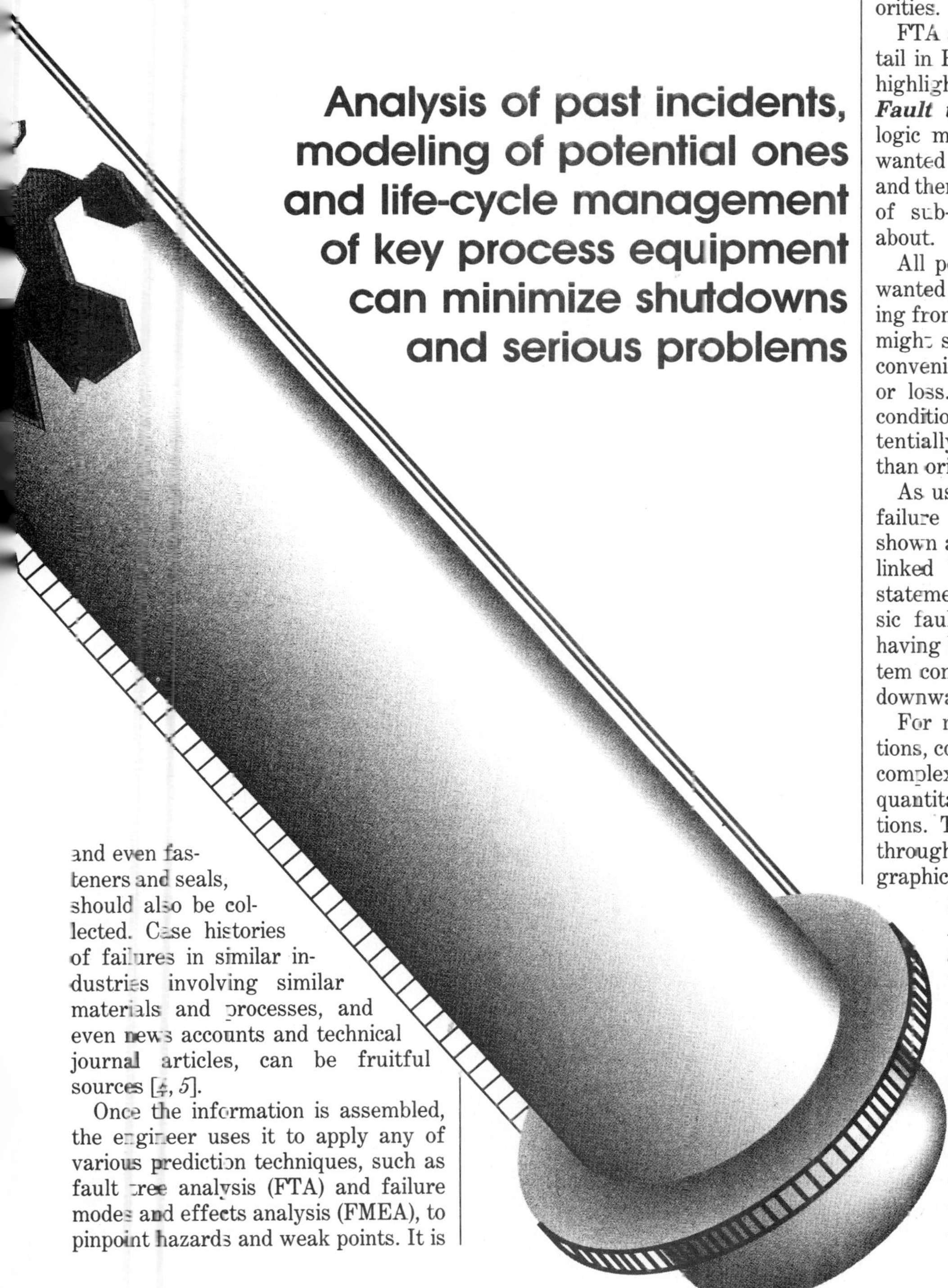

and even fasteners and seals, should also be collected. Case histories of failures in similar industries involving similar materials and processes, and even news accounts and technical journal articles, can be fruitful sources [4, 5].

Once the information is assembled, the engineer uses it to apply any of various prediction techniques, such as fault tree analysis (FTA) and failure modes and effects analysis (FMEA), to pinpoint hazards and weak points. It is usually preferable to conduct FTA before other predictive analyses, because it offers a good perspective of the entire system or sub-system and helps to identify critical areas and establish priorities.

FTA and FMEA are discussed in detail in References *6* through *10*. Their highlights are as follows.

Fault tree analysis is a deductive-logic model. It first assumes an unwanted event, such as a major failure, and then proceeds to identify conditions of sub-systems that could bring it about.

All possible contributors to the unwanted event should be included, ranging from the most serious to those that might simply cause an operational inconvenience without apparent damage or loss. Trivial-appearing events and conditions are often found to have potentially more-serious consequences than originally suspected.

As usually depicted graphically, the failure or other unwanted event is shown at the top of a network. This is linked below through various event statements and logic gates to more basic fault events, producing a "tree" having branches of sequences and system conditions spreading laterally and downward.

For many chemical-process applications, comprehensive fault trees can be complex, especially ones that employ quantitative probabilistic determinations. Their use can be made easier through computer models and network graphics packages prepared expressly for the purpose.

Failure modes and effects analysis is somewhat the reverse of FTA, in that it starts with basic elements of the system (i.e., components) and evaluates how each might contribute to system failure or some other unwanted event. It focuses closely on each system element so as to identify individual weaknesses and failure susceptibilities that, when corrected, will improve reliability of the system. The latter can be an entire plant, a process line, some critical component, or even a final product.

FMEA usually employs a tabular array like an accounting form, with a number of vertical columns. The backbone of FMEA is essentially a parts list, showing a breakdown into subassemblies and basic components (e.g., O-rings, seals, retainer pins and springs). Each part is identified by a number, plus its name and description. In the columns, the engineer notes possible failure modes and their causes, failure symptoms, part redundancies if any, consequences of the failure and their effects by varying degrees of severity, and probable occurrence frequency. Also included are brief comments on how to forestall or eliminate the potential problem or hazard.

The objective of FMEA is to consider and tabulate every item within a given system. The technique is most manageable if the engineer begins with a discrete system, then evaluates its subsystems, then looks at the subsystem components and their respective parts.

When carried out properly, FTA and FMEA are really never completed; they are always in a state of revision. This is because nothing in the chemical-process plant is static. Raw material grades and sources change, process conditions are modified, old product lines are discontinued and replaced with new ones, new regulations impose process and equipment modifications, labor considerations dictate logistical adjustments, and so on.

Life-cycle management

This approach focuses on the integrity of critical system components. It is important for at least two reasons. First, a predictive-analysis study itself may generate questions about integrity of equipment or systems in the plant. Alternatively, the engineer conducting the analysis may unfortunately make unwarranted assumptions about presumed integrity.

Such assumptions may be partly attributable to the traditional equipment-design concepts based upon safety factors, and on the fact that proof tests at prescribed intervals are still mandated by insurers and others. The assump-

This is a particularly fruitful tool. It has been said that we learn more from failures than we do from successes [*20*].

To be effective, analysis must be based on a strategy that includes definite procedures and objectives and tools, as outlined below. It also requires management support and an appropriate organizational structure. Ideally, the analyses should be conducted by an interdisciplinary team whose members possess experience directly related to various aspects of the system being studied.

The team studies the failures so that it can formulate courses of action to prevent recurrences. To do this, it must not only understand how the system or product was intended to function but also establish what combination of factors caused it to go wrong.

No two failures are exactly alike. This is mostly because virtually all failures, particularly the major ones, have multiple causes [*21, 22*]. Several conditions or events converge to precipitate the incident, sometimes suddenly. The more extensive and unexpected the failure, the more intricate the causal web surrounding it.

The table on the next page suggests the range of causes commonly associated with major failures. Although not all elements will be present in a given situation, note their breadth and their possibilities for interaction. Note also how very many are people-related.

Three pitfalls

The team should be careful that its study does not become an *inverted* investigation, i.e., one that starts with conclusions and works backward to assemble facts and align evidence to support those conclusions. Too many investigations are conducted along these lines. Often the method is adopted without conscious intent, or without awareness of its pitfalls.

Also to be avoided are *selective* investigations, which are inquiries having limited objectives and narrow focus. Selective studies zero in on some single feature of the occurrence, such as stress-corrosion at a welded joint, instead of addressing all of the causal factors. These kinds of investigations cannot be expected to provide all the information needed to prevent recurrences and improve reliability.

A different kind of caution should also be kept in mind. Major failures are uncontrolled and unexpected events. They create a disorganized situation in which ordinarily rational people may react illogically and unpredictably. Accordingly, there are risks associated with approaching and entering a failure area, and investigators must be continually alert to possibilities of secondary occurrences, explosions, or environmental hazards. It is advisable to be properly attired in the appropriate safety gear, and alert.

The steps to follow

The content and extent of failure investigations should be consistent with the magnitude of the occurrence and the likelihood of recurrences. The process details and the key issues will be specific to the particular incident. On the other hand, the basic objectives and procedures will not differ much from incident to incident.

The steps to follow are as follows. Note, however, that they are rarely carried out exactly in the sequence shown, because there are many unavoidable iterative loops in the procedure.

1 *Broadly characterize the problem*

The incident should be approached objectively, with the initial purpose of gathering a broad perspective that takes people, things and the surrounding environment all into account. A broad-view approach is desirable because the interrelationship of circumstances, causal elements and consequences is usually very complex, as noted earlier. Attempts to narrow the focus too early in the study can eliminate important information. Checklists can be useful in assuring that all relevant matters are addressed [*23, 24*].

2 *Document the observations* Contacts with people and things leave mental impressions. It is useful to document these permanently, in as much detail as possible. If litigation is involved, or is possible in the future, the investigator documenting the scene or

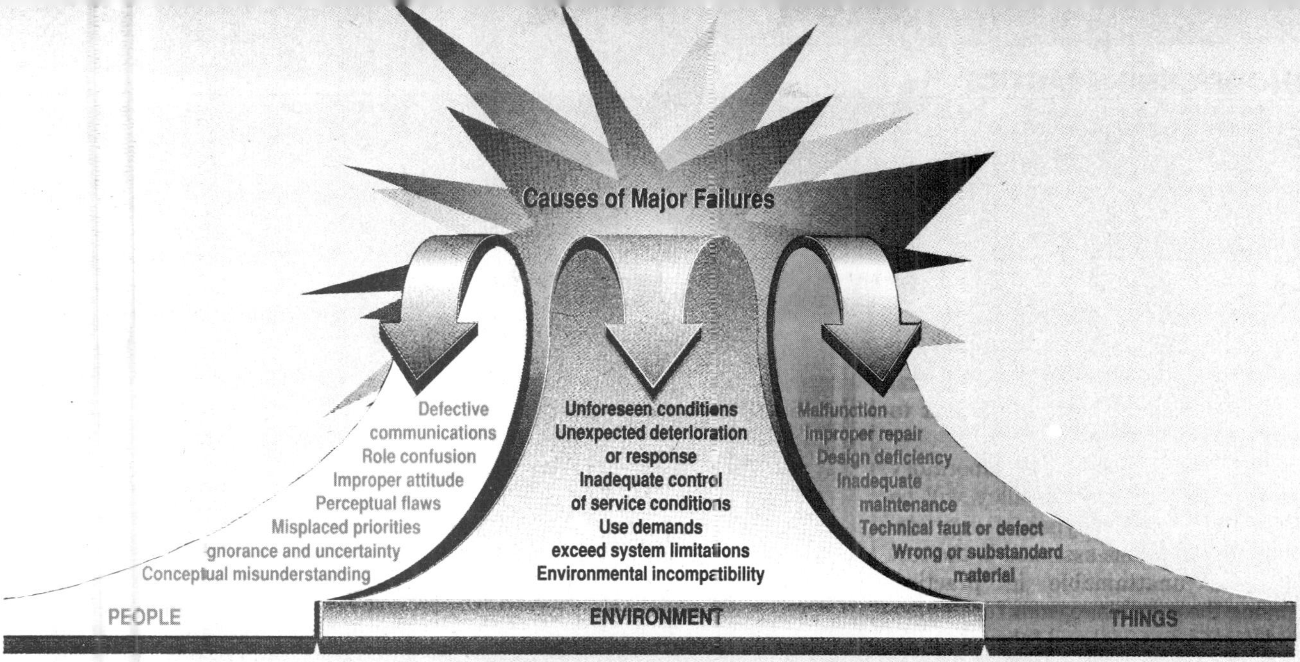

site should understand the factors that govern admissibility of evidence within the jurisdiction involved [*25*].

3 *Identify, collect and preserve evidence*

In the first place, it is important to keep in mind how wide the range of evidence can be. Depending on the situation, it may include not only physical items (e.g., samples of residual process material, damaged equipment) but also numerous kinds of data and documents: process and operating logs and flowcharts, raw material certifications, purchase requisitions, plant layouts, personnel records, design drawings, specifications and standards, statements of witnesses and other people involved, reports and memoranda, equipment-operating literature, product and sales brochures, patents, media accounts, and photos. The goal is to assemble enough material, of whatever kind necessary, to develop an unambiguous description of what occurred.

Physical items affected by the failure require special care. Before removing them as evidence, they should be photographed in the positions found, and these locations should be established with respect to fixed reference points. Collected materials should be handled as little as possible and placed in protective containers, or suitably covered to avoid contamination.

It is important to have a well-documented record of the origin and disposition of each sample or piece of evidence: the time, date, and place of its retrieval, and who retrieved it. And as the study progresses and other people take custody for testing or other purposes, each successive contact and recipient should be documented. An unbroken chain of custody is important.

Samples should also be taken of unaffected products or components, preferably from the same run or batch. These are useful in trying to understand how the faulty item may have deviated from its specifications.

4 *Examine the evidence*

It is difficult to generalize about this step, because the examination procedures and tests can be as varied as the incidents involved [*26-30*]. Before any tests are conducted, a game-plan acceptable to all parties should be devised to assure that the tests are performed in the proper order. Today's analyst has a vast collection of sophisticated techniques and instrumentation to assist in arriving at the correct conclusions [*31, 32*].

5 *Reconstruct what may have happened*

Once examinations and tests are completed, data are in hand, and results of evaluations available and assimilated, it is possible to reconstruct one or more failure scenarios. These are plausible sequences of events and circumstances that could have led to the occurrence.

It is essential to let facts, evidence, and test outcomes speak for themselves. The reconstruction must account for all established facts and be consistent with all hard and indisputable evidence. Care must be taken to avoid developing scenarios based upon absence of information.

Usually, the most direct and simple explanation consistent with available evidence is the correct one, even though many causative elements may have contributed. It is best to start a reconstruction on the basis of the most straightforward and obvious, then proceed to more-complex failure mechanisms and explanations only when facts and evidence absolutely require it.

6 *Determine why it happened*

Not only physical but also procedural causes must be considered. The extent of the consequences of the incident and range of damage can be affected by people's actions, responses, or failures to act or respond. This behavior may be a reflection of the overall system affected by the failure, and may require careful assessment.

As in the reconstruction step, all available facts and evidence should be accounted for and inconsistencies logically and reasonably explained. Conclusions based upon lack of information should be avoided.

7 *Formulate conclusions and recommendations*

If followed properly, the foregoing steps provide a sound basis for valid conclusions. These should be concise and objective and wholly supportable by the evidence. Scope and content of recommendations for preventing recurrences and improving the performance of the product, process, plant, equipment, or system should also reflect factual findings and be consistent with the conclusions.

8 *Report the results*

Unless the results and recommendations are clearly communicated to a receptive management or authority, the effort will be wasted.

tions ignore the fact that safety margins can be drastically reduced by undetected manufacturing defects, operation beyond design limits, unexpectedly aggressive environments, and unanticipated residual mechanical stress — especially within or next to welds.

For instance, even though laboratory tests show that a given alloy will be immune to various types of degradation, the conditions assumed in the test may be unattainable in practice. Among the possible reasons for this are undetected material and fabrication defects and resulting stress concentrations, compositional variability, localized alloy and microstructural heterogeneity, and variations in operating conditions.

The engineer should be alert to the need for life-cycle management at his or her plant. It is inconsistent to develop finely tuned analytical models for components and sub-systems of lesser consequence while, for example, a massive, expensive, and possibly old reaction vessel with enormous damage potential is considered only superficially — perhaps because nobody at the site really knows much about its integrity and continued suitability for service. The vessel may in fact be near the end of its intended design life, so questions should be raised about the need to retire it from service and replace it at considerable expense, or to continue its operation — and if so, for how long.

These questions are often complicated by other unknowns. For instance, the throughput of the vessel may have recently been increased beyond design ratings to achieve higher production. Or other changes in use may have been imposed that were not intended or expected by the original designers.

At least two life-cycle-management techniques developed to cope with such questions at nuclear power plants and other electric utilities can be adapted for use in the CPI. They focus attention directly on the actual hardware or system involved, aiming to estimate the degree of damage caused by previous service and the suitability for continued use (i.e., the remaining life). "Degree of damage" is defined as any reduction in a material's useful life.

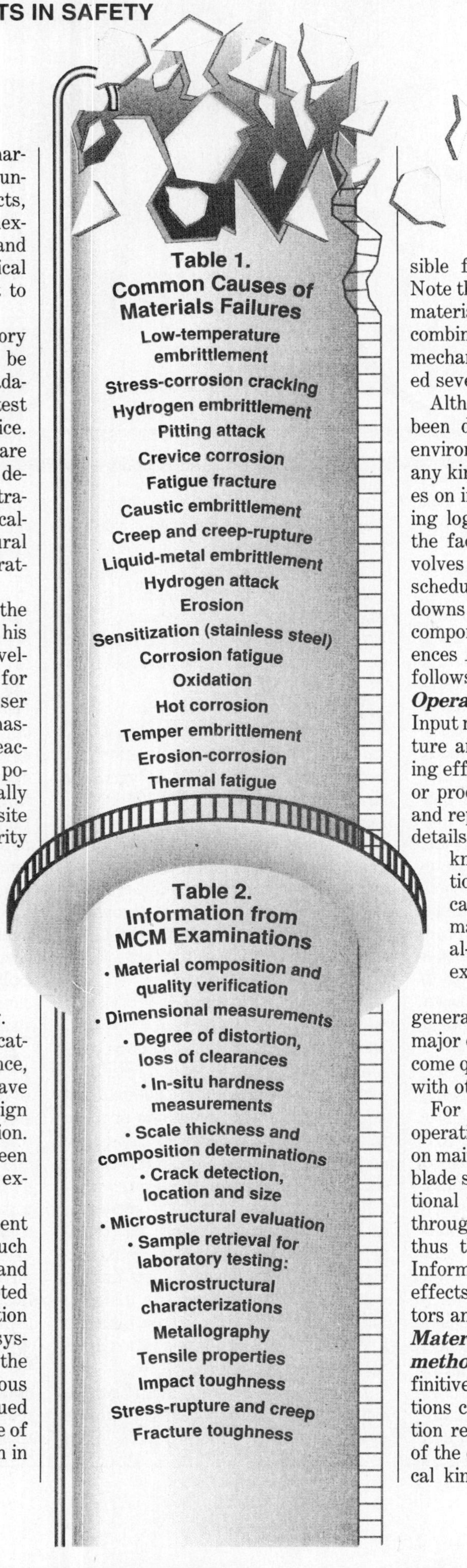

Table 1. Common Causes of Materials Failures

Low-temperature embrittlement
Stress-corrosion cracking
Hydrogen embrittlement
Pitting attack
Crevice corrosion
Fatigue fracture
Caustic embrittlement
Creep and creep-rupture
Liquid-metal embrittlement
Hydrogen attack
Erosion
Sensitization (stainless steel)
Corrosion fatigue
Oxidation
Hot corrosion
Temper embrittlement
Erosion-corrosion
Thermal fatigue

Table 2. Information from MCM Examinations

- Material composition and quality verification
- Dimensional measurements
- Degree of distortion, loss of clearances
- In-situ hardness measurements
- Scale thickness and composition determinations
- Crack detection, location and size
- Microstructural evaluation
- Sample retrieval for laboratory testing:
  - Microstructural characterizations
  - Metallography
  - Tensile properties
  - Impact toughness
  - Stress-rupture and creep
  - Fracture toughness

Illustrating the scope of such techniques, Table 1 lists a number of common failure mechanisms and adverse conditions that have been responsible for process industries failures. Note that many failures of engineering materials are caused by synergistic combinations of two or more of these mechanisms, sometimes with unexpected severity.

Although the two techniques have been developed for high-temperature environments, they can be modified for any kind of service. One method focuses on information derived from operating logs and other historical data for the facility in question. The other involves periodic examination during scheduled or unscheduled plant shutdowns when there is access to critical components. Details appear in References *11* through *13*; highlights are as follows.

Operating-data evaluation method: Input might include operating temperature and pressure, cycle time, operating efficiency, gas analyses, equipment or process modifications, maintenance and repair records, and original design details and specifications. Through knowledge of material and condition susceptibilities, the engineer calculates rough estimates of remaining life from standard material-test data and parameter extrapolations.

These qualitative estimates are generally too inaccurate for making major decisions. However, they can become quite informative when combined with other observations.

For example, satisfactory impeller operation in corrosive media depends on maintaining a protective scale on the blade surfaces. But an increase in rotational velocity (e.g., to achieve higher throughput) will remove the scale and thus threaten catastrophic corrosion. Information from failure modes and effects analysis could alert the operators and thus forestall a shutdown.

Material-condition-monitering method (MCM): This offers more definitive guidance. Periodic examinations can provide a wealth of information regarding the state and integrity of the critical system component. Typical kinds of data gathered appear in

Table 2. The information can be gathered via actual sampletaking, or amassed via techniques such as those discussed in this magazine's 1989-1990 series of articles on nondestructive testing, or both. (For related developments, see also "Holistic Maintenance," *CE*, April, pp. 30 - 35. And for information specifically related to control valves, see "Control Valves: Give Them Attention Before They Fail," *CE*, May 1990, pp. 90 - 95.)

Usually, MCM is employed for the most-critical and failure-susceptible locations, and multiple samples or data readings are taken. For proper sampling and meaningful evaluations, it is essential to understand the probable failure modes involved, their synergistic relationships, and their preferred sites.

Information gathered might include visual indications of distortion and wall thinning as well as less-obvious damage such as precipitation of brittle grain-boundary phases, early or advanced stages of stress-corrosion cracking, and intergranular cavity formation indicative of creep progression. Data like these can of course be combined with information from operating logs (as discussed above) and computations [*14-16*].

For optimal value, life-assessment methods must be applied periodically and on a routine basis. Comparing the most recent results with previous data will indicate trends in crack growth, and extent of environmental attack, strain accumulation, creep cavitation, and other forms of progressive material damage.

Going further

Preferably, the failure-avoidance strategy should include not only predictive models and life-cycle-management techniques but also real-time process monitoring [*17*]. While today it is impracticable for the typical CPI plant to monitor everything, indications from FTA and FMEA can pinpoint critical areas of the system that should be frequently checked. These high-risk areas are priority candidates for continuous monitoring.

Real-time monitoring can boost failure avoidance

Monitoring combined with real-time process control [*18*] promises especially worthwhile attractions for plants handling, for instance, extremely hazardous materials. And computer-integrated processing and manufacturing, now beginning to make a beachhead in the process industries ("CPI Likes CIM," *CE* Mar. 28, 1988, pp. 20,21), should offer even more attractive opportunities for failure avoidance [*19*]. ■

Edited by Nicholas P. Chopey

The author

Charles E. Witherell, 7426 Hillview Court, Pleasanton, CA 94588-4230, Tel. and FAX (510) 462-6460, is a consulting engineer in private practice, specializing in diagnosis, analysis and prevention of metallurgical failures. His 35 years of experience include associations with International Nickel (Research Manager), Eutectic Corp. (Vice President of R & D) and Lawrence Livermore National Laboratory (Principal Investigator). He received his B.S. in mechanical engineering from New Jersey Institute of Technology, and did his graduate study in physical metallurgy and law. He holds over 100 patents, and is author of the book "How to Avoid Products Liability Lawsuits and Damages," (published by Noyes), a member of several technical societies and a registered professional engineer.

References

1. England, S.M., "Opportunities in the Design of Inherently Safe Chemical Plants," *Advances in Chemical Engineering,* Vol. 15 (1990), pp. 73-135.
2. Florman, Samuel C. *Blaming Technology,* St. Martin's Press, New York, N.Y., 1981.
3. Swartz, E. M., "Hazardous Products Litigation," 2nd ed., The Lawyers Co-Operative Publishing Co., Rochester, N.Y. 1988, pp. 8-112.
4. Naumann, F. K., "Failure Analysis - Case Histories and Methodology," American Soc. for Metals, Metals Park, Ohio, 1983.
5. Engle, L. and Klingele, H., 'An Atlas of Metal Damage," Prentice-Hall, Englewood Cliffs, N.J., 1981.
6. Roberts, V. L. and Jones, P. L., Analytical Techniques in Product Safety, *J. of Products Liability,* Vol. 4, Pergamon Press, 1981, pp. 67-93.
7. Innis, C. L. and Hammond, T., Predicting Mechanical Design Reliability Using Weighted Fault Trees, in "Failure Prevention and Reliability," American Soc. of Mechanical Engineers, New York, 1977, pp. 213-18.
8. Gangadharan, A. C. et al., Computer Methods for Qualitative Fault Tree Analysis, in "Failure Prevention and Reliability," American Soc. of Mechanical Engineers, New York, 1977, pp. 251-62.
9. Lapp, S. A. and Powers, G. J., A Method for the Generation of Fault Trees, in "Failure Data and Failure Analysis: In Power and Processing Industries," PVP-PB-023, American Soc. of Mechanical Engineers, New York, 1977, pp. 95-101.
10. Kolb, J. and Ross, S. S., "Product Safety and Liability - A Desk Reference," McGraw-Hill, New York, 1980, pp. 115-47.
11. Viswanathan, R., "Damage Mechanisms and Life Assessments of High-Temperature Components," ASM International, Metals Park, Ohio, 1989.
12. Cane, B. J. and Williams J. A., "Remaining Life Prediction of High Temperature Metals," *International Materials Reviews,* Vol. 32, No. 5, 1987, pp. 241-262.
13. Woodford, D.A., "Design for Performance — A Predictive Framework for Life-Cycle Management," *J. of Metals,* Vol. 42, No. 12, December 1990, pp. 18-21.
14. Anderson, P. L. et al., "Monitoring and Modeling Stress Corrosion and Corrosion Fatigue Damage in Nuclear Reactors," *ibid.,* pp. 7-11.
15. Jones, R. H. and Friesel, M. A., "Using Acoustic Emission to Monitor Stress-Corrosion Cracking," *ibid.,* pp. 12-15.
16. Bruemmer, S. M., "Monitoring and Detecting Environmentally Induced Cracking," *ibid.,* pp. 16-17.
17. Wen, S., "An Intelligent Path to Quality — Process Monitoring and Control," *J. of Metals,* Vol 43, No. 1, January 1991, pp. 10-12.
18. "Frontiers in Chemical Engineering — Research Needs and Opportunities," National Research Council, National Academy Press, Washington, D.C., 1988, pp. 135-152.
19. Wu, B. P. F., "The Application of Expert Systems to Process Control," *J. of Metals,* Vol 43, No. 1, January 1991, pp. 13-16
20. Santayana, George, *The Life of Reason, Vol. 1, Reason in Common Sense,* 1905-1906.
21. Serrill, M.S., "Anatomy of A Catastrophe," *Time,* September 1, 1986, pp. 26-29.
22. Bignell, V. et al. "Catastrophic Failures," The Open University Press, Open University Educational Media, Inc. New York, 1977, reprinted 1978.
23. Witherell, Charles E., "How to Avoid Products Liability Lawsuits and Damages," Noyes Publications, Park Ridge, N.J., 1985, pp. 285-292.
24. "American Law of Products Liability," 3rd ed., Practice Aids, The Lawyers Co-Operative Publishing Co., Rochester, N.Y., 1987, supplemented, Sections 1-390.
25. Graham, M., "Handbook of Federal Evidence," West Publishing Co., St. Paul, Minn. 1981
26. Vander Voort, G.F., "Conducting the Failure Examination," *Metals Engineering Quarterly,* American Soc. for Metals, Metals Park, Ohio, May 1975, pp. 31-36.
27. Ryder, D. A. et al., General Practice in Failure Analysis, in "Metals Handbook," 9th ed., Vol. 11, Failure Analysis and Prevention, American Soc. for Metals, Metals Park, Ohio, 1986, pp 15-46.
28. Jur, T. A., "Microscopic Techniques in the Analysis of Mechanical Failures," *J. of Products Liability,* Vol. 1, No. 4, Pergamon Press, 1977, pp 265-85.
29. McCall, J. L. and French, P. M.," Metallography in Failure Analysis," Plenum Press, New York, 1978.
30. Colangelo, V. J. and Heiser, F. A., "Analysis of Metallurgical Failures," Wiley, New York, 1974.
31. Willard, H. H et al., "Instrumental Methods of Analysis," 6th ed., Van Nostrand, New York, 1981.
32. Gabriel, B. L., "SEM: A User's Manual for Materials Science," American Soc. for Metals, Metals Park, Ohio, 1985.

PPG Industries, Inc.

DON'T LEAVE

Roy E. Sanders and **James H. Wood**
PPG Industries, Chemicals Group

All of the effort expended in designing plant-safety systems is of little value unless accompanied by an adequate "prooftest" program and regular maintenance. These safety systems — consisting of such components as safety-relief valves, tank vents, critical alarms and protective isolation and shutdown devices — do not operate on a continuous basis. Rather, they are only called into service periodically, to warn of, or to prevent, conditions that could lead to plant accidents.

After just a few years of neglect, many protective devices and preventive process loops could become ineffective. Unfortunately, these safety-system failures may go undetected until a crisis occurs.

Determining the methods and frequencies of testing for critical instrument loops must be tailored to each plant's needs and resources. No matter how often testing is carried out, the facility must be committed to checking all plant-safety devices regularly. Testing programs must be impervious to all the other demands taxing a plant's resources, including key personnel changes. Periodic in-house audits should be carried out, to ensure that testing methods are effective.

The following recommendations were developed at the request of Industrial Risk Insurers (Hartford, Conn.) to highlight the elements and practices that are critical for ensuring the lowest practical risk at a given facility. The specific programs discussed in this article were developed for and by the Lake Charles, La. manufacturing complex of PPG Industries, Inc. (Pittsburgh, Pa.).

The 650-acre Lake Charles facility, situated on the Calcasieu River, contains 20 separate but interdependent chemical process units. The complex, in operation since 1947, employs about 1,600 people full time.

Almost all of the units produce large quantities of industrial chemicals on a continuous basis, with a very high onstream factor — nearly 95%. The main products are silica compounds, chlorine, caustic soda, hydrochloric acid, vinyl chloride monomer, methyl chloroform, and other chlorinated hydrocarbons.

On a typical day, huge quantities (about 3,000 gal/min) of saturated brine (NaCl) enter the complex, and more than 3,200 tons of chlorine are produced. In the humidity of southwest Louisiana, the salt, chlorine and hydrogen chloride gases circulating throughout the chlorinated-hydrocarbon processes attack equipment.

Particularly susceptible is equipment that has been improperly specified, installed, or maintained. Corrosion can damage plant-safety devices, leaving the plant vulnerable to industrial accidents. A rigorous testing program is intended to detect such damage.

The first half of this article details the development, and evaluation, of a test program for safety-relief valves (SRV). It takes into consideration the type of process operation (continuous vs. batch), the designed onstream factor, number of relief devices, corrosive nature of the chemicals processed, availability of qualified SRV repair shops, state laws, insurance recommendations and plant resources. Prooftesting methods and frequencies for complex, critical-instrument safety loops are discussed in the second half of this article.

The prooftesting program at PPG's industrial-chemical complex is provided as an example; individual programs will

Originally published February 1991

PLANT SAFETY TO CHANCE

differ on a plant-by-plant basis, and requirements will be different, as well, for low-volume batch processes used to manufacture pharmaceuticals and specialty chemicals.

In-house survey reveals flaws

PPG's Lake Charles complex has about 2,000 individual locations that are protected by safety-relief valves and conservation (pressure or vacuum) vents. SRVs range in size from ¼-in. threaded valves to 12-in. flanged valves, and are manufactured by 24 companies. Nearly 125 tank-conservation vents, which range in diameter from 3 to 24 in., are included in the test program.

Boiler-safety valves are not discussed here. Most states have strict specifications that should be followed for the safety testing of such devices.

Prior to 1974, each operating unit at Lake Charles handled its own relief-valve testing program, and there were no highly trained specialists in charge of it. In 1974, the PPG Lake Charles chemical plant began centralizing its efforts.

One technical employee was appointed to evaluate how effectively the existing safety-relief-valve program was operating, and to coordinate the implementation of improvements throughout the complex. The resulting coordinated effort was a great improvement over the previous system.

The results of the plant-wide survey of safety systems were alarming: A significant number of SRVs were failing to provide the security expected. Problems of corrosion, lack of training, lack or spare parts and inadequate testing procedures were blamed for this.

As well, the test-stand equipment, used to simulate operating conditions on valves in gas or vapor service was found to be providing an inadequate volume of fluid [3]. Hence, the testing method itself was flawed, compromising the accuracy of the test readings.

Certain valve designs and materials of construction in use at the facility were found to be unreliable in corrosive service. Before the system was centralized under the auspices of one employee, corrosion-related problems were not well documented, so problematic designs and materials of construction were often specified over and over again; there was no learning from past mistakes.

The in-house evaluation also found that, in general, small-volume test equipment are unreliable. Depending on them for safety-test results provided a false sense of security.

PPG's in-house auditor made several recommendations for the Lake Charles complex:

- Develop improved test methods
- Purchase adequate test-stand equipment
- Develop a detailed, centralized record-keeping system
- Base safety valve selection on design and materials of construction
- Provide more thorough training for in-house mechanics, who test and recondition the safety devices

PPG Industries, Inc.

Many safety systems in a plant are only called on periodically. Unless checked regularly, failures can go undetected — until a crisis occurs

This article is designed to help the novice, as well as the loss-prevention engineer and industrial-risk-insurance auditor to manage the total administrative effort of an SRV maintenance program.

Is your safety program safe?

The importance of pressure-relief valves cannot be overstressed, in terms of their role in ensuring plant safety. These "silent sentinels" [1] of the chemical process industries, often referred to as the "safety" or "safety-relief" valves are essential in helping to minimize industrial accidents caused by the overpressurization of pressure vessels and tanks.

Relief devices are very delicate instruments, and must be treated as such [2]. Their appearance is deceiving: Because they look like pipe fittings, they are often handled and stored in the same manner. This misconception is often the cause of significant abuse and damage.

Several questions must be answered when evaluating a plant's testing procedures for safety-relief valves.

- Does each vessel that is susceptible to damaging overpressure have a pressure-relief device in place?
- Are the relief devices well identified (Figure 1) with a visible number or a legible tag?
- Is each tag up to date, marked with the most recent test date?
- Is the adjustment screw sealed, to prevent adjustments or tampering by unauthorized personnel?
- Is there evidence of leakage from, or significant corrosion around the valve?

The connection between the safety-relief valve and the vessel it protects should be identified with a number or location tag. This will ensure that the SRV and vessel are properly paired again, after testing is complete.

Check the area in the vicinity of these safety-relief devices. Is there insulation, delicate equipment or walkways in the target area? Keep in mind that discharge from relief devices, operating at pressures greater than 15 psi, escapes at about 1,100 ft/s — nearly the speed of sound. The typical nozzle velocity of a shotgun is around 1,300 ft/s.

Inlet piping connected to the SRVs needs to be checked, too. If the fluid retained by the safety-relief device is

MEGAN REID

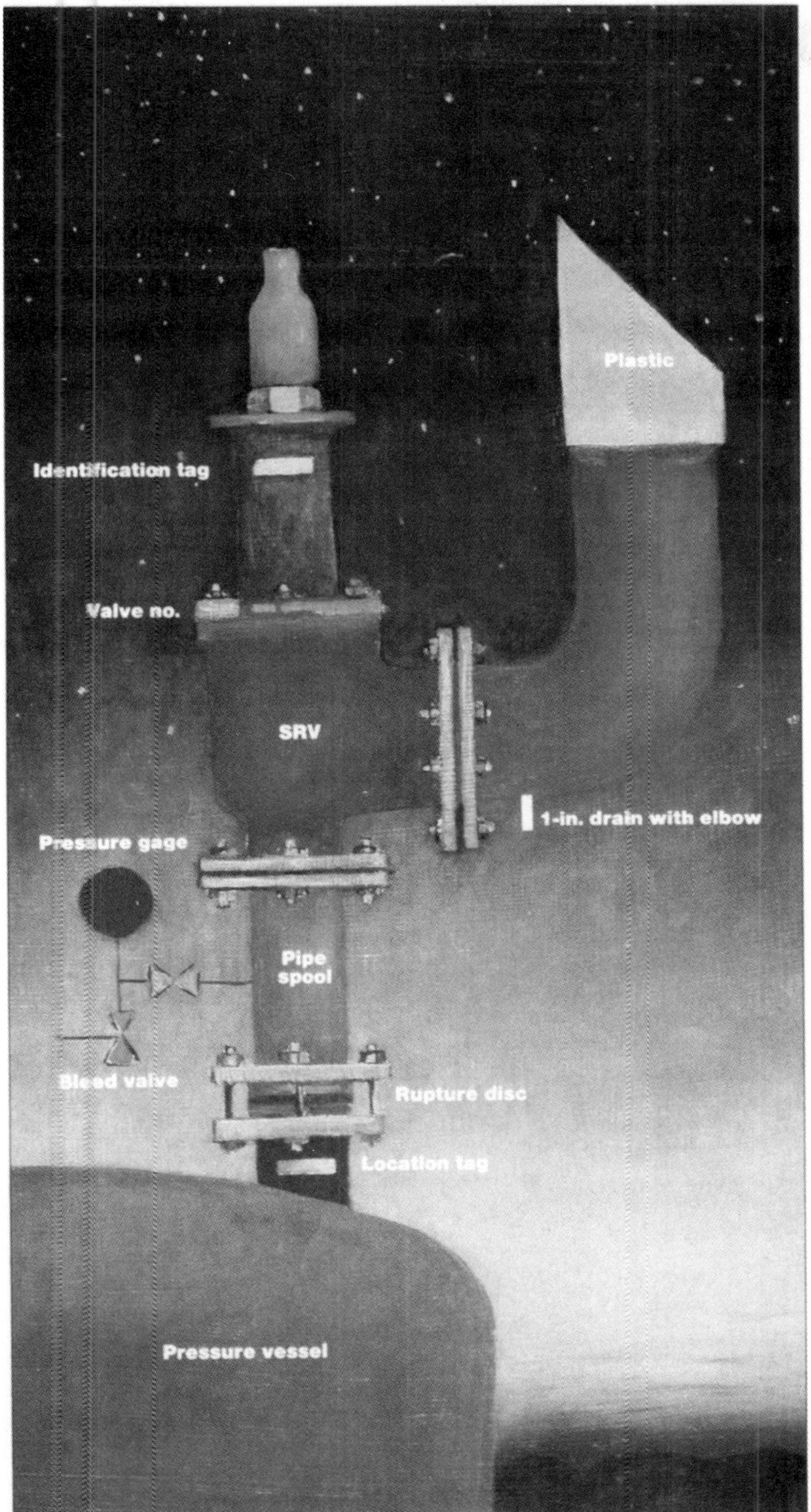

FIGURE 1. Properly identified safety-relief valves keep the critical information at your fingertips, and can save a lot of headaches down the road

corrosive, a corrosion-resistant rupture disc must be in place to prevent leakage. Any rupture disc in place must be in compliance with the ASME code.

Are there block valves below, or downstream of, the safety-relief valves? If so, are they "sealed" in the open, operational position? If not, tampering or accidental closure may occur, rendering the safety valves useless. Block valves or three-way valves should meet ASME Code requirements for pressure drop.

Each block valve in a safety-relief system should have a written "registration," approved by a supervisory technical-safety committee. Mechanics who remove safety-relief valves should report in writing on the condition (such as fouling, partial plugging and choking) of the inlet nozzle and discharge piping.

Next, observe all outlet piping. If outlet piping is the same diameter as the safety-relief valve, are there two or more elbows or long lengths? Such configurations may increase the risk of excessive back pressure.

If the safety-relief valve is in a service involving gummy or tar-like materials, process fluids can accumulate and clog or choke the discharge piping. Materials in SRVs that discharge to flare stacks or scrubbing systems may freeze, plug or solidify, and increase back pressure.

Are there rain caps on direct-discharge lines? These are helpful to keep contaminants from entering and clogging the discharge pipe.

The next step is to review the documentation of all SRV testing. Each operating unit should have a list of safety-relief devices. This list should include the set pressure and the last test date of each. Each plant must also develop a method to establish the frequency of testing.

Testing and repair procedures

Testing of SRVs relies on sufficient fluid volume or air pressure for dynamic testing. If simulated conditions do not provide adequate flow and volume, the test results will not be accurate. Valve performance should be carefully monitored under simulated process conditions, to ensure that an adequate volume of air is being used.

All process SRVs are pre-tested,

and the appropriate data on initial pop pressure and blow-down is recorded. This is carried out for all new valves, and for those removed from process service.

The safety valve is disassembled and examined, and all internal parts are blasted with glass shot, to remove rust and surface accumulations. The valve body is then sandblasted and repainted.

During this stage, all defective parts are replaced or reconditioned to the manufacturer's specifications. Metal seats are machined to the desired flatness. All moving or guiding surfaces, and all pressure-bearing parts are lightly lubricated [4].

Once reassembled, the valve is performance tested and adjusted to the correct set pressure, and the blowdown is set. All bellows-type and spring-loaded valves exposed to back pressure are back-pressure tested, to ensure the integrity of the bellows and gaskets on the exhaust side of the SRV [4].

Tamper-proof seals are added to discourage unauthorized or undetected field adjustments. Once testing is complete, a sheet-metal identification tag (usually of soft lead, which is easy to imprint yet impervious to corrosion) is attached to the SRV. It records the location number, set pressure and test date.

Valve openings are sealed with tape and cardboard to prevent the entry of foreign material during transit and storage, and lifting levers are tied down to protect the SRV en route to the user. The valve must be transported and stored in an upright position.

During testing, results are recorded on an audit form checklist (Figure 2). Information recorded for each SRV includes initial test pressure, and the condition of seats, stems, guides, springs, and inlet and outlet piping upon arrival at the plant. This information is sent to the appropriate area supervisor, and a copy is maintained in central SRV records [4].

PPG Industries, Inc.

Rigorous testing programs must be carried out regularly, and they must be impervious to all other demands taxing a plant's resources

Do-it-yourself testing

Many plants conduct their own safety-relief-valve testing (as opposed to using a factory-authorized testing service). To evaluate how effectively a program is working, plants conducting in-house testing should answer the following questions:

1. Are the repairmen factory adequately trained?
2. Does the plant have sufficient spare parts, equipment and a significant volume of fluid to simulate operating conditions, in order to adequately test the valves?
3. Does the plant have consistent testing and repair procedures in writing, and an up-to-date Quality Control Manual?
4. Are regular records kept and distributed, regarding the testing of valves and the existing conditions?
5. Is there a listing of low-pressure vacuum and pressure devices?
6. Are these devices regularly inspected?

How often is often enough?

PPG Lake Charles has established different test cycles for different types of process-safety equipment: Some is tested as frequently as every six months, some is tested as infrequently as every three years. SRVs operating in dirty, fouling or highly-corrosive conditions may require more-frequent testing to ensure reliable service.

State laws, corporate policies, insurance regulations, and any other appropriate mandates should be considered when establishing regular testing cycles. Additionally, any SRV that pops or discharges a corrosive fluid should be removed and reworked as soon as the discharge is detected.

At the Lake Charles facility, about 30% of the overpressure devices are tested annually; another 60% are tested every 24 months and the remainder are tested at other frequencies. In general, SRVs on the following types of equipment are tested at the corresponding intervals [3]:

- Positive-displacement pumps or compressors — every 12 months
- Vessels processing corrosive chemicals — every 12 months
- Process vessels with heat sources, including stills, kettles, hot-oil systems and low-pressure, refrigerated storage tanks — every 12 months
- Boilers — every 12 months (Louisiana law requires boiler SRVs to be tested, set and sealed annually)
- Process steam headers — every 24 months
- Storage vessels containing clean (non-plugging, non-corrosive) fluids — every 24 months
- Instrument-air manifolds within dehumidified control rooms — every 36 months
- Lubricating relief valves on enclosed systems, such as compressors, turbines and generators — every 36 months

Keep the records straight

A simple, sequential numbering system is acceptable for a small plant, such as a batch-chemical processing facility with only a few dozen SRVs. However, for the Lake Charles plant, a

SAMPLE SAFETY-RELIEF-VALVE TEST REPORT

SERVICE NO.:

_______ _______ _______ 46-_______
(DEPT) (DIV) (SEQ) (CAT) (SUF)

VALVE OUT OF SERVICE: 46- _______

VALVE INTO SERVICE: 46- _______

SERVICE DESCRIPTION: _______

DISPOSITION OF VALVE OUT, IF NOT RETURNED TO SERVICE:

☐ MOVED TO SERVICE NO. _______

☐ STORED AS SPARE ☐ SCRAPPED

REASON IF NO VALVE RETURNED TO SERVICE: _______

REQ. SET PRESS: VAC SET: BACK PRESS: COLD DIFF SET: OPER TEMP:

FIELD CONDITIONS:

RUPTURE DISC UNDER SRV: ☐ NO ☐ YES IF YES: ☐ GOOD ☐ LEAKING ☐ BLOWN ☐ NOT REPORTED

VESSEL OR PIPE NOZZLE: ☐ CLEAN ☐ DIRTY ☐ PARTIALLY PLUGGED ☐ COMPLETELY PLUGGED ☐ NOT REPORTED

TEST DATA ON VALVE NO. 46- _______

INITIAL TEST:

POP _______ PSIG

☐ EARLY

☐ OK (= ±10%)

☐ LATE

☐ NO TEST

IF NO TEST, EXPLAIN: _______

SPRING NO: _______

LUBRICANT: _______

FINAL TEST

SET PRESS _______ PSIG

VAC SET _______

BACK PRESS _______ PSIG

COLD DIFF SET _______ PSIG

TIGHT AT _______ PSIG

BLOW DOWN _______ PSIG

NOZZLE RING SET ☐ UP

_______ NOTCHES ☐ DOWN

GUIDE RING SET ☐ UP

_______ NOTCHES ☐ DOWN

VALVE CONDITION:

SEATS	STEM & GUIDES	SPRING
☐ GOOD	☐ GOOD	☐ GOOD
☐ FOULED	☐ FOULED	☐ RUSTED
☐ CORRODED	☐ CORRODED	☐ CORRODED
☐ ERODED	☐ GALLED	☐ CRACKED
☐ CRACKED	☐ FROZEN	☐ BROKEN

INLET: ☐ GOOD ☐ CORRODED ☐ DIRTY ☐ ERODED ☐ PLUGGED

OUTLET: ☐ GOOD ☐ CORRODED ☐ DIRTY ☐ ERODED ☐ PLUGGED

(ALL SAFETY VALVES MUST BE TESTED AS RECEIVED FROM OPERATING UNITS OR MANUFACTURER.)

VALVE DATA ON VALVE NO. 46- _______

NEW VALVE

MANUF. _______

SERIAL NUMBER _______

SET PRESS _______

OPER TEMP _______

EXISTING VALVE

MODEL _______

ORIFICE _______

VAC SET _______

BODY MATERIAL _______

CAP TYPE	LIFTING LEVER	TYPE CONSTRUCTION	O-RING
☐ PLAIN	☐ YES	☐ CONVENTIONAL	☐ YES
☐ OPEN	☐ NO	☐ BELLOWS	☐ NO
☐ PACKED		☐ BOILER	O-RING MAT'L.:
		☐ PILOT	_______

OUTLET : SIZE _______ IN. RATING _______ TYPE: ☐ SCRWD ☐ FLAT-FACE FLANGE ☐ RAISED-FACE FLANGE ☐ SOCKET WELD ☐ RING JOINT ☐ TONGUE AND GROOVE

INLET : SIZE _______ IN. RATING _______ TYPE: ☐ SCRWD ☐ FLAT-FACE FLANGE ☐ RAISED-FACE FLANGE ☐ SOCKET WELD ☐ RING JOINT ☐ TONGUE AND GROOVE

ACTION NOTE: _______

COMMENTS: _______

MECHANIC (SIGN) DATE W.O. NO.: SUPERVISOR (SIGN)

FIGURE 2. A formal safety checklist, such as this one used by PPG Lake Charles, should be customized for each facility

dual-numbering system is used, to keep up with 2,000 SRVs in service, and over 600 additional SRVs in storage. The devices kept in storage onsite are used as spares, or can be cannibalized for parts. Spare safety valve are a must [4].

In the dual-numbering system, one number refers to the valve type, and the other refers to its location and service (see Figure 2). The first number, identifying the equipment, is unique to the SRV, and is retired once the valve is taken out of service.

Equipment records include information on the orifice size, set pressure, manufacturer, model number, serial number, and inlet and outlet flange sizes and ratings.

Test results are entered into the mainframe computer according to each device's identification number. Personnel have "read" access to the records at over 500 terminals and personal computers throughout the complex.

The importance of proper identification of the safety-relief valve cannot be overstated. Without a strict numbering system, the risk is increased of incorrectly joining the wrong SRV to the wrong vessel after testing is complete.

For example, a 1-in. threaded valve can have an effective orifice area of from 0.06 to 0.314 in^2 (a 5-fold difference), yet both could fit the same installation. Similarly, a 1-in., threaded-type SRV set at 100 psi may have a capacity that is completely different from another valve with the same description.

Similarly, a 4-in. valve with 300-psi style flange could be equipped with either an "L" orifice (2.85 in^2) or a "P" orifice (6.38 in^2). The "P" orifice is 2.24 times larger than the "L" orifice.

The location portion of the numbering system indicates the SRV's specific site. Site records include a description of the location, temperature, set pressure, product and the presence or absence of a rupture disc.

Every other month, an inspection schedule is sent to the supervisor of each process unit. The form serves as a reminder of all of the SRV and conversation vent locations, recommended service and frequency of inspection, last test date and results (satisfactory, changed or unsatisfactory), and the correct set pressure, back-pressure, operating temperatures, and equipment number of each SRV and conservation vent. This schedule is intended to highlight those valves due for testing.

In addition, brief, statistical SRV Test Compliance Summaries are circulated to higher levels of management [4]. These summaries list the total number of SRVs in the area and the total number (and total percentage) that are overdue for testing. They serve as a periodic report card, signifying which process areas are in compliance, and which are delinquent.

Critical shutdown systems

In addition to the SRVs, numerous instrument loops provide critical safety, alarm and shutdown functions at the PPG Lake Charles complex. These are located on reactors, oil heaters, cracking furnaces, storage tanks, compressors, steam-heated vaporizers, kettles and stills, and flammable vapor detectors.

Such loops function in either an online or a standby manner; both types can fail. Failure of an online loop, such as a level-control valve, becomes known rather quickly when operation deviates from normal — either gradually or drastically, depending on the type of failure — and places a demand on the standby loop.

However, failure of a standby instrument loop, such as an alarm or safety interlock, will not become evident until a demand is placed on it to function. Problems developing in these loops must be discovered by periodic prooftesting.

A prooftest program cannot be left to someone's memory. It must follow a well-structured format to accomplish the essential steps, regardless of the myriad other activities and distractions that tend to absorb all the supervisors' and mechanics' time.

When PPG Lake Charles first initiated its's prooftest program, efforts to classify which safety devices were truly "critical" were not specific enough. Hence, the original program allowed too many instruments into the test system, which created a top-heavy burden. To prevent this from happening, the following information should be developed for critical loops:

- A listing of critical alarms and safety interlocks essential to safe plant operations
- Criteria by which the truly "critical instrument" system can be characterized as such, and discerned from other important (but not critical) systems

The test procedure for complex interlocks should be developed with engineering help, and, for consistency, should be documented in writing. Any proposed additions or deletions should be screened by a third party group or committee. An appropriate authorization path — through which changes in shutdown setpoints and instrument values can be made systematically and on the record — should be in place.

Test results must be recorded systematically, and must be easily accessible to all who need them for analysis. Defective or worn components should be identified and repaired immediately.

Block valves below SRVs require rigorous testing

Management must take an active role in the stewardship of this program. High levels of compliance with stringent test schedules should be urged and rewarded.

It takes years to develop and fine-tune a prooftest program. Time is required to define all of the instrument loops that need to be included and more time is needed to systematically collect or develop data sheets, containing basic operational information for each device or loop.

During the development stage, test methods need to be defined, and the appropriate personnel responsible for testing must be identified and trained. Finally, test frequencies must be decided.

Tamper-resistant measures should be in place for block valves upstream of SRVs

The Lake Charles facility has over 2,500 loops in its prooftest program. In many cases, the test frequency is determined by decision-tree engineering studies. Critical instrument loops appear on a prooftest schedule, which is distributed monthly to the operations and maintenance departments.

Prioritizing critical loops

Since some instrument loops are more critical than others, our instrument loops are assigned priorities:

- *Priority 1.* Critical instruments whose failure *would* either cause, or fail to inform of, situations resulting in accidental fire, explosion, uncontrolled release of dangerous materials, reportable environmental releases, or major property or production loss. The alarms assigned a "Priority 1" include those that have been mandated as such by: outside agencies; an in-house technical safety review committee; reliability studies; and specific alarms deemed critical by operations supervisors. All of these alarms are on a regular prooftesting schedule.

The Lake Charles facility has 1,295 "Priority 1" safety-instrumentation loops. These include alarms and trips on storage tanks containing flammable or toxic liquids, devices to control high temperature and high pressure on exothermic-reaction vessels, and control mechanisms for low-flow, high-temperature fluids on fired heaters. Other Priority 1 instruments include alarms that warn of flame failure on fired heaters, and vapor detectors for emergency valve isolation and sprinkler-system activation.

- *Priority 2.* Critical instruments whose failure *could* either cause, or fail to inform of, serious conditions involving environmental releases, property or production losses, or other non-life-threatening situations. These alarms are given a slightly lower priority, but are also prooftested on a regular schedule.

There are 1,346 "Priority 2" alarms at PPG Lake Charles. Examples include those alarms or trips on refrigeration compressors, rectifiers, cooling towers, kettles and stills, and those controlling power and instrument air.

- *Priority 3.* All other alarms. These non-critical devices are not on a regular prooftesting schedule.

Another method of classifying instruments into a hierarchical ranking is as follows [5]:

- Class I — Community-protection loops
- Class II — Employee-protection loops
- Class III — Major property- and production-protection loops
- Class IV — Minor property- and production-protection loops

Prooftest frequencies

Assigning prooftest frequencies for complex, safety-instrumentation loops requires "sound engineering judgment" for simple systems. For more complex, interlock systems, the frequency is a function of the tolerable hazard rates. For example, Du Pont Co.'s (Wilmington, Del.) Sabine River Works (Orange, Tex.) has 35,000 instruments in service [6]. Every safety interlock is tested on a periodic basis, usually every 12 months, following a detailed written test procedure.

Once a test frequency is established by Du Pont for a particular interlock, a significant review and multiple approvals are required to permanently remove the interlock or change its test frequency [6].

Many of PPG's high-pressure and high-temperature alarms are tested every six months. About half of the PPG Lake Charles Complex test frequencies have been developed using detailed reliability studies that consider the "hazard rate" (the acceptable probability of a process accident) and the "demand rate" (the number of times the critical alarm or shutdown function is required in service).

Within our chlorinated-hydrocarbon complex — which comprises a major portion of the Lake Charles facility, and produces eight different product lines — 26% of the loops are on a one-year test frequency, 42% are on a six-month frequency, 15% are tested every four months, 11% every three months and the remaining 6% at various other frequencies from weekly to every two years.

Administering the test program

Schedules for testing complex safety loops are distributed monthly in a packet to area supervisors. The packet includes a form listing the loops due in the current month (as well as those past due), the test frequency, and the authority requesting the test.

Accompanying the schedules are "working documents" for each loop due for testing. These documents contain the data necessary to perform the tests, such as set points and transmitter calibrations. The working documents are given to the mechanics who carry out the testing.

There are six groups of technical personnel responsible for prooftesting: electricians, critical-metering mechanics, refrigeration mechanics, analyzer-repair technicians, instrument-maintenance personnel, and area instrument mechanics.

Once the loops have been prooftested, the results are entered into the mainframe computer, along with the date tested and the condition found. All employees have "read" access to this data on many terminals throughout the complex, but only authorized personnel have access to the actual database.

Keeping track of other aspects of the system is also done by computer. For example, if a setpoint on a particular instrument needs changing, the request and approval (both approved by the area supervisor) are done at the computer terminal. Loops that are no longer in service can be deleted by the authorized personnel. New loops can be added by an individual designated by the technical safety committee, which reviews all new and modified installations.

Finally, monthly compliance reports are issued to top plant management. Particularly critical to the success of the program is management's full support and commitment to compliance with process-safety administration. ■

Edited by Suzanne Shelley

References

1. Braun, N., Silent sentinels: Lessons on pressure relief valve performance, *Valve Magazine*, Spring 1989, pp. 28-31.
2. Van Boskirk, B. A., Don't let your plant blow up — An overview of the elements of overpressure protection, paper presented at the joint meeting of Southeast Texas Section of the AIChE, and the Sabine-Neches Section of the Amer. Chem. Soc., April 1986.
3. Woolfolk, W., and Sanders, R., Process safety relief valve testing, *Chem. Eng. Progress*, March 1984, pp. 60-64.
4. Woolfolk, W. and Sanders, R., Dynamic testing and maintenance of safety relief valves, *Chem. Eng.*, October 26, 1987, pp. 119-124.
5. McMillan, G. K., Can you say 'Process Interlocks'?, *Tech Journal*, May 1987; also presented at the 1988 AIChE Loss Prevention Symposia, New Orleans, La., March 1988.
6. Barclay, D. A., Protecting process safety interlocks, *Chem. Eng. Progress*, February 1988, pp. 20-24.

The authors

Roy E. Sanders is superintendent of loss prevention at PPG Industries, Inc. (P.O. Box 1000, Lake Charles, LA 70602, Tel.: (318) 491-4322). He graduated from Louisiana State U. in 1965, with a degree in chemical engineering. Previously, Sanders held various technical positions in operations involving chlor-alkali products and chlorinated hydrocarbons. He serves on AIChE's Loss Prevention Symposia Committee. A registered Professional Engineer in La., he has authored articles on plant modifications, SRV testing and tank problems. Since 1984, he has taught a course, "Chemical Plant Accidents," for AIChE's Continuing Education Program.

James H. Wood is a senior design engineer in loss prevention at PPG Industries, Inc. (P.O. Box 1000, Lake Charles, LA 70602, Tel.: (318) 491-4792). He graduated from Ohio State University in 1976, with a degree in chemical engineering. Wood began his career in operations at an organics plant at PPG's Lake Charles complex, and became a loss-prevention engineer in 1979. He was instrumental in the development of an effective prooftest program for ensuring the reliability of critical instrument systems at the Lake Charles complex. Wood is co-author of a 1990 technical paper, "Stop Tank Abuse." ■

USE NITROGEN TO BOOST PLANT SAFETY AND PRODUCT QUALITY

Phillipe B. Laut and **David Johnstone**
Air Liquide America

Nitrogen offers a wide range of benefits during the handling and storage of raw materials and products in a chemical process plant. During blanketing, pressure transfer, purging and vapor condensation, nitrogen is used to exclude oxygen, which protects materials from oxidative degradation, and reduces the chance of fire or explosion when handling flammable products.

The following definitions are pertinent to any discussion of safety issues involving flammable liquids:

Flash point is the lowest temperature at which a liquid will produce enough vapors to support combustion. Some evaporation can take place below the flash point, but at a vaporization rate that is not sufficient to support combustion. *Ignition point* is the temperature at which a vapor mixture will ignite in the presence of a flame.

Fire triangle

Three elements are required for fire or combustion to occur: fuel, oxygen and an ignition source. Successful elimination or absolute control of any one prevents the possibility of fire or explosion.

Fuel sources for combustion include chemical vapors and dust particles. The storage of flammable liquids always creates potential fire hazards, particularly when the temperature of a volatile chemical rises above its flash point and its vapor concentration is in the explosive range.

With most flammable materials, there is a maximum and a minimum vapor concentration above or below which flame propagation will not occur, even when an ignition source is present. The lower explosive limit (LEL) and upper explosive limit (UEL) are usually expressed in terms of vol.% of chemical vapor in air (Table 1).

The explosive range for a given chemical is the region between the lower and upper explosive limits. It includes all concentrations of vapors in air at which flash and flame propagation will occur when ignited. As shown in Figure 1, mixtures that are too rich or too lean in fuel vapors will not ignite.

Many flammable chemicals have wide explosive ranges. For example, the range for hydrogen is 4 to 74 vol.%, which means that only a very lean or a very rich mixture of hydrogen in air is safe from an explosion if an ignition source is nearby.

To keep a mixture from becoming flammable, the vapor concentration in air must be kept outside the explosive range. This may be accomplished by increasing the concentration of chemical vapors above the UEL, or by displacing the chemical and decreasing its vapor concentration below the LEL. While it is possible to prevent such a mixture from burning, it is virtually impossible to accurately control these factors.

Ignition can be triggered by heat, sparks, static electricity, friction and improper grounding. It is not possible to exclude all ignition sources in a process facility. Even with proper design, grounding does not prevent internal static charges, and the potential for operator error is always present.

Oxygen is the third essential element. Unlike fuel and ignition sources, oxygen can be completely eliminated. While most chemical vapors need a minimum concentration of oxygen for combustion to occur — typically 10 to 15 vol.% — most dusts need only 8 vol.% oxygen for combustion. As long as the oxygen concentration is below this minimum oxygen for combustion (MOC), combustion cannot take place, even if the other critical factors are present. MOC values found in Column 5 in Table 1 correspond to the nose of the flammability curve in Figure 1.

The simplest and safest way to reduce the oxygen concentration is to dilute the chemical-air mixture in a vessel with an inert gas. Several gases are available (and discussed below), including combustion byproducts and carbon dioxide, but nitrogen is usually the most common and cost effective.

Preserve product quality

Many chemicals, polymers and food products degrade through reactions with oxygen or moisture in the air. Purging, pressure transferring and blanketing with nitrogen protects against material degradation. During blanketing, dry nitrogen is used to replace the air atmosphere inside a storage tank or reactor to minimize contamination and reduce the formation of undesirable and potentially hazardous reaction byproducts. By creating a slightly positive tank pressure, blanketing also minimizes air infiltration.

The risk of corrosion can also be reduced by using an inert atmosphere to exclude moisture. Discussed below are several other applications that are well served by low-cost nitrogen.

Emissions control. The use of a nitrogen blanket in the headspace of a vessel reduces vapor and product losses, which protects plant personnel and the environment. In blanketed storage tanks, vapors are vented only when the setting on a pressure-relief device has been reached, while in non-blanketed storage tanks, potentially hazardous vapors may escape continuously through evaporation.

Sparging. Very often, product streams reaching a reactor vessel or storage tank contain considerable quantities of dissolved oxygen, which can be more detrimental to product

Originally published June 1994

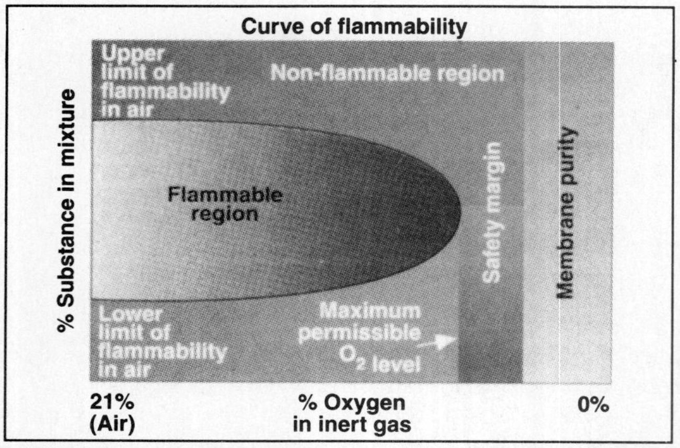

FIGURE 1. For flammable liquids or gases, combustion can occur only if the relative amounts of fuel vapor and oxygen in the mixture fall within limits, which are defined by the upper and lower explosive limits and the maximum permissible oxygen level (shown as the nose of the curve). A curve of flammability, such as this one, can be plotted for any chemical

CARLA MAGAZINO

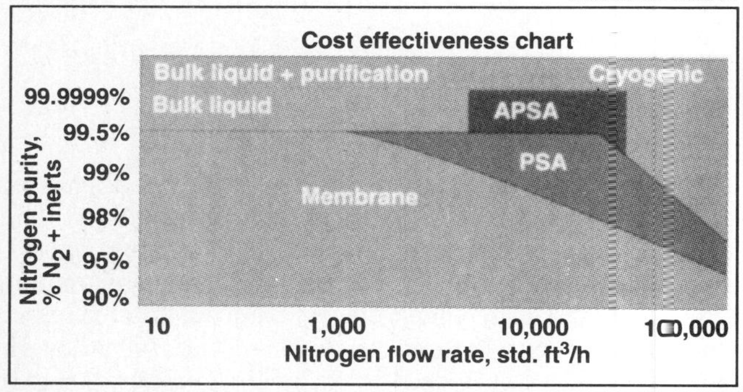

FIGURE 2. Each nitrogen-generating technology is the most economical over a specific range of flowrates and purities

quality than head-space oxygen. Sparging systems dissolve or infuse nitrogen into liquids, to strip dissolved oxygen.

Pressure transfer. In many cases, pumps are not suitable to move products. These include applications where several products must be transferred concurrently, when product degradation or contamination from pumps is likely to occur, when transfer must be done between different locations in the facility, or during handling of corrosive or abrasive chemicals. Nitrogen consumption is a function of the density of the liquid to be moved, length and configuration of the transfer pipe, desired liquid flow rate, and differential liquid levels between transfer points.

Vapor recovery. When supplied in cryogenic form, nitrogen is also useful as a refrigerant to condense and recover chemical vapors for either environmental or economic reasons. Such systems use cryogenic nitrogen to recover chemical vapors as they are vented from trucks, rail cars or other vessels.

A shell-and-tube heat exchanger is typically used as the vent condenser. Liquid nitrogen passes through the tubes, while solvent-laden air from a truck or vessel is ducted to the shell side of the heat exchanger. The solvent vapors cool and condense into a liquid holding tank. The remaining "clean" vapor stream contains little residual vapor, and can be vented to a proper location. The nitrogen exhaust can then be re-used elsewhere in the facility.

A given recovery system can be used for a variety of chemical vapors simply by adjusting its operating temperature. Such systems are especially well suited for operations in which significant volumes of chemicals with large venting requirements are handled every day.

In terminals where nitrogen is supplied in bulk as a liquid and then gasified onsite, a vent-condensation system requires little extra nitrogen or energy. This is because the vapor-recovery device does some of the work of the vaporizer normally used to convert the merchant nitrogen from a liquid to a gas.

Operators have several sources to choose from, to reduce fire hazards and protect raw materials and finished products from oxygen attack

Inert-gas supply options

By definition, an inert gas is nonflammable, chemically nonreactive and noncontaminating. For most industrial uses, the inert gas must also be nontoxic and free of moisture. Discussed below are the most commonly used inert gases:

Combustion gases. These inert gases, formed as byproducts from combustion of natural gas, are inexpensive but have several disadvantages. While they are theoretically composed of 79% nitrogen and 21% carbon dioxide, variable residuals of other byproducts, including hydrogen, oxygen, carbon monoxide, hydrocarbons, and water are always present. Accordingly, treatment steps, including filtration and drying are usually required prior to use.

Carbon dioxide. Because CO_2 is soluble in many liquids, it may react with moisture to form unwanted carbonic acid, which can react with the products being blanketed or cause equipment

FLAMMABILITY CHARACTERISTICS OF COMMON GASES AND LIQUIDS						
PRODUCT	BOILING POINT TEMPERATURE °F (1)	FLASHPOINT °F (2)	AUTO-IGNITION TEMPERATURE °F (3)	FLAMMABILITY LIMIT IN AIR VOL.% (4)		MAXIMUM OXYGEN CONTENT VOL.% (5)
				LEAN (LEL)	RICH (UEL)	
Acetaldehyde	70	36	365	4.1	55	12
Acetone	133	0	1,000	2.6	12.8	11.6
Acetylene	-119	–	571	2.5	81	–
Allyl chloride	113	-25	737	3.3	11.1	12.6
Ammonia	-28	–	1,204	16	25	15
Benzene	176	12	1,044	1.4 (at 212°F)	7.1	11.2
1,3-Butadiene	24	–	804	2	11.5	10.4
Butane	32	–	761	1.9	8.5	12.1
Butyl acetate	259	79	789	1.7	7.6	11.5
1-Butene	20	–	723	1.6	9.3	11.4
2-butene	34	–	615	1.8	9.7	11.7
N-butyl formate	225	64	612	1.7	8	12.4
Carbon disulfide	115	-22	212	1.3	44	5.4
Carbon monoxide	-310	–	1,128	12.5	74	5.6
Cyclopropane	-27.4	–	928	2.4	10.4	11.7
1,1-Dichloroethylene	99	-5	856	5.6	11.4	10
2,2-Dimethylbutane	121	-54	797	1.2	7	12.1
Ethane	-127	–	959	3	12.5	11
Ethanol	173	-55	793	4.3	19	10.6
Ethyl acetate	171	28	798	2.2	11.5	11.2
Ethyl bromide	101	–	952	6.7	11.3	14
Ethyl chloride	54	-58	966	3.8	15.4	13
Ethylene	-152	–	842	3.1	32	10
Ethylene oxide	56.4	<0	804	3	100	–
Ethyl ether	94	-49	356	1.9	48	–
Ethyl glycol	273.2	100.4	460	1.8	14	10.7
Ethyl formate	130	-4	851	2.7	13.5	10.4
Ethylglycol acetate	312.8	134.6	716	1.7	–	11
Gasoline	–	-49	996	1.4	7	–
Gasoline (octane 60)	–	-45	536	1.4	7.6	11.6
Gasoline (octane 92)	–	–	734	1.5	7.6	11.6
Gasoline (octane 100)	–	-36	853	1.4	7.4	11.6
Heptane	209	25	433	1.2	6.7	11.6
Hexane	156.2	-7	453	1.2	7.5	11.9
Hydrogen	-422	–	1,085	4	75	5
Isobutane	11	–	864	1.8	8.4	12
Isopropanol	179.6	70	860	2	12	12
Isopropyl ether	154.4	-18	830	1.4	21	10
Methane	-263	–	999	5.3	14	12.1
Methanol	151	52	867	7.3	35	9.7
Methylethylketone	176	26	960	1.8	12	11.4
Methylisobutylketone	242.6	73.4	858	1.4	7.5	12
Methyl acetate	135	14	935	3.1	16	10.9
Methylamine	19.4	–	806	4.9	20.7	10.7
Methyl butene	87.4	<20	–	–	–	11.4
Methyl chloride	-11	–	1,170	10.7	17.4	15
Methyl formate	89.6	-2	853	5.9	20	10.1
N-butanol	244.4	95	649	1.4	11.2	11.3
Pentane	97	<-40	588	1.5	7.8	12.1
Propane	-44	–	871	2.2	10	11.4
Propylene	-54.4	–	770	2.4	10.3	11.5
Propylene oxide	93	-35	869	2-2.1	21.5-22	~10
Toluene	230	39.2	996	1.4	7	9.1
tert-Butylamine	112	16	716	1.7	8.9	<11
Vinyl chloride	9	–	882	4	22	9
Xylene	284	85	867	1.1	7	8

1. Boiling point temperature at standard atmospheric pressure
2. Flashpoint is the minimum temperature at which vapors of a combustible liquid will be ignited by a flame in certain experimental conditions
3. Autoignition temperature is the minimum temperature at which a product will spontaneously oxidize in air
4. Flammability limit is the volume percentage of combustible gas in air such that below the lean limit (LEL) or above the rich limit (UEL), the mixture is considered nonflammable
5. Maximum oxygen content is the oxygen percentage in a combustible gas mixture below which the mixture is non-flammable

TABLE 1. Flammability characteristics are provided for some gases and liquids commonly found in chemical process facilities

Compound	o_s	Measured LEL	(o_s)LEL	Maximum permissible oxygen
Acetaldehyde	2.5	4.1	10.3	12
Acetone	4	2.6	10.4	11.6
Allyl chloride	4	3.3	13.2	12.6
Ammonia	0.7	16	12	15
Benzene	7.5	1.4	10.5	11.2
1,3-Butadiene	5.5	2	11	10.4
n-Butane	6.5	1.9	12.4	12.1
Butene	6	1.7	10.2	11.4
Carbon monoxide	0.5	2.4	2	5.6
Cyclopropane	4.5	2.4	10.8	11.7
2,2-Dimethyl butane	9.5	1.2	11.4	12.1
Ethane	7.2	3	10.5	11
Ethylene	3	3.1	9.3	10
Ethyl alcohol	3	4.3	12.9	10.6
Ethyl bromide	3	6.7	20.1	14
Ethyl chloride	3	3.8	11.4	13
Ethyl ether	6	1.9	11.4	10.3
Ethyl formate	3.5	2.7	9.5	10.4
n-Heptane	11	1.2	13.2	11.6
n-Hexane	9.5	1.2	11.4	11.9
Hydrogen	0.5	4	2	5
Isobutane	6.5	1.8	11.7	12
n-Butyl formate	6.5	1.7	11	12.4
Isopropyl ether	9	1.4	12.6	10
Methane	2	5.4	10.8	12.1
Methanol	1.5	7.3	10.9	9.7
Methyl acetate	3.5	3.1	10.9	10.9
Methyl formate	2	5.9	11.8	10.1
Methylamine	2.2	4.9	11	10.7
Methyl chloride	1.5	10.7	16	15
Pentane	8	1.5	12	12.1
Propane	5	2.2	11	11.4
Propylene	4.5	2.4	10.8	11.5
Vinyl chloride	2.5	4	10	10

Table title: MAXIMUM PERMISSIBLE OXYGEN LEVELS, VOL. %

TABLE 2. Shown are the maximum permissible oxygen levels for a variety of chemicals, calculated from Equation 4. For most of those shown here, there is good agreement (± 10%) between the calculated [(o_s)LEL] and the measured values for maximum permissible oxygen content. Note, however, that Eqation 4 does not yield good results for hydrogen or ethyl bromide

corrosion. And, because CO_2 cannot be economically produced onsite, its price is strongly influenced by geographic and market conditions.

Nitrogen. Nitrogen is the most commonly used blanketing gas. It is considered non-reactive except under extreme temperatures and pressures, and its low solubility keeps it from contaminating most liquids. In addition to being colorless, odorless, tasteless, and non-toxic, nitrogen has a very low dew point and is a poor conductor of heat and electricity.

Nitrogen can be delivered continuously through pipelines, supplied from a dedicated cryogenic plant onsite, from periodic truck or rail delivery, or it can be generated onsite in a non-cryogenic system, using hollow-fiber membranes or pressure swing adsorption (PSA).

In all but the latter, nitrogen is produced using traditional cryogenic separation, whereby air is fractionally distilled to yield nitrogen and oxygen. This nitrogen is then liquefied for easier handling and transportion and is ultimately re-vaporized as needed.

The best supply option is chosen on a case-by-case basis. If customers are near a pipeline network, large quantities of inexpensive gas are typically available from centralized production units. In other cases, volume requirements may justify construction of a dedicated cryogenic plant.

However, most nitrogen is delivered as a liquid by cryogenic trucks and stored in onsite cryogenic storage vessels. Usually, such customers need only moderate volumes of nitrogen — in the range of 100,000 to 20 million standard cubic feet (scf) per month.

Customized vendor software is often used to optimize the selection of an onsite nitrogen system

Over the last few years, more and more operators in the chemical process industries have been turning to non-cryogenic production methods. Although both membrane and PSA systems produce nitrogen with some residual oxygen — anywhere from 0.1 to 10%, depending on design specifications — these slightly lower nitrogen purities are still adequate for many applications.[2] In most cases, switching to the newer non-cryogenic technologies has demonstrated considerable savings — up to 50% of delivered costs.

PSA systems involve passing compressed air through activated carbon adsorbent beds. This technology has been commercially available for the last twenty years. Purities up to 99.9% can be cost effective relative to liquid nitrogen, although nitrogen purities up to 99.95% are easily attained with a PSA system.

The adsorbent technology is relatively mature, and has been performing at the top of its capabilities for nearly 20 years, so there is not much room for performance improvements. However, the simpler and more efficient membrane-based systems use relatively new technology, and have only been commercially available for the last five to seven years. The introduction of new polymers continues to improve membrane performance. As a result, the older PSA technology continues to occupy a smaller segment of the non-cryogenic nitrogen market.

Air separation by hollow-fiber membranes begins by compressing air to 175 psig or more, and then filtering the air to remove moisture, oil and other contaminants. Hundreds of thousands of polymer fibers are packed into small modules, through which the treated air circulates. The oxygen molecules diffuse radially out of the fiber module faster than do the nitrogen molecules. Nitrogen molecules travel axially inside the hollow membrane fibers, and eventually exit the membrane vessel for delivery to the process.

A new cryogenic nitrogen-generation technology called APSA[1] produces gaseous nitrogen at purity levels of 99.9999% purity without the complicated use of add-on, hydrogen-based deoxidation units. Mid-range users, requiring 6,000 to 30,000 scfh are best served by these systems, which offer economies of scale that reduce overall manufacturing costs. Onsite systems typically offer the following:

Cost savings. Figure 2 shows that over a wide range of nitrogen purities and flowrates, non-cryogenic nitrogen gas is produced at a lower cost than liquid nitrogen.

Capital savings. Aside from a small number of companies that sell nitrogen-generation systems to the end user, most gas vendors establish long-term contractual agreements with their clients. In such an arrangement, the vendor typically owns and operates the equipment, and the customer pays a monthly fee that is proportional to the capacity of gas produced at the specified product purity and flowrate. The benefit is that the user company does not have to approve a major capital expenditure, nor is it responsible for the maintenance and performance of the equipment. When handled through a service contract, routine upgrades to satisfy changing consumption needs or evolving technologies are easily accommodated by the gas supplier.

Reliability. Onsite systems require no dedicated personnel and can run unattended. Most of the major industrial gas vendors offer similar service contracts. Typically, the supplier provides all necessary preventive maintenance to ensure maximum reliability. Operational problems can be immediately detected by a telemonitoring system, so that field technicians can be dispatched on a 24-hour basis. Keeping a backup tank of liquid nitrogen onsite (or cylinders, for smaller systems) ensures that supply will continue uninterrupted, even during unplanned downtime.

Safety and quality. Onsite non-cryogenic nitrogen purity can be maintained at the optimum level to minimize costs without compromising

[1] APSA is a patented process from Air Liquide.

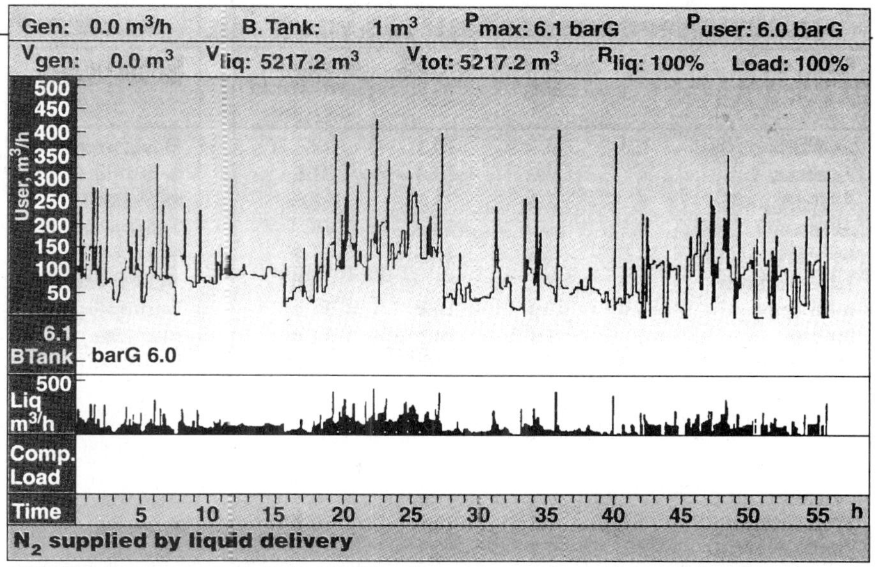

FIGURE 3A (above) and 3B (right). Shown here are two competing nitrogen-sourcing schemes for a single user, over a 56-h period. Figure 3a shows a simple case, where the nitrogen is supplied from a liquid source, hence, there is no compressor load and no buffer capacity. Figure 3b shows a case where nitrogen is generated onsite. The profile shows how the buffer supply is used to ensure adequate capacity. To meet peak demand, liquid nitrogen is supplied from a backup source

explosion and fire safety, or causing undesired or dangerous oxidation reactions. Another benefit is reduced plant traffic by a reduction in liquid tanker deliveries. And, awkward and potentially dangerous handling of cylinders is reduced.

NOMENCLATURE	
Gen	Nominal capacity of the membrane generator
B.tank	Buffer tank volume
P_{max}	Maximum pressure allowed in the buffer tank
P_{user}	Nitrogen pressure delivered to the user
V_{gen}	Total volume consumed from the membrane source
V_{liq}	Total volume consumed from the liquid source
V_{tot}	Total volume of nitrogen consumed (V_{gen} + V_{tot})
R_{liq}	Ratio of V_{liq} to V_{gen}
Load	Percent of the time the membrane system is online
User, m^3/h	Instantaneous nitrogen flowrate consumed by the customer
B. tank, barG	Pressure inside the buffer tank
Liq., m^3/h	Demand pattern from the backup liquid nitrogen supply
Comp. load	Compressor load (0=off; 1=on)

Cost optimization

Choosing the right onsite system is a function of desired flowrate, purity, consumption pattern, electricity cost, location and other site-specific factors. When using an onsite nitrogen generator, users have two ways to ensure that an adequate supply of nitrogen is always available. A liquid nitrogen backup supply is recommended to provide capacity needed for peak shaving, emergency shutdowns, special high-purity applications, or cryogenic cooling needs. In some cases, a buffer vessel is also added, to capture and store excess nitrogen that is generated but not consumed. Not only does this reduce wasted capacity, but it allows operators to meet peak demand at a lower cost than by using liquid nitrogen.

Three components are critical when evaluating onsite system costs. In every case, the objective is to minimize the sum of these three costs:

- Capital for the air compressor, generator, and liquid nitrogen tank. These are proportional to the system's unit capacity, and are reflected in the monthly facility fee
- Energy for the compressor, whose size is defined by pressure needs
- Liquid nitrogen backup supply, and storage capacity of the buffer system

A compromise must be found between the sizes of all the components. For example, undersizing one component (such as liquid nitrogen supply) would minimize its cost but may drastically increase the cost of the other components (such as the need of a large, expensive buffer tank). This will impact the cost to produce nitrogen onsite.

To optimize the design, a detailed ni-

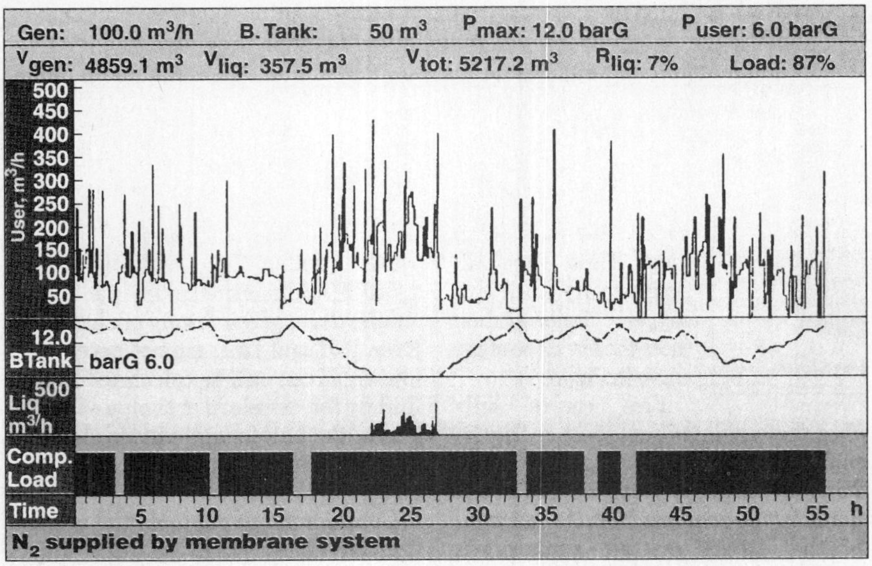

trogen flow profile must be recorded under normal operating conditions. A computerized flow-monitoring-and-recording skid is installed to collect performance data from an existing system.

If the nitrogen requirement is new, the plant should start on liquid nitrogen to obtain accurate consumption data. Once such demand data have been collected, specialized software can be used to analyze flow patterns and facilitate system sizing.

Figures 3A and 3B show competing sourcing options for a given user's flow profile. During analysis with the modeling software, the user is allowed to vary the volume of the buffer tank and size of the nitrogen-generating system. The goal is to determine the optimum combination of design parameters that meets the user's consumption pattern with minimal consumption of nitrogen from the relatively costly backup source for liquid nitrogen.

By plotting several critical process variables (Figure 4), one can visualize the thought process that gas suppliers use when designing systems to produce nitrogen at the lowest unit cost. Assuming that the required nitrogen pressure is fixed, the surface, S, represents the unit cost for nitrogen, and is produced by plotting the following:

$$S = f(G,B)$$

where:
G = relative generator capacity
B = buffer tank capacity

The low points on the surface show the lowest cost at the optimum combination of design parameters.

Flowrate instrumentation. Precise flow measurement is the single most important factor in developing an accurate consumption profile. Three types of devices are available:

Orifice plates are relatively inexpensive, but are less accurate and less flexible than competing methods. They must be sized separately for each new set of operating conditions, including pressure, temperature and pipe size. They have a relatively low turndown ratio — on the order of 4:1.

Pressure drop across the orifice plate is proportional to flowrate and is transmitted to a differential pressure recorder, which includes a totalizer that records the total volume of gas that has passed through the orifice.

Thermal–convective mass flowmeters use insertion probes that are accurate for several pipe sizes. These are connected to a flow computer and transmitter, which produces digital readouts of instantaneous and total flowrates.

These devices consist of two thermocouples with a heat-generating source in between. Heat dispersion, measured by a downstream thermocouple, is proportional to flowrate.

Once installed, these mass flowmeters are much more flexible than orifice plate meters. Unlike competing devices, mass flowmeters require no further system shutdowns when changing conditions would require recalibration or adjustments. They have a very high turndown ratio, up to 300:1, which ensures accurate measurements over a very broad range of flowrates.

Vortex flowmeters are the most commonly used, thanks to their ease of use, high reliability (with no moving parts) and fairly good turndown ratio (on the order of 14:1). These devices have small hurdles or bumps that create rotational movement in the gas. They measure the linear motion of the rotating gas, which is proportional to flowrate. Vortex flowmeters combine the low cost and simplicity of an orifice plate with the higher versatility of a thermal-conductive mass flowmeter.

Optimizing system design

In addition to sizing software, many industrial gas suppliers have developed software that helps clients to meet the quality objectives discussed below:

FIGURE 4. The unit cost of nitrogen produced onsite can be optimized by varying the size of the components and the buffer system, unlike that of liquid nitrogen, for which the delivered cost per 100 scf is fixed. The surface of this 3-D plot represents the relative gas cost (cost/total flow), which accounts for the rental cost and pricing structure of a typical service contract

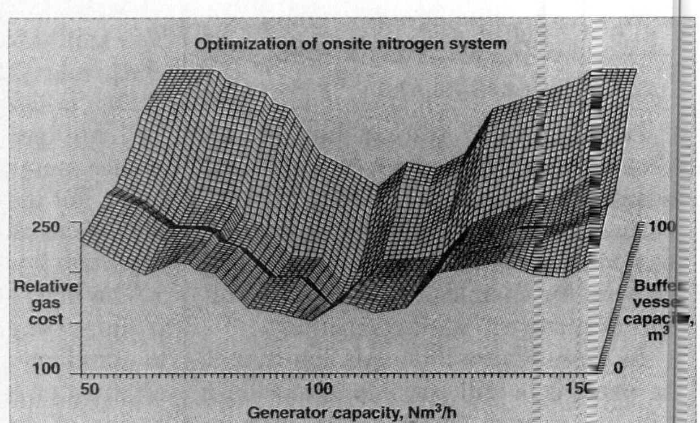

Combustibility and flammability. Historically, operators did not have to worry about the purity of nitrogen, because fractionally distilled nitrogen is ultrapure by design. Today, however, users of cheaper, non-cryogenic nitrogen must accept slightly higher amounts of residual oxygen. It is therefore the duty of both the user and the supplier to decide exactly how much oxygen a process can withstand without compromising safety and quality.

Doing so is not always easy, even though many critical values, including UEL, LEL, and maximum permissible

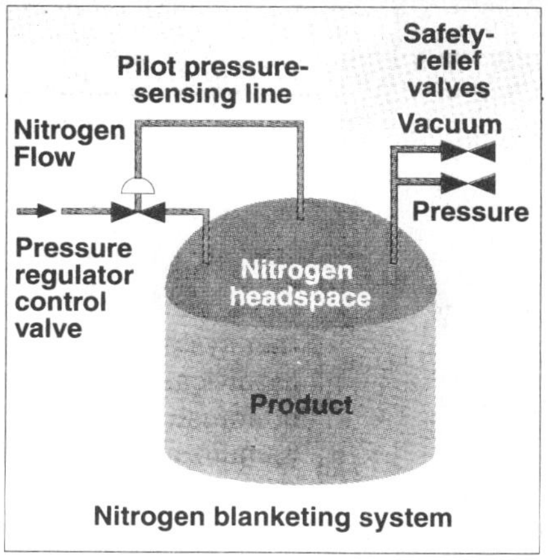

FIGURE 5. All of the components needed for a nitrogen-blanketing system to protect tank contents from fire or explosion are shown here

oxygen levels (described at the beginning of this article) can be found in published references. For chemicals with no known values, the critical parameters must be estimated.

For convenience, computer software has been developed to calculate maximum safe oxygen levels based on theoretical formulas. The calculations use stoichiometric oxygen values for the individual components that make up the studied molecules.[5] Using this software, it is possible to reasonably predict the maximum safe percentage oxygen (stoichiometric oxygen required for combustion) for some combustibles.

The software follows this generalized combustion equation, and requires as input the number of atoms of carbon, oxygen, nitrogen, sulfur, hydrogen and chlorine in the chemical compound:

$$C_cH_hO_oCl_dN_nS_s+[c+n+s+(h-a)/4-o/2]O_2 \longrightarrow cCO_2+aHCl+nNO_2+sSO_2+[(d-a)/2)]Cl_2+[(h-a)/2)]H_2O \qquad (1)$$

Unfortunately, testing has shown that this equation cannot be used in all cases.[2] In general, it is best suited for carbon-, hydrogen- and oxygen-bearing compounds; however, it does not seem to work for chemicals that include sulfur- or nitrogen-based compounds.

In cases where this equation cannot be used, it is still possible to calculate the maximum permissible oxygen level, especially for chlorinated or aminated products. For example, if the studied hydrocarbon is methylamine, the combustion reaction would be:

$$CH_3NH_2 + 2.25\ O_2 = CO_2 + 2.5\ H_2O + 0.5\ N_2 \qquad (2)$$

The calculated adiabatic temperature (T_{adia}) is the lowest temperature for which the combustion reaction is complete at the LEL. When the combustion reaction (Equation 2) is considered complete at the LEL point, any excess oxidant will simply behave as an inert. Thus, the LEL does not vary much with oxygen concentration (note the low slope of the curve in Figure 1).

Fire occurs only when the amount of heat produced by the combustion reaction is equal to or larger than the amount of heat absorbed by the inerts, by the products from the reaction, by the excess reactive components (which behave as inerts) and by any endothermic decomposition products. Therefore, the following can be written:

$$(H)(LEL)=100(C_p)(T_{adia}-T_i) \qquad (3)$$

where:

H=heat of combustion for the organic chemical, cal/mol

LEL=lower explosive limit for the organic chemical in air at 25°C and 1 atm pressure, vol%

C_p= heat capacity of nitrogen = 8cal/mole/°C

T_{adia}= adiabatic flame temperature for the organic chemical, °C

T_i = initial temperature of the organic chemical = 25°C

For well-known chemicals, H and LEL are can be found in published reference materials, so T_{adia} can be calculated. For instance, for methylamine, H = 234 kcal/mole and LEL = 4.9%. Thus, Equation 3 yields T_{adia} = 2,642°F.

The LEL corresponds to the concentration of oxygen required to produce a complete combustion reaction of the aminated product. This concentration is determined by the stoichiometric factor (o_s) of the oxidation reaction. This concentration is by definition the MOC, since below it, there will not be enough oxygen to maintain the 2,642°F needed for flame propagation.

The following relationship is thus true:

$$MOC=LEL(o_s) \qquad (4)$$

For methylamine, the MOC is then 4.9(2.25) = 11%. Table 2 shows comparisons of calculated and published values.

For more complicated amines, the calculation method described above yields similar T_{adia} values, around 2,700°F, and demonstrates that T_{adia} is similar for a given family of chemicals. Even if H and LEL are not necessarily known, H can still be calculated by first finding the standard enthalpy of formation of the basic mono-atomic elements present during combustion, and then using the Hess Law.

Consider as an example the calculation for dimethylethylamine (DMEA). The calculation for DMEA yields H=690 kcal/mole. Once H and T_{adia} are known, the unknown LEL can be found. Using Equation 3, the LEL for DMEA = 1.7% ± 2%.

Now that H, T_{adia} and LEL are known for DMEA, o_s can be calculated using the combustion equation:

$$DMEA + o_sO_2 = 4CO_2 + 5.5\ H_2O + 0.5\ N_2 \qquad (5)$$

where:

$DMEA = (CH_3)N(C_2H_5)$

$o_s = 6.75$

Using Equation 4, the MOC can then be calculated for DMEA: 1.7(6.75) = 11.5% ±1.4%. This exercise illustrates that for DMEA — even factoring in a 50% safety margin — nitrogen produced onsite is sufficient, as long as it contains no more than 6% residual oxygen (having a purity of 94%).

To summarize, if LEL is known, it is

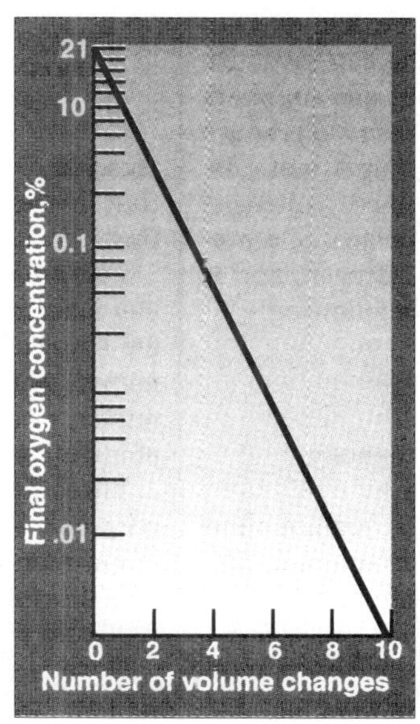

FIGURE 6. Sweep purging is used to replace the original atmosphere (air) inside a vessel with nitrogen. This relationship, shown using purge nitrogen of cryogenic purity (99.999% pure), shows the number of complete volume changes that are needed to achieve a desired final oxygen concentration, irrespective of the vessel volume. Using membrane-generated nitrogen of slightly lower purity shifts the curve slightly to the right

[2] Information source: Y. Torre, Research Technologist, Air Liquide Research Center, Les Loges, France; tel.: 33-1-3067-6260

easy to find MOC using Equation 4. If LEL is not known, it can be estimated starting with T_{adia} for the family of products to which the chemical belongs (aminated, chlorinated or others), and calculating the combustion heat. Table 2 shows maximum permissible oxygen calculated using Equation 4 for a variety of chemicals. In nearly all cases, the calculated numbers show very close agreement with known values, which underscores the validity of this method.

Tank blanketing. A typical software program used to estimate tank-blanketing requirements needs the following data: Tank contents, capacity and dimensions, the type of roof on the tank, average fluid level, liquid discharge and fill rates, flash point of the contents, average storage temperature and daily temperature changes, and the pressure-relief setpoint. The program then calculates vapor volumes, daily breathing and monthly working losses, total venting and inbreathing requirements, total monthly nitrogen usage, and the time and volume required to empty each tank.

Tanks can be blanketed by a continuous flow of nitrogen, or by maintaining a constant positive pressure in the vessel. Typical blanketing operations require relatively low pressures (0.5 to 2 in. water column).

A typical blanketing system includes the following components (Figure 5):

- A regulator that controls flow into the tank and provides nitrogen at adequate pressure on demand
- A pressure-relief device that protects the tank from over-pressurization during normal filling or breathing conditions, or a rupture disk that protects the tank from over-pressurization during emergencies
- A vacuum-relief device that keeps the tank from drawing a vacuum during pumping out of tank contents, or from breathing losses

Working losses occur during transfer operations, since the pressure in the tank changes as the product volume changes. All components must be sized for the largest anticipated pumping rates for product addition and withdrawal. Similarly, breathing losses occur as vapors expand or contract due to temperature variations.

Purging. This operation is used to reduce the concentration of oxygen, water, or other unwanted chemicals and vapors in transfer lines, railcars, vessels, towers and storage tanks. Vessels must be purged with an inert gas when they have been open to the air, cleaned with steam or water, or when their chemical contents are to be replaced with a different material.

If the purge gas is lighter than the original gas, it must be added near the top of the vessel, and the exit should be near the bottom, as far away as possible from the purge gas entrance. If the purge gas is heavier than the original gas, it should be added near the bottom and exit near the top, as far from the purge gas entrance as possible.

Common purging methods

Choosing the most efficient purging method can significantly reduce both nitrogen consumption and operating time. Each has specific design and equipment requirements. In all cases, the use of warm nitrogen can eliminate liquid pooling at the bottom of vessels.

Dilution. In this, the simplest purging method, nitrogen is continuously added and vented to dilute the atmosphere inside a vessel and lower the concentration of the unwanted compounds. Dilution requires efficient mixing within the vessel and is appropriate for spherical or squat cylindrical vessels. It requires little internal piping, and works best when the nitrogen inlet and outlet points are widely separated.

Figure 6 shows the relationship between the desired oxygen concentration and the number of volume changes required during dilution purging. For example, a nitrogen purge to replace air in a vessel can reduce the original oxygen concentration from 21 vol.% to less than 1 vol.% with seven full volume changes. Thus, total nitrogen required would be seven times the headspace capacity. After any purging operation, an analysis of the remaining gas is recommended to prove that the needed oxygen level has been reached.

Displacement. In this process, nitrogen moves in a plug-flow pattern through the vessel, ideally with a uniform cross-sectional concentration. Gas velocity must be low to minimize turbulence and mixing.

Displacement purging is typically used in long, narrow vessels, and can be extremely efficient, because theoretically only one vessel volume is required for a complete purge. Pipelines and transfer lines are also well suited for displacement purging. Of all the methods mentioned, this one uses the least nitrogen but takes the most time.

Pressure. This process is effective for railcars and vessels with unusual geometries that can withstand moderate internal pressures. Nitrogen is added until the pressure in the vessel reaches a preset point. Then a vent is opened, allowing the pressure to drop to the original level. The pressure inside the vessel is cycled between two different points to enhance mixing. This technique uses more nitrogen but works relatively quickly.

Vacuum. This cyclic process combines vacuum and purging dilution. Drawing a vacuum on the vessel reduces the concentration of the contaminant gas. The vessel is then repressurized with nitrogen and the cycle is repeated. Each purging method has certain maintenance, equipment, time and volume requirements. Selection of the best one will impact nitrogen conservation and efficiency. ■

Edited by Suzanne Shelley

The authors

Philippe B. Laut is the national program manager for the Floal family of onsite nitrogen-generation systems for Air Liquide America (5230 S. East Ave., Countryside, IL 60525; tel.: 708-579-7708). Prior to joining Air Liquide in 1991, he spent four years as an applications-development engineer for GE Plastics and Dow Plastics. Laut holds a B.S. in mechanical engineering from France's Paris XIII University, an M.S. in materials sciences from Montpellier II University, and a Master's degree in business administration from Indiana University (Bloomington). He also holds the degree of European Engineer from the European Federation of Colleges of Engineers, and has authored several papers related to the use of onsite systems within various industries.

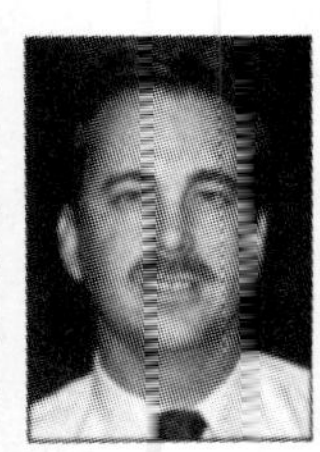

David W. Johnstone is an application engineer at Air Liquide America (12800 West Little York, Houston, TX 77041; tel: 713-896-2261). He received a B.S.Ch.E. from Texas Tech University, and is a member of AIChE and TAPPI. He has worked as a process engineer in the chemical industry and has several years experience in water treatment chemical sales and service. Johnstone is currently responsible for developing and marketing industrial gas applications for the chemical and refining industries in Tex., Okla., La., Ark. and Miss.

Section III

Quantitative Approaches to Risk Management

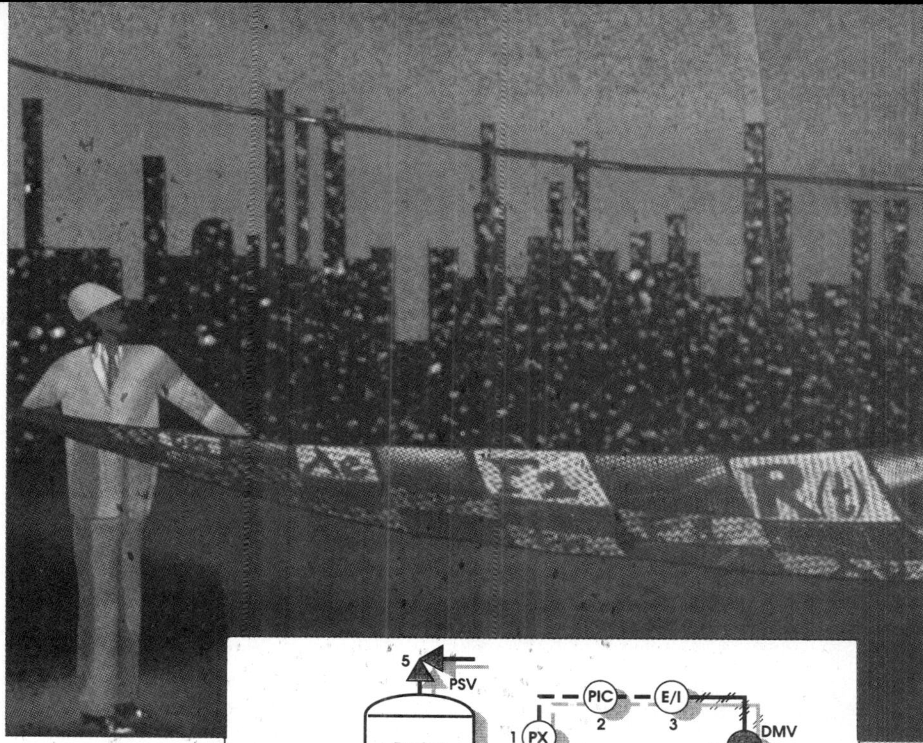

A NEW WAY OF QUANTIFYING RISKS

H. Wade Goodner
Consultant

For years, engineering designers have relied on qualitative techniques to determine the safety of process systems. These simple methods — based mostly on experience and intuition — are adequate to determine the potential failure modes of a process, but fall short for quantifying the relative risks of a failure (*1*).

In contrast, Fault Tree Analysis (FTA) is perhaps the most comprehensive quantitative method for assessing process risks. However, it's use has been largely limited because it is an overly complex and mathematics-intensive methodology.

But now, FTA has been refined and extended into a new technique that's much simpler to use. Called Hazard Assessment by Risk Analysis (HARA), the method clearly illustrates the safety and operational weaknesses of process systems.

HARA can be used for estimating the process risks of total systems. The technique is fully adaptable to any manufacturing or data-processing system whose component lives to failure are exponentially distributed. Part I of this article describes the basic FTA methodology on which HARA is based, and Part II — to be published in November — demonstrates the HARA technique itself.

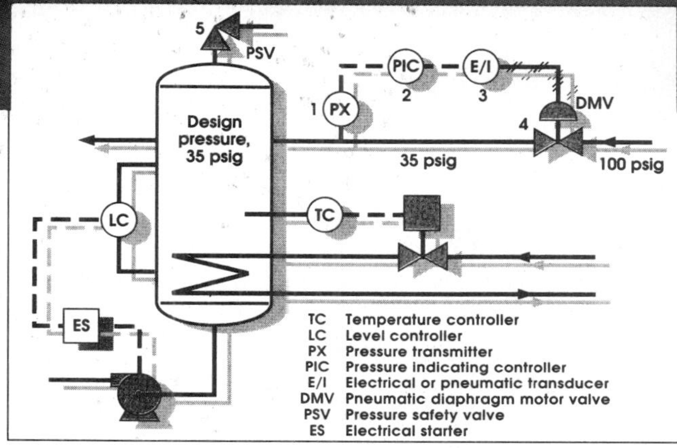

Figure 1. Quantitative, rather than qualitative, methods are needed to assess the risks posed by process systems, such as the one on the right

Streamlining the analysis

FTA consists of two main steps for estimating process system risk. First, users determine the minimum combinations of failed process components for a given failure mode of a system. Then, these combinations — called minimum cutsets — are analyzed with probability equations to estimate a system's failure frequency for a specific loss.

While the mathematical procedure has been much simplified with the publication of Fussell's approximations [*2*], what's been missing is an easy-to-use method of representing Fault Tree diagrams. Traditionally, Fault Tree diagrams are used to determine minimum cutsets, but these figures can quickly become very cumbersome. For example, they use many graphical notations, such as OR, AND and INHIBIT gates, to represent different events. As a result, the diagrams can be confusing because of the resulting muddle of symbols with various meanings — sort of like getting lost among the "trees in the forest" [*3*].

HARA replaces Fault Tree diagrams with Failure Logic diagrams that simplify the process of determining minimum cutsets (Figure 2). In contrast to the Fault Trees' complex notations, the Failure Logic diagram is drawn as a simple piping schematic in which the components are shown as valves.

Here, a failed component is a closed valve, and any minimum combination of closed valves that prevents an imaginary fluid from flowing from left to right across a diagram depicts a system failure. Because of the diagram's simplicity, the minimum cutset — the combination of closed valves that causes a system failure — can be easily and accurately determined.

As with FTA, HARA uses minimum cutsets to calculate a time-dependent failure rate for the system. This failure rate is then compared with acceptable risk-frequency criteria — such as from

Originally published October 1993

MARY JO MAZZELLA

An easy-to-use technique extends traditional Fault Tree Analysis for assessing process system risks

industry — over the same period, giving an indication of the safety level.

HARA has been successfully used for 10 years to evaluate new process designs. It has also been used to resolve differences between engineering contractors and their clients over problematic designs and specifications.

For example, HARA pinpoints the ineffectiveness of process safety devices that are added as a "just in case" measure. Often, these devices have no beneficial effect because they are part of a series chain of components whose overall strength is determined by its weakest link, most often a dominant, low-reliability device. System safety is worsened by such add-ons.

What's acceptable risk?

The essential purpose for quantifying risk is to answer the question, "Is this process safe for the people exposed to it?" However, this cannot be done until one determines what "safe" is, and what the criteria for a "safe" process are. In HARA, a "safe design is one that does not expose personnel to a risk that exceeds the average risk for the industry, or the risk of an 'Act of God' for a community" [4].

Generally, all activities have outcomes that are either a success or a failure. For each attempt of an activity, there is no guarantee of either success or failure. However, for each attempt, there is always a risk of failure and this, undeniably, is an essential characteristic of its safety level.

Some activities, of course, have greater risk of loss, such as of life or money, than others. But believing that zero risk exists or can be accomplished is a dangerous fallacy that precludes adequate system safety. However, risks can be cut to acceptable levels, as is implicitly accepted with regard to such everyday activities as traveling in a car or walking across a thoroughfare.

A hazard assessment of an engineering design estimates its inherent risks, and then evaluates them against some specified criteria to determine this design's acceptability for safety. If the estimated risks do not exceed the criteria, the design can be considered acceptable with regard to its operating risks; if, however, the risks are excessive, the design must be changed to meet the specification for maximum allowable risk.

Therefore, risk can be classified as either acceptable or unacceptable by its comparison with preset risk criteria. Note: For the purposes of identifying hazards, *a hazard is considered solely as an unacceptable risk.*

While risk has many qualitative definitions, the only objectively meaningful one is quantitative; it is based on the estimated loss caused by a failure event over a period of time, or that resulting from a number of trial events. So, to quantify industrial risk levels, parameters of human losses as a number of fatal or disabling injuries per 10^8 work-hours are employed.

An activity's risk is described as the product of the losses (consequences) from the failure of an activity and estimated frequency of failure over a given period of time:

$$\text{Risk} = \text{Frequency} \times \text{Consequences} \quad (1)$$

It is essential to note that while an event's certainty cannot be precisely predicted, its probability of failure can be reasonably estimated. Consequently, risk — not certainty — should be the parameter of interest by which system safety is evaluated. Thus, if it can be shown that a given process system does not expose personnel to a risk level that exceeds an industry's annual average (over the most recent 10 years), the process must be considered an acceptable safe design.

In performing a risk analysis of a process system, the probability of system failure, $\bar{P}_s(t)$, is calculated over a period of time using probability theory. The assumption here is that failure times for basic components are exponentially distributed. This permits the calculation of a system's time-dependent failure rate, which can then be compared with accepted time-based risk criteria. The equation for $\bar{P}_s(t)$ is:

$$\bar{P}_s(t) = 1 - P_s(t) \quad (2)$$

where, $P_s(t)$ = System's probability of success (reliability)

This implies that a system essentially operates with a series of unannounced component failures until a combination of such individual component failures (minimum outset) causes the overall system to break down. In HARA, a system's probability of failure, $\bar{P}_s(t)$, is determined by conventional reliability analysis methodology for various operating time periods (to a maximum of one year, or 8,760 hours). This generates the system's probability of failure over a year, assuming no preventive maintenance (PM) actions are taken over the period and that the system has not failed.

The reason for taking a one-year time period as a baseline is because the system's performance is determined by comparing its probability of failure, $\bar{P}_s(t)$, with $\bar{P}_a(t)$, the annual maximum acceptable probability of failure. The latter figure is based on a one-year average of industry losses.

Next, the two probabilities of failure are converted into failure rates. This yields more-useful parameters for selecting system components and for setting preventive maintenance schedules that can help achieve these acceptable

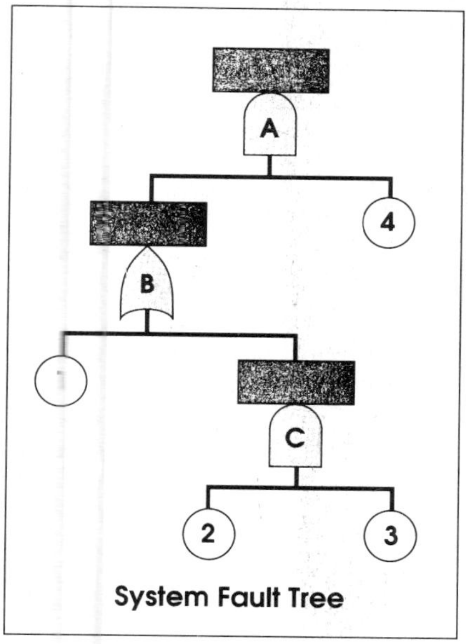

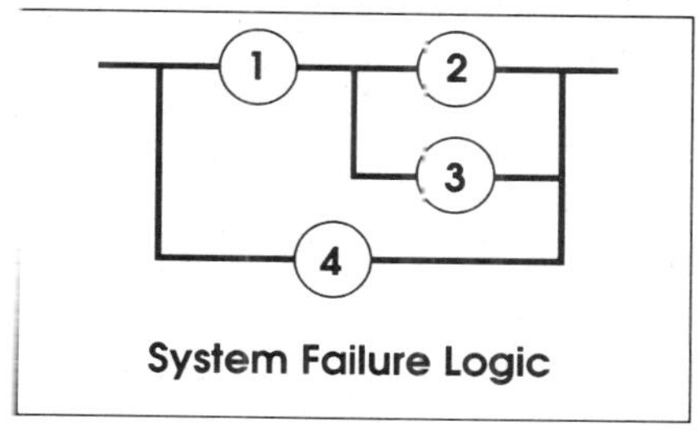

Figure 2. As can be seen, the Fault Tree (left) is far more complex than the Failure Logic diagram (right) for the same process system

risk levels. With specific failure data and appropriate probability equations with a failure model, a reasonable estimate of a component's probability of failure over a period of time can be made. Whenever this probability of failure approaches a specified maximum level (the maximum risk criterion), the component can undergo preventive maintenance or be replaced (Figure 2).

Average Annual Fatality Rate, or AAFR. Annual nonvehicular fatality data, together with the associated employee working hours, are published for the petroleum and petrochemical industry by the American Petroleum Institute (API; Washington, D.C.). An average annual value for fatalities per 10^8 exposure workhours, determined from 10 years of data, are used as the AAFR for these industries [*5*]. (The same can be done for any industry.)

To establish a criterion for acceptable risk, the AAFR is used as the maximum allowable risk level for fatal injuries to which employees may be exposed during their working hours. (*Use of the term "maximum allowable risk" does not imply that any certainty of a fatality is acceptable, but that criteria must be selected against which operating risk can be compared in order to evaluate a system's safety level.*)

Assuming that plant personnel are exposed to 10 potentially fatal events during their work shift, this AAFR is reduced by a factor of 10 for single-event applications. This ensures that the total employee exposure for all 10 events does not exceed the maximum allowable value. The current, 10-year-average AAFR value is 5E–8 hours^{-1}, which, when reduced by a factor of 10 gives 5E–9 hours^{-1} as the maximum allowable death risk for a single event.

Maximum Fatality Risk Frequency. For a single event, the Maximum Acceptable Fatality Risk Frequency (F_1) for a process system failure must not exceed 10% of the AAFR if the process is to be considered safe. To determine the value of this frequency, the estimated risk for the conditions of human exposure is equated to one-tenth of the historical AAFR. As a result, the expected risk for a failure must not exceed the 10-year average. If a single event's estimated risk is less than or equal to AAFR/10, then:

Estimated Risk = 1/10 × (5 fatalities/100,000,000 exposure workhours)

$F_1 \times P \times E$ = 5E–9 hours^{-1}

$F_1 = (5E - 9)/(P \times E)$ **(3)**

where, F_1 = Maximum Acceptable Fatality Risk Frequency, hours^{-1}

P = Percent of exposed persons fatally injured (a qualitative estimate)

E = Workhours of exposure to the process/Total operating hours of the process

Maximum Acceptable Disabling Injury Risk Frequency. A similar equation can be derived for the Maximum Acceptable Disabling-Injury Risk Frequency (F_2). However, this is based on a different industry value than that used for defining the maximum allowed frequency.

According to the National Safety Council's (Washington, D.C.) *Accident Facts,* there are about 200 disabling injuries for each fatality in a non-vehicular work exposure. The number is the average of the figure for disabling injuries over the past 10 years. Using this ratio, a maximum allowed frequency for a disabling injury risk is:

F_2 = (1E–6)/(P×E) hours^{-1}) **(4)**

where, F_2 = Maximum Allowable Disabling Injury Risk Frequency, hours^{-1}

P = Percent of exposed persons injured (a qualitative estimate)

E = Workhours of exposure to process/Total operating hours of process

Equations (3) and (4) suggest the current criteria of five deaths and 1,000 disabling injuries per 10^8 working hours as the total maximum allowed levels for risks to personnel. These estimates form the basis for evaluating a system's risk levels, numbers that can be used as upper limits that process risks must not exceed for adequate system safety.

P and E values are estimates made by the risk analyst from information about the failure event under investigation. P is simply a reasonable estimate as to the percentage of personnel injuries (fatal or disabling) that may result from exposure of workers to the event.

Values of P range from 0 to 1 in increments of 0.25. If personnel are in close proximity to the event, the P value may be 0.75 or 1. However, if they are at a greater distance or are protected by barriers, the P value may be taken as 0.25 or 0.50. If there is no probability of fatalities or disabling injuries because of the event, the P value is 0 and there is no risk of human loss.

E is the personnel exposure factor that is based on the number of people and periods of time they are exposed to the event. It is the ratio of exposure workhours to the system's operating hours over a repeated uniform period (shift, day, week and so on) of the system. The E value can range from 0 to more than 1 — even as high as 10, 25 or 50 — and can be defined as personnel exposure intensity factor.

The maximum allowed frequencies for both equations are simply a function of PE (probable injuries), and will have a range from zero to values exceeding one, i.e., they can reach 10, 50, 100 or more for single-event failures. A specific system design's Maximum Allowable Failure Rate — as determined by either Equation 3 or 4 — will vary with locations having different PEs. The variation occurs because the values of F_1 or F_2 will be a function of the number of people exposed at each location, even though the maximum risk level is held to a single, constant value for all locations.

Figure 3 for equations (3) and (4) shows the maximum allowable failure rate (F_n) for both fatal and disabling injuries for PE values ranging from of 0.01 to 100. For PE values equal to or greater than 10, the maximum allowable failure rate is reduced by a factor of 5. This constitutes an additional safety factor for personnel faced with high-risk exposures, such as in offshore platforms and process control rooms, that have greater densities of people.

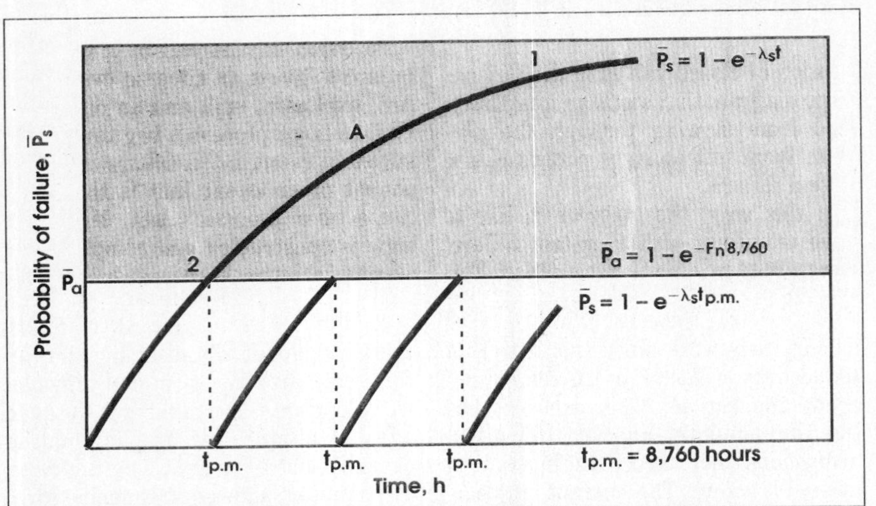

Figure 3. Here, a plot of $\bar{P}_s = 1 - e^{-\lambda_s t}$ (curve A), equation (2), with $P_s(t) = e^{-\lambda_s t}$ is shown for a component or system. This indicates the values of $\bar{P}_s$ for a specific system's failure rate (λ_s) at various time values. The horizontal limit for $\bar{P}_a = 1 - e^{-F_n t}$, where $F_n = F_1$ or F_2, as calculated by Equation 3 or 4, indicates the maximum acceptable probability of failure for the time period t = 8,760 hours for a specific risk. It is obvious that $\bar{P}_s$ is greater than $\bar{P}_a$ at point 1 for the one year period; however, $\bar{P}_s$ is equal to $\bar{P}_a$ at some lesser time, point 2. This would be the approximate preventive maintenance time for these cases

The single-event risk for a large number of personnel losses is $^1/_{10}$ that of the AAFR. This frequency for a fatality risk (1E–10 $hours^{-1}$) for 10 or more people is slightly less than that of the frequency of deaths in the U.S. caused by lightning, cataclysmic and other so-called "Acts of God" [6]. Any value less than this is felt to be meaningless.

Fault Trees and Failure Logic

Generally, a process system failure requires at least two — and, sometimes, as many as four, five or six — components to be in a failed state at the same time. (This is different from when the components *all fail at the same time*, which is the result of a cause originating outside the process system.)

For example, if a standby component has failed, its failure is not known until it is routinely inspected or called upon to act during a process upset. A good example is a relief valve protecting a low-pressure vessel from a pressure-reducing regulator. As long as the regulator does its job, a relief valve in a "fail to operate" state is of no consequence. However, if the regulator fails to open or to respond (for example, if it is in a "stuck" position) to a pressure increase at a time when the relief valve has already failed, the situation could be serious.

To evaluate a system, one can at first assume that the annual turnaround period for a continuously operating system is 8,760 hours, since this is the crucial period for discerning risk acceptability against annual criteria. However, systems utilizing high-failure-rate equipment can never successfully operate with acceptable risk over a year's period without some form of maintenance.

As a result, one should avoid operating a process system beyond what can be ascertained as an acceptable probability of failure. This can be accomplished by using the most reliable equipment in a favorable design configuration, together with the necessary preventive maintenance programs to restore any unreliable equipment to an "as-new" condition.

HARA currently uses Failure Logic

diagrams together with non-repair probability of failure theory. In Figure 2, the Fault Tree and Failure Logic are equivalent failure logics for a given system. However, only the Failure Logic can be viewed as a fluid-flow diagram that represents a process system whose operating success is certain whenever fluid can flow from the left to right through the diagram. The Failure Logic diagram has been patterned to resemble a piping system, since that's a concept that most process engineers are familiar with.

An adverse event caused by a system failure in any of the process characteristics of pressure, flow, temperature, level and so on can be represented by certain minimum number of component combinationational failures (minimum cutsets) of items 1 through 4. In the Failure Logic diagram, each component is represented as a closed or open valve. If a component fails, then the valve is shown as "closed." Therefore, any combination of closed valves in the Failure Logic diagram that stops an imaginary fluid from flowing through the diagram from left to right constitutes a system failure.

In this way, the diagram is like a valve manifold, with a system failure represented by a no-flow condition. For example, in Figure 2 there is a system failure if items 1 and 4 are in a failed state simultaneously, or if items 2, 3 and 4 have failed. However, there is no system failure if only items 1, 2 and 3 have failed.

To begin an analysis of a system's failure to avert an adverse event — a fire, explosion, spill and so on — the Failure Logic process is begun with the initiating event, i.e., a failure in a component or an event that is the initial cause of a process upset. Next, the logic is constructed and completed by placing component loops in parallel with the initiating event for each equipment loop that is specified to control, correct or shutdown the process before a more serious condition occurs. These parallel loops contain series as well as parallel components if standby or redundance is built into the system.

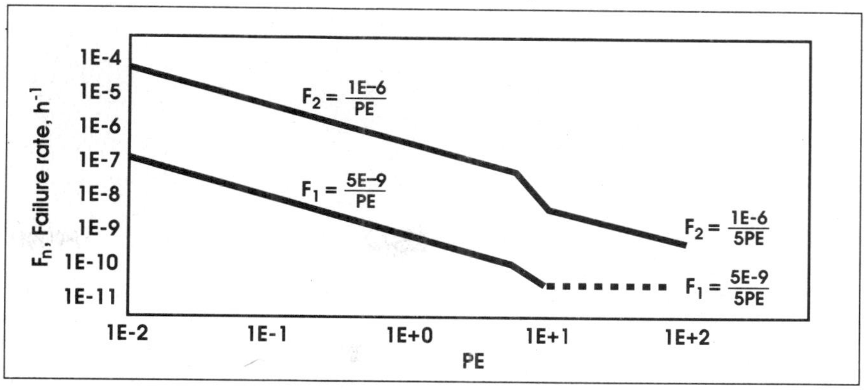

Figure 4. The maximum allowed failure rate (F_n) vs. the probable injuries are shown for both fatal and disabling events

Before selecting minimum cutsets, the initial Failure Logic diagram is reduced by combining components wherever this is legitimate. In certain circumstances, one can combine failure rates for components and, separately, their probabilities of failure. This reduces the number of components and facilitates a lower number of minimum cutsets. The reduction is needed to calculate an equivalent failure frequency that more accurately approximates the system's value for time period (t):

$$\lambda = 1 - R(t)/\int_0^t R(t)dt \qquad (5)$$

where, λ = System time-dependent failure rate

$R(t)$ = System reliability equation

A significant advantage of the Failure Logic over the Fault Tree is its ability to accept systematic, "ladder-like" listings of the secondary safety systems. These systems are added as parallel loops to the initiating event loop (process control failure) in the same order as they would be actuated to mitigate a failure.

The Failure Logic diagram can also distinguish between various subsystems, such as redundant vs. standby safety systems, which need to be treated differently when determining the minimum cutsets. In Fault Trees, the differences are not as obvious. ■

Edited by Jayadev Chowdhury

The author

H. Wade Goodner (106 Brittany Park, Anderson, SC 29621) is a consultant for risk analysis of process systems for engineering design contractors, chemical firms, and insurance underwriters. He retired last year as senior engineering specialist of reliability engineering at Phillips Petroleum Co. He holds 7 patents. He is a member of IEEE (Risk Soc.), Soc. of Risk Analysis, and Instrument Soc. of America. He has a BSME from the U. of Alabama, and is in Who's Who of America's listings of Science and Engineering, and Finance and Industry.

References

1. Kletz, T. A., Evaluating Hazards, *Chem. Eng.*, Apr. 1, 1985, p. 59.
2. Fussell, J. B., How to Hand Calculate System Reliability and Safety Characteristics, *IEEE Transactions of Reliability*, Vol. R-24, No. 3, August 1975.
3. Johnson, R., Fault Tree Faults, *AIChE International Conference on Hazard Identification and Risk Analysis, Human Factors, and Human Reliability in Process Safety*, Jan. 15–17, 1992.
4. Goodner, H. W., Risk Analysis of Human Exposure to Systems, *AIChE International Conference on Hazard Identification and Risk Analysis, Human Factors, and Human Reliability in Process Safety*, Jan. 15–17, 1992.
5. Goodner, H. W., A Basis for Development of Risk Based Industrial Standards, *Probabilistic Safety Assessment and Management (PSAM) International Conference*, Beverly Hills (Calif.), Feb. 4–7, 1991.
6. National Safety Council, *Accident Facts*, issues from 1981 to 1991.
7. Goodner, H. W., Risk Management of Engineering Designs, article to be published in the *Encyclopedia of Chemical Processing and Design*, M. Dekker, Inc., New York.

Figure 1. The PIC and PSV, shown as part of the failure analysis schematic (the orange section of the diagram), are the only components that can lead to an overpressurization of the system

STEVE HAIMOWITZ

A NEW WAY OF QUANTIFYING RISKS

PART 2

H. Wade Goodner P. E.*
Consultant

Part 1 of this article discusses the basic methodology of Fault Tree Analysis and of a new FTA-based technique, Hazard Assessment by Risk Analysis (HARA)*. This second part illustrates the use of the HARA technique, together with some examples.

The failure of a device or of a group of components can result in an adverse event, such as a fire, explosion, spill or release. In a given system or group of devices, the failure time of any particular component cannot be exactly determined. However, reasonable estimates of the device's probability of failure ($\bar{P}$) can be made from failure rate data and appropriate failure models.

Equipment failure rates, lambda (λ, h^{-1}), are available in the literature. These values are used in the exponential probability-of-failure equation to calculate the probability of failure of devices, $\bar{P}$:

$$\bar{P} = 1 - e^{-\lambda t} \quad (1)$$

Equation (1) is used in cases where a single device or simple system is started up and allowed to run for a given time period (t), during which time it does not undergo any repair or adjustment. Specifically, the equation calculates a system's probability of failure over a given operating time period (t).

Such a calculated probability of failure by itself does not provide sufficient information to evaluate a system's risk. What is needed, in addition, is a value for the maximum probability of failure ($\bar{P}_r$) that is considered acceptable for that system. Then, by comparing the derived failure rates of $\bar{P}$ and $\bar{P}_r$ for the period, one can evaluate the system's risk for acceptability.

Before continuing, it is important to point out some significant differences in the parameters based on the word "failure." For example, the predominant failure parameters are distinctively different from each other, and are calculated using different formulae.

Failure rate, λ, is the estimated number of failures per a device's *operating* unit of time. It is calculated by dividing the number of failures of a like quantity of operating components by the cumulative number of operating-time units of the components. These are random failures, and are not ones resulting from operation outside the unit's specifications or because of design or other faults. The number of failures are considered to be a constant value that is independent of time (this does not include idle or non-operating time).

Rate of failure, or ROF, is the rate or number of failures for a *clock* unit of time of one cycle of uptime plus down time to restart, over as many cycles as one wishes to examine for an extended analysis. It is generally used for announced component failures that undergo repair without a system failure. ROF is equal to λ when the system operates without repair, and is less than λ if the system undergoes repair (provided that ROF and λ are expressed in the same units).

Probability of failure is a decimal value ranging from 0 to 1 that represents the confidence level that a system will fail over a time period. The probability of failure is also used to represent

* Part 1 — which also gives the author's biography — p. 61.

MARYJO MAZZELLA

HARA assesses the operational weaknesses of safety devices, process systems and plants

a fraction of an undesirable number of events. Its symbol is $\bar{P}$, F(t) or $\bar{R}$.

A system's Failure Logic diagram for an adverse event resulting from a process upset (or other circumstance) is built as series and parallel combinations of symbols. Each symbol represents the failure of a control component or of a human operator.

Analyzing system failures

In a two-component system, if only one of the two must fail to cause an adverse event, then both components are placed in "series" logic. However, if both components must be in a failed state to cause system failure, then the components are placed in "parallel" logic to each other.

Most components in instrumentation loops are in series failure logic. In contrast, two sensors are in parallel logic if they measure the same process condition point (temperature, pressure and so on), and their outputs — one being in standby or voting logic — input to the same controller in series with them.

The building of a Failure Logic diagram is started with an initiating event symbol for a probable equipment failure or process condition upset (mode). Next, loops are constructed in parallel with the initiating event. The loops consist of components in series or parallel, or both, to represent failures of human, process control or safety systems that are installed to prevent the development of an adverse event.

Once all the potential failures for equipment and human reactions are represented in the Failure Logic diagram, the next step is to determine the minimum cutsets†. The minimum cutsets are gotten by listing all the possible combinations of the minimum number of component failures, each including the initiating event, that will cause an adverse event.

For the purposes here, this article will take a maximum of four component failures together with an initiating event, giving a total of five events, as a minimum cutset. Only in cases with large failure rates do a greater number of events significantly influence a system's failure rate.

Next, equations are used to determine a system's equivalent failure rate (λ_e) over a year of operation (8,760 h, without maintenance). If this value is equal to or less than the set criterion, F_1 or F_2, the system's design has acceptable risk. (See Part 1, p. 116, for descriptions of F_1, the Maximum Acceptable Fatality Risk Frequency, and F_2, the Maximum Acceptable Disabling-Injury Risk Frequency.)

However, if the system's risk is not acceptable, then additional measures must be taken to achieve acceptable risks. In other words, the system must be redesigned using additional or more reliable components, or incorporate redundant components, or the system must have increased maintenance.

Which operating system?

The method employed to estimate the failure frequency of a system due to a process condition is determined by the system's operating mode. Typically, operating systems are categorized as active-parallel standby, inactive-parallel standby, and batch. Batch systems are essentially active-parallel standby, but here they are treated as a separate case because of their short and, sometimes, infrequent operation.

All systems are evaluated on an annual basis. This is because the industrial averages used for setting the "maximum and acceptable" risk levels for process systems are calculated from losses that are reported on an annual basis. These maximum acceptable risk levels are compared to a system's estimated operating risk to determine the system's overall acceptability.

Active-parallel standby system. The failure analyses of such systems utilize minimum cutsets that are taken from a Failure Logic diagram for a specific failure scenario, such as over- or under-pressure, or over- or under-filled. Each cutset has a maximum of five events. The equations and required data are:

Probability of success of component (n), P_n:

$$P_n = e^{-\lambda_n t_n} \quad (2)$$

where, λ_n = failure rate of component, hours^{-1}

t_n = operating period between maintenance, hours

n = component number (1,2,3 . . . n)

Or, $P_n = (1 - \bar{P}_n)$ (3)

where, $\bar{P}_n$ = component's probability of failure

Probability of success of cutset (k), P_k:

$$P_k = 1 - \Pi \bar{P}_n \quad (4)$$

where, k = cutset number (a,b,c . . . k)

Failure rate of cutset (k), hours^{-1}:

$$\lambda_k = (\ln P_k)/-T \quad (5)$$

where, T = time period, 8,760 hours

The system's equivalent failure rate λ_e, hours^{-1}:

$$\lambda_e = \Sigma \lambda_k \quad (6)$$

Inactive-parallel standby system. These systems are characterized by an idle (inactive) or secondary (standby) component that is energized upon failure of the primary system. These systems usually contain only three components: a primary operating unit, a detection and startup device to turn-on a secondary unit, and a secondary unit or correcting device to compensate for the primary unit's failure.

An inactive-parallel system can be autonomous or be a part of a larger

† A minimum cutset, defined on p. 57 of Part I, is the smallest combination of component failures with the initiating event that will cause a system failure.

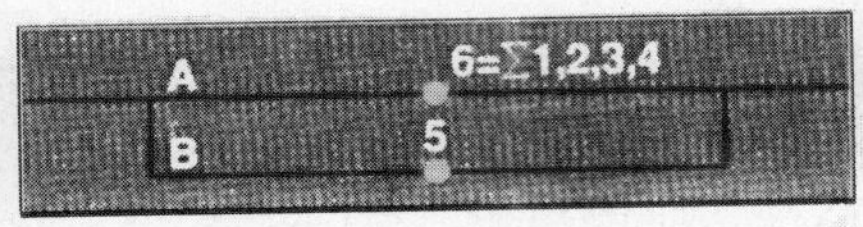

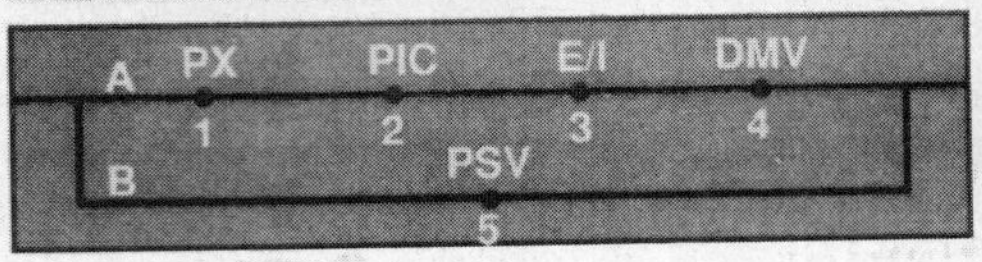

Figures 2–5. Series failure logic for vessel overpressurization is shown in Figure 2 (top, left), and overall failure logic in Figure 3 (bottom, left). Summing the failure rates yields Figure 4 (top, middle), and simplification provides a single failure rate in Figure 5 (top, right)

active-parallel system. If the inactive system is part of an active one, then the inactive system's failure rate (λ_u) becomes a replacement component in the active system. Then, the system's λ_e is obtained by the active-parallel method:

Secondary system's probability of success, P_{SB}:

$$P_{SB} = e^{-\lambda_2 t_2}(1 - \bar{P}_{ST}) \qquad (7)$$

where, λ_2 = secondary system's failure rate, hours^{-1}

t_2 = secondary system's required operating period before primary is restored, hours

$\bar{P}_{ST}$ = secondary system's probability of failure to start

Secondary system's failure rate, λ_{SB}:

$$\lambda_{SB} = (\ln P_{SB})/-8{,}760 \qquad (8)$$

Primary's probability of success, P_1:

$$P_1 = e^{-\lambda_1 t_1} \qquad (9)$$

where, λ_1 = primary system's failure rate, hours^{-1}

t_1 = primary system's intended operating period, hours

Unit's probability of success, P_{unit}:

$$P_{unit} = P_1 + [\lambda_1/(\lambda_1 - \lambda_{SB})](P_{SB} - P_1) \qquad (10)$$

Unit's failure rate, λ_u

$$\lambda_u = \ln P_{unit}/-t_1 \qquad (11)$$

Batch operations. The procedure to determine the failure rate (λ_e) of batch systems is done in two steps. The first one determines the system's probability of failure per batch using only equipment that is in operation from the beginning to the end of the batch cycle. In step two, λ_e is calculated.

Step one allows for the legitimate assumption that component failures detected during the batch can be corrected before the next batch. Equipment with long-term preventive maintenance cycles, such as relief valves, are not considered in this phase, but later.

First, the calculation is carried out as if the system is an active-parallel one with or without any inactive-parallel subsystem for a period of one batch. Then, a failure rate (λ_y) is determined for a year's operation by multiplying the failure rate for a single batch by the annual number of batches.

This failure rate (λ_y) is used with the standby equipment (those that were not previously considered as part of the active-parallel system) to calculate λ_e for a year of operation (8,760 hours). Note, λ_y is equal to λ_e in cases where no component's preventive maintenance period exceeds that of the batch period.

Step 1: Probability of a system's success per batch, P_{BAT}:

$$P_{BAT} = \Pi P_k \qquad (12)$$

where, P_k = probability of cutset success a,b . . . k

$$P_k = 1 - \Pi \bar{P}_n \qquad (13)$$

where, $\bar{P}_n$ = component's probability of failure

Probability of failure per batch, $\bar{P}_{BAT}$:

$$\bar{P}_{BAT} = 1 - P_{BAT} \qquad (14)$$

Probability of failure per year, $\bar{P}_y$:

$$\bar{P}_y = n\bar{P}_{BAT}(1 - \bar{P}_{BAT})^{n-1} \qquad (15)$$

where, n = batches per year

The estimated failure rate for a year's operation without components having preventive maintenance cycles greater than the batch period, λ_y:

$$\lambda_y = \ln(1 - \bar{P}_y)/-8{,}760 \qquad (16)$$

Step 2: λ_y is now combined with devices, such as relief valves, that have preventive maintenance periods longer than that of the batch cycle. The analysis to determine λ_e for the whole system is now treated as an active-parallel standby system using Equation (6).

Risk analysis procedure

Hazard assessments of risks for single components or full systems are carried out using the equations for active-parallel, inactive-parallel or batch operations, together with the equations for F_1 and F_2. Thus, a system's risks to humans are determined and evaluated against predetermined criteria, with the most severe case being the determinant for an acceptable system design. This is done as follows:

1. First, list — on a line-by-line basis — the principal process equipment (such as vessels, reactors or rotating equipment) and their modes of failure, whether by overpressure, overfilling, stopped flow or overtemperature. For each failure mode, construct a Failure Logic diagram that includes the initiating event and its associated preventive or mitigating equipment.
2. For each failure mode, list the failure effect and its consequences, for example, rupture of a vessel and the nature of human or financial losses.
3. For single-component failures, list the failure rates (h^{-1}) as given by sources of acceptable failure-rate data. For multiple-component systems with parallel or series failure logic, determine the equivalent failure rate (λ_e) for each failure mode for the total system. Also find the appropriate equations.
4. Obtain information from the operating group to determine estimates for probable human injuries and fatalities for each failure-mode event. Calculate F_1 or F_2, or both (the lowest value of F_1 or F_2 prevails as the maximum acceptable failure frequency).
5. For each failure mode, if $\lambda_e \leq F_n$, the system's design has acceptable risk and no hazard is considered to exist for the given conditions. If $\lambda_e \leq 2F_n$, corrective measures might be taken on a discretionary level. But, if $\lambda_e > 2F_n$, the system's design is considered to have excessive risk, and corrective measures are recommended.

The construction of a Failure Logic diagram for a particular failure mode is made easy by first completing a representative schematic, such as Figure 1, of the system. Then, by analyzing Figure 1, one can see that, for instance, one failure mode of the low-pressure vessel is rupture due to overpressurization.

This overpressurization will occur if these events happen at different times without correction: 1) the downstream pressure increases because of reduced

COMPONENT	CODE	COMPONENT DESCRIPTION	FAILURE RATE, h^{-1}	PREVENTIVE MAINTENANCE
1	PXPI	Pneumatic pressure transmitter	6.0E-5	8,760
2	CTL1	Pneumatic controller	2.3E-5	8,760
3	E/P1	Transducer	6.4E-5	8,760
4	DMVP	Pneumatic control valve	6.5E-5	8,760
5	PSV3	Relief valve, fail to open	6.0E-5	8,760

CUTSET	COMPONENT	EQUIVALENT FAILURE RATE, h^{-1}	
1	1,5	2.45E-07	The system's equivalent failure rate (λ_e) = 8.7E-07 h^{-1}
2	2,5	1.09E-07	
3	3,5	2.57E-07	
4	4,5	2.60E-07	

TABLES 1 and 2. If the series components are not summed up, the four cutset analysis results in a high equivalent failure rate for the system (Table 1, above). Summation yields a single cutset and the correct failure rate (Table 2, below)

COMPONENT	CODE	COMPONENT DESCRIPTION	FAILURE RATE, h^{-1}	PREVENTIVE MAINTENANCE
1	PXPI	Pneumatic pressure transmitter	6.0E-5	8,760
2	CTL1	Pneumatic controller	2.3E-5	8,760
3	E/P1	Transducer	6.4E-5	8,760
4	DMVP	Pneumatic control valve	6.5E-5	8,760
5	PSV3	Relief valve, fail to open	6.0E-5	8,760
6	SUM	Component sum (1+2+3+4+5)	2.12E-4	8,760

CUTSET	COMPONENT	EQUIVALENT FAILURE RATE, h^{-1}	
1	6,5	5.06E-07	The system's equivalent failure rate (λ_e) = 5.1E-07 h^{-1}

or stopped flow; 2) the pressure control loop (PIC) "fails open" or "fails to respond" to the increase in downstream pressure; 3) and the pressure relief valve (PSV) is in a "failed closed" state.

Although the vessel may have various attached systems for control of other process parameters, such as level or temperature, in this particular case only those components pertinent to pressure and its control are shown on the schematic. Therefore, a failure analysis schematic of the vessel in Figure 1 due to an overpressure rupture would contain only the pressure control system (PIC loop) and the overpressure prevention device (PSV). This is shown in orange in Figure 1.

Whenever the failure of any single component of a multicomponent system results in a process upset or accident, the components are placed in a linear, series configuration. In Figure 1, since high-pressure-control failure by any of the four components could cause vessel overpressure, they are placed in series logic in Figure 2.

In this configuration, any component failure — represented as a closed valve condition in the line — would stop the flow of an imaginary fluid from left to right. This represents the loop's failure to control the pressure.

On the other hand, if a process failure requires two or more components to be in a failed state at the same time, the components are placed in a parallel logic configuration. For example, the PSV and PIC loop must both be in a failed state at the same time for the vessel's pressure to rise dangerously. (Note that this doesn't mean that the components should fail at the *same time* — the cause of failure in such simultaneous failures is from outside the system and doesn't involve random failures of components.)

Rules of simplification

In constructing the full Logic Failure diagram, Figure 3 is the initial logic using all five components of Figure 1. However, for it to be properly used, it must be reduced to a simpler diagram.

There are some specific rules that must be followed to correctly simplify loops of the initial logic diagram into one with a more manageable logic, one that yields a reduced number of minimum cutsets. These rules are fundamental to reliability engineering methods, and are necessary to correctly solve for a system's overall failure rate or its probability of failure:

1. Failure rates (λ) of loop components in series with each other may be added together to obtain a loop failure rate.

$$\lambda_{loop} = \Sigma_{i=1}^{n} \lambda_i \quad (17)$$

2. Failure rates (λ) of a loop's components in parallel must *never* be multiplied. While the solution for a single-valued time dependent λ requires the use of Equation (5) from Part 1, the HARA method for an active-parallel system is closely equivalent.

3. Some systems have component probabilities of failure ($\bar{P}$) instead of failure rates (specifically, human error is represented in this fashion). An equivalent system or loop probability of failure ($\bar{P}_{loop}$) for a series of components having probabilities of failure ($\bar{P}_i$) is:

$$\bar{P}_n = 1 - \prod_{i=1}^{n} (1 - \bar{P}_i) \quad (18)$$

where, i = Individual components probability of failure

n = Number of components in series

Π = Symbol for "product of"

4. Generally, in a logic diagram, failure rates in series with probabilities of failure are not combined. If multiple failure rates and probabilities of failure exist in a series, each is summed separately to obtain a two-component logic loop in series — one for failure rate and the other for probability of failure — using Equations (17) and (18).

Having carried out these simplification steps, it is now possible to reduce Figure 3 to a simpler logic. Invoking Rule 1, the failure rates of components 1,2,3 and 4 are summed, to yield 6 in parallel with 5, as shown in Figure 4.

Taking minimum cutsets (6,5) and solving by the active-parallel procedure results in a single-valued component failure rate, λ_e, as shown in Figure 5. λ_e is the time-dependent failure rate of the total system for a specific time period (t). This is the estimated system failure rate that is to be compared with criteria for determining if the system's risk is acceptable.

This example serves to illustrate a purpose other than demonstrating logic construction: The need for simplified logic. Cutset reduction is important whenever a specific component appears frequently in various cutsets and when the product of the component's failure rate and preventive maintenance period is close to or exceeds 1.0. (Such reduction is not needed when the product is much smaller than 1.0.)

Reduction of the cutsets is necessary

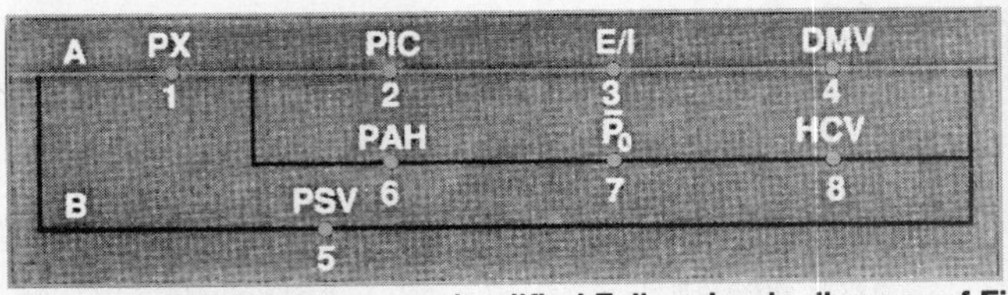

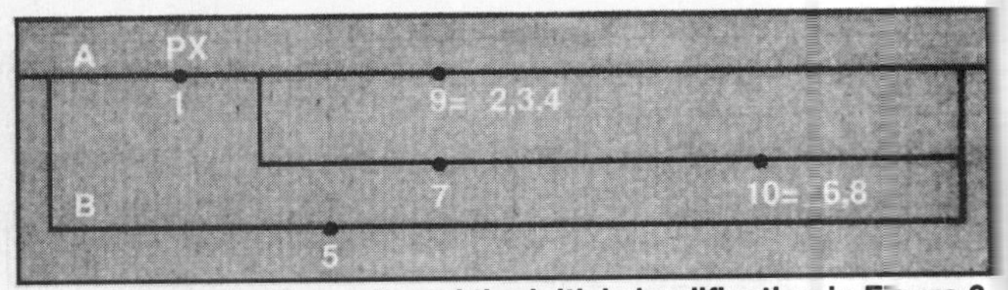

Figures 7 and 8. The non-simplified Failure Logic diagram of Figure 6 is shown in Figure 7, and the initial simplification in Figure 8

to prevent the generation of unrealistically high values that are clearly in error. For example, if the above example is worked out using the active-parallel method but without summing the series components of Figure 3, the minimum number of cutsets will be four, and the λ_e will be 8.7E–7 h^{-1}.

This is incorrect because the system's overall failure rate is greater — when it should be smaller — than either of the failure rates for the two parallel loops. (A greater value can be expected only if the loops are in series.)

The correct answer, λ_e = 5.1E–7 h^{-1}, is gotten by applying the same calculation procedure to Figure 4, which can be confirmed using Equation (5) in Part 1. Computer calculations for analyzing four and one cutsets are shown in Tables 1 and 2, respectively.

Completing the analysis

Before one can determine a system's safety, it is necessary to obtain personnel exposure data for the system. These data are needed to define the criteria of what constitutes a "safe" process design for a given set of exposure conditions.

For instance, let us assume for the system in Figure 1 that four persons are exposed to the operating vessel for one hour each during an eight-hour shift. Let us also assume that the casualties resulting from an explosion would be 50%. In this case, the maximum acceptable risk frequency (F_1) from Equation (3) in Part 1 is:

F_1 = (5E–9)/[(0.5)×(4/8)] = 2E–8 h^{-1}

Thus, under this criterion, the design is not safe (because λ_e = 5.1E–7 h^{-1} > F_1 = 2E–8 h^{-1}), and in need of corrective measures. In other words, the design is a hazard (unacceptable risk) to operating personnel because it results in a risk exposure that is greater than the industry average. Measures to reduce the risk to acceptable levels can include increased maintenance of the PIC loop, or replacement of the PIC with a more-reliable regulator.

Some applications specify an "alarm and operator" system that permits an operator to manually close a separate valve to prevent the vessel from rupturing. However, this will not substantially improve the system's safety, because it is unlikely that an operator would attempt to close a normal block valve, especially if it is close to the motor valve or the vessel. In fact, it is unreasonable to expect this action in view of the speed with which a vessel can overpressurize and explode.

However, as a practical exercise for adding safety components let's see what improvement one can expect with an operator loop. This is illustrated in Figure 6, which is Figure 1 modified to show the additional components, including a high-pressure alarm, an operator and a hand-operated control valve. Figure 7 is the non-simplified Failure Logic diagram of Figure 6, and Figure 8 is an initial simplification of the orange section of Figure 6.

Further simplification of Figure 8 might be necessary because the product of the 9's failure rate (1.52E–4 h^{-1}) and 8,760 operating hours exceeds 1.0. Thus, repeated use of this product in the cutsets will cause an inflated answer for the system's overall failure rate. Since 9 in Figure 8 appears in two minimum cutsets that are added to obtain the system's failure rate, Figure 8's failure rate is greater than the true value from Equation 5 in Part 1.

This becomes obvious on calculating the system's failure rate using Figure 8's logic, which yields a value of 6.1E–7 h^{-1}. The value indicates that adding the operator loop actually worsens the design — but this is simply not true.

However, simplifying Figure 10 to a two-event cutset using appropriate reliability equations and solving results in an λ_e value of 4.3E-7 h^{-1}. Thus the addition of an operator loop makes no significant improvement in the original system's failure rate of 5.1E-7 h^{-1} ■

Edited by Jayadev Chowdhury

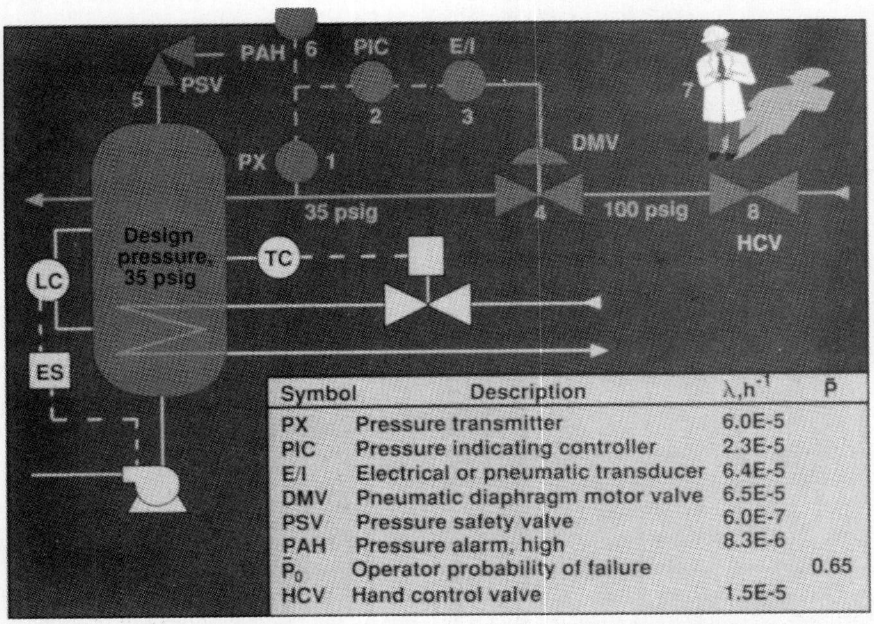

Symbol	Description	λ, h^{-1}	$\bar{P}$
PX	Pressure transmitter	6.0E-5	
PIC	Pressure indicating controller	2.3E-5	
E/I	Electrical or pneumatic transducer	6.4E-5	
DMV	Pneumatic diaphragm motor valve	6.5E-5	
PSV	Pressure safety valve	6.0E-7	
PAH	Pressure alarm, high	8.3E-6	
$\bar{P}_0$	Operator probability of failure		0.65
HCV	Hand control valve	1.5E-5	

Figure 6. An operator loop introduces an alarm, operator and hand-control valve into the process schematic (components 6, 7 and 8). Despite these additions, the system's overall risks are not substantially improved

Other sources

1. Shooman M. L., *Probabilistic Reliability: An Engineering Approach,* McGraw-Hill, New York, 1983.
2. McCormick, H. J., *Reliability and Risk Analysis,* Academic Press, New York, 1978.
3. Lee, F. P., *Loss Prevention in the Process Industries,* Butterworth, London, 1980.
4. Henley, E. J. and Kumamoto, H., *Reliability Engineering and Risk Assessment,* Prentice-Hall, Englewood, N.J., 1981.

V. Anthony Ciliberti
The Lubrizol Corp.

RANKING EQUIPMENT CRITICALITY

The payoff is in improved production, yield and quality

In the last two years, the chemical process industries (CPI) have been spending a lot of time and money to comply with new process safety regulations.* To a great extent, these efforts center on reducing hazards to people and the environment by ensuring the mechanical integrity of equipment.

In actual practice, these compliance efforts yield very little, if any, return on investment. Simply put, money spent on safety does not directly translate into increased profits.

In contrast, time and money expended on process-related – as opposed to safety – issues have a definite payback. For example, programs to ensure the mechanical integrity of equipment critical to a plant's operation do increase a plant's output, yield and product quality and, thus, earnings. Therefore, it is worthwhile for plants that are addressing the safety hazards of process equipment to also address the question of how critical such equipment is to the smooth running of the processes themselves.

In starting up a program, the logical step is to first evaluate the most critical equipment. To do so, however, it is necessary to rank the equipment based on some criteria, such as their relative importance to a process.

A technique for making such rankings of process equipment is based on hazard and process criteria. The rankings should be performed by teams that are thoroughly familiar with the process being evaluated – a good choice being a plant's Process Hazard Analysis (PHA) group. Since the team's familiarity with a process is highest at the end of a PHA study, the best time for ranking the criticality of equipment is toward the tail end of such safety evaluations.

Six easy steps

The procedure for ranking equipment criticality ranking system consists of:

1. Determining the Hazard Criticality Rating (HCR) of various process equipment using Table 1 *(Continues)*

* The new process safety laws and the techniques for addressing them are covered on p. 22.

HAZARD CRITICALITY RATING

1. ASSESS THE PROCESS CHEMICAL HAZARD (PCH) BY:
- **Determining the NFPA ratings (N) of process chemicals for:**
  Health, Fire, Reactivity hazards
- **Selecting the highest value of N**
- **Evaluating the potential for an emissions release (0 to 4):**
  High (RF=0): Possible serious health, safety or environmental effects
  Low (RF=1): Minimal effects
  None (RF=4): No effects
- ***Then, PCH = N - RF.*** (Round off negative values to zero.)

2. RATE OTHER HAZARDS (O) WITH AN ARBITRARY NUMBER (0 TO 4) IF THEY ARE:
- **Deadly (4), if:**
  Temperatures >1,000°F
  Pressures are extreme
  Potential for release of regulated chemicals is high
  Release causes possible serious health safety or environmental effects
  Plant requires steam turbine trip mechanisms, fired-equipment shutdown systems, or toxic- or combustible-gas detectors*
  Failure of pollution control system results in environmental damage*
- **Extremely dangerous (3), if:**
  Equipment rotates at > 5,000 rpm
  Temperatures > 500°F
  Plant requires process venting devices
  Potential for release of regulated chemicals is low
  Failure of pollution control system may result in environmental damage*
- **Hazardous (2), if:**
  Temperatures > 300°F;
  Extended failure of pollution control system may cause damage*
- **Slightly hazardous (1), if:**
  Equipment rotates at > 3,600 rpm
  Temperatures > 140°F or pressures > 20 psig
- **Not hazardous (0), if:**
  No hazards exist

3. SELECT THE HIGHER VALUE OF PCH AND O AS THE HAZARD CRITICALITY RATING

*Equipment with spares drop one category rating. A spare is an inline unit that can be immediately serviced or be substituted by an alternative process option during the repair period.

TABLE 1. The Hazard Criticality Rating (HCR) is determined in three steps

Originally published October 1994

2. Determining the Process Criticality Rating (PCR) using Table 2
3. Determining the Process and Hazard Criticality Ranking (PHCR) of equipment using Table 3
4. Generating a criticality list by ranking equipment using their alphanumeric PHCR values
5. Determining priorities for developing maintenance programs. The alphabetic part of the PHCR value determines the priority:

Category A	Highest priority
Category B	High priority
Category C	Medium priority
Category D	Low priority

6. Correlating critical parameters to PHC rankings to determine inspection frequencies. Tables 4 and 5 illustrate the procedure for rotating equipment

PROCESS CRITICALITY RATING	
Essential (4)	The equipment is essential if failure will result in shutdown of the unit, unacceptable product quality, or severely reduced process yield
Critical (3)	The equipment is critical if failure will result in greatly reduced capacity, poor product quality, or moderately reduced process yield
Helpful (2)	The equipment is helpful if failure will result in slightly reduced capacity, product quality or reduce process yield
Not critical (1)	The equipment is not critical if failure will have little or no process consequences

TABLE 2 (above). The Process Criticality Rating (PCR) ranks how important equipment is to the smooth operation of the plant on a scale of 1 to 4

PHC RANKINGS					
Process Criticality Rating	Hazard Criticality Rating				
	4	3	2	1	0
4	A44	A34	A24	A14	A04
3	A43	B33	B23	B13	B03
2	A42	A32	C22	C12	C02
1	A41	B31	C21	CD11	D01

Note: The alphanumeric PHC value is represented first by the alphabetic character for the category (for example, category A is the most critical while D is the least critical). The first numeric portion represents the Hazard Criticality Rating, and the second numeric part the Process Criticality Rating.

TABLE 3 (above). The Process and Hazard Criticality Ranking is got from the process and hazard criticality ratings

CRE VALUES				
PHCR	Mean time between failures, months			
	0–6	6–12	12–24	>24
A	a1	a2	a3	a4
B	a2	b1	b2	b3
C	a3	b2	c1	c2
D	a4	b3	c2	d1

TABLE 4 (above). The Criticality and Repetitive Equipment (CRE) values for rotating equipment with various MTBFs

MAINTENANCE CYCLES Frequency, days				
CRE	7	30	90	360
a1, a2	VM	LT		
a3, a4		VM	LT	
b1, b3			VM	
c1, d1				VM

VM: Vibration monitoring
LT: Lubrication sampling and testing

TABLE 5 (above). Predictive maintenance frequencies for rotating equipment based on their CRE values

TABLE 6 (below). A typical spreadsheet for ranking equipment criticality

SPREADSHEET FOR CALCULATING EQUIPMENT PHCRS											
Equipment Number	Equipment Description	NFPA rating H	F	R	RF	PCH	Other	HCR	PCR	PHCR
TKO	Tank	4	4	0	0	4	0	4	4	A44
TKO	Tank	4	4	0	1	3	3	3	4	A34
PU1BFW	Pump	2	0	0	0	2	1	2	3	B23

The proof of the pudding

Once all the equipment is ranked by criticality, the teams involved can generate priority lists. These lists can then be used to ensure the mechanical integrity of critical equipment by prioritizing predictive and preventive maintenance programs, inventories of critical spare parts, and maintenance work orders in case of plants upsets.

The PHCR system has been incorporated into a mechanical-integrity program at Lubrizol's Deer Park, Tex., plant. At present, all equipment – fixed, rotating and instrumentation – are evaluated using the method.

At first, the hazards posed by the plant's different operating units are ranked and prioritized based on a PHA. These rankings are then used to determine the order in which the hazards need to be addressed. When these PHA approach completion, team members evaluate the equipment in each operating unit using the PHCR system.

In preparation for the PHCR evaluation, a spreadsheet containing all equipment and associated information is assembled. Typically, this information is compiled from historical data in the maintenance computer and downloaded into a spreadsheet (Table 7).

The PHA team determines a risk reduction factor (RF), "other" hazards (0), and a process criticality ranking (PCR). The length of time for completing the studies varies from two to six hours, depending on the size of the operating unit. The following example illustrates the technique for ranking the criticality of a boiler feedwater pump (PU1BFW in Table 6).

1. From Table 1, N = 2, the highest of the NFPA ratings of 2,0,0, and the risk reduction factor (RF) = 0, since there is the potential for serious burn from hot water. The value for the Process Chemical Hazard (PCH) = 2, because N – RF = 2. The rating of Other Hazards (O) = 1, because of the high temperature. Thus, the Hazard Criticality Rating (HCR) =2, the higher numerical value of PCH and O
2. From Table 2, the Process Criticality Rating (PCR) = 3, since the pump is essential for plant operation (but an inline spare is available)
3. From Table 3, the Process and Hazard Criticality Ranking (PHCR) = B23
4. Since the pump is in Category B, it is "high priority" equipment
5. Assume the pump has a mean time between failure (MTBF) of 10 months. From Table 4, CRE = b1
6. From Table 5, the unit requires vibration monitoring every 90 days ■

Edited by Jayadev Chowdhury

The author

V. Anthony Ciliberti is a maintenance engineer at The Lubrizol Corp. (51 Tidal Road, Deer Park, TX 77536-0158; Tel: 713-479-2851; Fax: 713-884-5314). He graduated from Texas A&M in 1987 with a B.S. in chemical engineering. He is a member of the International Maintenance Institute, and is a treasurer of the Soc. of tribologists and Lubrication Engineers.

INCINERATOR RISK ASSESSMENTS: CHANGE IS IN THE AIR

Due to stringent land disposal restrictions, incineration is often a choice technology for hazardous waste management. However, public concern over the potential health risks associated with incineration has prompted the U.S. Environmental Protection Agency (EPA; Washington, D.C.) to reevaluate the use of it to destroy hazardous waste.

On May 18, 1993, EPA Administrator Carol Browner announced that the agency would be conducting a major overhaul of existing waste-combustion rules.[1] This initiative calls for establishing new, more-stringent permit conditions, which would require risk assessments for boilers and industrial furnaces (BIFs) and RCRA[2] hazardous-waste incinerators currently operating under interim status. EPA predicts that as many as 25% of affected facilities could be forced to close as a result. EPA has also stated[3] that the public deserves to have broader opportunities for participation in the risk-management process.

While human-health risk assessments have been part of the permitting process for most incinerators, those previously required for BIFs were limited to analysis of inhalation pathways. In its proposal, EPA has changed the scope, to include not only inhalation of emissions, but analysis of indirect exposure pathways, too (Figure).

This new risk assessment will be mainly concerned with dioxins, furans, and metals. However, any chemical whose toxic effect can be magnified in the food chain will require assessment as a result of the proposed initiatives.

[1]The May 18 announcement, entitled Draft Strategy Press Release (EPA Doc. #530-F-93-012), details Browner's re-examination of the Agency's current hazardous-waste-management progam, and outlines the main points of the proposed initiatives. To obtain it, call the RCRA-Superfund hotline: (703) 412-9810.

[2]Resource Conservation and Recovery Act.

Originally published June 1994

How are these risks assessed?

A risk assessment is a numerical model that estimates the potential risks posed by a facility or an emission source. Many compounds that are fed into an incinerator are known to be hazardous. Others, such as dioxins, are formed during the combustion process.

The goal of a risk assessment is to determine which compounds pose the greatest risk, how they may reach the public, who may be affected, and in what manner. Performing one for regulatory compliance typically involves the following four steps.

Hazard identification. All compounds handled or produced by the combustion unit must be evaluated, to determine if they should be studied in greater detail. Key compounds are usually decided by the waste-feed characteristics, results of a trial burn, risk-assessment precedents, availablity of supporting toxicity information and the potential for bioaccumulation.

Toxicity assessment. To determine if exposure to the compounds may create adverse human health effects, a numerical estimate of the toxicity of each constituent must be obtained. Such data can be found by consulting EPA, or state or local regulatory agencies.

Exposure assessment. This step evaluates all potential exposure pathways and determines the amount of chemicals that exposed individuals could take into their bodies. Historically, EPA has required assessment of the Maximum Exposed Individual (MEI). This method makes the following assumptions: On a daily basis, the exposed individual breathes air, ingests and dermally contacts soil near the facility, consumes drinking water, fish, beef and dairy products from local sources, and eats fruits and vegetable from a local garden — all of which contain the upper-bound estimates of compound concentrations contributed by the facility. For simplicity, it has been assumed that all exposures occur at the same physical location (which is called the receptor), and that the MEI resides and works at this same outdoor location for an entire lifetime.

[3]In its November 15-17, 1993, National Roundtable on Hazardous Waste Minimization and Combustion Workshop (Washington, D.C.).

Meanwhile, an assessment of a "reasonably maximally exposed" individual evaluates "high end" as opposed to "maximum" exposures. As an example, for the RME individual, one might assume that 10% of his or her food intake is produced at the closest farm, instead of the MEI assumption that all consumed food is produced at the location that is maximally affected by the facility for every single pathway.

Prior to EPA's move toward accepting an RME individual as the preferred basis for risk assessment, the only way for a facility to present a more realistic view was to include a separate analysis of "average" exposure. Such analysis could model risk using present average conditions for the specific facility and its immediate environment, instead of generic reference values.

STEVE HAIMOWITZ

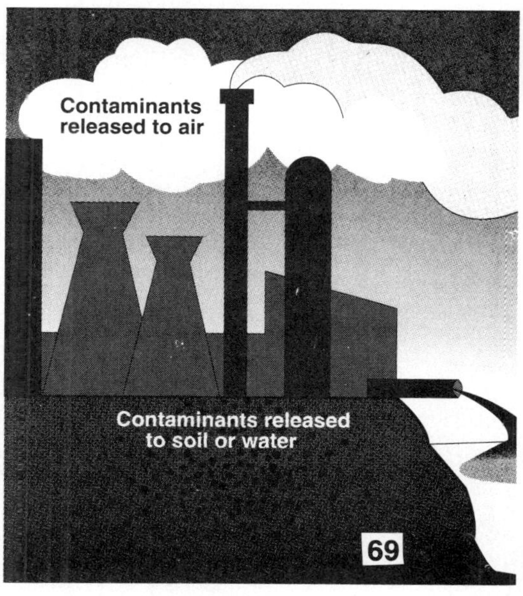

Proposed guidelines may change the way assessments are conducted, and widen the scope for affected facilities

Douglas Smith,
Allison Yuhas and
Brian Magee
ENSR Consulting and
Engineering

ENSR Consulting and Engineering

Risk characterization. This step evaluates both carcinogenic and non-carcinogenic risks. For the former, it specifically estimates the likelihood (over and above the background cancer rate) that an exposed individual at the specified location could develop cancer in his or her lifetime, as a result of exposure to chemicals of concern.

The second evaluation (hazard index) indicates the risk of non-cancer health effects, such liver or kidney damage. Risk characterization combines all the exposure and toxicity information, to identify the potential risks. The results are compared with risk-management criteria to determine their acceptability.

Recent proposed changes

On September 24, 1993, EPA issued a draft memorandum titled "Implementation of Exposure Assessment Guidance for RCRA Hazardous Waste Combustion Facilities (EPA 530-F93-029)." On the same date, the agency released another draft document, entitled "Addendum to the Methodology for Assessing Health Risks Associated with Indirect Exposure to Combustion Emissions." This document, revised on November 10, 1993 (EPA/600/AP-93/003), serves to update the previous guidelines for combustor risk assessment, which were released in 1990 (EPA/600/6-90/003).

Discussed below are some areas where incineration risk assessment guidelines may change. However, since the documents cited above provide the first view of proposed policy directives, the final policy could change from that presented here.

Multi-pathway risk assessments. As stated, EPA now intends to include evaluation of all relevant exposure routes for all chemical emissions, with new emphasis on chemicals that may accumulate in the environment (Figure). While consideration of these routes will provide a more complete picture of potential risk, the specification of additional receptor types and locations, and analysis of more complicated exposure routes will undoubtedly increase the cost of conducting the risk assessment — to $200,000 or more.

Stack emissions estimates. EPA is now recommending that risk assessments, especially for interim status units, consider actual operating data under worst-case operating conditions. EPA recommends that trial burns be conducted on actual waste in addition to surrogate waste, and that a variety of test conditions be used. Dioxin testing may be required for all facilities. For units under development, emissions data from facilities with similar combustion and air-cleaning technologies should be used until actual operating data are available.

Other emissions sources. Currently, most risk assessments address stack emissions only. However, the September 24 memorandum states that risk assessments should include emissions from all sources ". . including activities such as storage, blending, and handling of wastes fed to the combustion unit itself, as well as storage and handling of combustion residues (such as flyash, bottom ash and quench water) generated by the combustion facility." If additional assessments are required, it is possible that non-stack sources would have to be assessed in a supplementary analysis.

Exposure scenarios. The Fall 1993 draft policy endorses a move from the MEI approach to the RME concept of risk assessment. While the MEI approach is considered a worst-case analysis, it should also be noted that the RME concept was developed to represent the high-end exposures that several different types of receptors could potentially experience.

The EPA draft policy proposes to an-

A multi-pathway risk assessment sums the risks from all environmental pathways — including ingestion of produce, meat, cow's and mother's milk, and fish — that could affect a person's total risk from chemical exposure. It is then up to management to decide whether predicted levels are "acceptable," or whether mitigation measures should be taken to diminish potential impacts

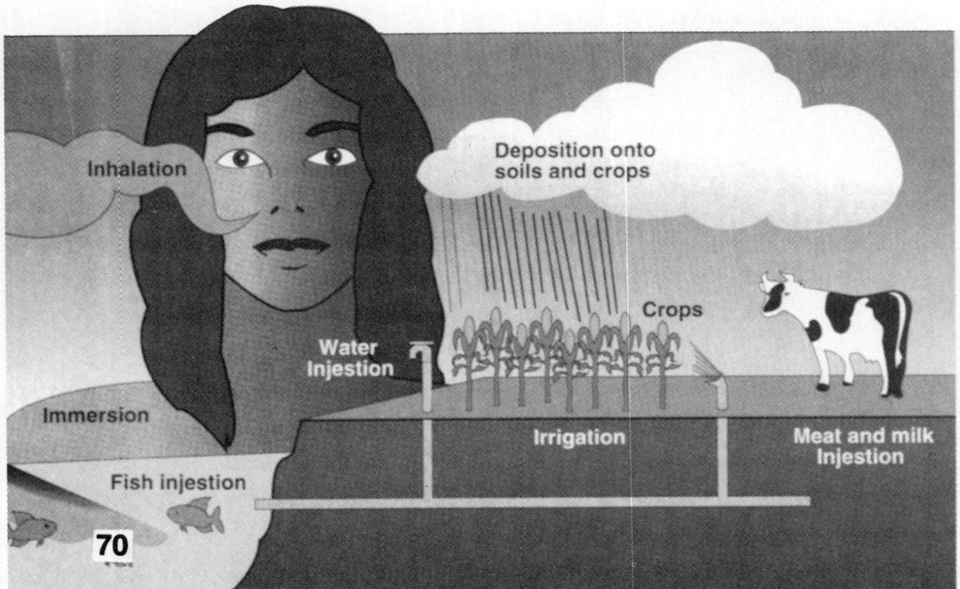

alyze receptors with different potentials for exposure. For instance, it would consider various receptors, such as a nearby resident, farmer, school child, backyard gardener and plant worker. Although this will call for more analysis, facilities may benefit from the improved realism of the results, particularly when it comes to interfacing with the community.

Population risks. Not all areas surrounding a facility have risks as high as those estimated for the MEI. Even the RME estimates are only representative of people living in specific receptor areas.

Now, EPA wants risk assessments to characterize the variation in risks for different geographical areas in the vicinity of a facility. One way to do this is to draw hypothetical, concentric rings around the facility, and estimate the number of people and the attendant risks from the facility's emission sources within each zone.

One concern, however, is that any presentation of a hypothetical number of cancer cases will give the impression that the risk assessment is an actual *predictor* of human health effects within a given proximity to a facility. In reality, such assessments are primarily planning decision tools, which do not predict specific health outcomes for people living near incinerators.

Exposure duration. The new EPA policy recommends less-than-lifetime exposure durations, such as 30 years. This is a departure from the previous standardized guidelines, which assume a 70-year exposure duration. Although this proposed change is not consistent with many state and regional guidelines currently applied to incinerators, if adopted, it will provide more realistic estimates of risk.

Wet-deposition modeling. According to the recent EPA documents, wet-deposition modeling (which accounts for the deposion of chemicals by precipitation onto the ground, water or plant surfaces) will be required for all new risk assessments. Most risk assessments now rely on dry-deposition modeling because of uncertainties associated with wet-deposition modeling.

Inclusion of wet-deposition modeling will increase risk assessment costs. And, in nearly all cases, the stated impact per unit of emission will rise. In addition, the location of maximum impact will often differ from that defined by dry-deposition modeling.

Dioxin-vapor modeling. There is a new requirement to model dioxin vapor uptake into plants. Up to now, risk assessors generally assumed that all dioxin is either absorbed by, or adsorbed onto, particles. This new modeling approach will affect beef, produce and dairy exposure pathways, and could raise the cited risks to as much as 100 times higher. Mass-balance considerations indicate that the EPA approach in this realm is scientifically flawed, so this aspect of the proposed guidance needs careful evaluation.

Acceptable risk. According to EPA, the "acceptable" or target risk for the sum over all pathways of exposures to all carcinogenic chemicals is $1\text{x}10^{-5}$. If this threshold is formally adopted, it would make the RCRA risk guideline consistent with criteria now used in many states. In the previous BIF rule, the sum of inhalation risks for four metals (arsenic, cadmium, chromium and beryllium) could not exceed $1\text{x}10^{-5}$. Thus, this target represents a tightening of the BIF regulations.

At high dose levels, most chemicals can produce some form of toxic effect, whether or not they are also potentially carcinogenic. For the non-carcinogenic effects of chemicals, EPA recommends that background exposures be considered. EPA suggests allocating 75% of the allowable exposure dose (as represented by a hazard index) of a chemical to sources other than the combustion facility, and the remaining 25% to the facility. The sum would have to be $\leq$ the published reference dose levels.[4]

This approach will clearly pose problems for many facilities. In particular, it will be a challenge to attain this goal for mercury because of its very low allowable exposure levels, and potential for bioaccumulation in the food chain.

Conduct your own assessment

In its September 24 memorandum, EPA indicated that it intends to perform the risk assessments itself or in conjunction with the facility. However, as more experience is gained by facility owners and their consultants, EPA indicates it will delegate the task to these parties under close supervision.

Because of different assumptions and methodologies, risk assessments carried out by EPA can differ from those prepared by the facility itself. By performing your own assessments, you will be able to anticipate agenc results and explain any apparent problems, which may be crucial to keeping your facility operating. In addition, a precise, multi-pathway risk assessment may provide the basis for negotiations, determine if site-specific approaches are needed, and ascertain probable emissions limits. It will also identify trial-burn requirements, determine environmental controls that are likely to be needed, and help communicate realistic risks to the community.

If proposed revisions to incinerator risk-assessment guidelines are enacted, more receptors and more scenarios would need to be considered. As an operator, you can prepare for any potential changes by collecting operating data, attending or monitoring roundable meetings held by federal and regional EPA offices, and working to understand how the changes may affect existing units or proposed projects. ■

Edited by Suzanne Shelley

[4]These reference levels can be found in the BIF Rule, or in EPA's Integrated Risk Information System (IRIS), an online database that contains health risk and EPA regulatory information on 400 chemicals and is mounted on the National Library of Medicine's Toxicology Data Network (Toxnet). For information, call (513) 569-7254.

The authors

Douglas Smith is principal environmental health scientist in the Risk Assessment Dept. of ENSR Consulting and Engineering (35 Nagog Park, Acton, MA 01720; tel.: 508-635-9500). He has 23 years of experience in risk and consequence analysis, and holds a B.S. in physics from Franklin and Marshall College, an M.S. and Sc.D. in environmental health sciences from Harvard School of Public Health.

Allison L. Yuhas is a risk assessment specialist with ENSR. She holds a B.A. in human biology from Brown University, and has four years of experience in risk-assessment analysis. Her project work has included multi-pathway risk assessments and exposure models for large combustion facilities and Superfund sites.

Brian Magee is a senior toxicologist with ENSR. He has 18 years of experience conducting multi-pathway risk assessments, and using those data to negotiate more favorable permit conditions with regulatory agencies. He holds a Ph.D. in toxicology from Mass. Institute of Technology, an M.P.A. in science and technology policy from the U. Washington, an M.S. in chemistry from U. California (San Diego) and a B.S. in chemistry from U. Virginia.

HOW RELIABLE IS YOUR SAFETY SYSTEM?

Lawrence V. Beckman
Triconex Corp.

Engineers are faced with many options in selecting instrumented protective systems. These safety units are designed to detect trouble spots in a process and then trigger proper action to safeguard the associated equipment and personnel in the plant.

The engineer needs to decide whether the system should be fail-safe or fault-tolerant. A few other options are: Will the overriding consideration of safety be reconciled with that of production? Given the reliability of programmable logic controllers (PLCs), how about installing a redundant* system made up of two or more PLCs in parallel? Or should a microprocessor-based system specifically designed for safety applications be used?

In the chemical process industries (CPI), redundant microprocessor-based safety systems are gaining in popularity because of the increasing reliability and favorable unit price of these electronic components. The use of redundant PLCs is fraught with both installation and operational problems, such as wiring of redundant inputs and outputs, implementation of algorithms for fault diagnosis, and incorporation of failure-mode scenarios in the application software.

Fail-safe vs. fault-tolerant

In weighing the pros and cons of redundant microprocessor-based safety systems, however, keep in mind that a fail-safe system may run counter to the goal of minimizing operating cost and loss of production. An emergency shutdown system that is fail-safe, for example, is designed to detect an unusual rise in pressure or temperature in a chemical reactor system, and safely shut off the entire process, or the sources of pressure or temperature, before equipment damage or personnel injuries can occur.

Note that in the fail-safe mode of operation, the entire process may be shut down upon detecting a fault, not a failure, in the safety system. Thus the major disadvantage of a fail-safe system is that unnecessary shutdowns or false alarms may be triggered by a

*Some experts prefer the term "replicate" and its derivatives, such as duplicate, triplicate, quadruplicate and so on because redundancy implies more than what is necessary.

Originally published January 1992

fault or malfunction in a single component of the protective system.

If the safety system is constantly "false tripping" the process, well-intentioned operators may disable the system to prevent the recurrence of such nuisance events, and thus defeat the purpose of installing such a system at all. Under these circumstances, what is even worse is the false sense of security in the minds of engineers who come to believe that they do have a safety system in place.

Fault-tolerant safety systems avoid false trips by sustaining the operation of a process even in the presence of a fault. Bear in mind that a failure is defined as a system's loss of ability to perform its intended function. A fault, meanwhile, is not a loss of function; it implies an impairment of some level of redundancy. Therefore, all fault-tolerant systems are necessarily redundant by incorporating two or more microprocessors, for example, in parallel.

Most fault-tolerant configurations allow online repairs in order to quickly recover the safety integrity of the system from the faulted operating state. For example, if a dual-redundant system is operating with a single microprocessor because the other unit is malfunctioning, it is only one step away from complete failure. To maintain a safe distance from failure, most fault-tolerant systems used in the CPI are triple-redundant, providing three independent levels of redundancy.

Another advantage of a triple-redundant design is the accommodation of a "voting" system. If two out of three microprocessors, for example, detect a problem, a decision can be made on the basis of the majority vote. The voting system also can single out the malfunctioning legs) in the system. This capability is a key factor in enabling the system to self-diagnose a fault, and

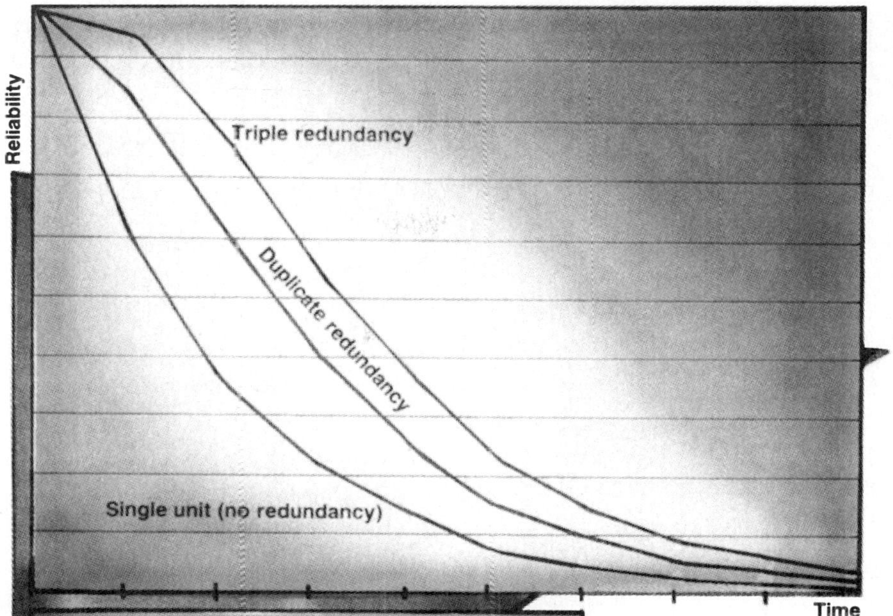

FIGURE 1. Increasing the degree of redundancy boosts reliability of a safety system

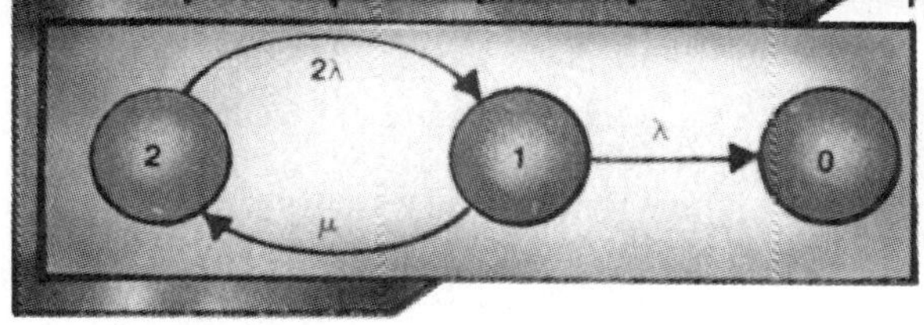

FIGURE 2. A fault-tolerant safety system existing in state "1" can be left alone to deteriorate to "0" state or brought back to no-failure ("2") state via the repair path

Quick, yet rigorous, calculations assess the chance of process safeguards failures

thereby improving its safety integrity.

Two other factors in favor of a fault-tolerant system are its enhanced reliability and availability. Reliability of a system is defined as the probability that the unit will perform its intended function for a stated period of time — say, a year. Availability is the percentage of time over which a system is capable of performing its intended function during a given time period.

Reliability in redundancy

Exponential distribution, which is applicable to situations where a failure can occur at any point in time, is the probability distribution most often employed to quantify reliability. In the absence of additional information, it is assumed that the failure rate (λ), expressed as the number of failures per year, is constant. The reliability (R) of a single microprocessor is then given by

$$R = e^{-\lambda t} \tag{1}$$

where t is time, and e is the base of the natural system of logarithms (having numerical value of about 2.718). It is clear that as t approaches infinity, the reliability goes to 0. Also, the higher the failure rate, the faster the drop in reliability.

For a parallel system with n identical redundant units in active operation, the reliability turns out to be

$$R_{system,1} = 1 - (1 - R)^n \tag{2}$$

For a system with two redundant units ($n = 2$), for example, we obtain

$$R_{system,2} = 2e^{-\lambda t} - e^{-2\lambda t} \tag{3}$$

The time interval between consecutive failures of a system is called mean time to failure (MTTF). For a dual-redundant system, $MTTF_2$ can be obtained from its reliability by integrating Equation 3:

$$MTTF_2 = \int_0^\infty R dt = \int_0^\infty (2e^{-\lambda t} - e^{-2\lambda t}) dt = 1.5/\lambda \tag{4}$$

With no redundancy, $MTTF_1$ of a single system would be $1/\lambda$. By adding one level of redundancy, we have increased $MTTF_2$ for the system by 50% with respect to $MTTF_1$. Adding a third unit gives

$$R_{\text{system},3} = 3e^{-\lambda t} - 3e^{-2\lambda t} + e^{-3\lambda t} \quad (5)$$

$$MTTF_3 = 11/(6\lambda) \quad (6)$$

We have again increased $MTTF_3$ of the system, but by a smaller percentage than the first level (22% vs. 50%). Thus the greater the number, n, of redundant units, the larger $MTTF_n$. However, there is a diminishing rate of return as the number of units increases (Figure 1).

The above results apply only to full redundancy in that all identical units must fail in order for the safety system to fail to function. The notation (3-2-1-0) is used to designate such systems, where "3" implies all three units are active, "2" indicates two out of three units are functioning, and "0" state means system failure.

There may, however, exist a form of partial redundancy in which only one unit is allowed to fail. In other words, if the system is made up of three identical units, two units are required for the functioning of the system. Such systems are designated by (3-2-0). The corresponding reliability is given by

$$R_{3\text{-}2\text{-}0} = 3R^2 - 2R^3 \quad (7)$$

Redundancy with repair

The reliability and availability of a system increases significantly if a failed unit can be repaired on line. The increase is attributable to the fact that the window of vulnerability is decreased by an amount equal to the interval of time between the instant the fault is first diagnosed, and subsequently repaired to restore the system to fault-free status. This time interval is called the mean time to repair (MTTR), and is equal to the reciprocal of the repair rate, μ.

Let us now define the mean time between failure (MTBF), and availability (A) of the system as follows

$$MTBF = MTTF + MTTR \quad (8)$$

$$A = MTTF/MTBF \quad (9)$$

For a dual-redundant system of two identical units operating in the fully active redundancy (2-1-0) mode

$$MTTF_{2\text{-}1\text{-}0} = \mu/(2\lambda^2) \quad (10)$$

where $\mu = 1/MTTR$, and $\mu >> \lambda$.

Keep in mind that Equation 10 has been obtained by using a theoretical framework called a Markov model of the system. The model assumes that the system can exist in all possible states from fully operational to completely failed. In a dual-redundant system, for example, there are three possible states:

- Both units operating ("2" state)
- One unit failed ("1" state)
- Both units failed ("0" state)

The probability of the system being in state "2" at time $t = 0$ is 1, and that of being in state "0" as t approaches infinity is likewise 1. These two values of the initial and final probabilities, in essence, specify the boundary conditions for the Markov model of the system. The transition steps from one state to another are shown in Figure 2, where the mean frequency of either unit failing, and thereby placing the system in state "1," is given by 2λ. The single-failure state ("1") can, of course, be brought back to the no-failure ("2") state via the repair path at a mean frequency of μ.

Now consider a triple-redundant (3-2-1-0) system consisting of three identical units. For fully active redundancy (that is, a single unit is required for operation), $MTTF$ of the system can be calculated as follows

$$MTTF_{3\text{-}2\text{-}1\text{-}0} = \mu^2/(3\lambda^3) \quad (11)$$

Compared to $MTTF_{2\text{-}1\text{-}0}$ of a dual-redundant system (given in Equation 10), $MTTF_{3\text{-}2\text{-}1\text{-}0}$ offers an improvement of $2\mu/(3\lambda)$. Because $2\mu/(3\lambda) > 1$, this reflects the enhanced availability of the triplicate system.

Consider the case of partially active redundancy (3-2-0), where two of the three units are required for operation of the system. It can be shown that

$$MTTF_{3\text{-}2\text{-}0} = \mu/(6\lambda^2) \quad (12)$$

From Equation 12, it is clear that $MTTF_{3\text{-}2\text{-}0}$ is smaller than $MTTF_{2\text{-}1\text{-}0}$. Therefore, the requirement of partially active redundancy imposes a severe restriction on the operation and availability of this triplicate system.

Periodic inspection pays off

If repair can be initiated upon diagnosis of a fault, this is of course the recommended course of action. However, there are situations where the repairs of the faulty units must be done at regular intervals only. This is especially true with unattended equipment that is inspected at periodic intervals (often called the proof test interval). A complete failure occurs if the system's redundancy is not adequate to sustain multiple faults during the proof test interval.

It turns out that a system's $MTTF$ is a function of the proof test interval. It is best to keep the proof test interval as small as possible in light of the fact that the failure rate of a typical system increases with time. It can be shown that a small decrease in this time interval can yield a significant increase in $MTTF$, and a much higher level of safety integrity.

Frequent proof testing is even more important if general-purpose PLCs are used in implementing the safety system. Because a typical PLC does not provide automatic testing and fault-diagnosis, faults may occur and go undiagnosed for an extended period of time. These undiagnosed or unrevealed (also known as latent or covert) faults may impair the system to the extent that it may be incapable of performing its intended safety functions.

Equations 10, 11 and 12 can be used to compute the periodically repaired system's $MTTF$, by adjusting the repair period to reflect the average downtime (ADT) as follows

$$ADT_{1\text{-out-of-}2} = T/2 \quad (13)$$

$$ADT_{1\text{-out-of-}3} = T/\sqrt{3} \quad (14)$$

$$ADT_{2\text{-out-of-}3} = T/2 \quad (15)$$

where T is the proof test interval, and the subscript 1-out-of-2 indicates that the dual system requires only a single unit to operate. Similarly the subscript 1-out-of-3 suggests that the triplicate system requires only a single unit to operate (Figure 3), while 2-out-of-3 implies that two of the three units are required for the system to continue operation (Figure 4).

Substitution of the respective value of $1/ADT$ (from Equations 13, 14 and 15)for μ in Equations 10 11 and 12 produces the following $MTTF$s for the corresponding systems

$$MTTF_{1\text{-out-of-}2} = 1/(\lambda^2 T) \quad (16)$$

$$MTTF_{1\text{-out-of-}3} = 1/(\lambda^3 T^2) \quad (17)$$

$$MTTF_{2\text{-out-of-3}} = 1/(3\lambda^2 T) \quad (18)$$

The system's availability can be approximated as follows

$$A = 1 - U = 1 - \lambda \times MDT \quad (19)$$

where U is the system unavailability, MDT is the mean downtime defined as follows

$$MDT = T/(n - k + 2) \quad (20)$$

where k out of n units are required for the system to operate.

On substituting the appropriate values of n and k, we obtain the system availabilities from Equation 19

$$A_{1\text{-out-of-2}} = 1 - \lambda^2 T^2/3 \quad (21)$$

$$A_{1\text{-out-of-3}} = 1 - \lambda^3 T^3/4 \quad (22)$$

$$A_{2\text{-out-of-3}} = 1 - \lambda^2 T^2 \quad (23)$$

Note that the triplicate system operating in 3-2-1-0 mode, has the highest availability ($A_{1\text{-out-of-3}}$). Interestingly, a dual system operating in the 2-1-0 mode has a higher availability ($A_{1\text{-out-of-2}}$) than the triplicate system ($A_{2\text{-out-of-3}}$) operating in the 3-2-0 mode.

Varieties of failure modes

Systems can fail in ways not directly related to the control system itself. Such systems are disabled by failures in the sensors, actuators and wiring, for example. Limiting the present discussion to faults within the control system, note that they can be both the overt and covert variety. Whether a particular fault will be placed into the overt or covert class depends on the diagnostic capabilities of the system. In other words, if the internal diagnostics of the system are comprehensive enough, then almost all faults will be overt or revealed to the system.

Systems with comprehensive internal diagnostics are said to have a high coverage factor. Most systems designed for safety applications have coverage factors ranging from 96% to 99+% (indicating that 96% to 99+% of the faults are self-diagnosed). Of course, the internal diagnostics of systems with triplicate or higher redundancy can always be enhanced by invoking the voting capability of the diagnostic components.

The proof test interval can be extended only if the coverage factor of the internal diagnostics is reasonably high. In fact, where the potential for human error in conducting the proof test is also high, extension of the proof test interval is often well justified.

However, no system in existence today offers 100% fault coverage. Considering that the most dangerous type of fault in a safety system is an unrevealed fault that could cause the system to fail, proof testing should be used to detect and repair this type of fault as well. The coverage factor of the diagnostics in redundant systems is crucial in maintaining a high level of system integrity, that is, keeping the system operating in fault-free state. One can also take advantage of the online repair capability to further enhance system integrity.

The ability of the safety system to respond to a process demand (that is, when called on to act) depends on its reliability. The percentage of time, during a given time interval, in which the safety system is either in a failed state, or unable to shut down the process in the event of an emergency is its fractional downtime (FDT; also known as unavailability).

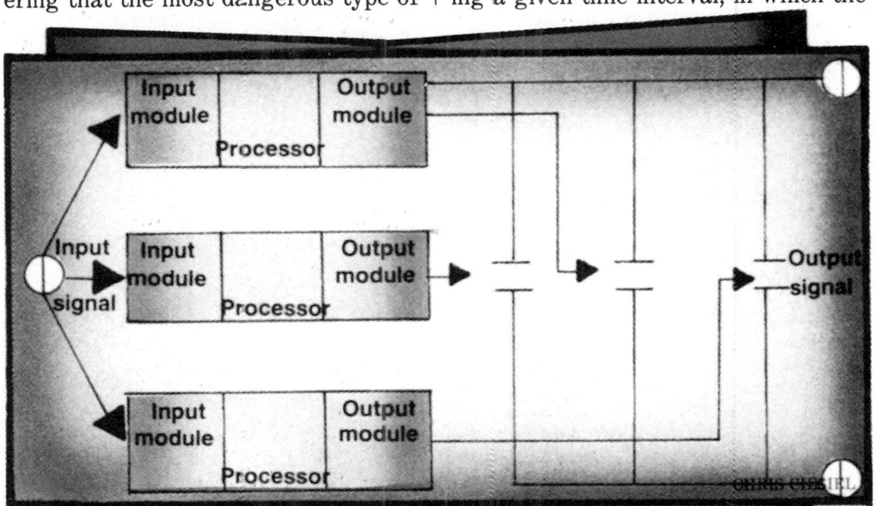

FIGURE 3. A triplicate safety system (top) operating in the 1-out-of-3 mode requires only a single unit to keep functioning

FIGURE 4. When in the 2-out-of-3 mode (bottom), a triplicate system needs at least two of the three units to operate

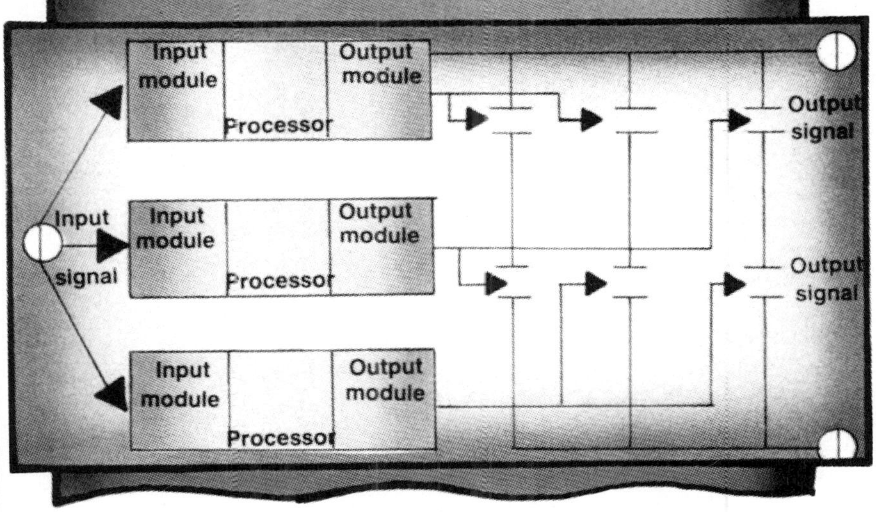

$$FDT = U = H/D \quad (24)$$

where H is the hazard rate, which is the reciprocal of the mean time between hazards (MTBH), and D is the demand rate, that is, the rate at which the safety system is activated by a process demand. Rearrangement of Equation 24 gives

$$MTBH = 1/(D \times FDT) \quad \textbf{(25)}$$

Let us compare the *MTBH* of dual- and triple-redundant systems featuring equivalent coverage factors. Suppose the systems exhibit an undetected failure rate of 0.18 failures/yr (one undetected fault every 5.56 years). For a proof test interval of three months (0.25 yr), and a demand rate of 0.5 emergencies/yr (once every two years), we can calculate the MTBH for the dual system as follows

$H_{1\text{-out-of-2}} = D \times \lambda^2 \times T^2/3$
$= 0.5 \times 0.7\ 10^{-3} = 3.5 \times 10^{-4}$
$MTBH_{1\text{-out-of-2}} = 2{,}857$ yr

For the triplicate system, we obtain

$H_{1\text{-out-of-3}} = D \times \lambda^3 \times T^3/4$
$= 1.15 \times 10^{-5}$
$MTBH_{1\text{-out-of-3}} = 86{,}956$ yr
$H_{2\text{-out-of-3}} = D \times \lambda^2 \times T^2$
$= 1.05 \times 10^{-3}$
$MTBH_{2\text{-out-of-3}} = 952$ yr

Assuming that $D \times T < 0.2$, a triplicate system running in the 3-2-1-0 mode increases the MTBH by a factor of 30 as compared to the dual system operating in the 2-1-0 mode. Also note that the increase is by a factor of 90 when compared to a triplicate system running in the 3-2-0 mode.

Bear in mind that what is an acceptable hazard rate in one situation may not be so under a different set of circumstances. Today's technology may allow an order of magnitude of 10^{-5} for the hazard rate. However, it may turn out to be too expensive to maintain the hazard rate at this level. Therefore, be skeptical of lower hazard rates quoted by an equipment vendor.

Another factor that can adversely affect the hazard rate is the occurrence of "common mode" failures. These result from the failure of two or more independent items due to a common cause, such as power supply to the system. Common-mode failures may also come to pass because redundant items are never completely independent. They may suffer from the same design or manufacturing faults; they may all be affected by fire or mechanical damage; or they may be maintained

Triconex Corp.

Online repair of a fault-tolerant system allows the user to replace a malfunctioning module without process interruption

by the same person who makes the same mistakes.

Although common-mode failures are quite rare in practice, they do occur, and therefore must be considered in a comprehensive failure analysis. The common-mode failure is typically modeled by adding an element in series with the redundant (parallel) elements in order to reflect the potential for this type of failure. Because the added element is in series configuration, it can adversely affect the reliability prediction. To avoid producing misleading results, this series element must be handled judiciously.

A comprehensive program of failure-mode analysis and fault-injection testing can dramatically reduce the possibility of common-mode failures in a process plant. Safety system vendors are also placing major emphasis on these procedures to validate the integrity of their systems.

Software reliability

A microprocessor-based safety system requires some internal programs as well as application software. The internal programs supplied by the vendor as part of the system include the operating system, self-diagnostics, and so on. This embedded software (often called firmware), along with the system hardware, is likely to be thoroughly tested and validated by the supplier, and as such is highly reliable.

The reliability of the application software is a user responsibility, and it must be tested and validated by the user to ensure that it is free of faults, particularly latent faults (bugs). The available standards for quantifying software reliability are not generally accepted, and new ones are only now in preparation. If comprehensive diagnostics are not provided by the vendor, the application software will be required to include input-output diagnostics and voting capability, thus increasing the program's complexity dramatically.

However, most application software used in safety systems is quite simple, and can be easily verified. Because of the use of a graphic language, such as ladder logic diagrams, the potential for human error is greatly reduced.

In order to minimize the likelihood of common-mode failures in the application program, a technique called diversity can be employed, whereby the programming is performed by two or more teams using different design, coding and testing methodologies. The parallel development effort is intended to detect errors by cross-comparison of test results during validation of the software. Keep in mind, however, while this concept is good in theory, the actual implementation in redundant systems is somewhat impractical, and rarely used anywhere except in military applications.

Given that the application program is an integral part of the safety system, it plays a critical role in the system's reliable operation. Prior to commissioning the system, the application software must be formally documented, and integrated testing should be performed to ensure proper operation.

Maintaining system integrity

The safety system should be designed in such a way as to simplify maintenance of both the hardware and software. For example, there must be provision for quick access to the faulty hardware components. Also, there must be visual indication for the plant operator, identifying the individual fault in the safety system. Repair procedures should be simple, to keep the repair time short.

The software in microprocessor-based systems offers unmatched flexibility both in function and testing. Program changes can be tested offline and brought back online without disabling the safety system. The software can monitor the repair being done on sensors and actuators to ensure that proper procedure is followed.

The economics of redundant microprocessor-based safety systems are becoming increasingly more attractive as the unit price of microprocessors keeps falling. As a result, engineers can now afford to limit the use of dual systems to fail-safe applications, where reduced availability and false trips are not a primary concern. Triplicate systems are, of course, the most costly alternative, but they offer the highest level of safety and availability.

The ideal design is the one that satisfies the safety issues while addressing production as well as operational considerations. The design of the system should also match the hazards associated with the process under control, and thereby counterbalance the effects of a failure of the safety system. A method of providing high levels of integrity is to eliminate all single points of failure within the system.

Analysis of cost over the life cycle of the system indicates that triplicate systems increase safety and production by dramatically reducing the occurrence of false trips. In certain circumstances, the payback for the redundancy can be achieved in a year. However, given the cost of incremental improvement in system integrity, levels of redundancy beyond triplicate systems are very difficult to justify. The optimal mode of operation is the fully redundant 3-2-1-0 mode, and the best policy is not to operate a safety system in the single, or "1," state for a long period of time.■

Edited by Gulam Samdani

The author

Lawrence V. Beckman is vice president of marketing at Triconex Corp., 15091 Blake Parkway, Suite D, Irvine, CA 92718; tel. (714) 768 - 3709. He began his career with Monsanto Chemical Co., and then worked at IBM, TANO, and Texas Instruments. A member of the Instrument Soc. of America SP84 safety committee, he has participated in several corporate seminars on safety and protective systems. He received his B.S. in 1964 and Ph.D. in 1969 both in chemical engineering, at Tulane University.

CHRIS CI[illegible]IEL

ESTIMATE TWO-PHASE FLOW IN SAFETY DEVICES

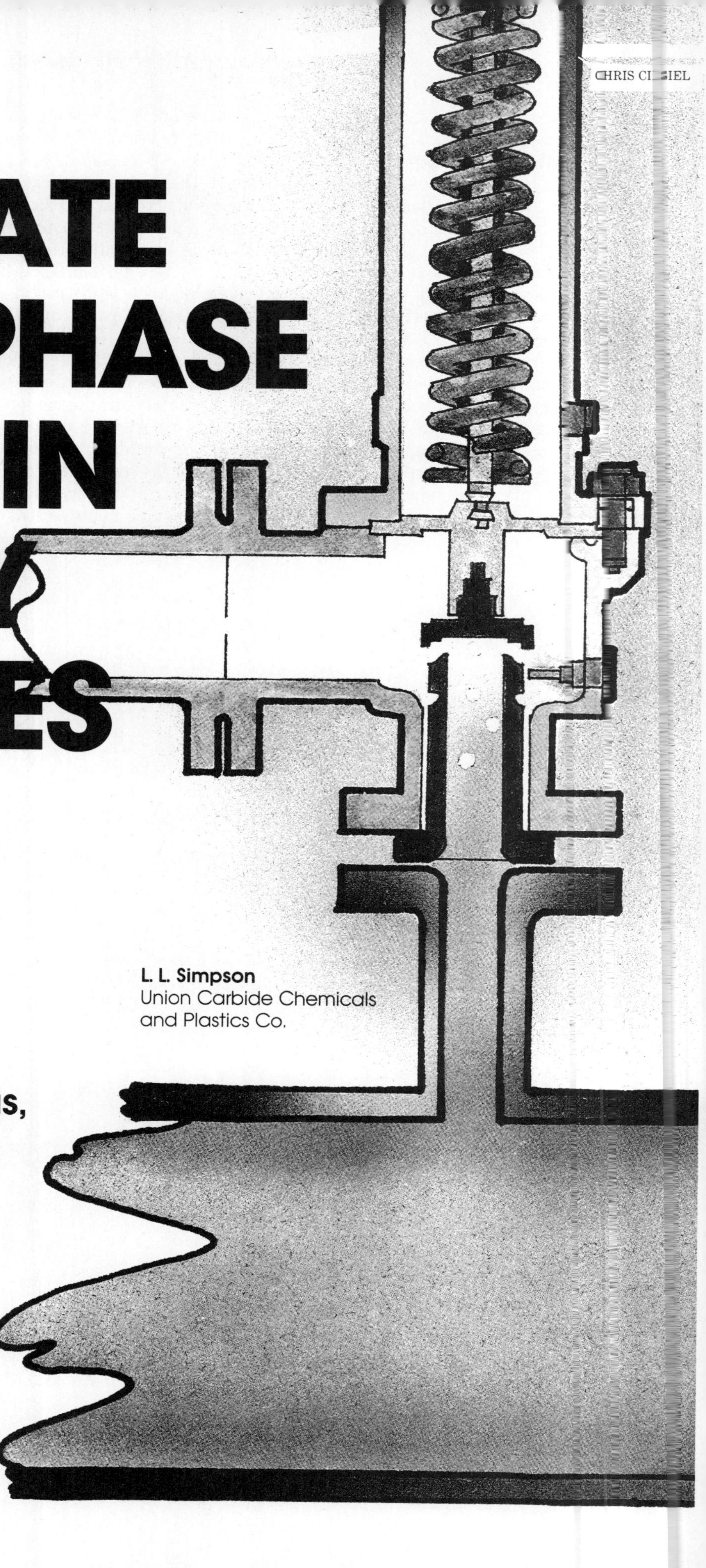

L. L. Simpson
Union Carbide Chemicals and Plastics Co.

A flexible, yet rigorous, approach computes gas-liquid flow through relief valves and rupture disks

Originally published August 1991

With an increasing awareness of safety in the chemical process industries has come a greater sophistication in the analysis of safety issues. One area where this particularly evident is in the study of two-phase flow in safety devices.

Virtually all CPI plants have pressure vessels containing both gases and liquids, and two-phase flows can develop in these systems as the result of fires, runaway reactions and other process upsets. Of particular concern are events involving the flashing of hot liquids across safety devices, because the mass flowrate will be much less than if the liquid were cold and nonvolatile.

Process engineers must decide whether or not to size a safety device for two-phase flow when designing a system. Most two-phase safety calculations involve ideal-nozzle concepts similar to those employed for single-phase gas and liquid computations. (Ideal here is synonymous with frictionless, or constant, entropy, i.e., isentropic.)

However, two-phase calculations are far more complex, involving many more sets of assumptions for describing flow in ideal nozzles. For example, ideal nozzles of different lengths can exhibit very different flow behavior, and various models are needed to help understand such behavior.

Over the years many techniques have been proposed to estimate two-phase flow [*1-8*], but none of them has found complete acceptance in industry. Generally, the methods lack the flexibility, rigor or computational accuracy needed to model the wide range of applications that one encounters in processing chemicals.

In recent work, researchers have focused on simple equations and all-encompassing charts for sizing safety devices [*6-8*]. But even these can be too difficult for the occasional user and the underlying technology does not lend itself for easy representation by charts and equations.

An algorithmic approach

Fortunately, two-phase-flow computations employing a generalized algorithm can be easily solved with personal computers. This new procedure estimates the ideal-nozzle mass velocity (mass flowrate divided by the minimum throat-flow-area) and the throat pressure.

The method can utilize any of the common models that have been used for this purpose. Calculational errors are typically less than one percent, and rarely exceed four percent, even at the thermodynamic critical point.

The tool is highly flexible and can deal with a wide range of model assumptions. The technique handles virtually any fluid, from subcooled liquids to superheated vapors, and fixed gas-fractions to multicomponent mixtures. Fluids that exhibit incompressible or critical flow, or that are subject to flashing, condensing or nonideal gas behavior can also be dealt with.

Further, users can estimate the exit thrust (or reaction force) on a safety device. This thrust, determined at the exit plane of the device, is the pressure multiplied by the area plus the mass flowrate times the velocity.

The best starting point for calculating two-phase flows is the volume-based energy equation:

$$G^2 = \left(\frac{\int_{P_r}^{P_t} -2[xv_g + (1-x)v_f]\mathrm{d}P}{[(xv_g/S + (1-x)v_f)^2 (xS^2 + 1 - x)]_t} \right)_{max} \qquad (1)$$

Here, G, P, S, v, and x are the mass velocity, pressure, slip ratio at the throat of the nozzle, specific volume, and weight-fraction gas, respectively, and the subscripts f, g, max, r and t refer to liquid, gas, maximum, reservoir, and the throat of the nozzle. The slip ratio is the gas-to-liquid velocity ratio.

This equation is integrated along an isentropic path from the reservoir pressure toward the downstream pressure until a maximum mass velocity is indicated. Except for nonflashing liquids, the maximum will occur at a throat absolute pressure that is roughly between 50% and 95% of the reservoir's absolute pressure.

Equation (1) is the framework for developing most of the models that have been proposed for analyzing two-phase flow through ideal nozzles. These include the Homogeneous Equilibrium Model (HEM), the Frozen Equilibrium Model (FEM), the Homogeneous Nonequilibrium Model (HNEM) and the Moody Model

The equation is general and devoid of assumptions. However, it is also cumbersome and computationally expensive to employ in this form, because the integral requires numerous isentropic phase-equilibrium calculations — except for the FEM — to resolve.

WHAT ABOUT PIPE FLOW?

There is no equivalent of the generalized ideal-nozzle equation for pipes, partly because the theoretical framework covering piping can be based on enthalpy, mechanical energy, momentum or Bernoulli-type equations. Nonetheless, the technique described here can be used — regardless of how the piping equation is formulated — if the equation is cast in the form of Equation (1). The homogeneous model of Benjamin and Miller (*12*) is probably the most widely used equation for two-phase flow through piping:

$$G^2 = \left(\frac{\int_{P_r}^{P_t} -\frac{\mathrm{d}P}{v}}{\ln(v_t / v_r) + (N+1)/2} \right)_{max}$$

This equation varies slightly from the original in that the velocity pressure coefficient N (including pipe friction) is increased by one to account for an entrance from a reservoir. This technique is much simpler than combining a pipe equation with a nozzle equation.

The quality of pipe-flow estimates is poorer than corresponding ideal-nozzle calculations, partly because N values, generally obtained from incompressible single-phase tests, are being applied to complex model-dependent cases. When applied to cases where the friction factors are dependent on Reynolds numbers, such as laminar flow, attention must be paid first to the integration step, to ensure orderly changes in N.

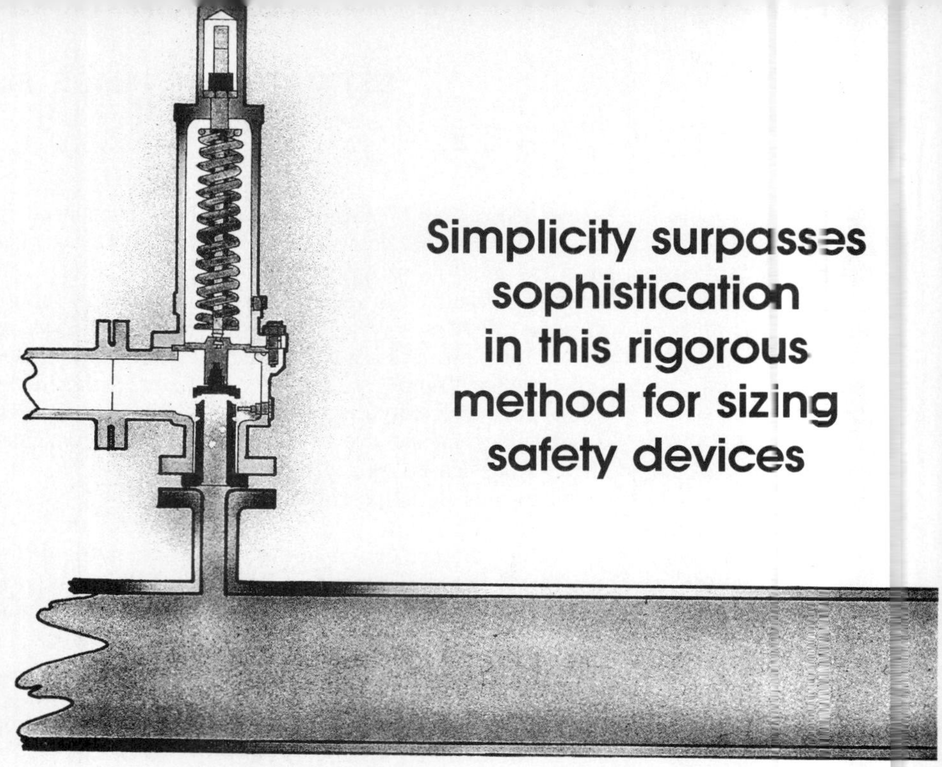

Simplicity surpasses sophistication in this rigorous method for sizing safety devices

As its name implies, HEM refers to a situation where the gas and liquid are at equilibrium inside the nozzle, with both phases flowing at the same velocity. With FEM, the phases also flow at the same velocity, but there is no change in the weight fraction of the gas in the nozzle; here, the flowrate can be determined by upstream physical properties.

One way to transform Equation (1) into a nonequilibrium model is to reduce the weight fraction of the gas throughout the nozzle so that the liquid does not flash as much as predicted by an equilibrium assumption. Equations (2) and (3) provide a means of reducing the flash fraction. Limited testing indicates that these simple empirical expressions will give results comparable to the Henry-Fauske model [4].

$$x_{ne} = x_A + x^2 - x_A^2 \quad \textbf{(2)}$$

$$0 \gtreqless x_{ne} \gtreqless x \quad \textbf{(3)}$$

Here, the ne subscript refers to a nonequilibrium condition. To use Equation (2), merely replace the x value in equation (1) with x_{ne}.

Moody-type models are based on equilibrium; however, theoretical or empirical correlations are used to relate the gas-to-liquid velocity ratio in the throat of the nozzle to the phase density ratio. A useful variation is a model with a fixed throat-velocity ratio.

Some simplifications

Fluid may enter a nozzle as a single phase (subcooled liquid or superheated vapor) or as two phases. The single-phase equations are well known and can be solved analytically or numerically; these solutions can be coupled to the methods that follow. In the saturated or two-phase regime, the following approximations greatly facilitate the solution of Equation (1) without introducing significant error:

$$x = a + bP \quad \textbf{(4)}$$

$$0 \gtreqless x \gtreqless 1 \quad \textbf{(5)}$$

$$1/v_g = cP^{1/n} \quad \textbf{(6)}$$

$$1/v_f = d + eP \quad \textbf{(7)}$$

FEM requires Equation (6) only, and both HEM and HNEM utilize Equations (4) through (7). This set of equations adds approximations but not assumptions to Equation (1). Equation (4) can be thought of as an interpolation scheme. Many fluids tend to exhibit such linear behavior; however, the equation works well even when the behavior is nonlinear.

Equation (6), for estimating the specific volume of the gas, is the most easily justified since it is of the same form as if the gas were ideal; n is an average isentropic coefficient. Because the liquid density tends to be relatively constant, Equation (7) is a good approximation.

The constants in Equations (4), (6) and (7) can be estimated two ways. First, the values can be based on the upstream saturation pressure and a rough estimate of the nozzle throat pressure. Alternatively, conditions in the nozzle can be projected from known behavior at the upstream saturation pressure. These techniques will be called the two-point interpolation procedure and the one-point projection method, respectively.

With two-point interpolation, the constants a, b, c, d, e and n are fixed from the x, v_g and v_f values at two different specified pressures, P_A and P_B. P_A is taken to be the saturation or bubble-point pressure when the fluid entering the nozzle is subcooled; it is the upstream pressure when the entering fluid is saturated or a two-phase mixture. Equation (8) is used to determine P_B.

$$P_B \cong 0.8P_A - 0.3x_A \quad \textbf{(8)}$$

P_B is intended to be a pressure in the vicinity of the critical throat pressure. The fact that it is not always a good estimate is unimportant because results are insensitive to P_B. As a result, values that depart from that calculated by Equation (8) are permissible. If the estimate from Equation (8) is less than

Nomenclature

C_{pf}	Liquid heat capacity
G	Mass velocity
n	Isentropic coefficient
P	Pressure
S	Slip ratio
T	Absolute temperature
v	Specific volume
x	Weight-fraction gas
a, b, c, d, e	are constants
λ	Latent heat

Subscripts

f	Liquid
g	Gas
max	Maximum
r	Reservoir
ne	Nonequilibrium
t	Throat of nozzle condition
A	Upstream saturation condition
B	Throat pressure

the back pressure, P_B may be equated to the back pressure.

The one-point projection technique utilizes the Clausius-Clapeyron equation to evaluate b in Equation (4). From an elementary manipulation of the thermodynamic equations, it can be shown that:

$$b = -TC_{pf}(v_g - v_f) / \lambda^2 \quad (9)$$

where C_{pf}, T and λ are the liquid heat capacity at constant pressure, absolute temperature and latent heat, respectively. Together with v_f, they are assumed to be constant and are evaluated at the saturated reservoir conditions. The equation also assumes that:

$$(\delta v_f/\delta T)_p \ll C_{pf}(v_g - v_f)/\lambda \quad (10)$$

A further, more subtle, assumption is that the single-component form of the Clausius-Clapeyron equation can be used to describe mixture behavior. Such a technique is questionable because latent heat does not have a rigorous thermodynamic definition. The coefficient n is set equal to one when the one-point projection method is combined with HEM. It is the isentropic coefficient when using the FEM.

The simple choice

This one-point projection technique with HEM is akin to the so-called Equilibrium Rate Model (ERM). In the range where the ERM is applicable, these two methods will yield results of similar accuracy if the most advanced form of the ERM [*6*] is used. However, neither technique is as good as the two-point interpolation method.

Nor is the ERM a distinctly different model. Rather, it approximates the HEM model and provides acceptable convenience and accuracy under many, but not all, circumstances.

With either approach for evaluating the above constants, the integral in Equation (1) can be solved directly for a subcooled liquid that enters the nozzle over a pressure range that varies between that of the upstream pressure and the saturation pressure. For calculations in the two-phase region, it is necessary to divide the pressure difference $P_A - P_B$ into small increments and to solve the equation at each intermediate pressure until a peak G value is obtained or the back pressure is reached.

There is a natural tendency to want to use more-complex technology than that represented by the above equations. For example, one might suspect that quadratic equations with three data points would work better than linear equations. Direct integration of Equation (1) is also tempting because it permits the use of a numerical minimization scheme.

Such temptations should be avoided. The calculational errors associated with this method are much smaller than the uncertainty associated with specific models. In addition, the procedure is free from the calculational quirks that often develop by using more-complex methods. Indeed, in this case, simplicity surpasses sophistication.

HEM is the most widely used method for two-phase safety-device sizing. It does not always yield the best estimate, but it usually gives a good estimate. Often, the predicted flowrates are somewhat lower than experimental results. As more data become available and computerized methods become more common, models with $S > 1$ are likely to become more prevalent.

Nevertheless, we will focus our attention on HEM. Equation (1) becomes:

$$G^2 = (v_t^{-2} \int_{P_r}^{P_t} -2\, v\, dP)_{max} \quad (11)$$

where,

$$v = xv_g + (1-x)\, v_f \quad (12)$$

An example

We will size a rupture disk to relieve 400,000 lb/h of 102 psia condensate saturated at 100 psia. Pressure downstream of the rupture disk is near atmospheric. In accordance with the above rules, P_A = 100 psia and P_B = 80 psia. From standard steam tables [*9*], we estimate the weight fraction flashed from 100 psia to 80 psia along an isentropic path. The input data is summarized in Table 1:

TABLE 1. Gas-liquid properties

	A	B
Pressure, psia	100	80
Weight-fraction gas	0.0	0.0179
Gas density, lb/ft³	0.2257	0.1828
Liquid density, lb/ft³	56.37	56.91

Let us illustrate how the constants are evaluated by using Equation (6), to estimate c and n.

$$0.2257 = c\ 100^{1/n}$$
$$0.1828 = c\ 80^{1/n}$$
$$n = \ln(100/80)/\ln(0.2257/0.1828) = 1.0585$$
$$c = 0.2257/100^{1/1.0585} = 0.002911$$

Similarly, it can be shown that a = 0.0895, b = –0.000895, d = 59.07 and e = –0.027. The solution of Equation (11) is displayed in Table 2.

TABLE 2. The maximum mass velocity occurs at 97 psia

P	x	V_g	V_f	v	$-\Sigma(v_{av}DP)$	G^2
102	0.0		0.01774	0.01774	0.0	0.0
101	0.0		0.01774	0.01774	0.01774	112.7
100	0.0	4.431	0.01774	0.01774	0.03548	225.4
99	0.000895	4.473	0.01773	0.02172	0.05521	234.1
98	0.001790	4.516	0.01772	0.02577	0.07896	237.8
97	0.002685	4.560	0.01771	0.02991	0.10680	238.8
96	0.003580	4.605	0.01771	0.03413	0.13882	238.3
95	0.004475	4.651	0.01770	0.03843	0.17510	237.1

The mass velocity is estimated from this table.

$$G = \sqrt{(238.8)(144)(32.174)} = 1{,}052 \text{ lb/s/ft}^2$$

P = 97 psia at the nozzle throat. The flow is choked at 97 psia and, like a gas, the mass velocity does not increase with back pressures less than this choked value. The numerical integration in the last column utilizes a trapezoidal technique.

This mass-velocity result is 1.6% less than the value obtained from detailed, multiple flash calculations (1,068 lb/s/ft²). One published shortcut equation [*10*] yielded G = 1,512 lb/s/ft², which is a 42% error.

The rupture-disk flow coefficient given in the ASME Code [*14*] is 0.62. Accordingly, the required rupture-disk area is:

(144)(400,000)/[(3,600)(0.62)(1,052)]
= 24.5 in.^{2}.

A rupture disk installation with 6-in. standard-weight pipe that has an internal cross-sectional area of 28.89 in.^{2} is the logical choice for this installation.

A second example illustrates the one-point projection method. We will estimate the flowrate of saturated liquid ammonia through a safety valve with a code-certified area and coefficient [*15*] of 3.17 in.^{2} and 0.858, respectively. When the valve is fully open, the ammonia temperature and pressure are 100°F and 211.9 psia. Physical properties are $C_{pf} = 1.177$ Btu/lb °F, $l = 477.8$ Btu/lb, $v_g = 1.419$ ft^3/lb, and $v_r = 0.02747$ ft^2/lb.

The constant b is estimated from Equation (9), where,

$b = -(100 + 459.67)(1.117)(1.419 - 0.02747)/[(5.404)(477.8)^2]$, or -0.0007436.

The 5.404 entry converts Btu to ft^3 psi. Other constants are determined to be $a = 0.1576$, $c = 0.003326$ and $n = 1$. A calculation similar to that suggested in the above table gives $G = 1{,}755$ lb/s/ft^2. The safety valve's mass flowrate is: (1,755)(3,600)(3.17/144)(0.858), or 119,000 lb/h.

Results from the techniques described here, using both HEM and slip models, have been compared to many different sets of experimental data, including those given in [*1, 3, 4, 5, 11, 13*]. They also have been compared to results from other calculation techniques.

As mentioned earlier, mass velocities from the two-point interpolation scheme are within a percent or so of those obtained by the detailed method, using multiple flash calculations in place of linearized equations. Calculational errors in the method are highest for saturated liquids, particularly those at elevated pressures.

However, the HEM mass-velocity estimate for steam at its thermodynamic critical point is within 4.2% of the true value. Other "two point" estimates are even closer to the best values.

More experimental data is needed on two-phase flow through safety devices. This will help to determine flow coefficients and model applicability, and enable a better understanding of stability limits and the effects of inlet pressure drop and back pressure. The behavior of safety devices in highly viscous two-phase flows is also a fertile field for research. ■

Edited by Jayadev Chowdhury

References

1. Palmer, M. E., Sallet, D. W. and Wu, K. F., ASME publication 78-WA/HT48, December, 1978.
2. Starkman, E. S., Schrock, V. E., Neusen, K. F., and Manelly, D. J., *Journal of Basic Engineering,* June, 1964.
3. Sallet, D. W., Somers, G. W., Gühler, M., Irvine, C. L., Popp, M., *Pressure Relief Valve Performance Study,* Unpublished Report, 1984.
4. Henry, R. E. and Fauske, H. K., *Journal of Heat Transfer, Trans. ASME,* May 1971.
5. Moody, F. J., *Journal of Heat Transfer, Trans. ASME,* February 1965.
6. Leung, J. C., *AIChE Journal,* Vol. 32, No. 10, 1986.
7. Leung, J. C. and Grolmes, M. A., *AIChE Journal,* Vol. 34, No. 3, 1988.
8. First, K. E. and Huff, J. E., *Plant/Operations Progress,* Vol. 8, No. 1, 1989.
9. *ASME Steam Tables,* American Soc. of Mechanical Engineers, 1967.
10. Fauske, H. K., *Plant/Operations Progress,* Vol. 4, No. 3, 1985.
11. Chisholm, D., Chapter 18, in *"Handbook of Fluids in Motion,"* (ed. by Cheremisinoff, N. D. and Gupta, R.), Ann Arbor Science, 1983.
12. Benjamin, M. W. and Miller, J. G., *Trans. ASME,* October, 1942.
13. Sozzi, G. L. and Sutherland, W. A., *Non-Equilibrium Two-Phase Flow,* ASME, 1975.
14. *ASME Boiler and Pressure Vessel Code* Section VIII, Div. 1, ASME, NY, 1983.
15. *Pressure Relief Device Certification,* The National Board of Boiler and Pressure Vessel Inspectors, Columbus, Ohio, 1987.

The author

L. L. Simpson is a Corporate Fellow in the Central Research and Engineering Technology department of Union Carbide Chemicals and Plastics Co. (Technical Center, Box 8361, South Charleston, WV 25303). His work includes plant design and startup, and for the past 25 years, a specialization in fluid dynamics. He has also been managing the Heat Transfer and Fluid Dynamics group for 15 years. Mr. Simpson has written many articles on various aspects of fluid dynamics, including pipe sizing, heat-exchanger tube failure, and mixing. He has a BSChE from Purdue University and an MSChE from West Virginia University. He is a registered professional engineer in West Virginia and has been active in AIChE.

SIZING RELIEF VENTS FOR RUNAWAY REACTIONS

Jasbir Singh
Hazard Evaluation Laboratory

The hazards of runaway reactions and unstable materials concern a wide range of industry segments including pharmaceutical, plastics, fine chemicals and petrochemicals. High pressures generated by these reactions can lead to explosive rupture of process vessels. To guard against such overpressurization, the most common protection device is a relief vent.

The term "runaway" refers to the fact that the rate of reaction increases exponentially with temperature, and that the chemical reaction itself is exothermic bringing about the unabated increase in temperature. Frequently, the rate of reaction doubles for every 10°C rise in temperature. Generally,

$$\dot{r} \propto e^{-E/RT} \tag{1}$$

where $\dot{r}$ is the rate of reaction (kmol/s), E is the activation energy (J/kmol), R is the gas constant (8314 J/kmol-K) and T, the absolute temperature (K).

Chemical reactors, storage tanks and most other process vessels are normally fitted with a bursting disc or a relief valve which, in the event of accidental overpressurization, opens to relieve the pressure and thus prevent equipment damage. The size — that is, effective orifice diameter — of the relief device has to be selected so that it can vent at a sufficiently high rate to compensate for the rate of pressurization. Thus, sizing involves two broad steps:

- Determination of the rate of pressure rise
- Calculation of the vent diameter large enough to cope with the rise

The rate of pressure rise is dependent on the cause of the venting incident normally some instrument failure or human error, and the specific nature of any chemical reaction. Thus the first step, irrespective of the equipment type is to look at all possible causes and ascertain the maximum rate at which the pressure may rise. This defines the so-called worst credible case under runaway conditions.

An example of a runaway reaction is the exothermic esterification of methanol with acetic anhydride. The rise in temperature with time for this system is shown in Figure 1. Note the characteristic manner in which the temperature rises very slowly at first, then accelerates rapidly. This is seen more clearly in Figure 2 where the rate of rise is plotted against temperature. The shape of the curve is characteristic of kinetically simple runaway reactions.

The pressure rise also shows a trend very similar to that of temperature. The rate of temperature rise, dT/dt (K/s) is easily converted to the heat generated by the reaction, $\dot{Q}$ (W/kg), using the relationship:

$$\dot{Q} = C_p \, dT/dt \tag{2}$$

where C_p is the specific heat of the liquid (J/kg-K). Thus Figure 2 can be translated into the heat output as a function of temperature, and the area under the curve gives the heat of reaction.

Almost invariably, the pressure rise in a piece of process equipment is due to the generation of vapor or gas. The safe relief sizing criterion therefore is:

Rate of vapor or gas removal $\geqq$ Rate of vapor or gas generation **(3)**

For example, if a closed vessel containing a liquid is heated at a rate Q (W) (say, by an external fire), then the rate of vapor generation $\dot{V}$ (kg/s) is given by,

Originally published August 1990

MEGAN REID

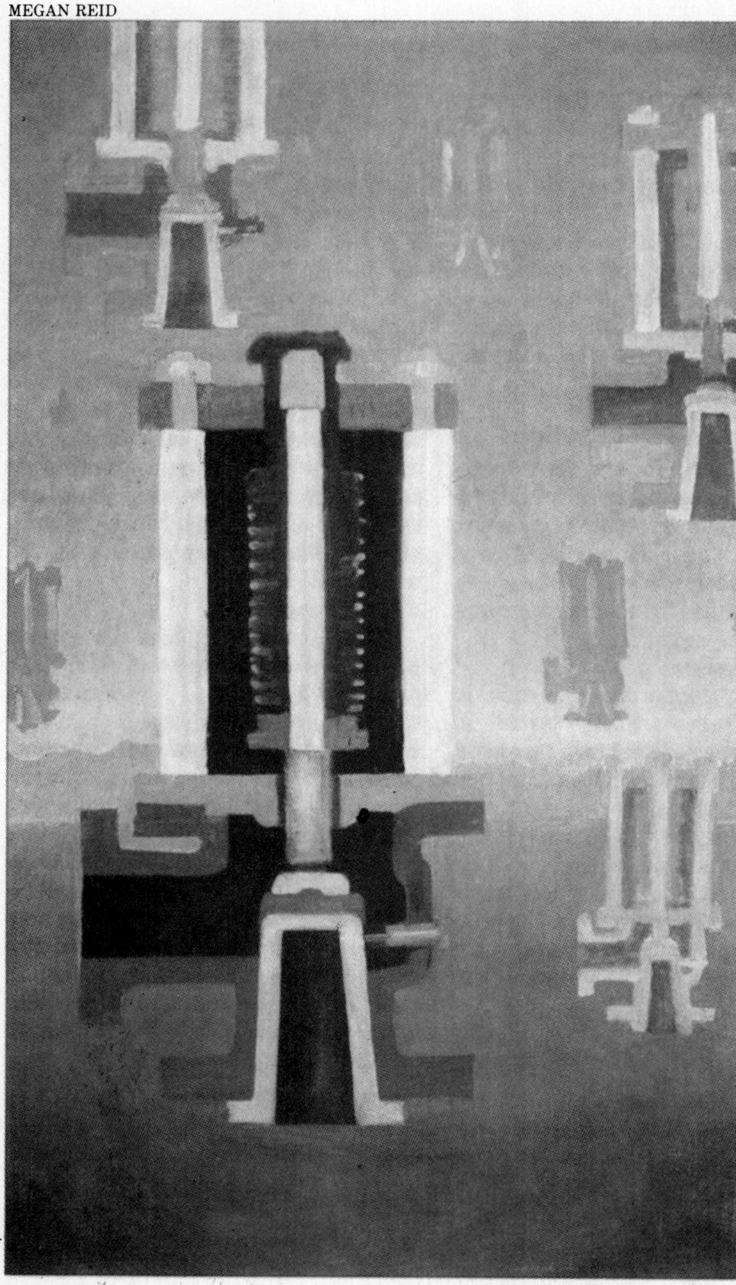

$$\dot{V} = Q/\Delta H_v \quad (4)$$

where ΔH_v (J/kg) is the latent heat of vaporization. If, on the other hand, the heat is generated by an exothermic chemical reaction (instead of fire), then:

$$Q = \dot{r}\ \Delta H_r \quad (5)$$

where ΔH_r is the heat of reaction (J/kmol). The value of $\dot{r}$ (and hence that of Q) is determined at the relief set point. It is assumed that evaporative cooling of the reactants will prevent any increase above this rate.

Given the rate of vapor generation, the traditional approach is to use a flow equation based on pure vapor and to ensure that the orifice is large enough to handle the rate given by Equation (4). Thus, for vapor flow,

$$G = C_d\, P\, [(\gamma/nRT) \times [2/(\gamma + 1)]^{(\gamma + 1)/(\gamma - 1)}]^{1/2} \quad (6)$$

where G is the mass flux through an orifice (kg/m²-s); C_d, the discharge coefficient; T, the vapor temperature (K); γ, the ratio of the specific heats at constant pressure and volume respectively; P, the upstream pressure (Pa); and n is the number of moles of gas per unit mass.

Equation (6) holds for sonic flow —

Simple calorimetric tests provide data for accurate calculation of vent diameter

FIGURE 1. Methanol and acetic anhydride reaction exhibits runaway characteristics. The temperature rises exponentially after a period of relatively small increase in temperature

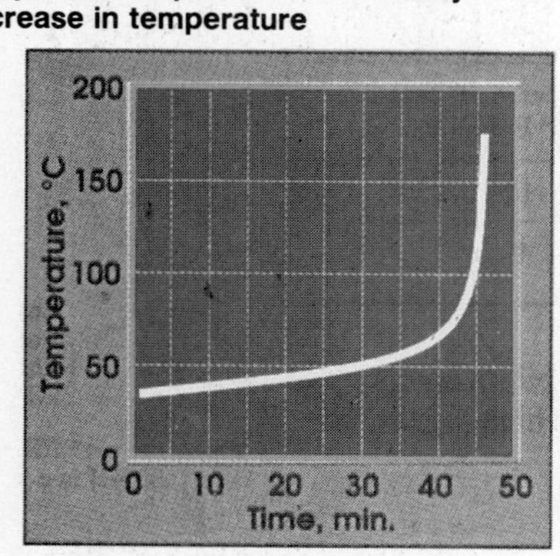

FIGURE 2. The rate of temperature increase (dT/dt) shows its peak for the runaway reaction between methanol and acetic anhydride

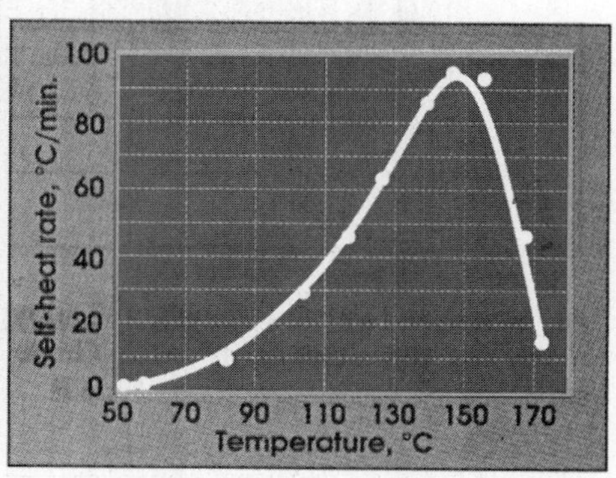

that is, if the ratio of the downstream to upstream absolute pressure is about 0.5 or less. Using this equation then the relief area, A (m^2), is given by $\dot{V}/G$

Limitations of general approaches

When the above approach is applied to reactive systems, a number of assumptions often break down. The relief-sizing criterion, as expressed by Equation (3) remains valid, but Equation (4) may not represent the total amount of gas being generated. If the reaction involves production of volatile gas for example, the rate is not directly related to the heat of reaction and latent heat. Thus, the gas rate needs to be determined in a different manner.

The rate of vapor generation is normally determined at the conditions when the vent opens. This rate is then used as the basis for vent sizing. Implicit is the assumption that cooling by latent heat after the vent opens will prevent any further temperature rise and hence prevent increase in the rate of pressure rise. This assumption is sometimes valid for venting in reactive systems, but not always. One certain case where it is not true is if the reaction generates a large quantity of noncondensible gas. The temperature continues to rise, and thus increase the reaction rate. The vapor rate for which the vent must be sized will obviously be higher than when the relief device first opened.

The final assumption — frequently invalid for reactive systems — is that the relief device can be sized to vent pure vapor. In practice, a vapor-liquid (two-phase) froth may have to be vented. The venting of the entrained liquid (in addition to the vapor) produces virtually no reduction in pressure, but its presence in the vent line reduces the vapor flow. Thus, the relief device sized for pure vapor will be too small for a two-phase mixture. Consequently, an order-of-magnitude underestimation in the area is possible.

Recognition of the above complex situation compared with the rather simple approach commonly used, led to the Design Institute for Emergency Relief Systems (DIERS) research program established in 1976 by the American Institute of Chemical Engineers (New York, N.Y.). This program produced both the designs for a testing device to assess chemical reactions and the methodology for sizing relief systems.

The use of DIERS methodology is becoming increasingly widespread. Detailed analysis of relief systems using this methodology, together with experimental testing, is now the accepted practice.

Types of runaway reactions

To size relief ports for a runaway reaction, it is necessary to classify the system according to the types of reaction products and their ability to remove heat during venting. Once this has been done, it is possible to select an appropriate vent-sizing equation.

1. Vapor-pressure systems

Here the system pressure is equal to the vapor pressure of the liquid. In such cases, the reactant will be cooled (due to the latent heat) as vapor is vented. At a sufficiently high vent rate, the heat of reaction can be balanced by the cooling effect, resulting in "tempering" of the reaction where no further temperature rise occurs. Since the pressure and temperature are directly related, the pressure will also cease to rise at this point.

Systems that behave in this way obey the Antoine relationship between pressure P and temperature T:

$$\ln P = A + B/T \qquad (7)$$

where A and B are constants. The methanol and acetic anhydride reaction discussed earlier is an example. The logarithm of the pressure against *(1/T)* follows a straight line as shown in Figure 3.

When a system of this type is vented, the pressure and temperature follow the trend shown in Figure 4, where an unvented runaway reactor and the same system under venting conditions are compared. Figure 4 also shows that when the vent opens, rises in both pressure and temperature slow down and then level off. Eventually the pressure will be reduced to ambient level as the reaction is cooled.

2. Gassy systems

These are cases where the system pressure is due entirely to the pressure of noncondensible gas, rather than the vapor pressure of the liquid. The gas is normally the product of decomposition. In these gassy reaction-mixtures, venting of the gas produces no noticeable cooling because the gas has no latent heat. As a result, the reaction temperature continues to rise during venting, and pressure rise can be arrested only by ensuring that the gas is vented at a sufficiently high rate. Thus, unlike the vapor-pressure systems, the pressure is controlled (and reduced) without cooling the reaction.

FIGURE 3. (below) Vapor pressure and temperature data confirm a tempered system for the methanol and acetic anhydride reaction

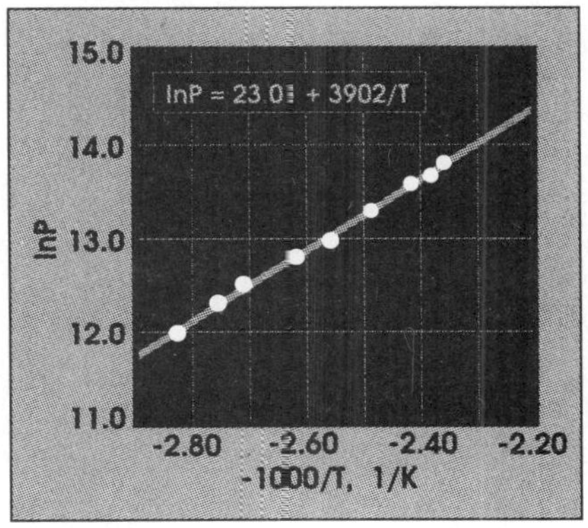

FIGURE 4. (right) Pressure and temperature characteristics for a tempered system with relief vents open and closed

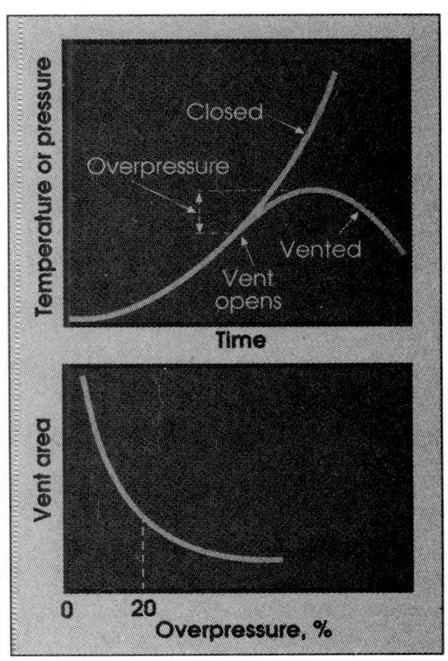

The pressure and temperature history for a reaction that has these characteristics is shown in Figure 5. The pressure and temperature rise until the vent is opened, when the pressure increase slows down but the temperature continues to rise. As the venting proceeds, the reaction rate continues to increase in line with temperature, but if the vent is made large enough, the pressure can still be kept under control.

In these systems, the rate of reaction governing the vent size will be higher than the one that governs when the vent opens. The pressure-temperature relationship for gassy systems is shown in Figure 6. The significant deviation from the Antoine relationship indicates effect of noncondensible gas.

3. Hybrid and complex reactions

There are many reactions that do not conform to the simple classifications above and therefore need to be considered differently. Among these is one common deviation: systems that have a significant vapor pressure and at the same time produce noncondensible gases — reaction systems often referred to as hybrids.

The behavior of such reactions during venting depends on the relative contributions of the vapor and noncondensible gas. An important case is a tempered hybrid. Here the vapor pressure contribution is large enough to provide evaporative cooling capable of preventing continued temperature rise after a vent opens. The relief of this type of system is therefore similar to that for a true vapor-pressure system (Figure 4) even though the pressure-temperature relationship will resemble a gassy system (Figure 6). If the vapor pressure is insufficient to temper the reaction (that is, prevent temperature rise after venting), then the system is treated as a gassy system.

There are no simple rules to decide when a hybrid reaction will temper; simple experiments can however provide the information, as discussed later.

Moreover, the nature of the chemical system can change during the course of venting. Consider, for example, a high-boiling reactant dissolved in a relatively low-boiling solvent. The pressure of the mixture is determined by the vapor pressure of the solvent and when the system starts to vent, it will behave like a tempered system (see Figure 3).

As the venting proceeds, the volatile solvent is stripped out leaving the high-boiling reactant behind. When the solvent is finally removed, the temperature starts to increase rapidly to the boiling point of the reactant. At the same time, of course, the reaction rate increases due to the increase in temperature and the change in composition.

Deviations can also result from changes in reaction kinetics and stoichiometry during the course of a venting incident. The classification of a reacting system can be influenced by a large number of variables. An experimental assessment should be carried out in each case to establish the true situation.

Vessel hydrodynamics

In addition to the thermal and kinetic aspects of a reaction, vent size is also influenced by the hydrodynamics within a vessel as this determines the composition of the vapor-liquid mixture vented. In general, the fluid entering the relief device can be considered in one of three categories (see Figure 7):

- All vapor with total disengagement between vapor and liquid
- Homogeneous two-phase mixture where the vapor and liquid are intimately mixed in the vessel, and the ratio entering the vent is equal to that within the vessel (zero disengagement)
- Two-phase vapor-liquid mixture in which the quantity of liquid entrained into the vent is less than that for homogeneous mixing, due to partial disengagement. Two disengagement models, namely bubbly and churn-turbulent flows are usually considered. The former regime relates to slightly less liquid disengagement (*i.e.*, more entrainment) than the latter.

Note that the two-phase or all-vapor behavior of a specific chemical mixture can only be verified through testing or plant experience. Theoretical prediction is not possible. The occurrence of

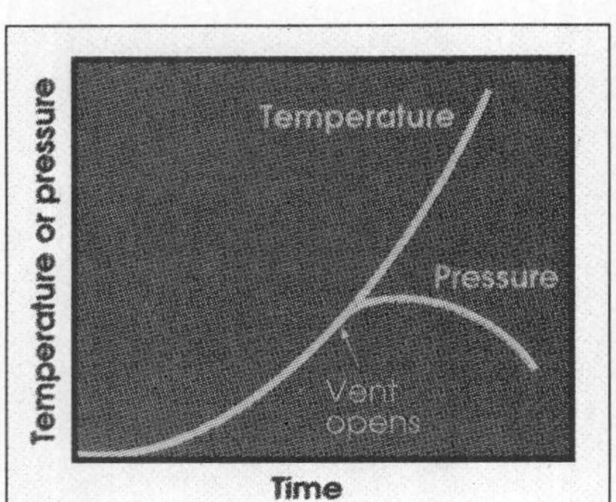

FIGURE 5. (below) Pressure and temperature characteristics for a gassy system with relief vents open and closed

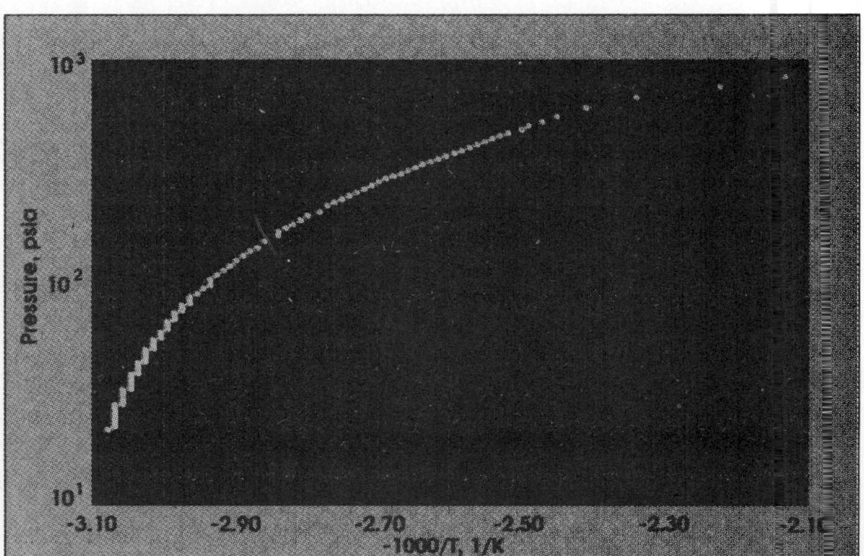

FIGURE 6. (right) Pressure-temperature relationship for a gassy system, with significant deviation from the Antoine equation (see Figure 3)

vapor venting is quite infrequent in reactive mixtures. In most cases, vapor venting leads to the smallest vent diameter.

Homogeneous venting is approached in many practical cases, and is frequently promoted by the inherent "foaminiess" of chemical systems, which tends to mask the behavior of discrete bubbles rising within a vessel. This foam can be produced by very low concentrations (ppm levels) of surface-active agents, and is quite stable under venting conditions. At present, the onset of this type of flow cannot be predicted from physical properties of the fluid. The assumption of homogeneous venting often (but not always) represents the most conservative relief size.

When foaming chemicals are vented, the superficial vapor velocity through the vessel has virtually no influence on the vapor-liquid ratio — even quite low velocities produce two-phase venting. This is in contrast to other substances where the vapor velocity is crucial in determining whether two-phase or all vapor venting occurs. Behavior of these non-foamy fluids is amenable to theoretical description when the fluids exhibit bubbly or churn-turbulent flow regime, as described below.

If a reaction vessel of cross section A (m^2) containing a non-foamy fluid is being vented, then the superficial vapor velocity $\dot{v}$ (m/s) at any liquid level h is given by

$$\dot{v} = (\dot{Q} h \rho_f A)/(\Delta H_v \rho_g A) = [\dot{Q} \rho_f/(\Delta H_v \rho_g)]h \quad (8)$$

where ρ_f and ρ_g are liquid and gas densities (kg/m^3) respectively, and $\dot{Q}$ (W/kg) is rate of heat generation per kg of total mixture. This shows that the velocity is proportional to liquid level.

The relationship between superficial vapor velocity and void fraction at which two-phase flow commences has been found [1] to be:

For bubbly flow,

$$\psi = \alpha (1 - \alpha)^2/[(1 - \alpha^3)(1 - C_0\alpha)] \quad (9)$$

and for churn-turbulent flow,

$$\psi = 2\alpha/(1 - C_0\alpha) \quad (10)$$

where α is the void fraction, C_0 is an empirical parameter (1.2 for bubbly flow and 1.5 for the churn-turbulent regime), and ψ is the dimensionless superficial velocity given by,

$$\psi = \dot{v}/U_\infty \quad (11)$$

where

$$U_\infty = [k(g\,\sigma\,(\rho_f - \rho_g))^{1/4}]/\rho_f^{1/2} \quad (12)$$

U_∞ = bubble-rise velocity in an infinite fluid (that is, in the absence of wall effects) (m/s)

k = 1.18 (bubbly) and 1.53 (churn-turbulent)

g = acceleration due to gravity (m/s^2)

σ = surface tension (N/m)

Thus if ψ is greater than the value given by Equation (9) or (10), two-phase venting will result. In the case of bubbly and churn-turbulent regimes then, two-phase venting begins at some high liquid level and eventually changes to all vapor as the level drops. Over the range of venting, the vapor-liquid ratio changes — starting at a low value and ending at 1.0 (all vapor). The variation in the weight ratio (so-called quality), χ, with ψ for churn-turbulent flow can be calculated [2] from:

$$\chi \sim (1 + \psi)/[(1 - \rho_f/\rho_g)\psi + ((\rho_f/\rho_g)(1 - \alpha)/(2\alpha))] \quad (13)$$

where α is the average void fraction in the vessel, based on liquid volume.

The relationship between ψ and the liquid level (or void fraction) dividing two-phase and vapor flow is plotted in Figure 7. This shows that if $\psi \geq$ 1 to 5, two-phase flow is ensured provided,

$\alpha \leq 0.6$ for churn-turbulent flow, and

$\alpha \leq 0.8$ for bubbly flow

Knowledge of the two-phase regime and the influence of superficial vapor velocity has important practical implications. In the case of a churn-turbulent system for example, it may be deduced that up to 40% of the vessel contents (mostly in the form of liquid) will be vented before vapor-venting begins. This has important downstream environmental consequences. Also, the fact that liquid entrainment can be influenced by the vapor velocity shows that vessel geometry (e.g., vertical versus horizontal) can be used to crucially affect the results.

Use of bench-scale calorimetry

The complexity of runaway reactions and two-phase flow can now be resolved relatively quickly due to developments in bench-scale adiabatic calorimetry, recently reviewed by Tharmalingam [3]. Such instruments typically consist of a thin-walled steel sample-cell containing a thermocouple and a magnetic stirrer. The sample cell is attached to a pressure transducer as shown in Figure 8. The container is surrounded by a number of zone heaters. The entire assembly is placed within a pressure vessel connected to a nitrogen supply.

As the sample temperature increases due to reaction, the zone heaters match

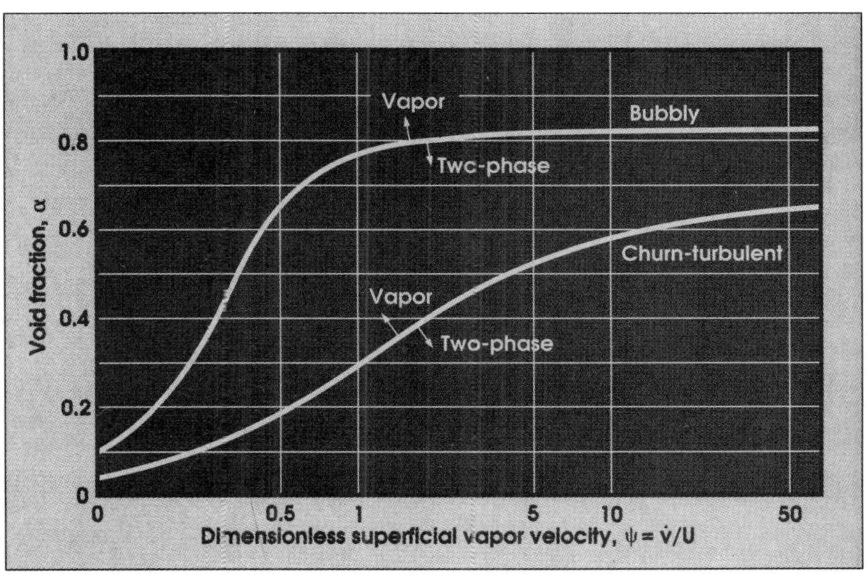

FIGURE 7. Change of venting regime with superficial vapor velocity

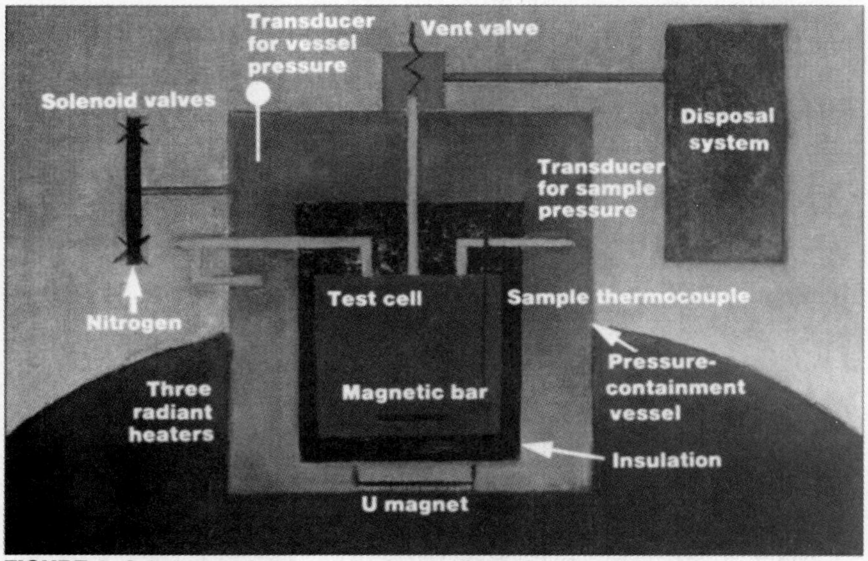

FIGURE 8. Calorimeter for the assessment of reactive materials

it exactly, thereby preventing any heat losses. The sample container is constructed from extremely thin sheet-metal to reduce the so-called phi-factor (ϕ):

$$\phi = 1 + (M C_p)_c/(M C_p)_s \tag{14}$$

where the subscript c refers to the container and s to the sample. Using a lightweight container, $\phi \sim 1.05$ (similar to that for largescale reactors) can be achieved.

This aspect is crucial to the use of the data for plant design without the need for sophisticated modeling or extrapolation. An example of the importance of the ϕ-factor is shown in Figure 9, where the temperature-time data for the same exothermic reaction is plotted with different values of ϕ [5]. This shows that the reaction was considerably slowed down from $\phi = 1.05$ to $\phi = 1.35$. More important, if ϕ is increased to 2.47, the exotherm is hardly visible.

FIGURE 9. Importance of ϕ-factor illustrated by amine-initiated cyanide reaction

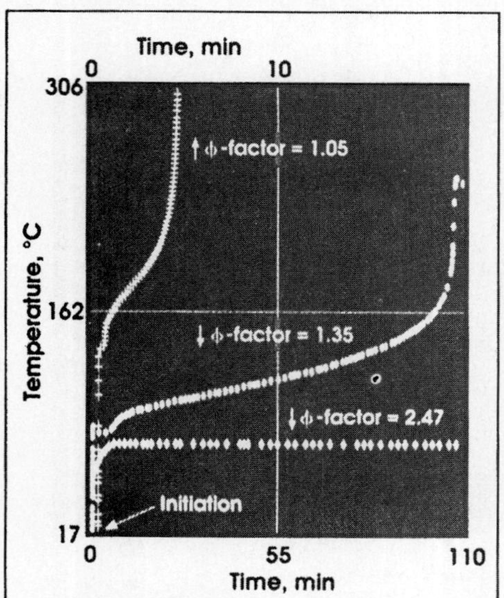

Clearly, thin-walled construction (which allows $\phi = 1.05$) limits the pressure to which a sample cell may be subjected. This limitation is overcome by balancing sample pressure with an equal external nitrogen pressure — a capability demonstrated during the DIERS program. The device can be used for different purposes — screening hazardous chemicals, investigating reaction kinetics, vent sizing, two-phase flow measurement and even assessing external quenching of vented reactants.

Vent-sizing equations

When a chemical system has been properly characterized by suitable experiments, and the relevant chemical and physical properties have been determined under the relevant conditions, it is often relatively straightforward to size the vent. The equations cited below are suitable only under the correct conditions, and therefore their limitations should not be overlooked. They are usually conservative, occasionally overly so. Equally, they may undersize if applied under the wrong conditions.

1. Sizing for tempered systems

Suppose a runaway reaction is vented when the temperature is T_s (corresponding to the relief set pressure P_s), and the rate of temperature rise is $(dT/dt)_s$. If the pressure is allowed to increase above the opening pressure (say, by 10 — 20%) to P_m, then suppose that at the maximum temperature T_m, the rate of temperature rise is $(dT/dt)_m$. For a vessel of volume V containing a mass M of reactants, the frictionless vent area A for homogeneous two-phase flow can be calculated [6] from:

$$A = M q /(G[((V/M)\, T_s \lambda \times (dP/dT))^{1/2} + (C_p\, \Delta T)^{1/2}]^2) \tag{15}$$

where q is the average rate of heat release between the vent-opening pressure and the maximum pressure, which can be approximated by $[(dT/dt)_s + (dT/dt)_m]C_p/2$; dP/dT is the slope of the vapor pressure vs. temperature curve (Pa/K); ΔT is the rise in temperature corresponding to the overpressure (i.e., $T_m - T_s$); λ is 1.0 for a vapor-pressure system; G is the two-phase flashing flow (kg/m^{2}s) which can be approximated by:

$$G = (dP/dT)\,[T_s/C_p]^{1/2} \tag{16}$$

It is possible to invoke the Clausius-Clapeyron approximation and eliminate dP/dT from both the above equations. Then the mass flux is obtained from:

$$G = (\Delta H_v / v_{fg})[T_s\, C_p]^{-1/2} \tag{17}$$

where v_{fg} is the difference between the vapor and liquid specific volume

2. Sizing for gassy systems

In general, the total amount of gas $\dot{F}_g$ plus vapor $\dot{F}_v$ generated during the reaction that has to be vented is given by:

$$\dot{F}_{tot} = \dot{F}_g + \dot{F}_v \tag{18}$$

where the rates are expressed in (m^3/kg-s). For systems with a low vapor pressure and those that produce large quantities of noncondensibles, $\dot{F}_v \ll \dot{F}_g$. If this gas is vented in the form of a homogeneous two-phase froth, the mixture density will be (M/V), and the required venting rate $\dot{F}_g (M/V)$, per kg of reactant. If the two-phase mass flux is G, then the vent area is given by:

$$A = [\dot{F}_g M^2]/[GV] \tag{19}$$

The two-phase mass flux can be estimated from incompressible Bernoulli's equation (using the two-phase specific volume):

$$G = C_1[2\Delta P\,(M/V)]^{1/2} \quad (20)$$

where ΔP is the pressure drop across the vent (Pa). Accurate estimation of the mass flux requires the use of a more complex set of equations [7].

If the gas generation rate is measured experimentally in a container of volume V_c at a temperature T_c calculated from the rate of pressure rise dP/dt in the container, then F_g is given by:

$$\dot{F}_g = (1/M_s)(V_c/P_p)\times (T_p/T_c)(dP/dt) \quad (21)$$

where T_p and P_p are the reactant temperature and pressure at the moment the gas generation rate is determined, and M_s is the mass of sample tested.

Since the temperature continues to rise after the vent has opened, it is important to ensure that the gas rate used for sizing is the maximum and not simply the value when the vent opens. T_p, P_p and dP/dt must be measured at this point of maximum gas rate.

The most conservative assumption is to use the maximum rate from a closed-vessel test. This may be overly conservative because the reactor may have been emptied (due to two-phase flow) before the maximum rate could occur. Therefore, care needs to be taken in ensuring a safe design while avoiding excessive conservatism.

3. Sizing for hybrid systems

The primary characteristic of the hybrid systems is that both vapor and noncondensible gas contribute to the total pressure, and thus they are neither entirely gassy nor vapor-pressure type. These systems are best divided into tempered and nontempered hybrids.

Vents for tempered hybrids can be sized using Equation (15) with a suitable value of λ, defined as the vapor pressure divided by the total pressure. For vapor-pressure systems, $\lambda = 1$, while for tempered hybrids, $0 < \lambda < 1$. If the vapor pressure of the reactants is known and the total pressure (i.e., including gas produced by reaction) is measured experimentally, then the ratio is λ. From Equation (15), it is clear that as noncondensible contribution increases, the vent area required is increased.

The calculation of ΔT in Equation

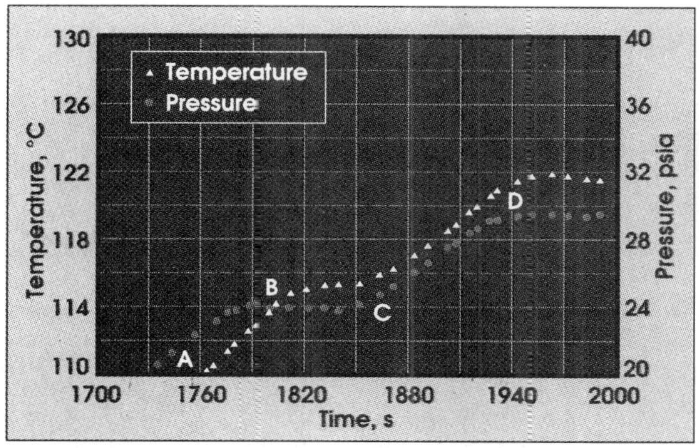

FIGURE 10. Calorimetric test results indicate tempered-reaction venting

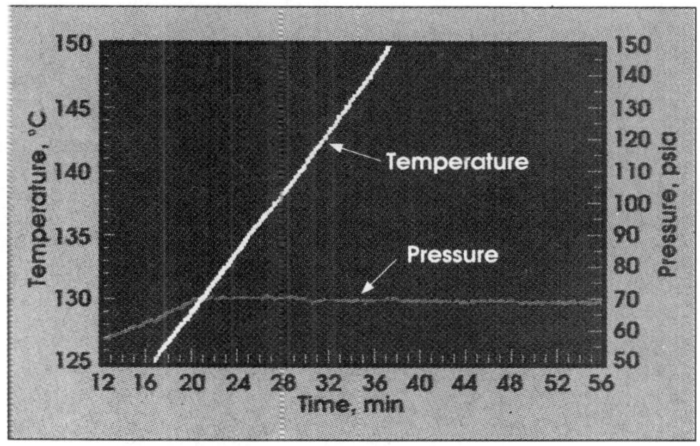

FIGURE 11. Calorimetric test results show gassy-reaction venting

ILLUSTRATIVE DESIGN

Sample calculation of vent size for a tempered runaway reaction is illustrated here using the experimental data presented in Figure 12. The method uses Equation (15). Homogeneous venting is assumed, which is normally considered conservative. (In the case of gassy reactions, all-vapor venting is sometimes more conservative. The only way to verify whether homogeneous venting occurs is to repeat the calculation using vapor and homogeneous venting, and use the largest diameter.)

Reactor parameters

Volume, $V = 5\ m^3$

Mass, $M = 4{,}000$ kg

Vent opening pressure = 10 bar (= 145 psia)

Temperature at set pressure, $T_s = 160°C$ (= 433 K)

Overpressure allowed above operating pressure = 10% = 1 bar (= 10^5 Pa)

Material properties

Specific heat, $C_p = 2{,}800$ J/kg-K

Slope of vapor pressure and temperature curve, $dP/dT = 15{,}000$ Pa/K (dP/dT can also be measured in the test)

ΔT for 10% overpressure $= 10^5/15{,}000 = 6.7$ K

Rate of reaction

Rate at set pressure, $(dT/dt)_s = 5.0$ K/min

Rate at maximum pressure, $(dT/dt)_m = 5.6$ K/min

Average rate = 5.3 K/min = 0.089 K/s

Calculation:

Two-phase mass flux, G

$= (dP/dT)(T_s/C_p)^{1/2}$

$= 15{,}000\ (433/2{,}800)^{1/2}$ kg/m²-s

$= 5{,}899$ kg/m²-s

Rate of heat generation $= \dot{Q}$

$= C_p(dT/dt) = 2{,}800\times 0.089$ W/kg

$= 248$ W/kg

Finally, using Equation (15):

$A = (4{,}000\times 248)/(5{,}899[(5\times 433\times 15{,}000/4{,}000)^{1/2} + (2{,}800\times 6.7)^{1/2}]^2)$

$= (9.92\times 10^5)/(90.1 + 136.97)^2 = 0.00326\ m^2$

Hence, vent diameter = 6.4 cm = 3 in (approximately)

The above figure is valid for an ideal nozzle. This must be adjusted to account for the actual type of relief device and downstream vent line.

(15) now presents more of a problem than for a vapor pressure system. It can be estimated from closed-vessel pressure-temperature data (as for a vapor pressure system), but this will underestimate ΔT and overestimate the vent size.

A better estimate requires a more detailed calculation, allowing for the fact that in a vented system, the noncondensibles will be vented, and therefore a larger ΔT will be available in order to achieve the same pressure rise. The maximum value of ΔT will correspond to the pressure-temperature slope of the vapor pressure line.

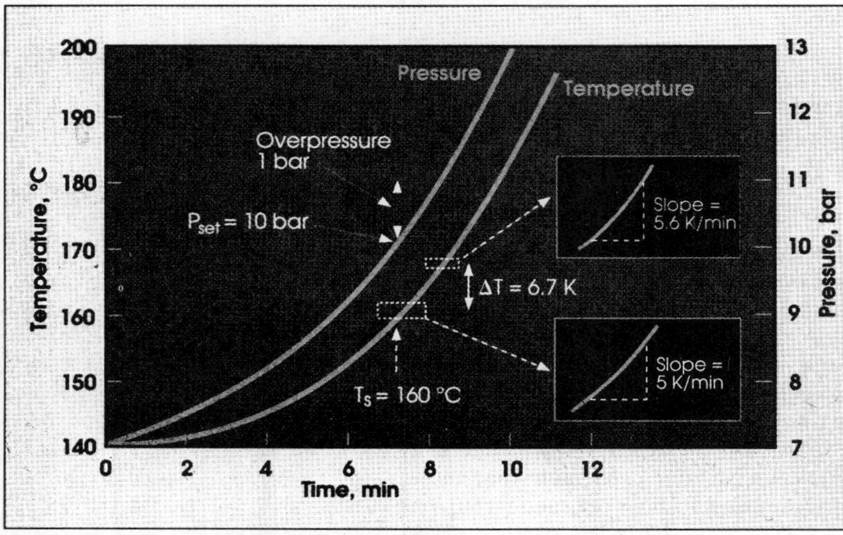

FIGURE 12. Calorimetric data used in the illustrative calculations for relief sizing

If the hybrid system is nontempered, the simplest approach is to use the gassy system equations. This will be somewhat conservative, since it allows no credit for cooling.

Data acquisition

Virtually all the data required by the vent sizing equations are obtained experimentally, including classification of reactions, using bench-scale equipment discussed earlier (Figure 8).

If a runaway reaction is simulated in the test cell, rate data of the type given in Figures 1 and 2 are obtained. If the pressure-temperature is then found to fit an Antoine equation (as in Figure 3), a vapor-pressure system is confirmed. If results such as those in Figure 5 are obtained however, the presence of gaseous components may be assumed.

In order to confirm tempering of a reaction during relief, it is necessary simply to allow the runaway reaction to occur in the test cell, and then open the vent at the required relief pressure and hold this pressure as venting proceeds.

If the temperature is found to remain constant (as in Figure 10), then the reaction is tempered, otherwise the result in Figure 11 will be obtained. In Figure 10, the pressure and temperature are seen to be increasing (A-B), and then the pressure is held constant at 24 psi (B-C). Immediately the temperature is also seen to turn around and remain steady at 115.5°C. Then, the pressure is deliberately increased to 29.5 psi (C-D); immediately the temperature increases to 121.8°C and levels off. This shows that the system is tempered, and that the boiling point is 115.5°C at 24 psi, and 121.8°C at 29.5 psi.

A similar experiment with a different chemical reaction is shown in Figure 11. In this case, while the pressure is held at 68 psi, the temperature continues to rise. This indicates a gassy reaction.

If further venting experiments are performed, using suitably selected vent diameters, it is possible to classify the reactions according to whether vapor venting or two-phase relief occurs. If the latter behavior is obtained, the experimental data will indicate whether homogeneous, bubbly or churn-turbulent modeling is relevant.

Software packages

Two software packages for relief sizing emerged from the DIERS program. They are Safire [8] and Deers [9] capable of running on mainframes as well as PCs. Both have been validated against the largescale tests conducted by DIERS, and are considered to be widely applicable.

Application of these computer codes to relief sizing problems requires a large amount of kinetic and physical property data. The programs are essentially reaction and vent-blowdown modeling aids. Once a system has been modeled and validated against experimental data, it is possible to use the model as a "simulator" to investigate relief under different conditions.

In general, the direct use of analytical equations together with experimental data is both efficient and adequate. Unless a system is too complex, or there are other reasons for modeling, software packages are not commonly used for relief sizing.

References

1. Fisher, H. G., DIERS Users Group Meeting, Philadelphia, Pa., Mar. 5-6, 1986.
2. Fauske, H. K., *Plant/Operations Progress*, Vol. 2, No. 1, Jan. 1983.
3. Tharmalingam, S., *The Chemical Engineer*, No. 463, pp. 33-41, Aug. 1989.
4. Singh, J., International Symp. on Runaway Reactions, Cambridge, Mass., AIChE, March, 1989.
5. Singh, J., *Polymers Paint Color Journal*, Sept. 20, 1989, pp. 127-129.
6. Leung, J. C., *AIChE Journal*, Vol. 32, No. 10, Oct., 1986.
7. Tangren, R. F., others, *J. Applied Physics*, Vol. 20, No. 7, pp. 637-645, 1949.
8. Grolmes, M. A. and Leung, J. C., 1985 AIChE Spring National Meeting, Houston, Tex., paper 56a.
9. Klein, H. H., *Plant/Operations Progress*, Vol. 5, No. 1, pp. 1-10, Jan. 1986.

The author

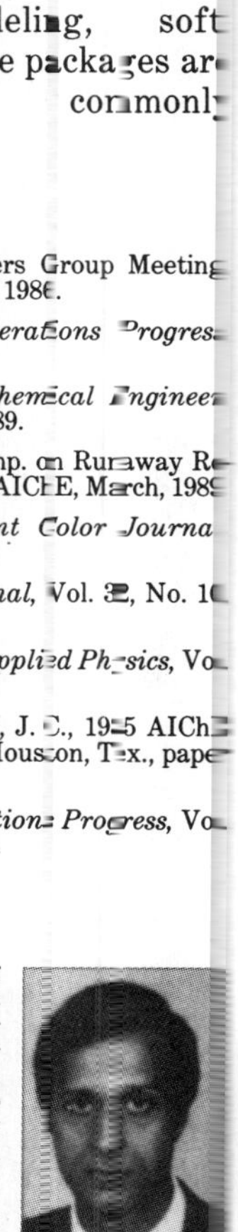

Jasbir Singh is the managing director of Hazard Evaluation Laboratory (Fire Research Station Site, Melrose Avenue, Borehamwood, Herts WD8 2SL, England; telephone: 081 - 905 1343), a company specializing in process hazards testing and consulting including development of calorimetry devices. His previous experience was in process design and hazard assessment gained in U.K., Netherlands and U.S.A. He received his B.S., M.S. and Ph.D. degrees, all in chemical engineering, from Imperial College (London University).

SAFER RELIEF-VALVE SIZING

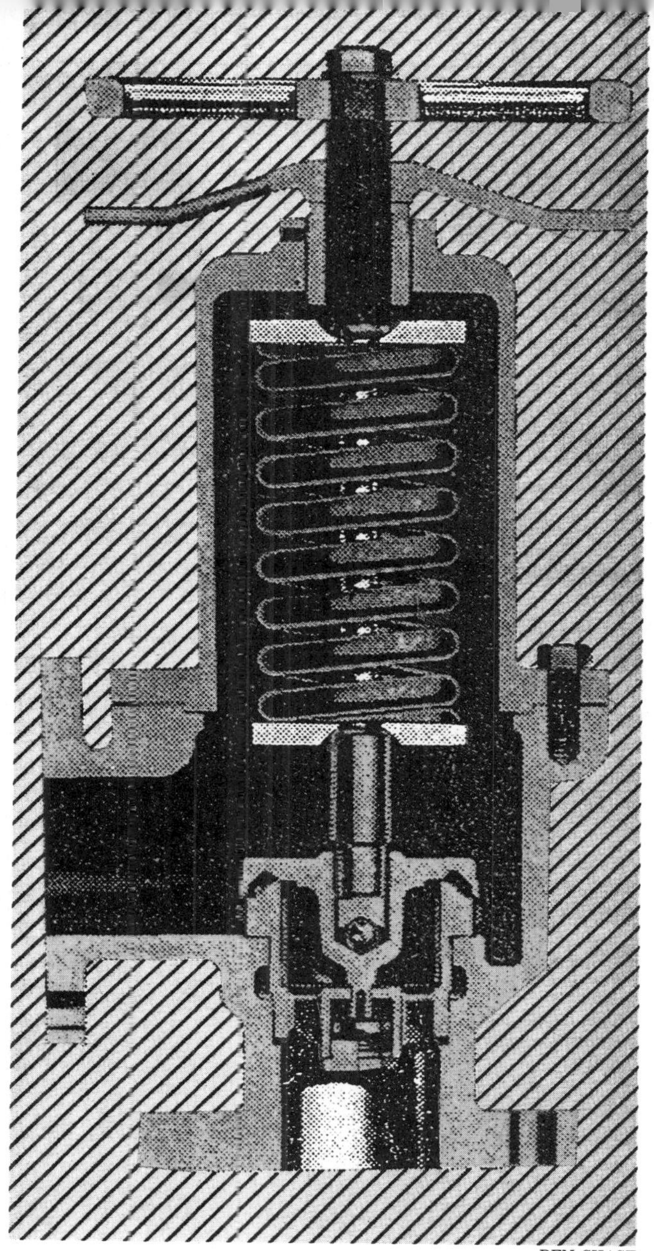

BEN CHASE

Wing Y. Wong, Syncrude Canada Ltd.

A rigorous calculation of the thermal relief requirement for process equipment is not practical because of the lack of data and the many assumptions involved. Because of this, a 3/4-in. by 1-in. pressure-relief valve, for example, might be installed without any assurance that it will adequately protect the system.

A method for calculating the relief volume and predicting the potential pressure increase caused by the thermal expansion of a liquid in piping and equipment blocked between process valves is described in this article. Examples illustrate the method. Also presented are criteria for determining where a thermal relief valve is not required as well as installation recommendations.

Expansion pressure

Thermal expansion can build up the pressure of a liquid contained in piping or equipment to a dangerous level. The main sources of heat for such thermal expansion include steam tracing, solar radiation, adjacent process fluid (such as a heat exchanger blocked in on the cold side but not on the hot side) and fire.

How much the pressures of some liquids will rise when blocked in a pipe and heated by 30°F are shown in Table I. Thus, only a 30°F rise in temperature can often boost the pressure of a liquid far beyond the design pressure of the equipment or pipe containing it.

A thermal relief valve is not required when:

- The equipment or pipe cannot be blocked in and heated during normal operations
- The blocked-in liquid cannot be heated higher than its ambient temperature
- Electric-heat tracing controls the liquid's temperature
- The pipe is above-ground and shorter than 100 ft
- One end of the line is blocked in with a check valve (the leakage is considered sufficient)
- The line is usually in service and can be vented and drained for a shutdown.

Protection methods

The following methods of protecting against overpressure are ranked in terms of reliability:

1. Installing a pressure-relief valve — The most common method of protecting piping and equipment against overpressure

2. Adding an open bypass around one

Size relief valves according to the potential pressure and the volume of the liquid blocked between valves that is likely to be released by thermal expansion

Originally published May 1989

of the block valves — May be suitable only if backflow through the bypass is acceptable

3. Placing a check valve around one of the block valves — An option only if leakage through the check valve can be tolerated. (During normal operation, the check valve is closed.)

4. Setting up a procedure that will ensure that the liquid is drained before the equipment or pipe is blocked in.

Basically, the maximum allowable working pressure of a vessel or pipe sets the operating pressure of the relief valve. To avoid inadvertent releases, this pressure should be as high as possible above normal operating pressure. It should also be less than, or equal to, the maximum pressure allowance for the weakest component in the system. The governing codes are the ASME Boiler and Pressure Vessel Code, Section I — pp. 67-73, Section VIII — pp. 125-136, and appendices M and 11; and ANSI Code B31.8 — Gas Transmission & Distribution Systems.

Calculating relief volume

The volume of liquid blocked in a pipe or equipment is considered a function of T and P:

$$dV = (\delta V/\delta T)_P dT + (\delta V/\delta P)_T dP \quad (1)$$

Divide by V:

$$dV/V = (1/V)(\delta V/\delta T)_P dT + (1/V)(\delta V/\delta P)_T dP \quad (2)$$

In the case of liquids, the partial derivatives of Eq. (2) are related to two commonly tabulated properties:

1. Volumetric expansibility, or cubical expansion coefficient A:

$$A = (1/V)(\delta V/\delta T)_P \quad (3)$$

2. Isothermal compressibility B:

$$B = -(1/V)(\delta V/\delta P)_T \quad (4)$$

Substituting Eqs. (3) and (4) into Eq. (2) yields:

$$dV/V = AdT - BdP \quad (5)$$

Both $(\delta V/\delta T)_P$ and $(\delta V/\delta P)_T$ in Eq. (2) are very small, as is evident from Figure 1, in which the isotherm lines on the left side for the liquid phase are very steep and closely spaced. This means that both A and B for liquids are very small — i.e., they are weak functions of temperature and pressure. Therefore, regarding them as constants over a small ranges of T and P will introduce only a slight error. Figure 2 shows that the isothermal compressibilities of some substances. Integrating Eq. (5) yields:

$$\ln(V_2/V_1) = A(T_2 - T_1) - B(P_2 - P_1) \quad (6)$$

Via Eq. (6), the pressure increase for a certain temperature rise of a liquid can be calculated, assuming that the volume of most liquids changes little with variations of T and P. This means that $V_2 = V_1$, which is valid because the volume of a blocked-in pipe or item of equipment can be assumed to not change with variations in temperature and pressure. Therefore, Eq. (6) becomes:

$$P_2 - P_1 = A(T_2 - T_1)/B \quad (7)$$

Calculate the pressure increase due to thermal expansion of a liquid by means of Eq. (7).

The required relief volume can be calculated via Eq. (6), assuming that the liquid is incompressible and the relief volume is not affected by a small change of the volume of the equipment or pipe with T and P. Thus, Eq (6) becomes:

TABLE I. Pressure increases of some liquids for 30°F temperature rise

Liquid	P, psia
Acetic acid	3,200
Acetone	3,260
Aniline	5,190
Benzene	3,860
n-Buty alcohol	2,590
Carbon tetrachloride	3,310
Methyl alcohol	3,900
Petroleum (sp.gr. = 0.8467)	2,340
Toluene	3,340
Water	1,100

TABLE II. Cubical expansion coefficients for selected liquids [1]

Liquid	$A \times 10^{-4}$, 1/°F*
Acetic acid	5.95
Acetone	8.26
Aniline	4.77
Benzene	6.87
n-Butyl alcohol	5.28
Carbon tetrachloride	6.87
Methyl alcohol	6.99
Petroleum (sp.gr. = 0.8467)	5.31
Toluene	6.11
Water	1.15

* (at 68°F)

TABLE III. Isothermal compressibility coefficients of selected liquids [1]

Liquid	Temperature °F	Pressure psia	$B \times 10^{-6}$ 1/psia
Acetic acid	77	1,360	5.54
Acetone	77	1,210	7.61
Aniline	77	2,670	2.76
Benzene	68	1,450-4,350	5.35
n-Butyl alcohol	63	120	6.12
Carbon tetrachloride	68	15-1,450	6.23
Methyl alcohol	32	15-7,250	5.37
Petroleum (sp.gr. = 0.8467)	162	15-220	6.31
Toluene	77	1,680	5.48
Water	77	15-7,250	3.15

$$\ln(V_2/V_1) = A(T_2 - T_1) \quad (8)$$

or,

$$V_2 - V_1 = V_1\{\exp[A(T_2 - T_1)] - 1\} \quad (9)$$

The error introduced by these correlations causes Eq. (9) to slightly overestimate the required liquid relief volume, but this conservative approach is acceptable based on general design practices.

Because $V_2 - V_1$ is very small, Eq. (8) can, for simplicity, be written as:

$$V_2 - V_1 = V_1[A(T_2 - T_1)] \quad (10)$$

For selected liquids, cubical expansion coefficients and isothermal compressibility coefficients are listed in, respectively, Table II and III. For liquid hydrocarbons, cubical expansion coefficients and isothermal compressibility coefficients are plotted against API densities in, respectively, Figure 3 and 4.

Example calculations

The first example involves calculating the relief volume resulting from the thermal expansion of blocked-in gasoil. A steam-traced 4-in. Schedule-40 pipe (4.026 in. I.D.) transports gasoil at 40 psig and 120°F, at which the density of the gasoil is API 13 deg. The pipe is designed for 160 psig, and its blocked-in length is 800 ft. At the normal operating conditions, the volume of the gasoil is:

$$V_1 = 800(4.026/12)^2(3.14/4) = 71 \text{ ft}^3$$

What will be the pressure of the gasoil in the blocked-in pipe if it is heated from 120°F to 150°F?

The pressure at 150°F can be calculated via Eq. (7). From Figure 3, the cubical expansion coefficient A is 0.00037. From Figure 4, the isothermal compressibility coefficient B is 0.0000053.

$$P_2 = 40 + 0.00037[(150 - 120)/0.0000053] = 2{,}134 \text{ psig}$$

Obviously a relief valve is necessary.

With the set pressure at 160 psig, the relief pressure would be 176 psig (i.e., 160 × 110%). Calculate the relief temperature with Eq. (7):

$$T_2 = [(B/A)(P_2 - P_1)] + T_1 = [(0.0143)(176 - 40)] + 120 = 122°\text{F}$$

Calculate the required relief volume via Eq. (10):

$$V_2 - V_1 = 71(0.00037)(122 - 120) = 0.0525 \text{ ft}^3 \text{ or } 0.4 \text{ gal}$$

In the second example, two heat exchangers in series are blocked-in on the cold (shell) side. Water on the cold side can easily be heated to 130°F by the hot-side fluid. The shell I.D.s are 58 in., or 4.83 ft; the tube O.D.s are 1 in., or 0.0833 ft; their lengths are 20 ft, and there are 1,236 of them in each exchanger. The shell's design pressure is 275 psig and the normal operating pressure is 84 psig. The normal temperature of the water is 100°F. Calculate the relief pressure, temperature and volume.

The volume of each shell cylinder = $(4.83^2)(3.14/4)20 = 367$ ft³. The volume of the two hemispheric heads of each exchanger = $0.2618(4.83^3)2 = 59$ ft³. The tube volume of each exchanger = $20(0.0833^2)(3.1415/4)1236 = 135$ ft³. Therefore, the total water volume in two exchangers = $(367 + 59 - 135)2 = 582$ ft³, or about 600 ft³, including the piping.

From Table II, the cubical expansion coefficient of water is about 0.000115; from Table II, its isothermal compressibility coefficient is about 0.00000315. Calculate the water pressure at 130°F via Eq. (7):

$$P_2 = [0.000115(30)/0.00000315] + 84 = 1179 \text{ psig}$$

Again, a relief valve is obviously needed.

For a set pressure of 275 psig, the relief pressure will be 302.5 psig. Calculate the relief temperature via Eq. (7):

$$T_2 = T_1 + B(P_2 - P_1)/A = 100 + 0.00000315(302.5 - 84)/0.000115 = 106°\text{F}$$

Calculate the relief volume with Eq. (10):

$$V_2 - V_1 = 600(0.000115)(106 - 100) = 0.42 \text{ ft}^3, \text{ or } 3.2 \text{ gal}$$

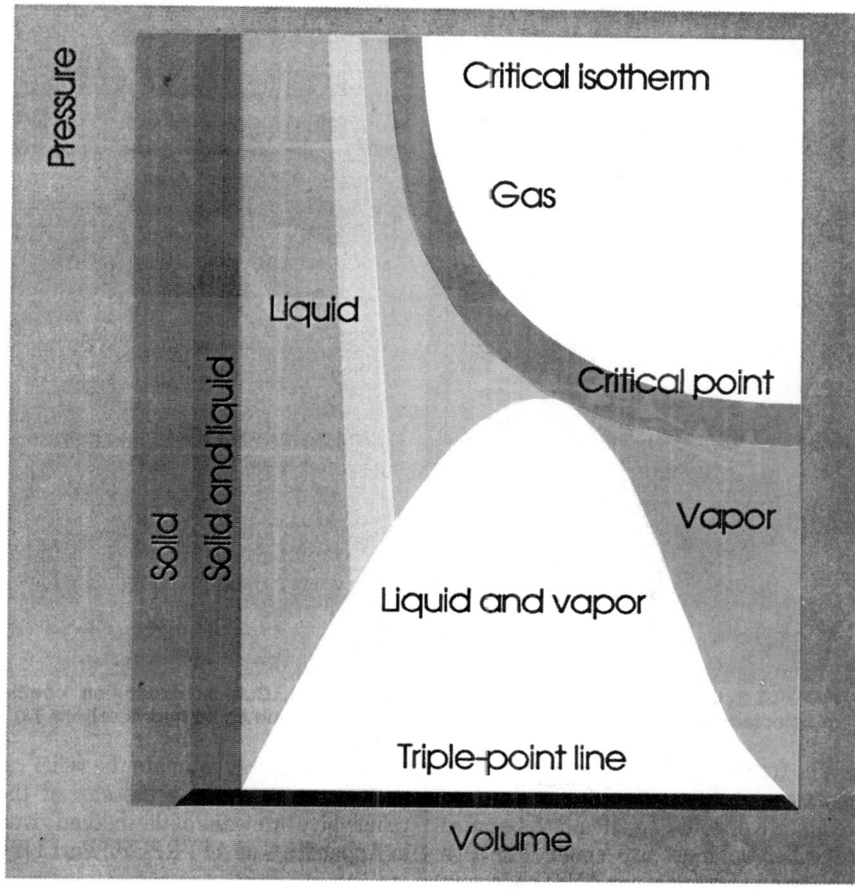

FIGURE 1. Representative pressure vs. volume diagram DIANE SARNO

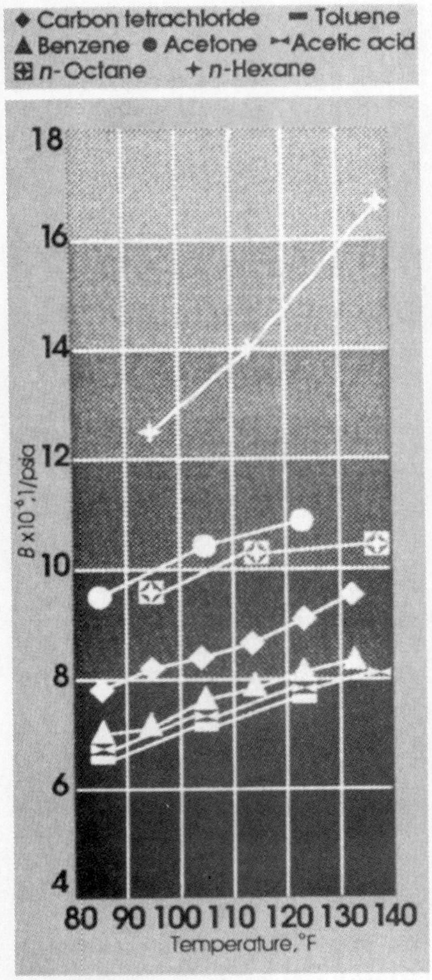

FIGURE 2. Isothermal compressibilities of selected liquids [3]

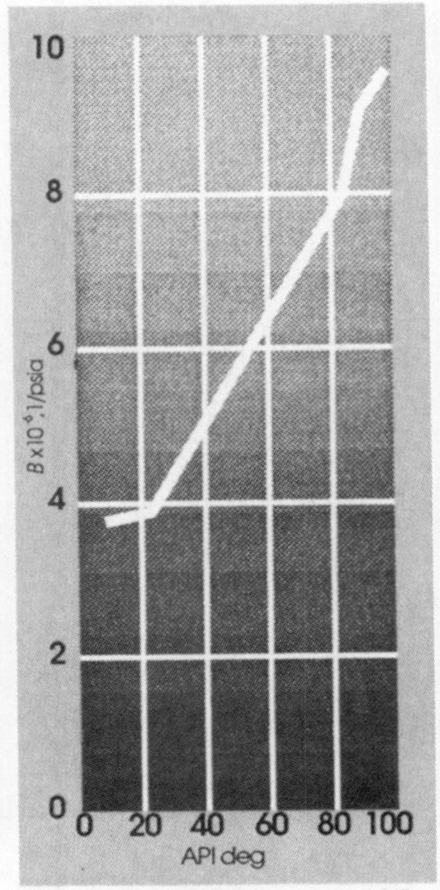

FIGURE 3. Cubical expansion coefficients of some liquid hydrocarbons [3]

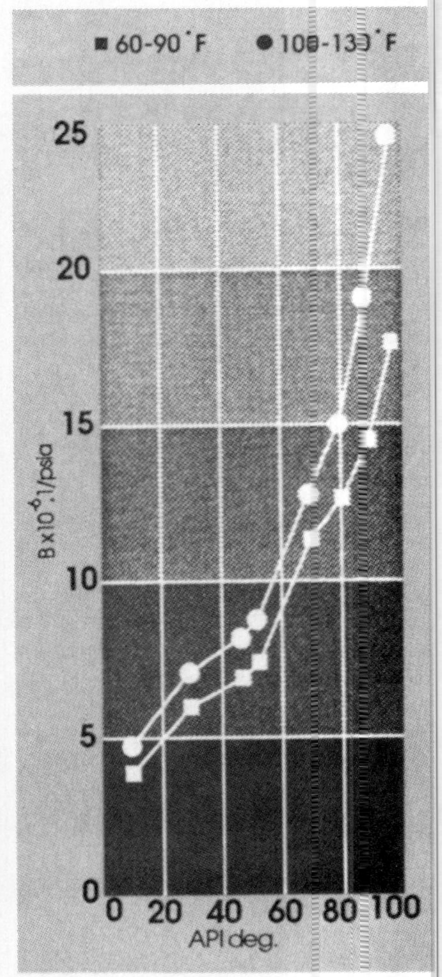

FIGURE 4. Isothermal compressibilities of some liquid hydrocarbons [4]

The foregoing examples show that for most cases one nominal 3/4-in. by 1-in. relief valve would be adequate because of relief volumes are small. For this reason and because, per ANSI code, any increase in volume would be offset by the allowable leakage through the block valves, a pipe of less than 100 ft does not need a relief valve (as previously noted).

If, however, there is a question (e.g., a pipe of large diameter, liquid-filled vessels or sizeable cold-side volumes of heat exchangers), first estimate the relief capacity, then determine the size of the relief valve by means of the method given in Appendix C of API RP-520, Part I [2].

Locating valves

Almost any point on a pipe or an equipment item could be suitable for installing a relief valve. The availability of a suitable discharge route and maintenance requirements may limit the options. A relief valve may be located on connecting piping rather than on the connected equipment if both belong to the same blocked-in system. For ease of service, valves should be located at the lowest elevation and at platforms.

If possible, a relief valve should discharge into a tank in which the particular liquid is stored. A liquid above its flashpoint, or above 600°F, or one that is toxic or flammable must be discharged into a closed system, such as flare header or an especially assigned vessel. ■

Nomenclature

A = Cubical expansibility, 1/°F
B = Isothermal compressibility, 1/psia
P_1 = Pressure at T_1, psig
P_2 = Pressure at T_2 or relief pressure, psig
T_1 = Temperature at P_1, °F
T_2 = Temperature at P_2 or relief temperature, °F
V_1 = Volume of a system at T_1 and P_1, ft^3
V_2 = Volume of a system at T_2 and P_2, ft^3

The author

Wing Y. Wong (237 Farrell Crescent, Fort McMurray, Alberta T9K 1L9, Canada) has been a senior process engineer and a senior applications engineer with Syncrude Canada Ltd. His previous experience has been in the area of process simulation, design and research with Shanghai Petro-chemical R & D Institute, Bechtel Canada Ltd. and Associated Kellogg Ltd. A recipient of an M. Eng. in process control from the University of Alberta and a B.S.Ch.E. from the East China Institute of Chemical Technology, he is a member of the Assn. of Professional Engineers, Geologists and Geophysicists of Alberta and the Canadian Soc. of Chemical Engineers.

Acknowledgement

The author wishes to thank J. Windhorst for reviewing the manuscript for this article.

References

1. Lang, N. A., "Handbook of Chemistry", 12th ed., McGraw-Hill Book Co., New York.
2. API RP-520, "Recommended Practice for the Design and Installation of Pressure Relieving Systems in Refineries, Part I — Design." American Petroleum Institute, Washington, D.C.
3. Perry, R.H., "Chemical Engineers Handbook," 6th ed., McGraw-Hill Book Co.
4. "Handbook of Chemistry and Physics," 66th ed., CRC Press, Inc.

Section IV

Minimizing Hazardous Exposure

HIGHLY TOXIC LIQUIDS MOVING THEM AROUND THE PLANT

Stanley S. Grossel
Hoffmann-La Roche, Inc.

Leaks and fugitive emissions are the two big safety and occupational-health concerns when pumping or otherwise transferring highly toxic liquids within a plant. With proper precautions against those two hazards, the engineer can safely choose from a broad array of equipment for making the transfer: centrifugal pumps, several varieties of positive displacement pumps, liquid- or gas-driven jet pumps, and gas-pressurization or vacuum-suction transfer systems. Of particular interest today are sealless pumps, having canned or magnetic drives.

This article focuses upon those mechanical features of such equipment that enhance safety with respect to toxic fluids.* It also covers selection criteria for various types of pumps and transfer methods, instrumentation to detect pump-component failure, methods for containing toxic liquids within the pumps and transfer equipment, materials-of-construction considerations, handling of leaks and emissions, and other safety considerations and engineering control techniques.

The importance of seals

Centrifugal pumps, the most common type of pump found in the chemical process industries, can move highly toxic liquid if equipped with appropriate seals and with safeguards against reduced flow. The sealing system should consist of either double-inside or tandem mechanical seals plus a seal-failure alarm. Double-inside arrangements employ two seals mounted back to back; the tandem consists of two facing the same direction. In both cases, the cavity between seals contains an innocuous buffer liquid.

A double mechanical seal that is protected by a buffer liquid circulating through a pressurized liquid system appears in Figure 1. If a seal fails, pressure on the reservoir will force the liquid either into the process fluid or out of the seal on the low-pressure side, and the liquid level in the reservoir will show an abnormal drop. The level switch on the reservoir should be not only connected to an alarm but also interlocked so as to shut the pump motor off. If desired, it can also be made to close the valve at the supply tank nozzle in the pump suction line.

To remove the heat of seal friction, a buffer-liquid recirculation system must include some sort of heat-transfer device. This can consist of a cooling coil within the reservoir or a jacket surrounding it, or jacketed piping, or a small heat exchanger. Conversely, it is sometimes necessary to heat the circulating buffer liquid, so as to control its viscosity or to prevent either it or the process liquid from freezing. Such provision is especially important for outdoor installations in cold climates.

A different buffer-liquid system for double mechanical seals appears in Figure 2. Here the liquid does not circulate through a reservoir but instead is furnished from a pressurized source, such as a process water system. A solenoid valve in the buffer-liquid supply line is interlocked with the motor for the process-fluid pump, so that the motor cannot be started unless the line is open. In addition, a pressure or flow switch is interlocked with the motor to shut it down if there is a loss of buffer liquid supply.

Tandem mechanical seals are used when no contamination of the process liquid by buffer liquid can be tolerated. In this arrangement, the latter is under lower pressure than the process liquid, usually at 1 atm. or only a little higher, so a failure of the inner seal will result in a rapid filling of the buffer-liquid reservoir with process liquid. The reservoir should be equipped with a pressure or liquid-level switch that will raise an alarm or shut down the process-fluid pump motor, or both.

Some centrifugal pumps sidestep the sealing problem altogether — by being sealless. These canned-motor and magnetic-drive designs are discussed later.

Keep the fluid flowing

With centrifugal pumps, it is also important to avoid operation with the unit shut off or at severely reduced flow, say below 50 to 60% of the flow that corresponds to the highest efficiency point for the pump. The instability conditions that crop up at such flowrates

*The overall mechanical design of such equipment is covered in a number of books. See References 1, 2 and 3.

Originally published April 1990

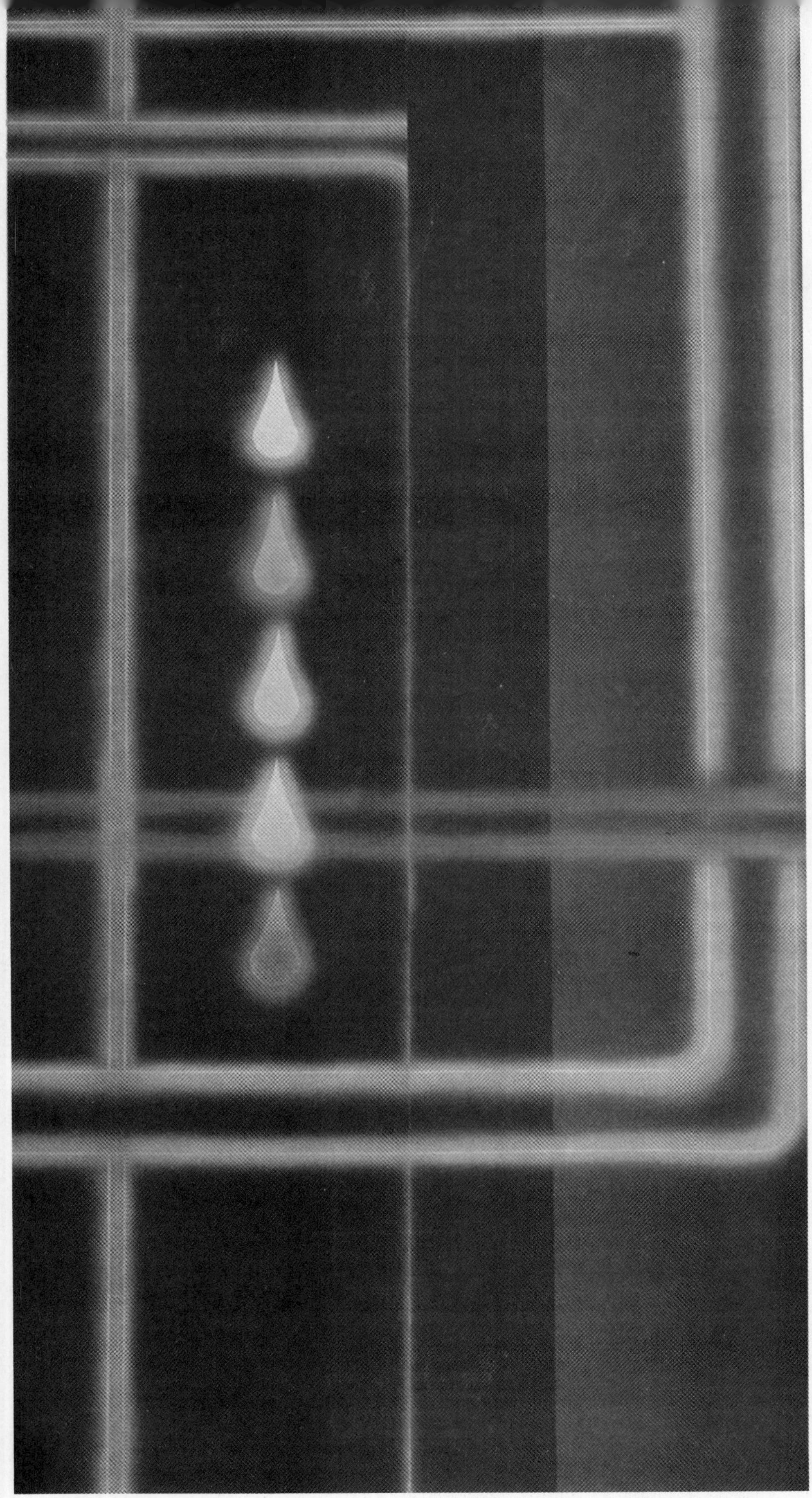

Most of the ordinary methods can be used with extraordinary care

can cause excessive vibration and damage to pumps, drivers, couplings, gears, piping or adjacent equipment, which could release the toxic fluid.

To avoid this problem, a minimum-flow recirculating line should be installed as a bypass in the discharge line from the pump, upstream of the block and check valves. This line should lead back to the pump supply tank if the latter is an adequate "heat sink" to absorb without undue temperature rise the heat added to the fluid while it passes through the pump. Otherwise, the bypass should be routed through an external cooler before returning to the supply tank or pump suction line. The minimum-flow-rate requirements for a given pump should be obtained from the pump vendor

If the liquids being pumped are not only toxic but thermally sensitive, it is also advisable to install a temperature sensor in the pump casing. This should be interlocked to shut off the pump motor when an excessive casing temperature is reached.

Vertical submersible pumps are preferred over other centrifugal pumps by some companies for transferring highly toxic liquids, as there is no suction piping from which a leak can occur. As in the case of an ordinary centrifugal pump, shaft sealing is required to prevent fugitive emissions, and double mechanical seals are recommended.

A number of vertical-pump designs, however, instead use packing to seal the shaft. In this case, a packing system that includes a buffer gas should be specified. To prevent toxic gas from leaking to the atmosphere, a buffer gas consisting of dry air or nitrogen is piped in between the lower and upper sealing element (see Figure 3). In case of packing failure, an interlock causes air to pressurize an emergency seal, which prevents toxic vapor leakage into the buffer chamber. A number of toxic liquids such as liquid chlorine, anhydrous hydrogen cyanide, phosgene and bromine are handled successfully by pumps with this kind of packing system.

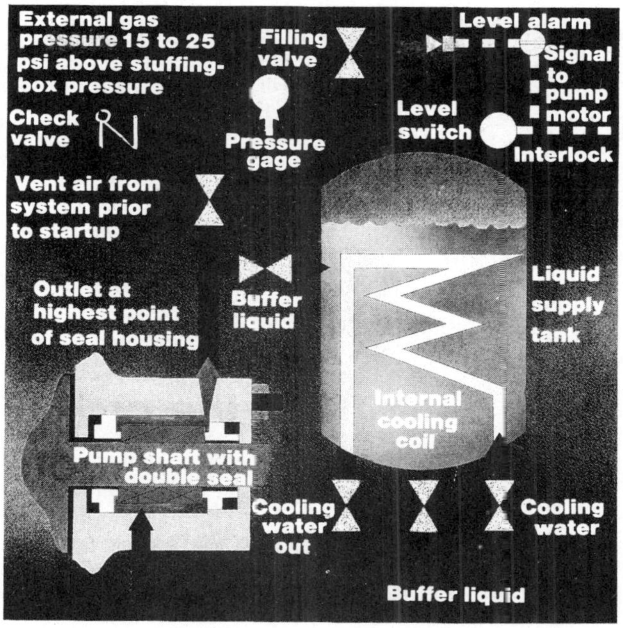

FIGURE 1. **Buffer liquid circulates between double mechanical seal (left) and pressurized reservoir. Upon seal failure, the buffer liquid (rather than the toxic process liquid) leaks, the liquid level in the reservoir drops and the pump motor shuts off**

The positive approach

Positive-displacement pumps are frequently the most economical choice for flows of 10 gal/min or less. Conventional rotating positive-displacement pumps (gear, vane, lobe, progressing cavity and screw type) can handle highly toxic liquid if outfitted with double or tandem mechanical seals. Furthermore, several manufacturers have recently introduced rotating positive-displacement pumps equipped with magnetic drives, which are inherently safer than the conventional type. These are discussed in the next section of this article.

Motor-driven diaphragm metering pumps and (for greater flows) air-operated diaphragm pumps are good choices for highly toxic liquids. Unlike centrifugal or rotating positive-displacement pumps with mechanical seals, they do not have a sealing device that can leak. Also, they cause less of a temperature rise at low flows.

For highly toxic fluids, pumps with double diaphragms should be specified. These allow the volume between the diaphragms to be monitored and alarmed, using instrumentation based on pH, liquid conductivity or pressure, depending on the fluid.

Most positive-displacement pumps require some relief device to protect against overpressure, which can be caused by a closed discharge valve or a restricted discharge line. External relief valves are usually used — it is easy to see when they open, they can readily be set at any pressure, they remain open only long enough to relieve pressure (and so minimize process-fluid loss), and they will dissipate pump heat if the valve discharge is not piped directly to the pump suction. It is recommended that the relief valve discharge instead be piped back to the suction-side supply tank.

Internal relief valves are commonly installed in hydraulically driven diaphragm pumps, as protection in case the external relief fails to work. A major drawback is that the internal relief does not dissipate heat. Even if the pump comes equipped with an internal relief valve, it is strongly advisable also to provide an external one when pumping highly toxic liquids. In the case of slurries, a rupture disk is usually used because a valve might plug up.

An external relief device should be located as close as possible to the pump discharge. It should be upstream of any block valve.

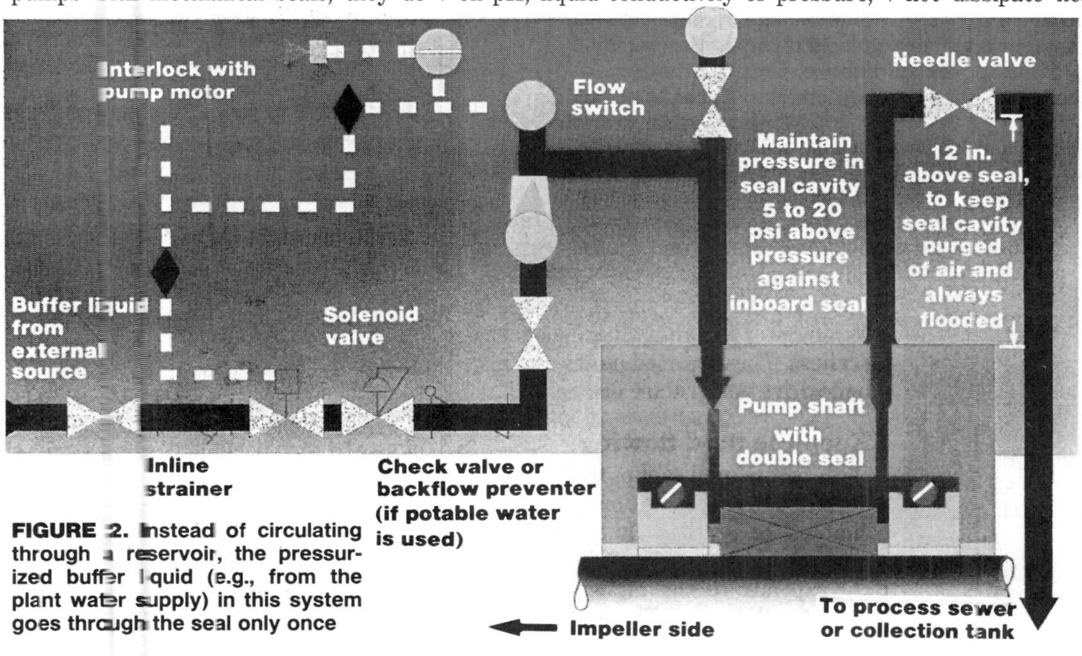

FIGURE 2. Instead of circulating through a reservoir, the pressurized buffer liquid (e.g., from the plant water supply) in this system goes through the seal only once

Sealless pumps

The canned-motor and the magnetic-

Diagrams: STEVE HAIMOWITZ

drive pumps are becoming more and more popular for handling highly toxic liquids. Each is driven by a magnetic coupling between the pump and an external rotating motor: the magnets are attached to the pump shaft and the motor shaft, with a non-magnetic shield between them. Magnetic-drive pumps use permanent magnets. Canned-motor pumps use electromagnets in the external motor stator to drive the internal motor rotor, the shield between stator and rotor being called a can. For more on design features, applications, and problems of sealless pumps, see References 4 and 5.

Because these pumps have no seals, there is less opportunity for leaks and fugitive emissions. But there are two "weak points" that potentially may interact to cause serious leakage. These are the containment shell or can, and the bearings.

The containment shell is usually fairly thin so as not to impede the magnetic field any more than necessary. Thus if the rotor bearings — which are internal and lubricated by the pumped fluid — wear enough to cause the rotor to rub against the can, the latter may rupture, causing uncontrollable loss of the fluid being pumped.

Innovative design features from a number of pump manufacturers address this problem. Some pumps have an outer shell for secondary containment if the can ruptures. At least one manufacturer equips this secondary-containment shell with a sensor well to accommodate a moisture-sensing probe, which can be interlocked with an alarm or with the motor to shut it down. Another vendor offers as an option a mechanical seal between the radial ball bearings and the secondary-containment housing.

A monitor to detect bearing failure is available on some sealless pumps. So are temperature probes.

A number of operating precautions for sealless pumps are particularly important when the process fluid is highly toxic. For instance, low-boiling liquids may flash when circulated through the internal bearing and rotor assembly, resulting in vapor binding of the pump when the vapor-liquid mixture is returned to the lowest-pressure zone at the back of the impeller.

With canned-motor pumps, this can usually be prevented by providing a reverse-circulation system to return some fluid back to the suction tank. With magnetic-drive pumps, a reverse-circulation system cannot be as easily installed because the can is blinded on the back side by the external magnetic drive. However, a small bypass stream sometimes can be taken off the discharge end, cooled externally, and injected into the can area.

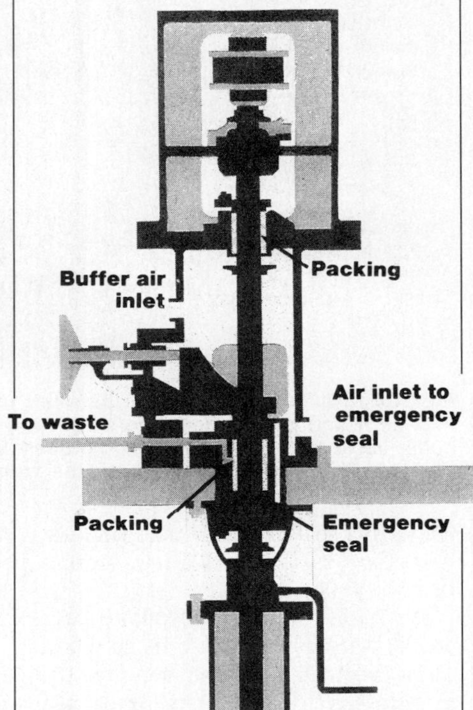

FIGURE 3. To prevent leakage of toxic vapors rising from impeller (not shown), vertical pump includes upper and lower packing, buffer gas between them, and an emergency seal that can be pressurized

Another possible cause of failure is the presence of solids in the pumped fluid. These can block small passages in the internal lubrication system. Various pump manufacturers use different techniques to handle this problem. Examples include use of self-cleaning strainers, and provision of a piping arrangement similar to those for flushing mechanical seals.

Special care should be taken to avoid running sealless pumps dry, which can result in worn bearings and internal rotor galling or ripping open of the can. Some manufacturers offer an electric-current-sensitive device to detect a dry running pump.

Jet power

Highly toxic liquids can readily be transferred by liquid- or gas-driven jet pumps, which rely on the momentum of one fluid to move another. Although the eductor nozzle in these pumps gradually erodes, their lack of moving parts makes them very reliable.

They have been used in such diverse applications as circulating nuclear-reactor coolant, pumping fuel within aircraft and spacecraft, and augmenting thrust in propulsion systems [6]. In the continuous manufacture of trinitrotoluene (TNT), they have been employed to transport water-insoluble explosive mixtures through the various steps of a nitration process, thus avoiding the risks that conventional pumps pose to friction-sensitive explosives [7].

Jet pumps have low efficiencies on the order of only a few percent and the ordinary versions develop low head. This limits their usefulness to, for instance, transferring liquid from one vessel to another nearby, with the pipe run being relatively short.

Furthermore, they dilute the fluid being pumped by mixing it with the motive fluid. However, if the latter is air or nitrogen, it can readily be disengaged from the process fluid in the downstream receiver. Also, in some situations a fluid substance that would have to be added to the pumped material in the immediately downstream operation can serve as the motive fluid.

Using pressurized gas...

Liquids can be transferred by means other than pumping. This option entails use of pressurized gas or of vacuum. Such an arrangement has neither eroding nozzles nor moving pump parts.

At high enough pressure, air or nitrogen or (for some applications) other gas can force the liquid from one process vessel through connecting piping to another process vessel. This technique has been used widely in aerospace operations to feed liquid propellants from supply tanks to rocket thrust chambers. It is likewise common in the chemical industry — in fact, some chemical manufacturers even recommend that their drummed products be unloaded via gas pressurization. One example is acetaldehyde [8]. In this case the drums are of a special type, designed for 40 psig.

A typical gas-pressurization transfer system involving a process vessel is shown in Figure 4. The gas supply line to it is provided with a safety valve set at or below the maximum allowable working pressure of the vessel, to protect the latter against overpressure if the pressure-reducing valve fails. If a vessel being pressurized with nitrogen is in a closed room, the discharge line from the safety valve should be routed to a safe location outdoors, to prevent possible asphyxiation of personnel in the room if the valve discharges.

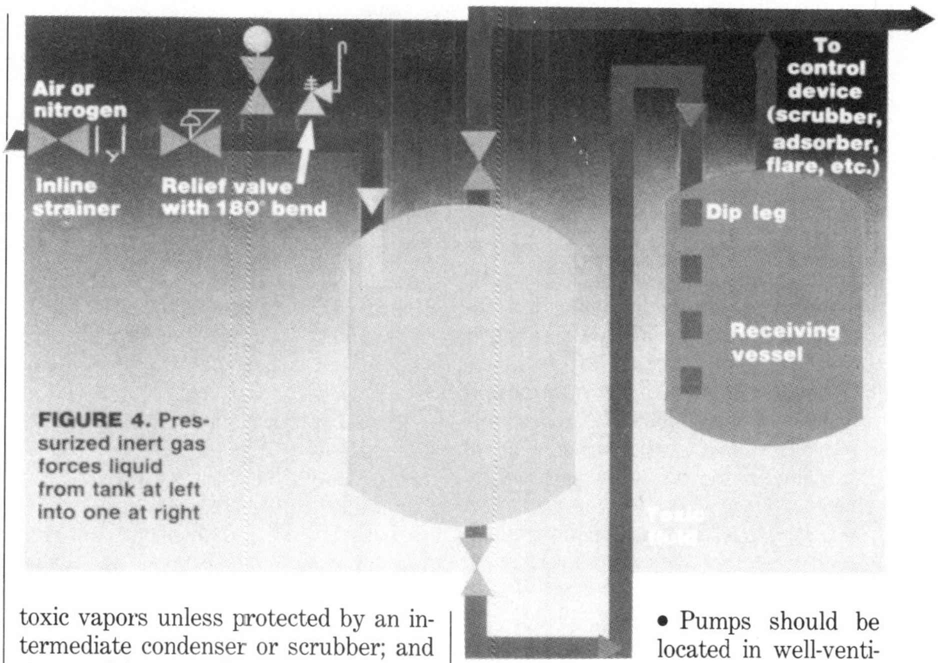

FIGURE 4. Pressurized inert gas forces liquid from tank at left into one at right

. . .or relying on vacuum

Unloading of drums of liquids by vacuum suction transfer is a common operation in batch chemical processes, and quite suitable for highly toxic fluids if proper precautions are taken. A typical arrangement for such a system appears in Figure 5.

This system is operated in the following sequence: Valve A is opened and the vacuum pump or ejector turned on, evacuating the process vessel until the desired vacuum is achieved. Then Valve A is closed and the vacuum pump turned off. Next, Valve B is opened, and the liquid sucked from the drum into the vessel. Then Valve B is closed, and nitrogen is admitted to the vessel to break the vacuum.

Transfer of highly toxic liquids this way should take place in an enclosure that is vented to a scrubber or adsorber. Within the enclosure, the drum should be emptied beneath an exhaust hood that will capture any vapor emissions from the drum bung hole, which likewise should go to the scrubber or adsorber. A scabbard, also vented to scrubber or adsorber, should be available to hold the discharge wand (dip pipe) after the drum has been emptied.

Vacuum suction transfers are limited to systems with short transfer lines, as the maximum available pressure differential is one atmosphere. Transferring of viscous liquids by vacuum suction may be difficult. Two other potential problems have to be kept in mind: The liquids in the vacuum source (pump oil, seal water, or stream-jet condensate) may become contaminated with highly toxic vapors unless protected by an intermediate condenser or scrubber; and the receiving vessel may be overfilled, which will contaminate the vacuum unless there is a trap in the line.

Widely relevant

The guidelines listed up to now relate to specific types of pumps or other transfer equipment. Also important when handling highly toxic fluids are a number of safety practices that are applicable across-the-board.

• Pumps should be located in well-ventilated open areas to prevent accumulation of toxic vapors. Furthermore, they should not be placed under pipe racks, partly because of possible ventilation problems but mainly to lessen the chance of fire complications if the toxic fluid is flammable
• A drain equipped with a trap should be provided near the pump, to carry any process spills to a collection sump or tank
• The foundations and supports for

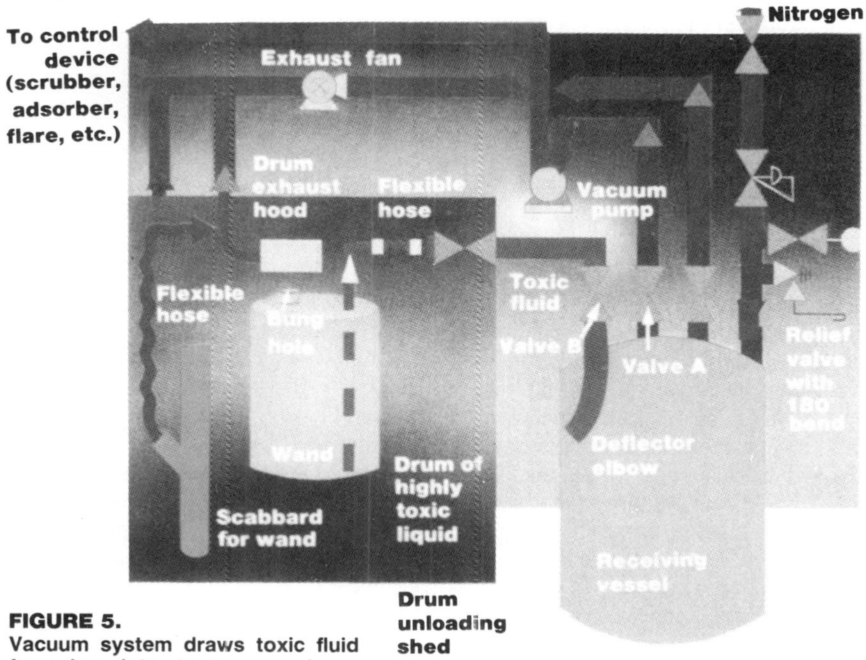

FIGURE 5.
Vacuum system draws toxic fluid from drum into storage vessel

pumps, and for their suction and discharge piping, should be fully able to withstand vibration and static and dynamic loads

- The engineer should consider installing remotely operated isolation valves at the supply-tank outlet nozzles in the suction lines to the pumps. If remote operation of a shutoff valve is thus provided, the valve must be interlocked to shut down the pump motor at the same time. This is especially important for sealless pumps, which cannot operate dry
- Air-operated diaphragm pumps should be connected to the process piping by flexible metal hoses instead of expansion joints, because these pumps vibrate

Protection of operating and maintenance personnel is vital when pumping or transferring highly toxic liquids. [*For more details, see the accompanying article.*] Maintenance workers should wear personnel protective clothing and respiratory protection equipment while servicing pumps handling such liquids

It is a good idea to install pump shrouds (hoods) to collect minor fugitive emissions, the shroud being vented to a scrubber or adsorber by means of an exhaust fan. The vacuum applied to the vent should generate an air flow of at least 200 ft/min linear velocity at the junction between pump and shroud [*9*]. One pump shroud mentioned in an unpublished NIOSH (U.S. National Institute of Occupational Safety and Health) report is designed so that half is permanently mounted whereas the other is hinged to fold upward; thus, while a maintenance worker has access to service the (running) pump, air currents continue to draw any emissions into the vent.

An active and continuous monitoring and maintenance program should be set up where highly toxic liquids are handled. For instance, all rotating shaft seals should be inspected visually by process operators at the beginning (or the end) of each shift. Visual inspection by a trained maintenance worker should take place weekly, or even daily. And the seal area should be monitored with direct-reading analytical instruments, so that the operator need not leave the vicinity to take the reading.

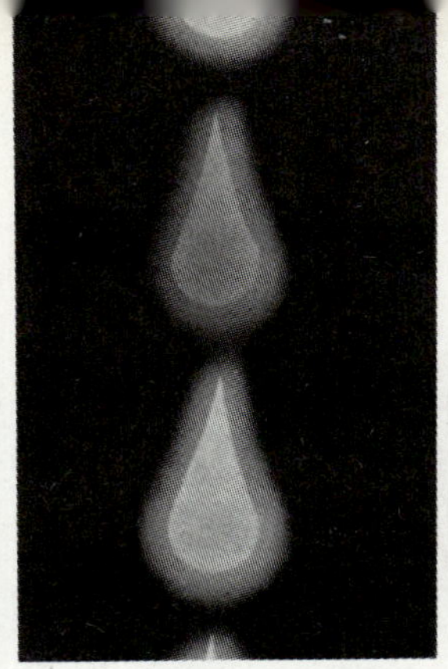

Handling toxic liquids requires active monitoring and maintenance

Materials of construction

Pumps made of brittle materials, such as cast iron, must not be used for highly toxic liquids, because of the risk of fracture. Even ones made of ductile iron should be avoided, because ductile iron can revert to cast iron if exposed to temperatures produced by flammable-liquid fires.

Normally, materials of construction for pumps and piping are selected for their resistance to corrosion. However, the engineer must also consider the effect that certain elements in alloys have on the stability of the specific liquid being pumped. For example, when pumping 54 wt.% (or higher concentration) monomethyl hydrazine at 160°F or higher, Type 304 stainless steel pumps should be used rather than Type 316, as the 3% molybdenum content of the latter catalyzes the decomposition of monomethyl hydrazine, which takes place accompanied by an outbreak of fire [*10*].

To avoid other such "booby traps," material safety data sheets and information on recommended materials of construction should be requested from the manufacturer of the chemical being pumped. Reference 11 also contains much information on catalytic effects of various metals on chemicals, resulting in hazardous conditions. Obviously, such conditions complicate whatever risks related to toxicity the process system may pose. ■

References

1. Karassik, I. J. and Krutzsch, W. C., "Pump Handbook," 2nd ed., McGraw-Hill, New York, 1986.
2. Poynton, J. P., "Metering Pumps: Selection and Application," Marcel Dekker, New York, 1983.
3. Henshaw, T. L., "Reciprocating Pumps," Van Nostrand Reinhold, New York, 1987.
4. Parkinson, G., Surging Interest in Leakproof Pumps, *Chem. Eng.*, June 1989, pp. 30 - 37.
5. Reynolds, J. A., Canned Motor and Magnetic Drive Pumps, *Chem. Proc.*, November 1989, pp. 71 - 75.
6. Nilavalagan, S., others, *Chem. Eng. J.*, Vol. 39, pp. 97-109 (1988).
7. Biasutti, G. S. and Camera, E., Safety Aspects of Continuous Nitration, *Loss Prevention*, Vol. 8, pp. 123-125 (1974).
8. Private communication from Hoechst Celanese Corp.
9. SRI International unpublished report to NIOSH, "Control Technology Assessment of the Pesticides Manufacturing and Formulating Industry," May 1980.
10. Olin Corporation brochure, "Monomethyl Hydrazine Handling and Storage."
11. Bretherick, L., "Handbook of Reactive Chemical Hazards," 3rd ed., Butterworth, Boston, 1985.

The author

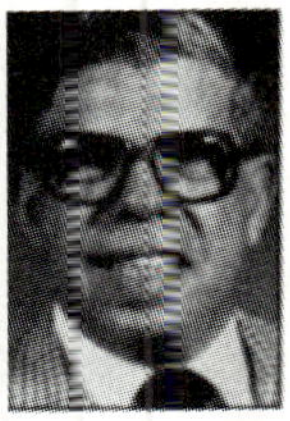

Stanley S. Grossel is Engineering Fellow in the Process Plants Engineering Dept. of the Fine Chemicals & Vitamins Div. of Hoffmann-La Roche Inc., 340 Kingsland St., Nutley, NJ 07110-1199, Tel. (201) 235-4251. Among his activities in AIChE (of which he is a Fellow) are: 1988 chairman of the Safety and Health Div.; past chairman and still active member of Loss Prevention Programming Committee, Health and Safety Programming Committee, and Solids Flow and Mechanical Separations Progamming Committee; overall Program Chairman for Safety and Health Div.; lecturer for continuing education course "Conveying of Bulk Solids"; chairman of CCPS Committee on Handling of Emergency Relief Effluents; member of Engineering Practices and Effects of Vapor Cloud Explosions and BLEVES committees. He is a member of the National Fire Prevention Assn. Explosion Protection Systems Committee involved with the NFPA 68 and 69 guides, and is U.S. Regional Editor of the *Journal of Loss Prevention in the Process Industries.* He holds a B.Ch.E. from City College of New York and an M.S.Ch.E. from Drexel Institute of Technology, and has done work toward a doctoral degree at Polytechnic Institute of Brooklyn.

Acknowledgments

I should like to acknowledge the following individuals and companies that provided me with technical information and advice of value in preparing this article: Herbert M. Fried (Hoffmann-La Roche, Nutley, N.J.), John A. Reynolds (Union Carbide Corp., South Charleston, W. Va.), Robert Hart (E. I. du Pont de Nemours & Co., Wilmington, Del.), Ali M. Nasr (Liquiflo Equipment Co.), Walt Bischoff (Sloan & Bischoff, Inc.) and William L. Drake (Smith-Koch, Inc.), plus Lawrence Pumps, Inc., Dickow Pump Co., Sundstrand Fluid Handling Unit of Sundstrand Corp., DEPA North America, Inc., Pulsafeeder Div. of Clevepak Corp., Ingersoll-Rand Co., and American Lewa, Inc.

HIGHLY TOXIC LIQUIDS

TEACHING OPERATORS TO HANDLE THEM

R. W. Gallant
Dow Chemical Co.

Operators who handle highly toxic fluids must pick up a variety of special skills and knowledge, such as stringent procedures for opening of lines and equipment, sophisticated methods for analytical instrumentation and sampling, and accommodation to bulky and inconvenient protective gear. But all such training must be based on the same principles as effective operator training in the more-general sense. Those principles provide the best starting-point for teaching toxic-fluid handling.

Basically, effective operator training focuses on what is to be done, why it is necessary, how it is done, why it is done that way, and how to recognize and respond to deviations from normal conditions.

Within this focus, instilling a responsible attitude is as critical as teaching the correct procedures. Supervisors must not only establish this attitude but lead in practicing it. And, they must make training a fundamental, top-priority task for their department and for themselves alike. Training must be ongoing, to keep operators up-to-date on changes in procedures. It should be be based on the following subjects:

- General plant and company rules
- Safety regulations and equipment
- The specific plant process
- Specific equipment: Purpose, function, troubleshooting
- Specific chemicals and reactions

Originally published April 1990

Safe plant procedures and worker protection thwart the toxic hazard

- Process control and instrumentation
- Emergency procedures
- Environmental regulations and procedures
- Laboratory procedures
- Non-routine procedures: Maintenance, line opening, others
- Alertness to unsafe conditions, and variations from normal operations
- Importance of asking for information or help when uncertain

Understanding the hazard

Within that framework, several major specifics need to be covered when highly toxic chemicals are present in the plant environment. In the first place, operators must thoroughly understand the toxic hazard(s) involved. This entails the answers to a number of questions:

Is the hazard acute (single-exposure danger), as with hydrogen cyanide and carbon monoxide? Or is it chronic (damage results from longer term exposures), as with carcinogens?

Is the toxicity concern mainly with inhalation, with ingestion or with absorption through the skin? Although the exposure problem is normally predominantly inhalation, absorption through the skin can be very serious — and this is the exposure mode that operators are especially likely to ignore.

What level of exposure is dangerous? Can operators detect the danger level by smell, as with chlorine and ammonia? Or must they learn to use specialized analytical instrumentation, as when handling ethylene oxide with its danger limit of 1 ppm or chloromethyl methylether at 100 ppb?

To understand the toxic hazard, the operator also needs to know how the toxic substances fit into the local plant setting. At what steps in the process are they present? Equally important, what are the specific risks of the chemicals migrating to a normally "safe" process stream or piece of equipment, due to leakage or other malfunctions? A related risk to be covered in the training is the unintended reaction of non-toxic chemicals to produce toxic products.

Operating safely

Once an operator has thus learned the nature of the hazards, he or she is in a better position to learn how to operate in the face of them. Safe procedures for day-to-day running of the plant will of course depend greatly on the particular processes and equipment that are involved. But two topics likely to be vital in any case are the taking of samples and the loading or unloading of equipment or vessels. Traditionally, the most extensive exposure of operators to toxics occurs during the sampling of streams. Loading or unloading, with its line connection and disconnection, is not far behind.

Good sampling procedures are specifically defined and designed to prevent operator exposure. Such procedures should cover not only in-plant sampling itself, but how samples are brought to the laboratory, how they are stored, and how they are disposed of. Operators must also be made familiar with the analytical instruments used to measure toxic chemical concentrations, as well as those to detect leaks. [*For information related to loading and unloading of toxic chemicals, see the accompanying article in this report.*]

Maintenance poses similar risks. Operators must be familiarized with written procedures (see discussion later in this article) for all line openings, vessel entries and maintenance tasks, and must be instructed to follow the steps on a checksheet for each such job. Normally, of course, such tasks also require specific authorization from the supervisor.

Relevant government regulations (such as those of the U.S. Occupational Safety and Health Administration, or the National Institute of Occupational Safety and Health) and the requirements they impose need to be thoroughly explained. Such record-keeping on personal exposure to toxics, ambient levels detected in the plant and related matters is not only important for satisfying government requirements and inspections but also useful in improving plant operation and efficiency. Operators must also be familiar with routine programs for monitoring industrial hygiene, by which the plant forestalls *new* releases of toxic substances.

Dress rehearsals

As regards personnel protective equipment, such as rubber suits or air packs, operators need not only thorough training in its use but a good understanding of when it is required. They must also be taught the importance of routinely inspecting such equipment.

Drills to practice evacuation procedures and communication during a toxic spill are essential, especially because protective equipment is often bulky or awkward. Realistic drills that include wearing all necessary gear can uncover a multitude of problems and inadequate training.

For instance, if operations in a control room are to continue after a nearby spill, the room must be equipped not merely for positive pressurization but also with air packs or fresh-air supply for the control board operators. Among other things, such a drill might dramatize to the operators the fact that verbal communication is impossible when wearing air packs (unless the equipment is of special design).

Operators need to know how to deal with an actual spill upon their own person or a fellow worker. Immediate correct action is often the difference between a serious problem and merely a near miss. Removal and disposal of contaminated clothing, including shoes, must be enforced. The training must cover procedures for getting the appropriate medical assistance quickly.

Delivering the message

The underpinning of any effective training program is a set of written procedures for all operations, and for handling non-routine events. It is impossible to maintain consistent operating procedures by word-of-mouth training.

The procedures should be detailed and clear enough that the operator can carry out all necessary steps without uncertainty. The document should also alert operators about things (e.g., signs of potential danger) to observe and clearly instruct them about when to take corrective action.

Table 1 is an example showing the

TABLE 1: Unloading Procedure for Ethylene Oxide

Particular Hazards Involved:

1. EO is highly reactive and flammable
2. Can cause chemical burns to eyes
3. Exposure by contamination of clothes can result in severe burn to skin
4. Short-term inhalation can result in vomiting and nervous system depression
5. Long-term inhalation can result in leukemia
6. Inhalation guideline is 1 ppm maximum (8-h average.)

Special Safety Equipment and Procedures:

1. MSA EO respirator when connecting, disconnecting and sampling
2. Monogoggles at other times
3. Rubber boots when connecting, disconnecting and sampling
4. If EO skin exposure occurs, affected area must be washed immediately and supervision notified
5. Repairing a leak requires a rubber suit, boots and supplied air

Fill in Appropriate Part of Check Sheet As You Proceed*

■ Check with Control Room Operator to be sure tank DH-1E will hold EO
Hazards: Could overflow DH-1
Precautions: Do not exceed 80% level in DH-1
Corrective Actions: Stop off-loading at 80%
■ Turn on blue light, set blue flags and derailer
■ Chock car and set brake
■ Connect bottom and top electrical ground
■ Catch EO sample through sample line
Hazards: Inhalation or skin exposure, fire
Precautions: EO respirator and rubber boots
Remarks: Know location of fire extinguishers, emergency trip valves and sprinkler system.
■ Have sample analyzed. Do not proceed until analysis is completed
■ Notify Control Room Operator that you are ready to unload
■ Check load valves and vent valves for leaks
Hazards: Inhalation or skin exposure, fire
Precautions: EO respirator, rubber boots
Remarks: DOT regulations require that the car be leak free. Immediately report any leaks to supervisor
■ Connect load and vent pipes
■ Open nitrogen to each connection and check for leaks

Note: This procedure contains 19 additional steps, omitted here.
*In the actual document, appropriate hazards, precautions, corrective actions are listed for each step. Shown here are only those that relate to toxic hazards. —*ED*

TABLE 3: Oral Test for Ethylene Oxide Sampling: Toxicity Aspects

1. What are the hazards of EO?
2. What are the allowable maximum exposure levels for EO (acute and chronic)?
3. How do you analyze for EO concentration? What is the level of detection for each of the analytical methods?
4. What safety equipment is required when sampling? Connecting lines? Repairing a leak?
5. What action do you take if you or a coworker are exposed to EO?
6. How do you check for leaks?
7. What contaminants increase the hazards related to EO?
8. What do you do if the flexible hose or safety valve develops a leak?
9. What do you do if the emergency alarm goes off?
10. What checks do you make on the air pack before using it?
11. What is the decontamination procedure for equipment before entering?
12. What is the procedure for sampling EO?
13. What do you do with EO samples after analysis is completed?

TABLE 2: Senior Operator Training Program

TOPIC	SUPERVISOR INITIALS/DATE
1. Toxins and Toxicity A. Definitions: Acute, chronic, TLV B. Types of exposure: Dermal, oral, inhalation C. Specific chemicals: Plant exposure hazard coding system D. What to do if exposure occurs E. What to do if a leak is detected	
2. Where to Find Specific Information A. Material Safety Data Sheets B. Computer system files C. Waste streams characterization sheets D. SARA Information Manual	
3. Hazard Communication Standard 29CFR 1910.120 (HAZCOM-"Worker Right to Know" Law) A. Film: "What's It All About" B. Contents and locations of HAZCOM Manual	
4. Review of Medical Surveillance Program A. Requirements B. Review of current personnel exposure data	
5. Handling Hazardous Waste A. Film on proper packaging, labeling, and storage B. Proper safety equipment C. Chemical hazards	
6. Spill Prevention, Control and Corrective Actions A. Review requirements of spill notification plan B. Emergency procedures C. Penalties for non-compliance	
7. Emergency Alarm System A. What activates the alarm system B. Evacuation procedures	

TABLE 4: Training Schedule for 1990

TOPIC	INSTRUCTOR
February	
Toxic Chemicals Classifications	Chuck Dodd
Hazardous Chemical Loading and Unloading	Donna Young
Plant Operations: Compressor Section	Phil Larson
April	
Air Pack and Respirator	Allan Bailey
Industrial Hygiene and Medical Surveillance	Medical Dept.
Valve Maintenance	Maintenance Dept.
June	
Plant Operations: Reactor Area	Allan Bailey
Analyzers for Toxic Chemicals	Chuck Dodd
Review of Past Potentially Serious Incidents	Phil Larson
August	
Plant Operations: Finishing Area	Donna Young
Instrumentation and Control: Module 1	Phil Larson
Emergency Procedures for Spills	Allan Bailey
November	
Instrumentation and Control: Module 2	Phil Larson
Vessel Entry Procedures for Toxics	Donna Young
Fire Fighting Training	Safety Dept.

procedure for unloading ethylene oxide from a tankcar. This chemical has a 1-ppm maximum inhalation level for chronic purposes and can cause acute problems due to single skin exposure. Note that key steps on the checksheet contain comments on potential hazards, precautions and corrective action, plus informational remarks in most of the cases. This detailed sheet becomes a fundamental element in the training, as well as in the subsequent operator testing.

Ideally, the training program will involve a combination of techniques, using films, classroom instruction, computer simulation, and hands-on training in the field with an experienced operator and supervisor. Such a program (with respect to toxic chemicals) for a senior operator in a specialty chemical plant appears in Table 2. Note that each step of the training requires the initials of the supervisor and the date completed.

It is important to measure the level of learning with written tests, oral tests, and field demonstrations of understanding. The purpose of the testing is to determine not only the operator's ability to carry out specific procedures but also his or her understanding of the reasons for them. It should also reveal what additional training steps are necessary.

Although the questions should not be tricky or confusing, they should among other things determine the operator's ability to deal with the unexpected. Table 3 shows the oral test for sampling a toxic stream.

The new operator should be teamed with an experienced operator until proficiency has been demonstrated. Also, a shift supervisor should make a point of personally making rounds with newer operators, to assure that they know the procedures and the process well. This is also an excellent time to establish a good working-relationship between operator and supervisor — and to ingrain the discipline of questioning anything that doesn't look normal and of asking questions when not certain of the correct procedure.

Incidentally, such rounds pay off in other ways for the supervisor. They provide a good training refresher, with the opportunity to relearn some specifics that haven't been dealt with for a while. And they may reveal that part of a written procedure has become inadequate or out-of-date.

Another key to good training, and to consistent operator performance, is a well-planned procedure for gathering data in the plant. A checksheet of observations made by the operator on the rounds is valuable to both operator and supervisor. It disciplines the operator to check the critical items. It also allows a supervisor to review current and past data to look for changes. Although much of this can be done by computer, the operator checklist remains a valuable operating and training tool.

Good training never ends

Training is a continuous process. Monthly or quarterly safety-and-operations meetings are important to explain new procedures, review current ones, discuss problem areas and operator ideas, and review the cumulative data on toxicity exposure.

In addition, periodic two- or three-day intensive "continuing education" sessions conducted by plant supervision are valuable techniques for refreshing knowledge, identifying areas where additional training is needed, updating written procedures, broadening operator understanding, and focusing on specific plant problems.

Table 4 is an example of a schedule for such an ongoing program. Note that it is planned out for the year, with specific people assigned to handle each topic. Many of these sessions may well include drills and simulations of the actual situation under discussion. ■

The author

Robert W. Gallant is vice president for materials management, capital and production planning, and hydrocarbons and energy supply in the corporate product department of Dow Chemical Co., 2030 Willard H. Dow Center, Midland, MI 48674; Tel. (517) 636-2099. Starting with Dow as a process engineer, he became successively a research group leader, a plant superintendent, a production manager, a manufacturing manager, and general manager of the Louisiana Div. of Dow Chemical U.S.A. He holds a B.S.Ch.E. from the U. of Florida. He has authored two technical books and contributed to two others, and has written numerous magazine articles, including several for *Chemical Engineering*. Among these was a November 1987 25-page feature report, "So you want to be a manager."

A PRACTICAL APPROACH TO COMPLIANCE WITH THE HON MACT

Mark M. Hultman
TRC Environmental Corp.

In today's highly regulated environment, most U.S. chemical manufacturing plants are already required to install devices to control air pollution, or take steps to otherwise reduce or eliminate emissions of hazardous air pollutants (HAPs). Nonetheless, operators throughout the chemical process industries (CPI) are bracing for the HON Rule — the Hazardous Organics National Emissions Standards, as promulgated by the U.S. Clean Air Act Amendments of 1990.

Determining whether or not the HON Rule applies to a given facility is not a simple matter, nor is compliance expected to come cheaply. This article will help manufacturers in the CPI to determine if the regulation applies to their facilities, and to assess what controls are required.

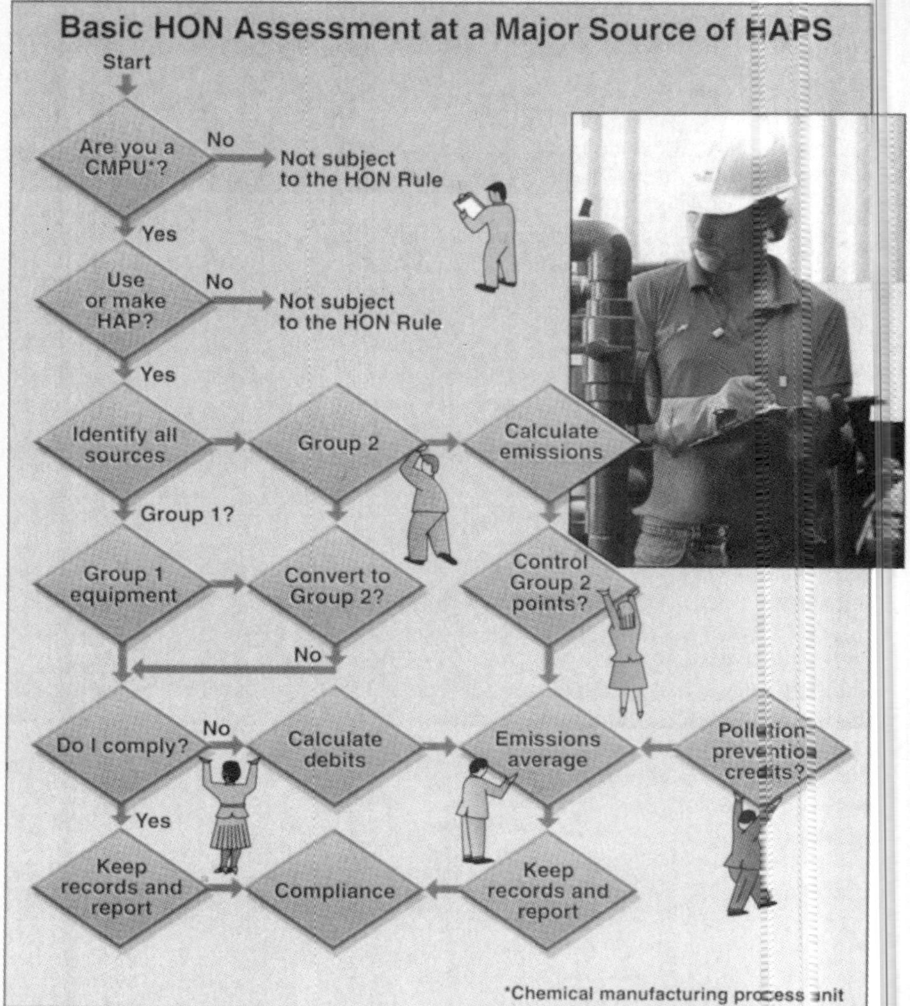

FIGURE 1. To assess the implications of the HON Rule on a given facility, operators must identify and evaluate all equipment in the regulated categories. This flow chart will help to guide these activities

An evolutionary process

Under the 1977 Amendments to the Clean Air Act, the U.S. Environmental Protection Agency (EPA; Washington, D.C.) made its first attempts to control hazardous air pollutants by promulgating seven National Emission Standards for Hazardous Air Pollutants (Neshaps). These regulations, codified, in 40 CFR 61, were seen as largely ineffective and incomplete.

Title III of the 1990 Clean Air Act Amendments (CAAA) greatly increased the control standards for hazardous air pollutants. The Amendments listed 189 toxic substances to be regulated (the so-called hazardous air pollutants or HAPs), and required a new approach to their regulation. Using this new approach, "source categories" of like or similar sources of HAPs are regulated similarly. One of these source categories is the Synthetic Organic Chemical Manufacturing Industry (Socmi).

Each source category will have its own new control standards, called Maximum Achievable Control Technology (MACT). The MACT standards are based on the maximum level of control that has been achieved in practice within the industry today.

For the Socmi source category, this standard is known as the HON (Hazardous Organic Neshaps). The HON Rule provides rigorous control and reporting requirements for 112 of the 189 HAPs listed in CAAA Title III. It is one of the most complex and pervasive rules ever issued by EPA, and by itself, covers more processes and pollutants than all previous EPA air toxics programs combined.

Who is subject to the HON Rule?

Not all Socmi sources are regulated under the HON Rule. For the HON Rule to be applicable to a Socmi operation, three basic criteria must be met. First, the Socmi source must be located at a facility that is considered a major source of HAPs.

In the rule, the term "facility" pertains to all the emission points on a contiguous or adjacent property under common ownership or control. Under the Clean Air Act Amendments, a facility is designated a "major source" if it has the potential to emit either 10 tons of one HAP, or 25 tons of any combination of the 189 HAPs annually.

The second condition for regulation under the HON Rule is that for major sources, the chemical manufacturing process unit (CMPU) must either manufacture or use one of the chemicals listed in Subpart F of the HON Rule (Table 1). Third, the CMPU must use as a reactant, or manufacture as a product, by-product, or co-product, one or more of the 112 HAPs listed in the HON Rule (Table 2).

The HON Rule identifies 385 CMPUs that are known to involve an organic HAP as either a product, byproduct, co-product, intermediate or reactant. When the HON was first proposed, it identified affected equipment as chemical manufacturing processes (CMPs). However, in the final rule this was changed to the CMPU designation to clarify exactly where the HON Rule applicability begins and ends. In the final rule, CMPU is defined by its primary product:

"CMPU means the equipment assembled and connected by pipes or ducts to process raw materials and to manufacture the intended product... includes air oxidation reactors and their associated product separators and recovery devices; reactors and their associated product separators and recovery devices; distillation units and their associated distillate receivers and recovery devices; associated unit operations; and any feed, intermediate and product storage vessels, product transfer racks, and connected ducts and piping. A chemical manufacturing process unit includes pumps, compressors, agitators, pressure relief devices, sampling connection systems, open-ended valves or lines, valves, connectors, instrumentation systems, and control devices or systems."

It is important to note that to be regulated under the HON Rule, *all three* criteria must be met. For instance, a facility that manufactures a chemical on the HON list (Table 1) and is a major source of one or more of the 189 HAPs may still not be subject to the HON Rule if its chemical process does not use or manufacture any of the 112 HAPs on the HON list (Table 2).

Similarly, a facility that has the potential to emit less than 10 tons/yr of any one HAP, or 25 tons/yr of any combination of HAPs, is considered a minor source and is exempt from the HON. If a facility can reduce its potential to emit below those thresholds — enough to achieve "synthetic minor" status by the compliance deadline — it will be exempt from the HON MACT.

For the moment, the manufacture of the HON chemicals by batch processing has also been exempted from the rule. However, a separate batch-processing rule may follow.

What is covered by the HON Rule?

The HON Rule consists of three subparts. Subpart F covers general provisions; Subpart G covers emissions standards for process vents, storage vessels, transfer racks and wastewater operations; and Subpart H covers emissions caused by equipment leaks.

This complex rule governs emissions from vents, storage vessels, transfer racks, wastewater operations and equipment leaks

By setting emission standards for process vents, storage vessels, transfer racks, and wastewater operations, the HON Rule is designed to reduce the emissions of 112 organic HAPs from the 385 CMPUs at both new and existing sources. Subpart H of the HON Rule specifies revised and more stringent standards for leak detection and repair of equipment leaks from valves, pumps, connectors, pressure relief devices, open-ended valves or lines, sampling connection systems, instrumentation systems, surge control vessels, bottoms receivers, and agitators. The provisions of Subpart H also apply to closed-vent systems and control devices used to control emissions from any of the listed equipment.

Compliance timetable

With limited exception, existing sources must comply with the provisions of the HON Rule by no later than April 22, 1997. The operating permit authority or EPA may grant up to one year of additional time to attain compliance for cases of extreme hardship.

Subpart H of the HON Rule defines five groups of sources under the equipment leaks section. For the first group of sources listed, compliance with Subpart H was required by October 24, 1994. Each of the remaining four groups of sources are required to comply in three month intervals beyond that date, with all subject sources in compliance by October 23, 1995. However, this deadline has been temporarily delayed for specific facilities while EPA tries to better define the complex and controversial term "potential to emit."

The HON Rule requires the categorization of process vents, storage vessels, transfer racks, and wastewater streams into "Group 1" sources (controls required) or "Group 2" sources (no controls required). The grouping requirements are discussed below.

A *process vent* includes any gas stream that is continuously discharged from an air-oxidation reactor, other reactor, or distillation unit within a chemical manufacturing process unit. For process vents to be subject to regu-

CRITERIA CHEMICALS FOR SOCMI OPERATORS[1,2]

- Acenaphthene
- Acetal
- Acetaldehyde
- Acetaldol
- Acetamide
- Acetanilide
- Acetic acid
- Acetic anhydride
- Acetoacetanilide
- Acetone
- Acetone cyanohydrin
- Acetonitrile
- Acetophenone
- Acrolein
- Acrylamide
- Acrylic acid
- Acrylonitrile
- Adiponitrile
- Alizarin
- Alkyl anthraquinones
- Allyl alcohol
- Allyl chloride
- Allyl cyanide
- Aminophenol sulfonic acid
- Aminophenol (*p*-)
- Aniline
- Aniline hydrochloride
- Anisidine (*o*-)
- Anthracene
- Anthraquinone
- Azobenzene
- Benzaldehyde
- Benzene
- Benzenedisulfonic acid
- Benzenesulfonic acid
- Benzil
- Benzilic acid
- Benzoic acid
- Benzoin
- Benzonitrile
- Benzophenone
- Benzotrichloride
- Benzoyl chloride
- Benzyl acetate
- Benzyl alcohol
- Benzyl benzoate
- Benzyl chloride
- Benzyl dichloride
- Biphenyl
- Bisphenol A
- Bis(Chloromethyl) Ether
- Bromobenzene
- Bromoform
- Bromonaphthalene
- Butadiene (1,3-)
- Butanediol (1,4-)
- Butyl acrylate (*n*-)
- Butylene glycol (1,3-)
- Butyrolacetone
- Caprolactam
- Carbaryl
- Carbazole
- Carbon disulfide
- Carbon tetrabromide
- Carbon tetrachloride
- Carbon tetrafluoride
- Chloral
- Chloroacetic acid
- Chloroacetophenone (2-)
- Chloroaniline (*p*-)
- Chlorobenzene
- 2-Chloro-1,3-butadiene (Chloroprene)
- Chlorodifluoroethane
- Chlorodifluoromethane
- Chloroform
- Chloronaphthalene
- Chloronitrobenzene (*m*-)
- Chloronitrobenzene (*o*-)
- Chloronitrobenzene (*p*-)
- Chlorophenol (*m*-)
- Chlorophenol (*o*-)
- Chlorophenol (*p*-)
- Chlorotoluene (*m*-)
- Chlorotoluene (*o*-)
- Chlorotoluene (*p*-)
- Chlorotrifluoromethane
- Chrysene
- Cresol and cresylic acid (*m*-)
- Cresol and cresylic acid (*o*-)
- Cresol and cresylic acid (*p*-)
- Cresols and cresylic acids (mixed)
- Cumene
- Cumene hydroperoxide
- Cyanoacetic acid
- Cyclohexane
- Cyclohexanol
- Cyclohexanone
- Cyclohexylamine
- Cyclooctadienes
- Decahydronaphthalene
- Diacetoxy-2-Butene (1,4-)
- Diaminophenol hydrochloride
- Dibromomethane
- Dichloroaniline (mixed isomers)
- Dichlorobenzene (*p*-)
- Dichlorobenzene (*m*-)
- Dichlorobenzene (*o*-)
- Dichlorobenzidine (3,3'-)
- Dichlorodifluoromethane
- Dichloroethane (1,2-) (Ethylene dichloride; EDC)
- Dichloroethyl ether (Bis(2-chloroethyl)ether)
- Dichloroethylene (1,2-)
- Dichlorophenol (2,4-)
- Dichloropropene (1,3-)
- Dichlorotetrafluoroethane
- Dichloro-1-butene (3,4-)
- Dichloro-2-butene (1,4-)
- Diethanolamine (2,2'-Iminodiethanol)
- Diethyl sulfate
- Diethylamine
- Diethylaniline (2,6-)
- Diethylene glycol
- Diethylene glycol dibutyl ether
- Diethylene glycol diethyl ether
- Diethylene glycol dimethyl ether
- Diethylene glycol monobutyl ether acetate
- Diethylene glycol monobutyl ether
- Diethylene glycol monoethyl ether acetate
- Diethylene glycol monoethyl ether
- Diethylene glycol monohexyl ether
- Diethylene glycol monomethyl ether acetate
- Diethylene glycol monomethyl ether
- Dihydroxybenzoic acid (Resorcylic acid)
- Dimethylbenzidine (3,3'-)
- Dimethyl ether
- Dimethylformamide (N,N-)
- Dimethylhydrazine (1,1-)
- Dimethyl sulfate
- Dimethyl terephthalate
- Dimethylamine
- Dimethylaminoethanol (2-)
- Dimethylaniline (N,N)
- Dintrobenzenes *(Not otherwise specified)*
- Dinitrophenol (2,4-)
- Dinitrotoluene (2,4-)
- Dioxane (1,4-) (1,4-Diethyleneoxide)
- Dioxolane (1,3-)
- Diphenyl methane
- Diphenyl oxide
- Diphenyl thiourea
- Diphenylamine
- Dipropylene glycol
- Di-o-tolyguanidine
- Dodecandedioic acid
- Dodecyl benzene (branched)
- Dodecyl phenol (branched)
- Dodecylaniline
- Dodecylbenzene (*n*-)
- Dodecylphenol
- Epichlorohydrin (1-chloro-2,3-epoxypropane)
- Ethanolamine
- Ethyl acrylate
- Ethylbenzene
- Ethyl chloride (Chloroethane)
- Ethyl chloroacetate
- Ethylamine
- Ethylaniline (*N*-)
- Ethylaniline (*o*-)
- Ethylcellulose
- Ethylcyanoacetate
- Ethylene carbonate
- Ethylene dibromide (Dibromoethane)
- Ethylene glycol
- Ethylene glycol diacetate
- Ethylene glycol dibutyl ether
- Ethylene glycol diethyl ether (1,2-diethoxyethane)
- Ethylene glycol dimethyl ether
- Ethylene glycol monoacetate
- Ethylene glycol monobutyl ether acetate
- Ethylene glycol monobutyl ether
- Ethylene glycol monoethyl ether acetate
- Ethylene glycol monoethyl ether
- Ethylene glycol monohexyl ether
- Ethylene glycol monomethyl ether acetate
- Ethylene glycol monomethyl ether
- Ethylene glycol moonooctyl ether
- Ethylene glycol monophenyl ether
- Ethylene gloycol monopropyl ether
- Ethylene oxide
- Ethylenediamine
- Ethylenediamine tetraacetic acid
- Ethyleneimine (Aziridine)
- Ethylhexyl acrylate (2-isomer)
- Fluoromethane
- Formaldehyde
- Formamide
- Formic acid
- Fumaric acid
- Glutaraldehyde
- Glyceraldehyde
- Glycerol
- Glycerol tri-(polyoxypropylene)ether
- Glycine
- Glyoxal
- Hexachlorobenzene
- Hexachlorobutadiene
- Hexachloroethane
- Hexadiene (1,4-)
- Hexamethylenetetramine
- Hexane
- Hexanetriol (1,2,6-)
- Hydroquinone
- Hydroxyadipaldehyde
- Isobutyl acrylate
- Isobutylene
- Isophorone
- Isophorone nitrile
- Isophthalic acid
- Isopropylphenol
- Linear alkylbenzene
- Maleic anhydride
- Maleic hydrazide
- Malic acid
- Metanilic acid
- Methacrylic acid
- Methanol
- Methionine
- Methyl acetate
- Methyl acrylate
- Methyl bromide (Bromomethane)
- Methyl chloride (Chloromethane)
- Methyl ethyl ketone (2-butanone)
- Methyl formate
- Methyl hydrazine
- Methyl isobutyl carbinol
- Methyl isobutyl ketone (Hexone)
- Methyl isocyanate
- Methyl mercaptan
- Methyl methacrylate
- Methyl phenyl carbinol
- Methyl tert-butyl ether
- Methylamine
- Methylaniline (*N*-)
- Methylcyclohexane
- Methylcyclohexanol
- Methylcyclohexanone
- Methylene chloride (Dichloromethane)
- Methylene dianiline (4,4'-isomer)
- Methylene diphenyl diisocyanate (4,4'-) (MDI)
- Methylionones (*a*-)
- Methylpentynol
- Methylstyrene (*a*-)
- Naphthalene
- Naphthalene sulfonic acid (*a*-)
- Naphthalene sulfonic acid (*b*-)
- Naphthol (*a*-)
- Naphthol (*b*-)
- Naphtholsulfonic acid (1-)
- Naphthylamine sulfonic acid (1,4-)
- Naphthylamine sulfonic acid (2,1-)
- Naphthylamine (1-)
- Naphthylamine (2-)
- Nitroaniline (*m*-)
- Nitroaniline (*o*-)
- Nitroanisole (*o*-)
- Nitroanisole (*p*-)
- Nitrobenzene
- Nitronaphthalene (1-)
- Nitrophenol (*p*-)
- Nitrophenol (*o*-)
- Nitropropane (2-)
- Nitrotoluene (all isomers)
- Nitrotoluene (*o*-)
- Nitrotoluene (*m*-)
- Nitrotoluene (*p*-)
- Nitroxylene
- Nonylbenzene (branched)
- Nonylphenol
- Octene-1
- Octylphenol
- Paraformaldehyde
- Paraldehyde
- Pentachlorophenol
- Pentaerythritol
- Peracetic acid
- Perchloromethyl mercaptan
- Phenanthrene
- Phenetidine (*p*-)
- Phenol
- Phenolphthalein
- Phenolsulfonic acids (all isomers)
- Phenyl anthranilic acid (all isomers)
- Phenylenediamine (*p*-)
- Phloroglucinol
- Phosgene
- Phthalic acid
- Phthalic anhydride
- Phthalimide
- Phthalonitrile
- Picoline (*b*-)
- Piperazine
- Polyethylene glycol
- Polypropylene gglycol
- Propiolactone (*b*-)
- Propionaldehyde
- Propionic acid
- Propylene carbonate
- Propylene dichloride (1,2-dichloropropane)
- Propylene glycol
- Propylene glycol monomethyl ether
- Propylene oxide
- Pyrene
- Pyridine
- p-tert-Butyl toluene
- Quinone
- Resorcinol
- Salicylic acid
- Sodium methoxide
- Sodium phenate
- Stilbene
- Styrene
- Succinic acid
- Succinonitrile
- Sulfanilic acid
- Sulfolane
- Tartaric acid
- Terephthalic acid
- Tetrabromophthalic anhydride
- Tetrachlorobenzene (1,2,4,5-)
- Tetrachloroethane (1,1,2,2-)
- Tetrachloroethylene (Perchloroethylene)
- Tetrachloropthalic anhydride
- Tetraethyl lead
- Tetraethylene glycol
- Tetraethylene pentamine
- Tetrahydrofuran
- Tetrahydronaphthalene
- Tetrahydrophthalic anhydride
- Tetramethlyenediamine
- Tetramethylethylenediamine
- Tetramethyllead
- Thiocarbanilide
- Toluene
- Toluene 2,4-diamine
- Toluene 2,4-diisocyanate
- Toluene diisocyanates (mixture)
- Toluene sulfonic acids
- Toluenesulfonyl chloride
- Toluidine (o-)
- Trichloroanaline (2,4,6-)
- Trichlorobenzene (1,2,3-)
- Trichlorobenzene (1,2,4-)
- Trichloroethane (1,1,1-)
- Trichloroethane (1,1,2-) (Vinyl trichloride)
- Trichloroethylene
- Trichlorofluoromethane
- Trichlorophenol (2,4,5-)
- (1,1,2-) Trichloro (1,2,2-) trifluoroethane
- Triethanolamine
- Triethylamine
- Triethylene glycol
- Triethylene glycol dimethyl ether
- Triethylene glycol monoethyl ether
- Triethylene glycol monomethyl ether
- Trimethylamine
- Trimethylcyclohexanol
- Trimethylcyclohexanone
- Trimethylcyclohexylamine
- Trimethylolpropane
- Trimethylpentane (2,2,4-)
- Tripropylene glycol
- Vinyl acetate
- Vinyl chloride (Chloroethylene)
- Vinyl toluene
- Vinylcyclohexene (4-)
- Vinylidene chloride (1,1-dichloroethylene)
- Vinyl (N-)-pyrrolidone (2-)
- Xanthates
- Xylene sulfonic acid
- Xylenes *(Not otherwise specified)*
- Xylene (*m*-)
- Xylene (*o*-)
- Xylene (*p*-)
- Xylenols (mixed)
- Xylidene

[1] Isomer means all structural arrangements for the same number of atoms of each element and does not mean salts, esters or derivatives.
[2] Chemical Abstract Service (CAS) numbers for each chemical can be found in the Federal Register, Vol. 59, No. 78, April 22, 1994, pp. 19462-19465.

TABLE 1. Several criteria must be met for operators to be regulated under the HON Rule. In addition to being designated a "major source" of organic HAPs, operators must manufacture or use one of these chemicals, which are contained in Subpart F of the rule

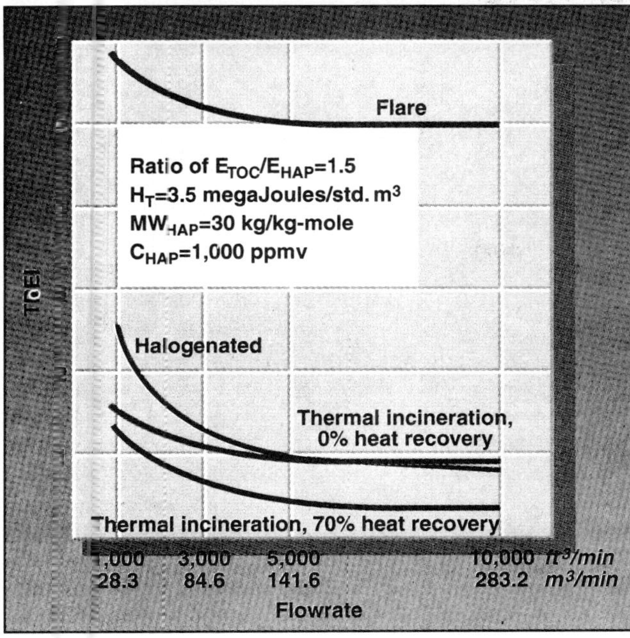

FIGURE 2. Using Equation 1, the TREI value was calculated for a low-concentration, low-molecular-weight vent stream. When TREI≤1, Group 1 controls are required. The lower the TREI value, the more cost-effective the control option. For this stream, a flare would not be justified. The use of thermal incineration with heat recovery would improve the cost-effectiveness of the approach

lation under the HON Rule, they must contain greater than 0.005 wt.% total HAPs.

Process vents typically include vents from distillate receivers, product receivers, product separators, and ejector-condensers, and include gas streams that are discharged directly to the atmosphere, or are discharged after diversion through a product-recovery device. However, the category typically excludes relief-valve discharges and leaks from equipment, which are regulated under HON Subpart H.

Once all process vents at a facility are identified, they must be categorized into Group 1 or Group 2, based on their flowrate, HAP concentration, and "total resource effectiveness index" (TREI). The TREI provides an indication of the cost versus the effectiveness of reducing emissions of a given HAP in a process vent stream, based on vent flowrate, emission rate of HAP, net heating value and corrosion properties.

A TREI value of 1.0 or less means that control is required and a value in excess of 4.0 means that control is not required. A TREI value that falls in between these values allows the vent to be designated a Group 2 vent until a detailed verification of the TREI is completed to assure that the characteristics of the vent do not change.

For process vents to be categorized in Group 1, the flowrate must meet or exceed 0.005 std. m^3/min, and the organic HAP concentration must be at least 50 ppm by volume (ppmv). Group 2 process vents are those that do not meet these criteria. In some cases, even when Group 1 values are exceeded, it may be possible to designate a process vent as a Group 2 by calculating its TREI, using Equation 1.

Additional details on calculating the individual terms listed below can be found in the process vents provision of the rule, Section 63.115 (d)(1,2,3) (Federal Register, Vol. 59, No. 78, April 22, 1994, pp. 19477-19478).

$$TREI = \frac{1}{E_{HAP}}[a + b(Q_s) + c(H_T) + d(E_{TOC})] \quad (1)$$

where:

TREI = TRE index value

E_{HAP} = Hourly emission rate of total organic HAPs, kg/h

Q_s = Vent stream flowrate, std. m^3/min, at a standard temperature of 20°C

H_T = Vent stream net heating value, megaJoules/std. m^3

E_{TOC} = Emissions rate of total organic compounds (minus methane and ethane), kg/h

a,b,c,d = Coefficients presented in paragraph (d)(3)(ii) and (iii) of Federal Register section noted above

Figure 2 shows the relationship of TRE to flowrate for a specific set of vent parameters. The parameters selected to prepare this figure represent a relatively low concentration of organic HAPs (1,000 ppmv) with a relatively low molecular weight (30 kg/kg-mole). Another simplifying assumption is that H_T is constant at 3.5 megaJoules/std. m^3, and that E_{TOC} is a direct ratio of 1.5 times E_{HAP}. Figure 2 shows that even for vent streams with relatively low flowrates and low concentrations of organic HAPs, the TRE index is below the 1.0 trigger value for Group 1 classification. The figure also shows that the cost-effectiveness of competing controls can be compared by calculating the TREI for each control device.

Storage vessels are grouped into Group 1 or Group 2 as a function of the capacity of the vessel, and the vapor pressure of the organic HAP stored in it. Table 3 shows Group 1 threshold values for new and existing vessels.

Transfer racks are classified into Group 1 if they load more than 0.65 million liters of liquid products whose organic HAPs have a rack-weighted, average, partial pressure of 10.3 kilopascals or more. Rack-weighted, average, partial pressure is calculated by taking the individual, maximum, true vapor pressures of products loaded at the rack, and weighting them by the volume of each liquid loaded.

Wastewater streams are also classified into Group 1 and Group 2 based on flowrate, concentration and compound, and the threshold values are different for new and existing sources. For instance, both new and existing process wastewater streams with an excess of 10,000 ppmw of HAP compounds are designated as Group 1 streams, irrespective of flowrate. Streams at new and existing sources whose flowrates are greater than 10 L/min and whose organic HAP concentration exceeds 1,000 ppmw are also classified in Group 1. Table 4 shows one control option for existing Group 1 wastewater streams.

At new sources, however, streams with a flowrate greater than 0.02 L/min, and an average concentration of 10 ppmw of one or more of the compounds listed in Table 5 are also designated as Group 1 wastewater streams. Group 2 streams (no controls required) include any wastewater stream that does not meet Group 1 designation criteria. All determinations of average flowrate and total volatile organic HAP concentration are made for the point of generation of the stream.

Reference control technologies

The setting of emission rate standards for each of the above portions of the SOCMI source is based on a "reference control technology" (RCT). Each portion of the SOCMI source has its own reference control technology, which is defined in Subpart G. These reference technologies represent the minimum baseline for the specific emission points at regulated Socmi sources.

- For process vents, the RCT requires a combustion device that will reduce organic HAP emissions by 98%, or achieve a maximum outlet concentration of 20 ppmv
- For storage vessels, the RCT specifies the use of an internal floating roof, an external floating roof, an external floating roof converted to an internal floating roof, or a closed-vent system to divert vapors to a control device achieving 95% reduction in organic HAP emissions. For the purpose of emissions averaging, these four technologies are considered equivalent
- For transfer racks, the RCT specifies a combustion device or recovery system that reduces organic HAP emissions by 98%, or achieves an outlet concentration of 20 ppmv
- For wastewater streams, the RCT specifies the use of specified controls, a steam stripper, or any of the specified alternative control measures; and a

The HON Rule is one of the most complex rules ever put on the books, and it covers more processes and pollutants than all previous EPA air toxics programs combined

TABLE 2. To be regulated under the HON Rule, major source operators must be makers or users of one or more of the chemicals listed in Table 1, and either use as a reactant, or manufacture as a product, byproduct or co-product, one or more of the criteria chemicals listed here

REGULATED POLLUTANTS CONTAINED IN SUBPART F[1,2,3]

- Acetaldehyde
- Acetamide
- Acetonitrile
- Acetophenone
- Acrolein
- Acrylamide
- Acrylic acid
- Acrylonitrile
- Allyl chloride
- Aniline
- Anisidine (*o*-)
- Benzene
- Benzotrichloride
- Benzoyl chloride
- Biphenyl
- Bis(chloromethyl)ether
- Bromoform
- Butadiene (1,3-)
- Caprolactam
- Carbon disulfide
- Carbon tetrachloride
- Chloroacetic acid
- Chloroacetophenone (2-)
- Chlorobenzene
- 2-Chloro-1,3-butadiene (Chloroprene)
- Chloroform
- Cresols and cresylic acids (mixed)
- Cresol and cresylic acid (*o*-)
- Cresol and cresylic acid (*m*-)
- Cresol and cresylic acid (*p*-)
- Cumene
- Dichlorobenzene (*p*-)
- Dichlorobenzidine (3,3'-)
- Dichloroethane (1,2-) (Ethylene dichloride; EDC)
- Dichloroethylether (Bis(2-chloroethyl)ether)
- Dichloropropene (1,3-)
- Diethanolamine (2,2'-Iminodiethanol)
- Dimethylaniline (*N,N*-)
- Diethyl sulfate
- Dimethylbenzidine (3,3')
- Dimethylformamide (*N,N*-)
- Dimethylhydrazine (1,1-)
- Dimethyl phthalate
- Dimethyl sulfate
- Dinitrophenol (2,4-)
- Dinitrotoluene (2,4-)
- Dioxane (1,4-) (1,4-Diethyleneoxide,
- 1,2-Diphenylhydrazine)
- Epichlorohydrin (1-Chloro-2,3-epoxypropane)
- Ethyl acrylate
- Ethylbenzene
- Ethyl chloride (Chloroethane)
- Ethylene dibromide (Dibromoethane)
- Ethylene glycol
- Ethylene oxide
- Ethylidene dichloride (1,1-Dichloroethane)
- Formaldehyde
- Glycol ethers[4]
- Hexachlorobenzene
- Hexachlorobutadiene
- Hexachloroethane
- Hexane
- Hydroquinone
- Isophorone
- Maleic anhydride
- Methanol
- Methyl bromide (Bromomethane)
- Methyl chloride (Chloromethane)
- Methyl ethyl ketone (2- Butanone)
- Methyl hydrazine
- Methyl isobutyl ketone (Hexone)
- Methyl isocyanate
- Methyl methacrylate
- Methyl *tert*-butyl ether
- Methylene chloride (Dichloromethane)
- Methylene diphenyl diisocyanate (4,4'-) (MDI)
- Methylenedianiline (4,4'-)
- Naphthalene
- Nitrobenzene
- Nitrophenol (*p*-)
- Nitropropane (2-)
- Phenol
- Phenylenediamine (*p*-)
- Phosgene
- Phthalic anhydride
- Polycyclic organic matter[5]
- Propiolactone (*b*-)
- Propionaldehyde
- Propylene dichloride (1,2-Dichloro propane)
- Propylene oxide
- Quinone
- Styrene
- Tetrachloroethane (1,1,2,2-)
- Tetrachloroethylene (Perchloroethylene)
- Toluene
- Toluene diamine (2,4-)
- Toluene diisocyanate (2,4-)
- Toluidine (o-)
- Trichlorobenzene (1,2,4-)
- Trichloroethane (1,1,1-) (Methyl chloroform)
- Trichloroethane (1,1,2-) (Vinyl trichloride)
- Trichloroethylene
- Trichlorophenol (2,4,5-)
- Triethylamine
- Trimethylpentane (2,2,4-)
- Vinyl acetate
- Vinyl chloride (chloroethylene)
- Vinylidene chloride (1,1-Dichloroethylene)
- Xylenes *(Not otherwise specified)*
- Xylene (*m*-)
- Xylene (*o*-)
- Xylene (*p*-)

1 For all listings above containing the word "compound" and for glycol ethers, the following applies: Unless otherwise specified, these listings are defined as including any unique chemical substance that contains the named chemical (i.e., antimony, arsenic) as part of that chemical's infrastructure.
2 Isomer means all structural arrangements for the same number of atoms of each element, and does not mean salts, esters or derivatives.
3 The Chemical Abstract Service (CAS) number for each chemical can be found in the Federal Register, Vol. 59, No. 78, April 22, 1994, p. 19465.
4 Includes mono- and di-ethers of ethylene glycol, diethylene glycol and triethylene glycol R-$(OCH_2CH_2)_n$-OR' where: n=1, 2 or 3; R=alkyl or aryl groups; and R'=R, H, or groups which, when removed, yield glycol ethers with the structure R-$(OCH_2CH_2)_n$-OH. Polymers are excluded from the glycol category.
5 Includes organic compounds with more than one benzene ring, and those with a boiling point greater than or equal to 100°C.

control device that reduces organic HAP emissions by 95%, or achieves a concentration of 20 ppmv for combustion devices. This includes all vapor streams vented from wastewater tanks, oil-water separators, containers, surface impoundments, individual drain systems, and treatment processes (including the steam stripper) managing wastewater.

The emission rate standard

The emission rate standard equation shown below is used to demonstrate compliance when using emissions averaging. Compliance with the allowable emission rates defined in Equation 2 can be demonstrated by verifying that the reference control technology has been applied; actual emission rates do not need to be calculated.

Determining the sum of the emission rates represented by each term in this equation is no small task, and new and existing sources are treated differently. For more details on calculating the individual terms listed below, see Subpart G of the HON Rule, Section 63.111 (Federal Register Vol. 59, No. 78, April 22, 1994, pp. 19525-19526).

$$E_A = 0.02\sum EPV_1 + \sum EPV_2 + 0.05\sum ES_1 + \sum ES_2 + 0.02\sum ETR_1 + \sum ETR_2 + \sum EWW_{1c} + \sum EWW_2 \quad (2)$$

where

E_A = Emissions rate allowed for the source, megagrams/yr

$0.02\sum EPV_1$ = Sum of the residual emissions from all Group 1 process vents, megagrams/yr

$\sum EPV_2$ = Sum of the emissions from all Group 2 process vents, megagrams/yr

$0.05\sum ES_1$ = Sum of the residual emissions from all Group 1 storage vessels, megagrams/yr

$\sum ES_2$ = Sum of the emissions from all Group 2 storage vessels, megagrams/yr

$0.02\sum ETR_1$ = Sum of the residual emissions from all Group 1 transfer racks, megagrams/yr

$\sum ETR_2$ = Sum of the emissions from all Group 2 transfer racks, megagrams/yr

$\sum EWW_{1C}$ = Sum of the residual emissions from all Group 1 wastewater streams

$\sum EWW_2$ = Sum of emissions from all Group 2 wastewater streams

GROUP 1 STORAGE VESSELS

CAPACITY, m^3	VAPOR PRESSURE, KILOPASCALS*
EXISTING VESSELS	
$75 \le$ capacity < 151	To be determined
$151 \le$ capacity	≥ 5.2
NEW VESSELS	
$38 \le$ capacity < 151	13.1
$151 \le$ capacity	≥ 0.7

* Maximum true vapor pressure of total organic HAP at storage temperature

TABLE 3 (above). These threshold values determine which storage vessels require a Group 1 rating — and additional controls

ORGANIC HAPS SUBJECT TO WASTEWATER PROVISIONS AT NEW SOURCES[1]

Allyl chloride	Phosgene
Benzene	Tetrachloroethylene (Perchloroethylene)
Butadiene (1,3-)	Toluene
Carbon disulfide	Trichloroethane (1,1,1-) (Methyl chloroform)
Carbon tetrachloride	Trichloroethylene
Cumene	Trimethylpentane (2,2,4-)
Ethyl benzene	Vinyl chloride (chloroethylene)
Ethyl chloride (Chloroethane)	Vinylidene chloride (1,1,-Dichloroethylene)
Ethylidene dichloride (1,1-Dichloroethane)	Xylene (*m*-)
Hexachlorobutadiene	Xylene (*p*-)
Hexachloroethane	
Hexane	
Methyl bromide (Bromomethane)	
Methyl chloride (Chloromethane)	

Note: This list of organic HAPs is a subset of Table 4.

TABLE 5 (above). New wastewater streams fall into Group 1 if the flowrate is greater than 0.02 L/min, and the average concentration of any of these compounds is ≥ 10 ppmw

TABLE 4 (below). For Group 1 wastewater streams, one compliance option is to reduce these organic HAPs by the percentage indicated

ORGANIC HAPS SUBJECT TO THE WASTEWATER PROVISIONS AND REQUIRED FRACTION REMOVED (FR) VALUES[1]

Compound	FR	Compound	FR
Acetaldehyde	0.95	Ethylidene dichloride (1,1-Dichloroethane)	0.99
Acetonitrile	0.62	Hexachlorobenzene	0.89
Acetophenone	0.72	Hexachlorobutadiene	0.99
Acrolein	0.96	Hexachloroethane	0.99
Acrylonitrile	0.96	Hexane	0.99
Allyl chloride	0.99	Isophorone	0.60
Benzene	0.99	Methanol	0.31
Benzyl chloride	0.99	Methyl bromide (Bromomethane)	0.99
Biphenyl	0.99	Methyl chloride (Chloromethane)	0.99
Bromoform	0.99	Methyl ethyl ketone (2-Butanone)	0.95
Butadiene (1,3-)	0.99	Methyl isobutyl ketone (Hexone)	0.99
Carbon disulfide	0.99	Methyl methyacrylate	0.98
Carbon tetrachloride	0.99	Methyl *tert*-butyl ether	0.99
Chlorobenzene	0.99	Methylene chloride (Dichloromethane)	0.99
Chloroform	0.99	Naphthalene	0.99
Chloroprene (2-Chloro-1,3-butadiene)	0.99	Nitropropane (2-)	0.98
Cumene	0.99	Phosgene	0.99
Dichlorobenzene (*p*-)	0.99	Propionaldehyde	0.99
Dichloroethane (1,2-) (Ethylene dichloride)	0.99	Propylene dichloride (1,2-Dichloropropane)	0.99
Dichloroethyl ether (Bis(2-chloroethyl)ether)	0.87	Propylene oxide	0.99
Dichloropropene (1,3-)	0.99	Styrene	0.99
Diethyl sulfate	0.90	Tetrachloroethane (1,1,2,2-)	0.99
Dimethyl sulfate	0.53	Tetrachloroethylene (Perchloroethylene)	0.99
Dimethylaniline (*N,N*-)	0.99	Toluene	0.99
Dimethylhydrazine (1,1-)	0.57	Toluidine (*o*-)	0.44
Dinitrophenol (2,4-)	0.99	Trichlorobenzene (1,2,4-)	0.99
Dinitrotoluene (2,4-)	0.38	Trichloroethane (1,1,1-) (Methyl chloroform)	0.99
Dioxane (1,4-) (1,4-Diethyleneoxide)	0.37	Trichloroethane (1,1,2-) (Vinyl trichloride)	0.99
Epichlorohydrin (1-Chloro-2,3-epoxypropane)	0.91	Trichloroethylene	0.99
Ethyl acrylate	0.99	Trichlorophenol (2,4,5-)	0.96
Ethylbenzene	0.99	Triethylamine	0.99
Ethyl chloride (Chloroethane)	0.99	Trimethylpentane (2,2,5-)	0.99
Ethylene dibromide (Dibromomethane)	0.99	Vinyl acetate	0.99
Ethylene glycol dimethyl ether	0.90	Vinyl chloride (Chloroethylene)	0.99
Ethylene glycol monobutyl ether acetate	0.76	Vinylidene chloride (1,1-Dichloroethylene)	0.99
Ethylene glycol monomethyl ether acetate	0.28	Xylene (*m*-)	0.99
Ethylene oxide	0.98	Xylene (*o*-)	0.99
		Xylene (*p*-)	0.99

1 Chemical Abstract Service (CAS) numbers for each chemical can be found in the Federal Register, Vol. 59, No. 78, April 22, 1994, p. 19545-19546.

TABLE 6. For process vents, operators must reduce total organic HAP emissions by 98 wt.%, or maintain a concentration of ≤ 20 ppmv. For widely used exhaust-control devices, this table shows the key parameters to be monitored and specific reporting requirements. Other equipment categories face similar levels of detailed reporting

MONITORING, RECORDKEEPING AND REPORTING

THERMAL INCINERATOR	CATALYTIC INCINERATOR	BOILER OR PROCESS HEATER*	FLARE
Firebox temperature[1]	Temperature upstream and downstream of the catalyst bed	Firebox temperature[1]	Presence of a flame at the pilot light
(Section 63.114(a)(1)(i))	*(Section 63.114(a)(1)(ii))*	*(Section 63.114(a)(3))*	*(Section 63.114(a)(2))*
1. Continuous records[2] 2. Record and report the firebox temperature averaged over the full period of the performance test — NCS[3] 3. Record the daily average firebox temperature for each operating day[4] 4. Report all daily average temperatures that are outside the range established in the NCS or operating permit, and all operating days when insufficient monitoring data are collected[5] — PR[6]	1. Continuous records[2] 2. Record and report the firebox temperature averaged over the full period of the performance test — NCS[3] 3. Record the daily average firebox temperature for each operating day[4] 4. Report all daily average temperatures that are outside the range established in the NCS or operating permit, and all operating days when insufficient monitoring data are collected[5] — PR[6] 5. Report all daily average temperature differences across the catalyst bed that are outside the range established in the NCS or operating permit — PR[6] 6. Report all operating days when insufficient monitoring data are collected[5]	1. Continuous records 2. Record and report the firebox temperature averaged over the full period of the performance test — NCS[3] 3. Record the daily average firebox temperature for each operating day[4] 4. Report all daily average firebox temperatures that are outside the range established in the NCS or operating permit and all operating days when insufficient monitoring data are collected — PR[6] * Units with a design heat-input capacity < 44 MW, and whose vent stream is not introduced with, or as, the primary fuel	1. Hourly records of whether the monitor was continuously operating and whether the pilot flame was continuously present during each hour 2. Record and report the presence of a flame at the pilot light over the full period of the compliance determination — NCS[3] 3. Record the times and durations of all periods when a pilot flame is absent or the monitor is not operating 4. Report the times and durations of all periods when all pilot flames of a flare are absent — PR[6]

[1] Monitor may be installed in the firebox or in the ductwork immediately downstream of the firebox before any substantial heat exchange is encountered. [2] The term "continuous records" is defined in Section 63.111. [3] NCS = Notification of Compliance Status, as described in Section 63.152. [4] The daily average is the average of all recorded parameter values for the operating day.

Note that there are eight sums in Equation 2. Two apply to each of the four types of equipment subject to regulation under the HON Rule. Thus, in essence, Equation 2 represents the total of the sums of the remaining emissions from all equipment categorized as Group 1 after the application of the reference control technology, plus the sum of the emissions from all Group 2 equipment.

The use of emissions averaging

In general, the HON Rule requires an overall reduction in HAP emissions of 98%. Emissions averaging would allow facilities to establish a system of emission credits and debits across its inventory of emission points, so that temporary, above-standard emissions from one unit could be used to offset below-standard emissions from another unit.

The HON allows operators to use a one-year averaging period, which may be helpful to chemical manufacturing facilities that make frequent operational changes, such as those that manufacture multiple chemicals in a CMPU during a given year. In such facilities, a moderate investment in emissions tracking might allow the operator to maintain operating flexibility without jeopardizing compliance status.

Emissions averaging may be used at existing facilities and may include a maximum of 20 emissions points. After calculation of the quantities of credits available for use in offsetting emission debits, a 10% discount factor is applied to the credits before using them in the averaging plan. However, if these credits are provided by pollution-prevention alternatives, no discount factor is applied and full credit is allowed.

Pollution-prevention alternatives include any process revisions that reduce emissions, rather than applying pollution controls to the exhaust stream. EPA also allows up to five additional points for inclusion in an emissions average if pollution-prevention reductions are used in the averaging plan.

To use emissions averaging, operators must determine that the "overcontrol" of one HAP to allow for the "undercontrol" of others does not present too much localized risk to the public, worker health or the environment. To make this determination, EPA is expected to provide a technical support document for use by the state regulatory agency in its approval of emissions averaging. This technical support document will take into account different toxicities of various HAPs, as well as differing dispersion and exposure characteristics of emission sources.

As facilities begin to gather the information required by the HON Rule, emissions averaging will become a more attractive option. In spite of its restrictions, such as earlier submittal dates for implementation plans and more frequent reporting, emissions averaging will be more cost-effective for many sources than retrofitting each piece of existing equipment to achieve compliance.

Monitoring and recordkeeping

All of the requirements for control of emissions from the points subject to the HON Rule require detailed records of operating parameters, emission rates, calculations, flows and other variables. The recordkeeping will depend on the compliance method or methods chosen.

The chosen compliance strategy will impact the level of detail required in recordkeeping, and therefore these requirements and their costs should be considered early on when determining a compliance strategy. Table 6 shows a detailed listing of the monitoring requirements for process vents, to demonstrate that they are in compliance with the 98 wt.% reduction requirement or 20-ppmv maximum concentration limits.

Subpart G contains reporting and recordkeeping requirements for subject sources. In general, five types of reports must be submitted to EPA:

1. Initial notification. For existing sources, this report was to have been

REQUIREMENTS FOR PROCESS VENTS

SCRUBBER FOR HALOGENATED VENT STREAMS*		ALL CONTROL DEVICES
pH of scrubber effluent	Scrubber liquid and gas flowrate	Flow diverted to the atmosphere from the control device
(Section 63.114(a)(4)(i))	*(Section 63.114(a)(4)(ii))*	*(Section 63.114(d)(1))*
1. Continuous records[2] 2. Record and report the pH of the scrubber effluent averaged over the full period of the performance test — NCS[3] 3. Record the daily average pH of the scrubber effluent for each operating day[4] 4. Report all daily average pH values of the scrubber effluent that are outside the range established in the NCS or operating permit and all operating days when insufficent monitoring data are collected[5]— PR[6]	1. Continuous records 2. Record and report the scrubber liquid-to-gas ratio averaged over the full period of the performance test — NCS[3] 3. Record the daily average scrubber liquid-to-gas ratio for each operating day[4] 4. Report all daily average scrubber liquid-to-gas ratios that are outside the range established in the NCS or operating permit and all operating days when insufficient monitoring data are collected[5]— PR[6]	1. Hourly records of whether the flow indicator was operating, and whether flow was detected at any time during each hour 2. Record and report the times and durations of all periods when the vent stream is diverted through a bypass line or the monitor is not operating — PR[6] **Monthly inspections of sealed valves** *(Section 63.114(d)(2))* 1. Record that monthly inspections were performed 2. Record and report all monthly inspections that show the valves are not closed or the seal has been changed — PR[6]

If all recorded values during an operating day are within the range established in the NCS or operating permit, such a statement can be recorded instead of the daily average. [5] The periodic reports shall include the duration of periods when monitoring data are not collected for each excursion as defined in Section 63.152(c)(2)(ii)(A). [6] PR = Periodic Reports as described in Section 63.152

submitted 120 days after publication of the Final Rule for the HON MACT for existing sources. Because the final rule was published on April 22, 1994, this initial notification was due to EPA by August 22, 1994.

For new sources, an initial notification must be handed in 180 days prior to commencement of construction. The initial notification must list the CMPUs that are subject to Subpart G, and the provisions that apply (process vents, storage vessels and so on). Details are not necessary at this stage, but a statement must be included as to whether the source expects to be able to comply on time.

COMPLIANCE OPTIONS FOR WASTEWATER TANKS

Capacity, m³	Vapor pressure, kPa	Control requirements
≤75	—	Section 63.133(a)(1)
≥75 and <151	<13.1	Section 63.133(a)(1)
	≥13.1	Section 63.133(a)(2)
≥151	<5.2	Section 63.133(a)(1)
	≥5.2	Section 63.133(a)(2)

TABLE 7. Specific control requirements for wastewater tanks vary as a function of tank capacity and vapor pressure

2. Implementation plan. Sources must file an implementation plan with their operating permit application. The plan must include details on how the source will comply with the provisions of Subpart G.

New sources must submit this plan with the initial notification. Existing sources that have not yet filed their operating permit application have two deadlines. For emission points not included in an averaging plan, the required submission date is 12 months prior to the compliance date (April 22, 1996). For points under an emissions averaging plan, the plan must be submitted 18 months prior to the compliance deadline (October 22, 1995).

3. Notification of compliance status. This notification is to be submitted 150 days after the source's compliance date. It must contain the emissions data for Group 1 points, and all necessary information to demonstrate compliance.

4. Periodic reports. In general these reports would be submitted semiannually. Quarterly reports, however, must be submitted for all points included in an emissions average. All periodic reports must include the information required to be reported under the recordkeeping and reporting provisions for each kind of emissions point. This would include performance tests, data that are beyond the acceptable ranges for specified time periods, and the results of periodic inspections.

5. Other reports. Other reports include reports of startups, shutdowns, malfunctions and repairs.

Get started

For Socmi sources that meet the major source threshold, the first step to complying with the HON MACT is to identify all equipment in the four categories of process vents, storage vessels, transfer racks, and wastewater streams. The initial listing should be all inclusive and identify all emission points. The question of applicability can then be answered for each by examining the applicable criteria. Figure 1 shows a step-by-step flow chart to organize these activities.

Once all subject equipment have been identified, each component should be assigned to either Group 1 or Group 2. Any vent identified as a Group 1 source should be examined further to determine if a process change or a TREI determination can move the emission point to Group 2.

Next, from engineering calculations or emissions test data, emissions from non-compliant Group 1 or Group 2 vents should be calculated. Any emission points used in calculating an emissions average should be identified. Once this phase of the assessment is complete, an initial understanding of compliance status will emerge.

To minimize the cost of compliance, it is of obvious importance to minimize the number of process modifications required, and to identify pollution-prevention alternatives. Maximum use of existing equipment is recommended. In some cases, this may call for the aggregation of vents, upgrades of existing controls, and additional emissions testing to verify compliance status.

For process vents, wherever possible, product-recovery devices and other techniques should be considered, where they may change the TREI of a process vent enough to move it from Group 1 to Group 2 status.

Before retrofitting storage vessels with floating roofs, it may be advisable to investigate the potential benefit of ducting emissions to an existing control device on site. This may help to minimize the impact of the HON Rule, since many facilities have already controlled tank emissions to minimize losses of raw material or product during transfer operations.

Transfer racks should include a vapor-balance system, which is designed to minimize the displacement losses of organic HAP vapors during the transfer of organic HAPs from storage vessels to transportation vessels. The use of sound loading practices and regular maintenance of equipment is also a must to minimize vapor losses.

The HON MACT strongly encourages pollution-prevention measures to deal with wastewater streams. Wherever possible, return water streams to the process, pipe them to approved incineration systems, and recycle the waste. All wastewater streams should be examined to determine the best way to control emissions. Otherwise costly steam stripping requirements will be mandated by the regulatory authority.

Most operators subject to regulation under the the HON Rule are also subject the Title V Operating Permit Program.* Any facility that is a major source of HAPs is required to obtain a Title V operating permit. Thus, by definition, anyone subject to the HON Rule will need an operating permit. A few states, including Florida, will require an operating permit of a HON source even if the result of application of the HON Rule makes the facility a minor source of HAPs. Many of these requirements are addressed in the HON Rule itself or in the General Provisions for 40 CFR Part 63.

The source startup, shutdown and malfunction plan will establish acceptable levels for process variables such as temperatures, flowrates, and periodic test measurements. Since continuous emissions monitoring systems are not available for most equipment sources subject to regulation by the HON Rule, regular monitoring of process variables will be necessary.

The plan should include sufficient detail to identify any unusual operating conditions that may occur and would cause a deviation from an established value of a monitored variable.

If these are made a part of the plan, they will not be considered "excursions" — a term used in the HON Rule to reflect a monitored variable that temporarily exceeds limits. Up to six excursions are allowed for the first semi-annual reporting period. Allowable excursions will be reduced by one per reporting period during each of the first three years of the HON Rule implementation. At the end of three years, only one excursion will be allowed during a reporting period.

The reporting requirements will allow compressed data to be submitted to fulfill reporting obligations. Systems using data compression do not record monitoring data at a set frequency; rather, they only record values that vary from previously recorded numbers by some set criteria. EPA has allowed data-compression systems to be used for HON Rule compliance for three reasons: To ease the burden of data storage and retrieval; to more accurately represent variations in operating parameters; and to quickly detect operating problems. EPA has established guidelines for such systems, and use of data compression requires prior EPA approval.

It is important to approach compliance with the HON MACT in a systematic manner to develop the most cost-effective and efficient compliance strategy. While the HON Rule has generally been judged as acceptable by the chemical process industries, it requires tremendous effort in terms of data gathering, analyzing and documentation. With just two years left to meet the April 1997 deadline for submission of the complete compliance plan, it's not too early to get started. ■

Edited by Suzanne Shelley

* For more on the Title V operating program, see Title V sends shock waves through the CPI, *Chem. Eng.*, January 1995, pp. 35-39, and Gearing up for Title V permits, *Environmental Engineering*, a supplement to *Chem. Eng.*, June 1994, pp. 14-24.

The author

Mark M. Hultman, P.E. is a principal consulting engineer in the Air Permitting and Compliance Engineering Group in the Windsor, Conn., office of TRC Environmental Corp. (5 Waterside Crossing, Windsor, CT 06095; tel.: 203-289-8631). Hultman received his B.S.Ch.E. from the University of Connecticut (Storrs). He is a registered professional engineer in five states, and is a member of AIChE.

THE BEST APPROACH TO ENVIRONMENTAL COMPLIANCE

Paul R. Ammann, Gayle S. Koch, M. Alexis Maniatis, and **Kenneth T. Wise**
Brattle/IRI

Regulations on the discharge of hazardous substances from chemical and other manufacturing facilities are getting stricter. Meanwhile, government agencies and public-interest groups are becoming more vigilant. Violations can mean substantial, even multimillion-dollar penalties, and capital and operating costs for compliance can be extensive. Given these high stakes, companies must take pains not merely to comply with applicable regulations and permits but to do so as cost-effectively as possible.

Compliance can almost always take several forms, differing significantly in cost and required time. Companies often overreact, installing much larger systems than necessary.

The picture becomes more complex when the firm not only faces current regulations but also expects different ones in the future. Another complexity arises when one or more compliance options can be designed (at additional cost) to achieve other, non-environmental, objectives.

Whatever their situation, companies should aim not merely to be cost-effective but to also avoid the reactive approach. They should instead aim for a more proactive mode, in which environmental compliance options actually play a role in the strategic direction of the firm.

The regulatory push

In the U.S., the main regulations that trigger non-compliance penalties for process plants are the Federal Water Pollution Control (Clean Water) Act, the Resource Conservation and Recovery Act (RCRA), and the Clean Air Act. The Clean Water Act sets the requirements for getting permits to discharge process water and stormwater. RCRA mandates permitting and reporting for waste handling and storage at operating facilities. The 1990 Clean Air Act Amendments include a timetable for chemical and other manufacturing plants to obtain permits, as well as to control emissions of about 200 hazardous pollutants over the next several years.

Permits issued by state or federal regulators define the concentrations and amounts of pollutants that may be contained in streams and gases discharged. Typically, they also specify the sampling methods, frequency of sampling, and the procedures and required frequencies (usually monthly) for reporting data.

Double trouble

Poorly designed strategies for complying with permits can, of course, result in permit violations and less-cost-effective investment. In addition, such strategies can also bring about excessive penalties for past non-compliance [*1*, *2*].

This latter effect comes about because the agencies assessing penalties for past non-compliance usually base their calculations on the money a company actually spends to bring a plant into compliance. In the absence of careful planning, these outlays may include investments made for reasons broader than compliance with the permit at issue, or simply may not be efficient solutions.

Three examples make the point. First, consider a company that spends $200,000 to build a treatment facility to achieve compliance as well as other objectives. Federal and state agencies alike will calculate the penalty for past non-compliance on the basis of that figure, even if the company could have achieved compliance by spending only $100,000.

Second, consider a manufacturer that had employed a chlorinated solvent to degrease metal parts. Solvent discharge exceeded the permit limit, so the plant engineer contacted vendors to seek compliance alternatives. The engineer first leaned toward installing a central incinerator for all of the process gas discharges from the plant. The capital cost was in the range of $1-2 million, and the incremental operating cost was about $200,000/yr.

But further evaluation revealed a novel method by which the residual solvent could be recaptured on the production line. The capital cost of this system was only about $150,000. And the recovery of solvent provided a significant return on the modest investment.

Third, a large chemical plant whose process water discharged into a municipal treatment plant was exceeding its permit limit. Several reactors at this facility produced chemical wastes, which were directed to a plant drain system and commingled with other, non-polluted streams.

The company built an end-of-pipe facility to treat all of the discharge, at a significant cost, which formed the basis of a large penalty. A retrospective analysis showed that compliance could have been achieved more simply with controls at the reactors, at a savings of 75% in capital cost, negligible added operating costs, and significantly lower penalties. ❐

Originally published February 1995

It makes more sense to be proactive than reactive

ENGINEERING ISSUES

The engineer can choose from three strategies for complying with environmental discharge permits: (1) apply the end-of-pipe approach, treating the overall plant discharge just upstream of the regulators' stipulated compliance point; (2) treat the pollutant source streams throughout the plant individually; or (3) modify or replace the overall process to avoid generating the pollutants in the first place. Each option has advantages and drawbacks.

The end-of-pipe approach is the most obvious, and probably most widely used. Centralizing all of the treatment equipment at one point, it simplifies compliance, at least conceptually. Unfortunately, it requires treating all the liquid and gaseous streams generated at the plant, including unpolluted ones, which requires much investment. Operating cost, on the other hand, may well be only slightly higher than for treating smaller flowrates.

The second option, source control, recognizes that in many plants, the pollutants arise at only a few points. Nevertheless, designing control alternatives for these sources requires analysis of the overall manufacturing process. And because exercising this option ordinarily leads to the installation of control systems at different locations in a plant, reagents must be supplied to each of these locations and the responsibility for managing the control systems must be divided among several operators.

The third option, changing the manufacturing process, does away with the need for environmental controls. But process change generally requires research and development, and lead-times of one or more years. It can also raise complications regarding the specifications for and marketplace acceptance of the plant product, because the latter is now being made via a different process. Investment in the process and product development, and in the new production facilities, often can only be justified by future business prospects and improved economics of manufacturing, and not by environmental compliance alone.

Nevertheless, modest process changes may be justified in some situations. For example, adding non-contact heating or cooling systems to chemical reactors to reduce the amount of liquids or gases discharged from a process often has a large return on investment while contributing to compliance. Such changes usually require only limited engineering study.

SEVEN STEPS

Finding the best compliance strategy for a given situation can be aided by following the sequence outlined in this section. To apply it properly, the engineer should be careful to take into account not only the immediate permit limitations, but also any anticipated future regulatory or business requirements.

1. Analyze pollutant sources

In most cases, regulators are concerned with not one but a number of pollutants or compliance parameters. These include not only standard items, such as stream pH or oil-and-grease content, but also the presence or concentration of substances that are specific to the particular operation. Such pollutants usually arise at certain points in the plant. The first step is to review the entire process to identify these points, as well as the paths the pollutants follow in leaving them.

For example, a batch reactor may give off an acidic waste gas. The acid compound might become picked up in not only an aqueous scrubber for the reactor offgases but also in a periodic rinse of the reactor. Both streams need to be identified, and their flows traced through the plant to their discharge point or points.

In applying this step, keep in mind that pollutants can take many forms. For instance, they may derive from unreacted feed materials, compounds produced by undesired side reactions, or materials such as particulates that are not removed from process streams.

DOCUMENTATION CAN HELP

No matter how conscientious their engineers and environmental managers, companies may at some point face noncompliance penalties. In such cases, careful documentation of the firm's cost analyses for past and future compliance can help minimize these penalties. Conversely, information that is not properly characterized and analyzed can lead to inappropriate and excessive penalty demands that needlessly complicate negotiations and litigation.

Before responding to requests from government agencies or citizens groups to provide information on compliance investments, company managers and engineers should take care to (1) analyze the compliance alternatives and to identify what investment items were (or will be) necessary to achieve compliance, (2) determine if some portion of the cost of these items contributed to achieving other objectives, (3) identify the incremental costs associated with compliance measures, (4) analyze and value properly the future costs that may be necessary for compliance, and (5) identify any incremental cost savings that are associated with compliance. Following these steps and taking the time to respond thoughfully to requests for information early can have a huge payoff in reduced penalties and litigation expenses later. ❐

2. Make a pollutant balance

Once the sources have been identified, the pollutant mass flows and concentrations at various locations in the plant should be determined. From a knowledge of the process and calculation of a simple material balance, one can in many instances then estimate the total flow of principal pollutants discharged into a common plant drain or stack. Sampling and analysis of various streams may be required in certain circumstances.

An audit in which detailed measurements are made on every polluted stream can aid the analysis. It is especially valuable when streams contain pollutants whose concentrations may vary significantly with slight changes in process conditions.

The engineer should draw a schematic diagram showing the main streams and the pollutants. Figure 1, for instance, shows the reactor referred to in Step 1. Gases it emits pass through an aqueous scrubber that collects acidic compounds, and the resulting small aqueous stream goes into a plant drain system.

The reactor is cleaned periodically with rinse water that, when discharged, is also acidic. The process drain system collects solutions from various sources, some free of pollutants, and discharges these solutions to an outfall.

3. Develop control schemes

The description of the waste-stream flows and compositions suggests sites for controls. In Figure 1, the pollutant arises from only one reactor unit. Treatment or control systems could be placed either at this unit or near the plant's discharge point. The choice, as in any situation, will determine the volume of wastes to be handled and, consequently, the investment and operating costs.

In this example, the pollutants are present in a combination of continuous and intermittent flows, and the concentrations are variable. During reactor operation or rinse, the pollutant-containing flow from the reactor unit (including the scrubber) to the drain is 10 to 45 gal/min. At the plant discharge point, the same pollutants are contained in solutions flowing at up to 175 gal/min. In this example, it seems sensible to control the pollutant discharge at the source, where the volumes to be treated are smaller.

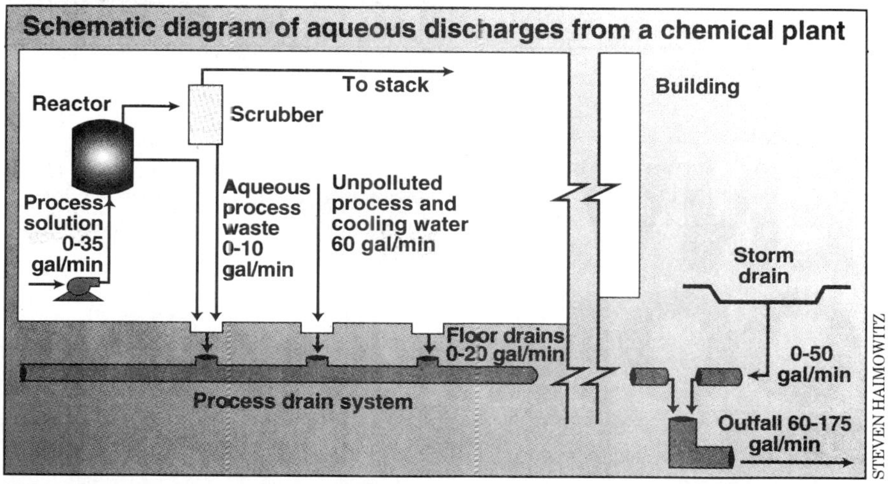

FIGURE 1. Aqueous solution, containing pollutants, that is discharged by the reactor-overhead scrubber can be treated (after it has become commingled with other liquid wastes) at the plant outfall, or it can instead be handled as-is at the process unit itself. In the example discussed in the text, the latter option proves to be the more cost-effective

4. Consider process changes

As already noted, major process changes require significant outlays of time and money, and usually require other justifications besides meeting permit requirements. On the other hand, minor process modifications that eliminate one or more waste streams may be feasible, and the engineer making a compliance analysis should be on the lookout for such possibilities.

Furthermore, if broader company strategies indicate that major process changes are likely to be made in the future, such intentions may influence the compliance strategy for the present. This point is discussed in more detail later.

5. Make a rough cost estimate

In most cases, the alternatives from Steps 3 and (if relevant) 4 utilize commercial equipment. So, the engineer can develop at least preliminary cost estimates.

In our example, the capital cost for the end-of-pipe system is estimated at $450,000. This is nearly three times higher than for the source-control system, because of the need for new infrastructure to collect the solutions, a flow-equalization system, and larger equipment to accommodate the higher flowrates.

Operating-cost difference between the options is estimated to be negligible; the major new cost is for chemicals, which is about the same in either case. Probably, no new labor would be needed for either alternative.

6. Is compliance assured?

As mentioned earlier, there is a basic difference between pollution control at the source (i.e., the process unit) and at one centralized downstream point. It concerns the responsibility for operating the treatment unit.

Treatment at the source implies that the control is the responsibility of the process-unit operator. Indeed, the control of the treatment module may be integrated into that for the process itself. In centralized treatment, the responsibility for consistent compliance is usually instead vested in a person who has that as his or her main (or sole) task.

The end-of-pipe system can cope with pollutants that arise from unforeseen sources. If there is any likelihood of occasional entry of such pollutants into the drain system, the end-of-pipe approach may be required to avoid violating the permit.

Engineers should not only be aware of such possibilities, but should also estimate the risks involved. A high risk of failures leading to non-compliance may justify a comparatively expensive treatment system.

7. Analyze the economics

Compliance alternatives may present considerably different trade-offs in investment, operating cost, and penalty exposure, so the comparative economics should be developed thoroughly. This usually takes into account any costs or savings associated with future environmental regulations.

Applying traditional economic-evaluation tools (e.g., Net Present Value analysis) to compliance situations involves considerations that do not regularly arise in conventional application of these tools. Accordingly, the rest of this article discusses the economic-analysis step in more detail. ❐

ECONOMIC ANALYSIS

There are three major steps in analyzing compliance options from an economic perspective. Adhering to them can help companies make better business decisions in selecting compliance systems, even without costly and complex engineering analyses.*

First, assess the alternatives from a total-company perspective, to uncover long-term opportunities that may be created or lost with each. Choosing a particular alternative today might create or eliminate important options for future compliance, or for changes in the manufacturing processes themselves.

Next, estimate the incremental cash flows associated with each alternative. As explained later, it is important to identify the incremental flows, rather than other common cost-engineering measures, such as allocated or accounting costs.

Finally, discount the cash flows to determine the present value of each alternative. This may require advanced option-valuation techniques, such as ones described in Reference [*3*].

* Many of the concepts discussed in this section may also be applied to valuing the "economic benefit" associated with delayed compliance with environmental laws. This concept becomes an element of penalty calculations for a wide range of regulations. For more details, see References [1] and [2].

Total-company perspective

Often, control technologies developed to satisfy one regulation can also help the plant to comply with other present-day ones, or to achieve compliance with stricter future regulations. Try to identify as many such options as possible. In our experience, options that provide valuable flexibility are often overlooked.

For example, one particular way for a plant to treat process wastewater might, at little additional capital or operating cost, also be able to handle small amounts of process waste that are now shipped offsite. Or, the same or a different option might also be able to meet future regulations concerning stormwater runoff.

The fast pace of technological change often creates future options that may be overlooked if the engineer thinks only in terms of today's technology. Solvents represent a good example.

Faced in the early 1980s with onsite treatment or offsite disposal of spent cleaning solvents, many companies installed costly onsite-treatment equipment, correctly expecting dramatic increases in offsite-disposal costs. However, more-knowledgeable engineers or consultants would have also taken into account the likely switchover to solventless, aqueous cleaners in the future. This may well have given the nod to offsite treatment, even though it was less attractive in the short run.

Once future options have been identified, a probability should if possible be assigned to each outcome. With these probabilities, the options and their expected cash flows can be incorporated directly into an expected-value analysis, using a probabilistic approach illustrated later. Our examples have been simplified to illustrate the concepts; in actual applications, these analyses can be extremely complex.

One must also consider the effect of each compliance alternative on product quality and manufacturing cost. This may best be done by staying in touch with the firm's product managers. They may be aware of expected changes in the manufacturing process, which might entail a different set of waste streams.

Incremental cash flows

Managers and engineers often look at project costs in accounting or traditional cost-engineering terms (for in-

EFFECT OF CHOSEN DISCOUNT RATE

Year	15% DISCOUNT					8% DISCOUNT				
		SYSTEM 1		SYSTEM 2			SYSTEM 1		SYSTEM 2	
	Present Value Factor $(1+r)^t$	Cash Flows ($)	Present Value ($)	Cash Flows ($)	Present Value ($)	Present Value Factor $(1+r)^t$	Cash Flows ($)	Present Value ($)	Cash Flows ($)	Present Value ($)
0	1.0000	10,000,000	10,000,000	6,000,000	6,000,000	1.0000	10,000,000	10,000,000	6,000,000	6,000,000
1	1.1500	600,000	521,739	1,250,000	1,086,957	1.0800	600,000	555,556	1,250,000	1,157,407
2	1.3225	612,000	462,760	1,275,000	964,083	1.1664	612,000	524,691	1,275,000	1,093,107
3	1.5209	624,240	410,448	1,300,500	855,100	1.2597	624,240	495,542	1,300,500	1,032,379
4	1.7490	636,725	364,049	1,326,510	758,436	1.3605	636,725	468,012	1,326,510	975,024
5	2.0114	649,459	322,896	1,353,040	672,700	1.4693	649,459	442,011	1,353,040	920,856
6	2.3131	662,448	286,395	1,380,101	596,656	1.5869	662,448	417,455	1,380,101	869,698
7	2.6600	675,697	254,020	1,407,703	529,208	1.7138	675,697	394,263	1,407,703	821,381
8	3.0590	689,211	225,304	1,435,857	469,384	1.8509	689,211	372,359	1,435,857	775,749
9	3.5179	702,996	199,835	1,464,574	416,323	1.9990	702,996	351,673	1,464,574	732,652
0	4.0456	717,056	177,245	1,493,866	369,261	2.1589	717,056	332,135	1,493,866	691,949
Totals		16,569,832	13,224,691	19,687,151	12,718,108		16,569,832	14,353,697	19,687,151	15,070,202

Proper selection of discount rate can be essential for choosing correctly between environmental-compliance options. In this example (discussed in the text), an inappropriate 15% rate favors System 2, whereas the more-realistic 8% rate gives the nod to System 1

stance, by including overhead expenses). Such approaches are inappropriate for evaluating compliance alternatives. Instead, as when evaluating any proposed investment, the analyses should be based on only the incremental cash flows associated with each measure: the cash outlays or savings that would occur as a result of each alternative, and that otherwise would not have been realized. Be sure to incorporate every incremental cost savings or outlay, including such not-obvious matters as materials savings or heat recovery.

Allocated costs need scrutiny, and should be excluded if they are not truly new cash costs that will arise as a result of the project being analyzed. In many instances, for example, little or no incremental labor may be required for a particular pollution-control system, even though labor might be allocated to the activity for internal accounting purposes.

The incremental analysis should lay out, year by year, the cash outflows and inflows of the alternative being considered. Typically, this includes an initial capital-cost outlay, annual operating and maintenance (O&M) costs, and material or other savings (if any). If the equipment is expected to have a salvage value or decommissioning cost, it should be included. The life of the equipment must also be considered, and outlays provided for replacement anticipated.

As discussed in the next section, the timing of each cash flow is important and care should be taken to place each outlay or receipt in the appropriate period. Enter the capital costs in the year they occur rather than allocating them over time as happens in many accounting systems.

Include the tax consequences of each investment. Among these are cash flows associated with depreciation or expense tax shields, and any investment or other tax credits.

Determining present value

To place cash flows that occur in different future years on a comparable basis, employ discount rates. Future cash amounts are discounted back to a given date (usually the present) via a discount rate that reflects the time value of money and the risk of the cash flows at issue. The relationship is the well-known present-value equation

$$PV[C_t] = C_t/(1 + r)^t$$

where $PV[C_t]$ is the present value of C_t, the cash flow in year t, and r is the annual discount rate; t is expressed in years from the date of the calculation. Present values obtained via this equation are tabulated in finance handbooks for a wide range of discount rates and time periods.

The net present value of an investment is the sum of all of the cash flows involved. This includes the initial outlay, annual O&M costs and any revenues or savings created.

According to a key axiom of modern finance theory, the appropriate discount or interest rate for a particular situation is the prevailing cost of capital for that kind of situation when its riskiness is properly taken into account [4]. Cost of capital is defined as the expected rate of return in capital markets on alternative investments having equivalent risk [5, 6]. While measuring the precise cost of capital for pollution control investments may be difficult, *it is generally likely to be considerably lower than the overall cost of capital for the company as a whole*. This is because pollution-control expenditures or savings are inherently low in risk.*

Ignoring this difference and thus using too high a discount rate has an important consequence. The resulting analysis will tend to understate the net present value of annual operating and maintenance costs. This will bias the decisions in favor of less-capital-intensive projects.

Consider this simple example. The ABC Company, reviewing its options for a new pollution-control system, has narrowed the choice to two. System 1 requires an immediate capital investment of $10 million and annual O&M costs of $600,000 in today's dollars. System 2 requires a much smaller capital investment, only $6 million, but has annual O&M costs of $1.25 million in today's dollars. For simplicity, assume that all O&M costs are incurred at the end of each period. Inflation is expected to be 2% annually over the ten year life of the project. The company's weighted-average cost of capital is 15%.

Using the present-value equation, the company determines the present value of the two cash flow streams as shown in the table on the previous page. The outlays required for System 1 have a net present value of $13.2 million, and those for System 2, $12.7 million. Because System 2 appears to have a lower overall cost, it would be chosen.

But the appropriate discount rate for this project is virtually sure to be lower, because the risk associated with pollution-control expenditures is lower than that for the company's expenditures as a whole. Table 2 demonstrates that if the same cash flows are valued using an 8% discount rate (arbitrarily chosen for this example), System 1 becomes the better choice, as the present value of its required outlays is now smaller.

While this example ignores tax effects, a more-complete analysis would consider the after-tax cash flows, including the firm's ability to write off annual O&M charges to expenses as well as the depreciation tax shields created by capital investments. Be sure that if after-tax cash flows are used, the discount rate is also calculated on an after-tax basis.

Elusive discount rates

Appropriate discount rates for environmental projects are hard to develop with any precision. If the engineer is unable to come up with an accurate rate, sensitivity analysis can be used to identify those situations in which the discount rate is critical to the decision.

In many cases, a single test can identify the most cost-effective solution without a precise discount rate. The first step is to calculate the discount rate that equalizes the net present values of the alternatives under consideration. Then, compare this calculated rate with the company's financial specialists' educated (though imprecise) guess as to the appropriate discount rate for environmental projects in general. If these two rates are close to each

* The company's cost of capital is typically a weighted average of the firm's after-tax cost of debt and equity. The cost of capital thus measures the riskiness of the cash flows received by the holders of the company's debt and equity (interest and dividends or earnings). These cash flows are typically quite volatile (risky) relative to the cash flows that are expected to be paid for pollution-control investments, and therefore require a higher discount rate. For more on cost of capital, and risk measurement, see [3].

other, then the right decision between the projects unfortunately remains unclear. But if the two rates are not close, then a comparison of the projects' net present values calculated at the "educated guess" discount rate should lead to a reasonably reliable decision. A more complex analysis will compare the net present values over a range of discount rates to identify the ranges in which the decision is sensitive to the choice of discount rate.

Don't overlook options

It is important to consider the future options that may be created or lost with a particular investment. Suppose the XYZ Company has narrowed its alternatives to two. System A requires an initial $4-million investment and an annual O&M outlay of $250,000 in today's dollars. System B has no initial capital cost but requires annual O&M spending of $750,000 in today's dollars.

System B (unlike System A) gives the company the chance to switch after three years to a new technology, System C. This system would require an additional initial investment of $1 million but is expected to require only $75,000 of annual O&M, both figures being in today's dollars.

Company engineers see a 50-50 chance that System C will be commercially viable and appropriate for XYZ's waste stream; they also know that if it is not available after three years, it will never become available. Inflation is expected to be 2% throughout the life of the 15-year project. The appropriate discount rate is assumed to be 8%. Again, taxes are ignored to simplify the example. Which system should the company install?

An effective way to analyze this kind of problem is to map out the possible outcomes [*7, 8*]. Figure 2 illustrates a decision tree for this problem. Each of the alternatives is laid out along a branch, and probabilities are assigned where uncertainty exists.

In this case, System A is represented along the left branch while Systems B and C are to the right. For clarity, System B is broken down into two components: System B1 represents the first three years of the project, while System B2 represents the final twelve years in the event that System C is not chosen or is unavailable.

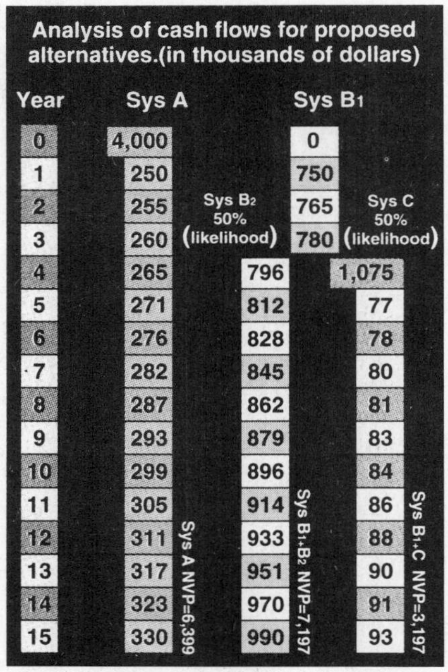

FIGURE 2. A graphic decision-tree approach (typified above, with present values omitted for simplicity) often aids the analysis of process situations entailing uncertainty. The findings of this analysis are summarized in the bar graphs below.

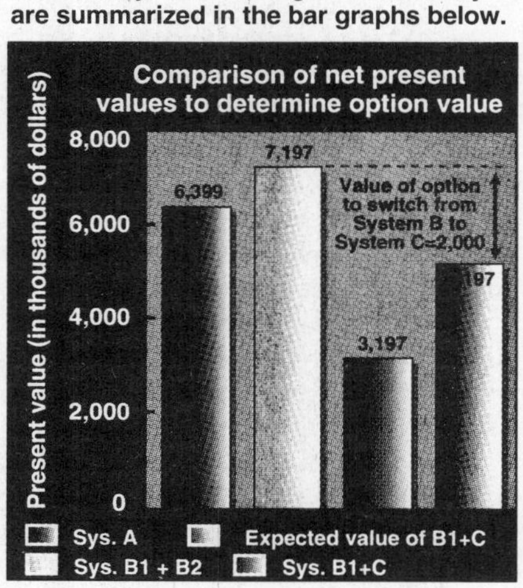

Each box shows the cash flow (outlay) for the given year. The flows have been corrected for inflation but, for visual simplicity, their present values are not shown.

A simple net-present-value (NPV) analysis of these data, ignoring the possibility of switching from System B to System C, reveals that the NPV of System A's outlays is $6.4 million while that of System B's is $7.2 million. This suggests choosing System A.

Now, however, take into account the possibility of making the switch. If the new technology, System C, becomes available and is appropriate for XYZ's waste stream, the NPV of the outlays for System B1 for three years followed by those for System C in the remaining twelve years is $3.2 million.

Since there is a 50% chance that System C will be implemented, the expected value of the combined System B1 and System C is {[50% × $7.2 million (if System C is not available)] + [50% × $3.2 million (if System C is available)]}, or $5.2 million. The value of the option to install System C after three years is therefore ($7.2 million − $5.2 million), or $2 million.

This example illustrates the most simple case, where investments and payoffs occur discretely and there is only one decision point. In practice, options may be much harder to value, especially if the costs involved are subject to large uncertainties, and decision trees can reach hundreds or even thousands of branches.

Furthermore, the option-valuation methods employed in practice are more complex than the approach shown here. Discussion of these methods is beyond the scope of this article. Instead, the aim here is to sensitize the engineer to the likely need for option-valuation in real-life situations.

Who pays for the R&D?

This second example illustrates a situation in which a delay in major capital investment in pollution control systems has overall economic benefit. Note, however, that the cash flows assume that the development costs of the future technology (System C) are borne by its vendors. In many other cases, the client company itself may have to invest in technology development.

To take account of the latter circumstance, the analysis could incorporate the estimated R&D outlays. By reviewing their present values, the engineer can assess the justification for this spending.

In summary, engineers evaluating compliance alternatives can face difficult decisions in selecting a system. Looking for creative solutions and preparing a careful economic analysis can often have a big payback. Projects should be evaluated from a companywide perspective, and important options associated with each system identified to achieve compliance with the most cost-effectiveness possible. ■

Edited by Nicholas P. Chopey

References

1. Wise, K.T., Ammann, P.R., others, EPA's 'BEN' Model: Challenging Excessive Penalty Calculations, *Toxic Law Reporter*, May 6, 1992
2. Wise, K.T., Maniatis, M.A., others, EPA's New 'BEN' Model: A Change for the Better?, *Toxic Law Reporter*, Feb. 24, 1993
3. Brealey, R.A., and Myers, S.C., "Principles of Corporate Finance," 3rd ed., Chap. 9, McGraw-Hill, New York, 1988
4. Brealey, R.A., and Myers, S.C., *op. cit.*, Chap. 9
5. Brealey, R.A., and Myers, S.C., *op. cit.*, Chap.7
6. Kolbe, A.L., and Read, J.A., Jr., with Hall, G.R., "The Cost of Capital," Chap. 2, MIT Press, Cambridge, Mass., 1984
7. Koch, G.S., Ammann, P.R., and Wise, K.T., Evaluating Environmental Costs: Accounting for Uncertainties, *Chemical Waste Litigation Reporter*, Vol. 27, No. 3, pp. 554-559, February 1994
8. Koch, G.S., Ammann, P.R., and Wise, K.T., Using Decision Analysis to Manage Environmental Costs, presented at 87th Annual Meeting of the Air & Waste Management Assn., June 1994, Cincinnati, Ohio

The authors

Paul R. Ammann is a Principal with Brattle/IRI, 30 Church St., Cambridge, MA 02138; Tel. (617) 864-7900. He has over 30 years of experience in the development, costing and application of environmental, chemical and metallurgical technologies. In the past decade, he has focused on environmental consulting in the areas of non-compliance matters, Cercla cost-recovery reviews, environmental-liability evaluation, R&D planning, and technical and market-feasibility analysis. Mr. Ammann holds B.S. and M.S. degrees in chemical engineering from the Massachusetts Institute of Technology.

Gayle S. Koch, also a Brattle/IRI Principal, is experienced in evaluating environmental liabilities in connection with settlements, insurance arbitrations, mergers and acquisitions, real estate transfers, litigation, and strategic planning. She has also conducted detailed market assessments for numerous new technologies. Ms. Koch holds a B.S. in chemical engineering and an M.S. in management science from the Massachusetts Institute of Technology, as well as a B.S. in that university's Science, Technology and Society Program.

M. Alexis Maniatis, an Associate with Brattle/IRI, has dealt with a wide range of environmental and financial issues, including work with companies involved in non-compliance penalty matters to identify cost-effective compliance strategies and calculate economic benefits. He has also helped companies evaluate ability-to-pay issues. Mr. Maniatis holds a B.A. in economics from Wesleyan University and a Master's degree in Public and Private Management from Yale University.

Kenneth T. Wise, a Principal of Brattle/IRI and director of the firm's environmental practice, has worked on environmental issues for 20 years. He has testified in litigation on recovery of hazardous-waste-cleanup costs, impacts of hazardous waste sites on property values, allocation of Superfund liabilities, and challenges to non-compliance penalties. He has also helped companies value insurance-recovery settlements and potential litigation risk. He holds a B.S. in physics from Harvey Mudd College and a Ph.D. in economics from the Massachusetts Institute of Technology.

PRODUCT STEWARDSHIP SHIFTS INTO

Gregory G. Bond,
The Dow Chemical Co.

The chemical process industries (CPI) have known about product stewardship for a long time. Recently, however, its importance has been growing rapidly in response to public and governmental concern about the effects of CPI products on people's health and the environment. The whole stewardship discipline is currently being redefined.

Initially, stewardship focused on potential hazards posed by the manufacture and handling of chemical products. Today, its scope is expanding to encompass the cradle-to-grave life cycle of a product, from sourcing of the raw material to disposal of consumers' end products. And environmental concerns are becoming a prominent component of the health and safety emphases of stewardship.

Developing a full-scale product stewardship program is a greater challenge than ever before. For example, the program is likely to call for change as more environmentally responsible methods are identified — not only in production and handling, but in all areas associated with CPI products.

Fortunately, with the growing challenge, many increasingly effective tools are also available for developing a stewardship program; this article discusses several of the most-important ones. Which combination you use will be determined by your product and its applications.

Responsible Care guidelines

The first tool in your kit is the industry-wide Responsible Care initiative. It has been adopted by the Chemical Manufacturers Assn. (CMA) to encourage all chemical producers to enhance their environmental, health and safety (EH&S) performance. Adherence to Responsible Care is a condition of membership in CMA. The initiative includes detailed guidelines and goals to help you develop a workable program for your company.

Included in the initiative are six Codes of Management Practice, designed to continuously improve manufacturers' EH&S performance, and standards for manufacturers to responsibly assist customers with their products. These codes address the issues typically examined by the public vis-à-vis chemical manufacturers and their clients.

These codes outline actions in specific operational areas, with the long-term objective of

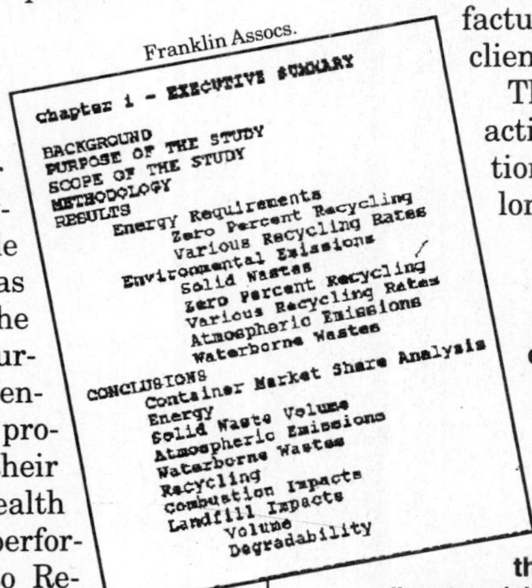

Franklin Assocs.

Chapter 1 - EXECUTIVE SUMMARY

BACKGROUND
PURPOSE OF THE STUDY
SCOPE OF THE STUDY
METHODOLOGY
RESULTS
Energy Requirements
Zero Percent Recycling
Various Recycling Rates
Environmental Emissions
Solid Wastes
Zero Percent Recycling
Various Recycling Rates
Atmospheric Emissions
Waterborne Wastes
CONCLUSIONS
Container Market Share Analysis
Energy
Solid Waste Volume
Atmospheric Emissions
Waterborne Wastes
Recycling
Combustion Impacts
Landfill Impacts
Volume
Degradability

FIGURE 1. Sample pages from life cycle inventory (LCI) documentation discuss the LCI purpose, scope, methodology and results. This particular document is 70 pages long (left, and throughout the article). Safe-handling literature (photo, top) is given to each product customer.

Source: Dow Chemical

FIGURE 2. The PSR procedure (right) hinges on interaction between the product management team and the review board

continuously improving company performance. Each code contains a number of management practices (106 in all) defining operational disciplines that CMA members must follow to achieve compliance. The six management codes are:

- Community awareness and emergency response (CAER)
- Pollution prevention
- Process safety
- Employee health and safety
- Distribution
- Product stewardship

Risk-management practices. These are practices that work interactively at each stage of a product's life cycle. They also extend the risk-management process beyond traditional boundaries to include suppliers, contractors, distributors and customers.

Product stewardship review

Another effective tool for the development of your program is the product stewardship review (PSR) process, which Dow has used for more than a decade (Figure 2). We have found PSR an efficient and effective means for identifying, examining and managing critical risks associated with new and existing products and services, and for continuously improving stewardship programs.

CPI companies are redefining how they care for their products ®

HIGH GEAR

The Product Stewardship Code brings the complete life cycle of each chemical product under scrutiny. Its intention is "to make health, safety and environmental protection an integral part of designing, manufacturing, marketing, distributing, using, recycling and disposing of [chemical] products." It also requires members of CMA to provide EH&S information about their products at any point during their chain of distribution. The Code's 12 management practices are grouped into three categories, which are discussed below.

Management leadership and commitment. This first category describes a series of practices that gives direction, provides resources, sets priorities and establishes responsibilities to create the foundation for successful product stewardship.

Information and characterization. These practices address the CPI's need to continually increase the body of knowledge that surrounds chemical products and their handling. The overall goal is to improve hazard identification and risk characterization at every step in a product's life.

PSR, a critical component of stewardship, provides an in-depth life cycle analysis of what is currently going on in each product business from the EH&S point of view. It also accurately assesses what is likely to go on in the next three to five years, what the product business group is doing to manage risks, and what it can and should do better. PSR analyzes the product's future only in the near term (other company processes study the longer term).

PSR board. A key to the successful implementation of PSR is the product stewardship review board (PSRB). In accordance with Management Practice No. 1 of Responsible Care's Product Stewardship Code, Dow's PSRB includes experienced senior managers from varied backgrounds. Board membership is cross-functional, including people both directly and indirectly associated with the product business under review.

Members of the board include representatives from public affairs, marketing, research and development, mammalian and environmental toxicology, legal services and manufacturing. Other functions, such as distri-

LCI GUIDELINES

When formulating your LCI report, use these basic rules:

- Use English or Metric units, but don't mix them
- Base data on a one-year average of performance
- Use average data (or a range of data) from individual plants, weighted by production rate
- Document the basis for all data. If data are estimated, reference the calculation method
- If there are coproducts, allocate these on mass balance; i.e., use mass units, not monetary value or moles. Allocate generated steam and power on an energy basis
- Specify electrical energy sources, and break these down by type (i.e., gas, coal, cogeneration, hydro or nuclear) to establish waste emissions and efficiencies. Contractors are especially well-suited to develop these data
- For constituents of products that are present at less than 1%, list and check the substances for unusual impact properties (e.g., toxicity). Although Dow includes data on substances present at less than 1%, these data should have minimal overall effect on your LCI, and can usually be omitted
- Specify water-supply source. Also specify usage amounts and costs for treatment, waste and energy, using industry-average values. Again, this data collection is usually handled well by outside contractors
- Identify and quantify waste and emission components

In the final LCI report, include the following elements:

- A statement of the purpose of the LCI
- The scope or boundaries of the inventory
- A process flow sheet, with indications of energy input and waste output
- The source of the data and time period covered
- The LCI data itself
- Assumptions, exclusions and limitations, with their reasons
- References for methods used in estimating fugitive emissions
- Copies of secrecy agreements
- Identification of report recipients

Finally, extensive peer review should be given to the LCI report. If you intend to issue the report publicly, consider additional outside peer review to improve the quality and credibility of the LCI. ❒

bution, may be included as necessary.

The board meets each month to review one or more groups of products. How frequently each product is examined depends on its level of hazard. The most-hazardous products are reviewed more frequently than others, about once a year, to ensure that appropriate attention is paid to risk management.

A major responsibility of the board is to assist line management with stewardship responsibilities. In addition, the board advises each product business team, helping it identify long-term issues and opportunities, so the teams can devise effective means of managing them. In the course of the PSR process, the board raises the business team's awareness of public-policy issues that should be factored into future planning.

PSR meetings have resulted in better EH&S practices and information sharing about product hazards (Figures 3 and 4). For example, some product-handling standards have become stricter. Board recommendations have included: increased industrial hygiene monitoring at customer locations, to ensure that potential exposures are being properly minimized; additional product toxicology testing; and changes in material safety data sheets (MSDSs) and product brochures.

In a few rare instances, a board meeting has determined that the risks involved in manufacturing a specific product outweighed the benefits. In these cases, the company discouraged certain applications for the product, or abandoned the product altogether.

FIGURE 3. Procedures for sampling chemical shipments from tank cars are one of the many topics spelled out in product stewardship literature

Dow Chemical

FIGURE 4. Chemical plant safety is a major consideration of a product stewardship program. Here, a warning sign alerts personnel to potential hazards in Dow's plant complex (Midland, Mich.)

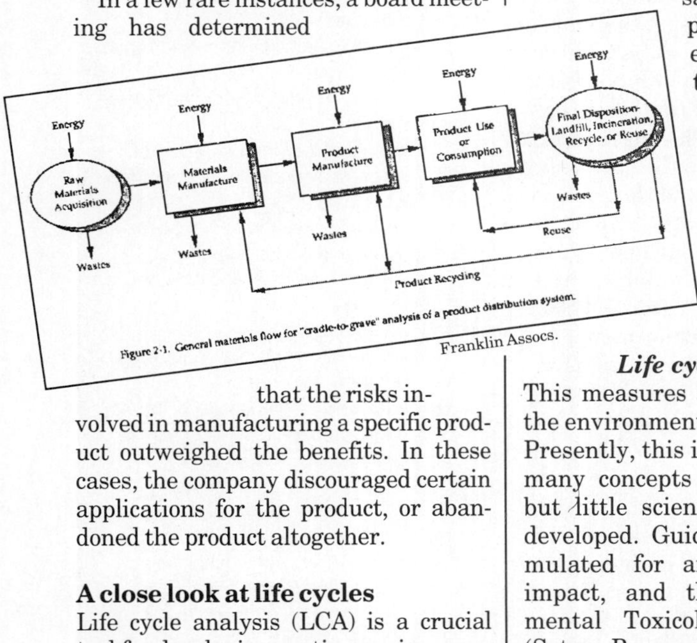

Franklin Assocs.

A close look at life cycles

Life cycle analysis (LCA) is a crucial tool for developing continuous improvement that can also be used to formulate a stewardship program. However, LCA is much more talked about than actually used in the CPI. Many companies are aware of the need for LCA, but don't know how to conduct it. Some analysts believe that it will take another five years before LCA is used properly and to its fullest extent in the CPI. A complete LCA consists of three separate and interrelated components. These are detailed below.

Life cycle inventory (LCI). Many people mistakenly call this LCA, but it is only the first step in the LCA process (Figure 5, p. 81). Traditionally, this is a close study of a product's transportation, storage, safe handling, health and safety profile and disposal. LCIs have now expanded to include the product's raw material sourcing, energy use and incorporation in consumer end products (Figures 1 and 6, pp. 78 and 82). The inventory focus depends on the product itself and its major applications.

Life cycle impact analysis. This measures a product's impact on the environment and on public health. Presently, this is a gray area in which many concepts have been proposed, but little science has actually been developed. Guidelines are being formulated for analyzing a product's impact, and the Soc. of Environmental Toxicology and Chemistry (Setac; Pensacola, Fla.) is establishing principles for life cycle impact

PRODUCT STEWARDSHIP DATABASES

Many useful databases are available to both producers and users of chemical products. The Chemical Manufacturers Assn. (CMA) provides some good services:

MSDSs. CMA has a database file of some 1.5 million MSDSs from member companies for use by the organization's Chemical Transportation Emergency Center (Chemtrec) and Chemical Referral Center. These documents are electronically transferred or scanned into the database as they are received from the chemical manufacturers, and are available in a graphic format.

In case of emergency, call the 24-hour hotline (800-424-9300) and receive a faxed MSDS in a matter of minutes. In non-emergency situations, call the Chemical Referral Center (800-262-8200), which is open from 9 a.m. to 6 p.m., Monday through Friday. A few MSDSs are restricted by the product manufacturer; in these cases, CMA refers the caller to the producer for service.

New database. CMA is planning a service to make MSDSs available on-line to your computer. This system, which will initially hold some 500,000 individual documents, is expected to start up in the third quarter of this year.

The U.S. EPA maintains several databases that are generally available to the public. These include:

Integrated Risk Information System (IRIS). This contains summaries of health-risk and EPA regulatory information on over 500 specific chemicals. It contains the EPA consensus opinion on potential, chronic human-health effects

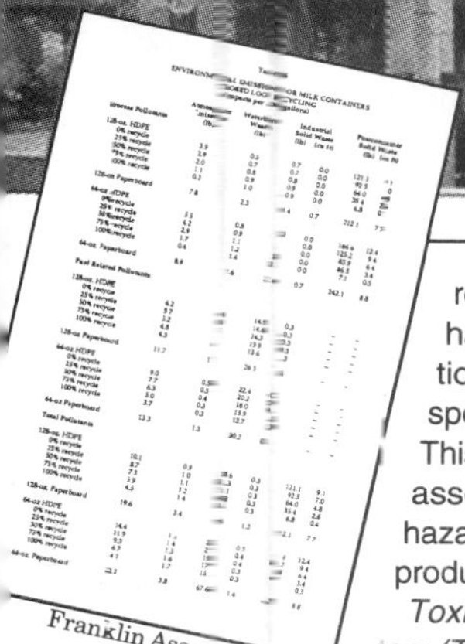

Franklin Assoc.

related to chemical-hazard identification and dose-response assessment. This can be useful in assessing the health hazard of a chemical product.

Toxic Release Inventory (TRI). This provides information on amounts of over 300 listed toxic chemicals that U.S. facilities release directly to air, water or land, or that are transported off-site. This information can help identify, among other things, chemical releases from downstream customers, and determine whether customers are using your products properly.

International Register of Potentially Toxic Chemicals (IRPTC). This is sponsored by the United Nations Environment Programme (UNEP). It contains information organized in "chemical data profiles" and covers a broad spectrum of subject areas related to hazard identification and risk assessment.

Several databases are available to help you develop a Life Cycle Inventory (LCI; box, this page). These contain inventories of individual materials, providing data on such topics as waste management, required energy, and sustainable levels of recycling. Among them are the Boustead Model (Dr. Ian Boustead, The Open University, E. Grinstead, England), Ecobalance (Rockville, Md.), and models developed by Franklin Associates, Ltd. (Prairie Village, Kan.) and Chem Systems (Tarrytown, N.Y.). ❑

analysis in North American industry.

Life cycle improvement analysis. This component is based on the results of the first two steps. It analyzes all the changes introduced into the product's sourcing, production, transportation, handling and disposal. This analysis must be performed as a continuing process. New information developed after changes are made must be fed back into the inventory and the impact analysis, to analyze and solve any problems generated by the changes.

At present, most LCA practitioners are carrying out only the first component, or at best the first and third, and introducing changes in processing and handling in response to the findings of the LCI. However, this situation is improving with the redefinition of product stewardship to include management issues for impact assessment.

Because state-of-the-art methodology for life cycle impact analysis is still incomplete, LCA is inadequate for definitively evaluating product risks and benefits. Managers should ensure that LCA findings are not used for propagandistic purposes, e.g., by environmental groups for "product bashing."

If you plan on undertaking an LCI, first consider what you intend to use it for. Ask these questions first: Are you willing to make major changes in your raw material sourcing, production methods, transportation, handling procedures or disposal, if necessary? Or do you plan only to improve your present practices, without making fundamental changes? Your intentions determine how you design your LCI.

According to Dow's "Global guidelines for life cycle inventory," an LCI starts with a statement of purpose. This statement defines how far into the past and future data must be gathered, how much detail must be developed, and who is the intended audience for the results. (Nat-

LIFE CYCLE INVENTORY (LCI) AND DATA FLOW

LCI coordinator collects data from (data collated in consistant format):

- Technology center
- EH&S department
- Power and utilities
- Plant
- Other sources

First draft LCI

- Attach example of data
- Review with LCI expert, plant engineers and technology center

Final LCI

- File report in technology center
- Fill out external questionnaire

First-draft data set

- Obtain consultant secrecy agreement
- Obtain approval of data set from manager and technology center, and correct if required

Final data set

- Send data set to panel chairman of LCI experts for review and computing

Consultant computing program

- Issue LCI report
- Have industry team review report

Industry LCI

- Database used by: authorities, institutes, universities and consultants to calculate LCAs on market applications

Dow-specific LCI

- Database used by Dow for product- and process-improvement analysis by plants and technology centers
- Dow internal LCAs on market applications
- Dow internal LCAs on research products and processes

Source: Dow Chemical

FIGURE 5. A life cycle inventory (LCI) includes exhaustive product data, which are rigorously examined

urally, an LCI for a stewardship program for a single product will require different parameters than an LCI for product comparison, where data relating to two or more different products or processes must be gathered.)

Next, determine whether the inventory information should be gathered by an outside consultant-contractor or by your own staff. Qualified consultants have some advantages. They can provide a uniform standard of rigor in the database they develop, particularly where your own product or process is being compared with a competitor's. Contractors may also be better able to obtain data for parts of the product life cycle that you and your plant staff cannot access in house.

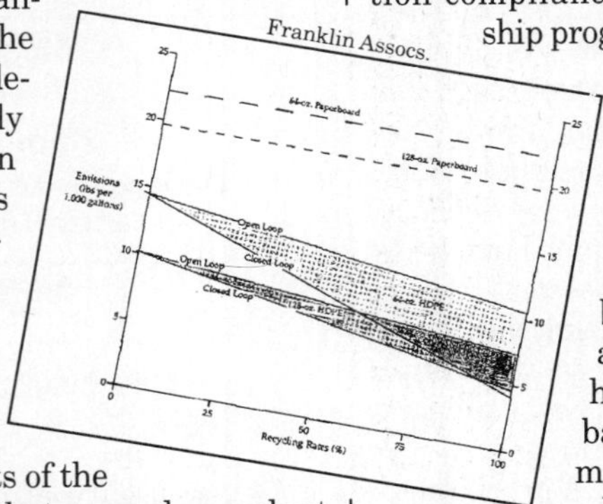

FIGURE 6. Sometimes called "resource and environmental profile analyses," LCI documents typically include flow charts and supporting data (above)

In addition, you may not have a sufficiently large staff to sacrifice people from regular production, sourcing or client service for such a time-consuming data-gathering process. Also, outside consultation can enhance the credibility of the findings among your target audiences.

Determining full costs

An item in the stewardship tool kit that holds great promise for the future is full-cost accounting. This includes several mechanisms to control all direct and indirect costs of a product or process, including the regulation-compliance costs, a full stewardship program, product emissions and energy consumption. The long-term goal of this discipline is to make sure these costs are traceable in the product price so that both the manufacturer and the customer can have a more-informed basis for the choices they must make.

Full-cost accounting is not yet adequately developed. So far, there is only limited basis for data, as is the case with the impact assessment component of LCA. Most practitioners are still at the preliminary phase of tracking indirect costs.

Industry must continue its dialogue in this area before full-cost pricing can be instituted. The main problem today is that any producer who raises prices to match the results of this accounting process will be at a competitive disadvantage, unless direct competitors are doing similar cost accounting.

Communicating with customers

The Responsible Care initiative stresses the importance of transferring product stewardship information to customers. In most companies, customer communication is principally conducted by field sales staff. Consequently, these people must be trained in product stewardship. A good stewardship program can give a producer a competitive edge, which salespeople can use in their customer contact.

Sales training. At Dow, all new field salespeople receive stewardship training as part of the company's commercial development program. In addition, they are periodically retrained at meet-

A PRODUCT-STEWARDSHIP CASE HISTORY

Ethyleneamines are used in many processes to make such products as fabric softeners, chelating agents, epoxy resins and fuel additives. The raw materials are furnished by Dow, and most sales are direct. Consequently, there is little need to review the stewardship processes of suppliers or distributors. The principal audience for all stewardship technology transfer is the chemical industry, since the product's major use is as a chemical reactant.

Ethyleneamines can be both corrosive and highly reactive. One aspect of product stewardship is to provide the customer with information on possible reactions. Because ethyleneamines are corrosive, they require astute handling procedures, for which Dow provides training. MSDSs and other safe-handling literature and videos are also provided to all customers.

Three product stewards are assigned to this product line; one each is responsible for activities in North America, Europe, and Latin America and the Pacific. The three work in concert with a chief steward, who is also responsible for several other product lines.

The ethyleneamines line does not yet undergo life cycle analysis (LCA). However, LCA concepts are applied when the product line is reviewed every three years as part of the company's PSR process. This process surveys every aspect of the product's health, safety and environmental impact, and determines what stewardship practices and tools are appropriate.

As a result of the recommendations resulting from the PSR process, Dow's product stewards are working with the American Industrial Hygiene Assn. (AIHA) to set workplace exposure limits on ethyleneamines. They are also developing validated analytical methods for industrial hygiene measurements and monitoring. Dow already has well-developed analytical methods for several of the more-hazardous ethyleneamines, but still needs to develop or refine methods for others. In addition, company scientists are studying the biodegradation of ethyleneamines as a means of modeling their fate in the environment.

The product stewardship program is communicated directly to ethyleneamine customers by the stewards and Technical Service and Development (TS&D) professionals. These are experts who ensure that clients' facilities are appropriately equipped to handle Dow products. For example, they may inspect a new customer's off-loading installations before delivering the first shipment of product and reinspect them periodically. They work with customers on ways to recycle used product.

In addition, TS&D workers give a safe-handling presentation with a question-and-answer session, to customer employees (if necessary, at all three shifts). They also leave behind a safe-handling video, which covers the material of the presentation, and distribute a literature piece giving recommendations on glove and respirator selection.

Finally, the TS&D staff's emergency telephone number, listed on the MSDS, can be called in case of an accident, such as a spill or emission. Experts, available 24 hours a day, talk customers through the handling of each accident. A steward can also be contacted for higher-level assistance during the emergency. ❐

ings of the sales districts and zones and through written communications, such as product stewardship bulletins.

Customer seminars. Dow holds regulatory seminars to update clients on developing regulations that govern specific products. These seminars provide a technical exchange and communicate the chemical producer's expertise in regulatory affairs, safe-handling practices and issues management. Customers want more than just a product: They want the supplier's expertise and support as well.

In many cases, customer seminars can be carried out through trade associations. For example, in the field of chlorinated solvents, the Center for Emissions Control (CEC; Washington, D.C.), which represents manufacturers, sponsors workshops for solvent users and cosponsors scientific meetings with the U.S. Environmental Protection Agency (EPA). The Chlorine Chemistry Council (CCC; Washington, D.C.) is planning similar activities for chlorine issues.

MSDSs. In addition, some chemical producers have other ways of making information on their products easily available to customers. Under the Occupational Safety and Health Administration (OSHA) Hazard Communication Standard in the U.S., and Canada's Workplace Hazardous Materials Information System, producers are required to supply their material safety data sheets (MSDSs), in their most-current updates, to distributors and customers. This practice is very labor intensive, requiring great quantities of paperwork, unless the documents can be made available electronically via database (box, p. 80).

One such database program is electronic data interchange (EDI), which was established by Dow several years ago as an American National Standards Institute (ANSI) standard and was first implemented between Dow and one of its major distributors, Van Waters and Rogers (VW&R; Seattle, Wash.). Today, Dow's MSDSs are stored in VW&R's system for transmission to regional offices as needed for further distribution to downstream customers. And, as the MSDSs are modified (Dow has a 30% annual revision rate), changes are automatically transmitted. Since the program was

DATABASE PRODUCTS

For more information on products and services, please circle the appropriate number on the Reader Service card

MSDSs	**625**
MSDS online	**626**
IRIS	**627**
IRPTC	**628**
Boustead Model	**629**
Ecobalance	**630**
Franklin Assocs.	**631**
Chem Systems	**632**

implemented by the two companies, it has expanded to other distributors and to customers who can receive the information electronically.

The role of product stewards

A particularly effective vehicle for technical exchange is the work of the product steward, who is responsible for the development and application of stewardship for a product or line. At Dow, for example, every product or line has at least one assigned steward, who works with a chief steward, of whom there are 12 in the company's North American operations.

The chief stewards are responsible for coordinating programs companywide. This includes seeking "synergies" among product lines, following long-term issues, such as toxicology studies, and managing industry-trade association issues. The stewards, on the other hand, work directly with customers and are responsible for transferring stewardship technology directly to the users of the products. Their concentration centers on a product's manufacture, transportation, handling and use by the customer.

A good deal of cooperation takes place among the stewards — not only among those assigned to the same line, but also with those working in other areas. On a global basis, the work of the stewards is coordinated by the company's global coordination team, which has representatives from each geographic area. This team's work is based on the philosophy that any stewardship principle learned in one area may apply elsewhere.

Distributor communications

For guidance on how to apply stewardship to distributors, refer to the activities of CMA or the National Association of Chemical Distributors (NACD; Washington, D.C.). Several main distributors are members of CMA and are therefore implementing Responsible Care. Many others belong to NACD, which requires all its members to adhere to its Responsible Distribution Process code. This code for chemical distributors is analogous to Responsible Care for chemical producers.

The most-recent development in NACD is a call for third-party auditing of members to assure their compliance with the Responsible Distribution Process. This mechanism, which aims at getting an unbiased assessment of distributor activities, is scheduled to start this year. Current members will be subject to third-party audit, while new members will be allowed five years of self-assessment to put all of their stewardship practices in place before auditing begins.

The assessment methods currently in use by NACD members have been developed in cooperation with Dow and other major suppliers. Over the past several years, Dow has conducted a distributor product stewardship survey program with its distributors. The procedures developed in this program provided the basis for the CMA's Distributor Self-Assessment protocol and NACD's Implementation Questionnaire. These will form the basis for third-party audits.

Stewardship: A holistic approach

Product stewardship is growing in importance to the CPI. The concept of cradle-to-grave stewardship, embracing all activities with a product, is the way of the future.

However, not all the components discussed in this article are necessary for every stewardship program. One mix of tools was used for Dow's ethyleneamine line (box, p. 82). Because of the complexities inherent in the stewardship concept, it is important for companies to examine all available tools and to use any that apply to their products. ■

Edited by Irene Kim

The author

Gregory G. Bond is global director of product stewardship for Chemicals & Performance Products, Hydrocarbons and Energy, The Dow Chemical Co. (Midland, Mich.). He also directs Chemicals Technical Service and Development and Environmental Affairs for Dow North America. He received a Ph.D. in epidemiology, a B.S. in cellular and molecular biology and a Master's in public health from the University of Michigan (Ann Arbor). The author of over 45 peer-reviewed journal articles on epidemiology research, he was elected a Fellow of the American College of Epidemiology in 1988. He has served as an advisor to various governmental agencies.

DEALING WITH LEAD IN THE WORKPLACE

STEPHEN WALSH

Kirk Sweetland
Environmental
Science & Engineering

New U.S. regulations target lead exposure for CPI construction and services

There is a new workplace safety standard on lead that will concern safety, health and environmental managers throughout the CPI. The standard was published in the May 4, 1993, Federal Register by the U.S. Occupational Safety and Health Administration (OSHA), and made effective on June 3, 1993. The new standard "applies to all occupational exposure to lead in all construction work in which lead, in any amount, is present in an occupationally related context."

Language in the standard clearly specifies a number of common CPI activities, including: petroleum and water tank repainting; underground storage-tank demolition; indoor and outdoor industrial facility maintenance and renovation; process equipment maintenance; and electrical cable splicing. Construction or retrofit projects throughout the CPI will inevitably be affected.

OSHA issued the new standard under Title X of The Housing and Community Development Act of 1992. Because of the conditions under which it was promulgated, it has the effect of a final OSHA standard (although the official version of that will not appear until much later). As such, the interim standard was not subject to the normal review and comment process typical of other OSHA standards.

Except for engineering controls (such as ventilation and containment systems), companies were required to comply with the standard at the beginning of last August. The engineering-controls deadline was October 1. States and territories with their own Federal-OSHA-approved occupational safety and health plans* were scheduled to complete a comparable standard November 4.

OSHA estimates the cost of compliance with the regulation at $750 to $3,600 per worker, per year, for a total annual cost of over $400 million. This figure assumes that a previous 200 $\mu g/m^3$ exposure standard was already being met.

Who is covered

The new standard essentially adapts the OSHA General Industry Standard for Lead (29 CFR 1910.62) and adds a new section, 1926.62, to Subpart D of Part 1926, covering employee protection requirements for construction workers exposed to lead. "Construction work" is defined by the standard as work for construction, alteration or demolition, including painting, salvaging and maintenance.

The new standard substantially changes the Permissible Exposure Limit (PEL) for lead concentrations in air to an 8-h time-weighted average (TWA) of 50 $\mu g/m^3$, and an "action level" of 30 $\mu g/m^3$, 8-h TWA. The action level mandates these employer duties: 1) performing the initial exposure assessment; 2) identifying compliance methods; and 3) instituting or installing worker protection practices and equipment.

Further, there is a list of worker tasks that presumes that workers are exposed to certain levels of airborne lead (Table). These tasks require specified worker protection until actual exposure levels are determined

The standard requires that workers with elevated blood lead levels be removed from work activities involving lead exposure. Specific benefits must be provided for employees who are so removed.

Many other administrative details will need attention. These include: sig-

*These states and territories are Alaska, Arizona, California, Connecticut, Hawaii, Indiana, Iowa, Kentucky, Maryland, Michigan, Minnesota, Nevada, New Mexico, New York, North Carolina, Oregon, Puerto Rico, South Carolina, Tennessee, Utah, Vermont, Virginia, Virgin Islands, Washington and Wyoming.

Originally published December 1993

HIGH-EXPOSURE TASKS		
1. < 500 $\mu g/m^3$; PF = 10	**2. > 500 $\mu g/m^3$; PF = 25**	**3. >2,500 $\mu g/m^3$; PF = 50**
Sites where lead-containing coatings or paint are present Manual demolition of a structure (example: dry wall) Manual sanding Heat gun applications Applications involving dust-collection systems	Use of a lead-containing motor, lead burning Surfaces with lead coatings or paint are present Rivet busting Power tool cleaning Clean-up activities where dry expendable abrasives are used Movement of abrasive blasting enclosures	Abrasive blasting Welding Cutting Torch burning

OSHA specifies certain tasks where the lead exposure level is assumed. This table shows the categories of tasks, along with the protection factor (PF) of the respiratory equipment needed

nage at work sites, training sessions, and access, for employees, to records.

Attaining an action level

The 30-$\mu g/m^3$ action level is determined by a test in which a worker is monitored for a full shift. Employee exposure at or above the action level for just one day a year triggers compliance action.

When the work activity is known to be above the action level, further testing is called for, including exposure tests and biological monitoring. The latter is primarily a blood test for the presence of lead and zinc protoporphyrin, a common analytical method. More-frequent and more-thorough exams are required when employees are exposed to the action level for more than 30 days per year.

Below the action level, the employer is not obligated to comply with most of the standard. However, retesting should be performed whenever there are changes in the equipment, process, controls, personnel or tasks that may result in employees' exposure.

The new permissible exposure limit (PEL) for lead in air has been reduced from 200 $\mu g/m^3$, 8-h TWA, to 50. The initial determination of lead exposure represents what would occur if the employee were not using a respirator. Alternatively, these initial determinations can be based on data from previous tests.

The standard assumes that the high-exposure tasks subject employees to lead levels above the PEL until demonstrated otherwise by the exposure assessment. Thus, employers are required to implement interim protective measures for employees until the exposure assessments are performed and to notify employees within five days of completion of an exposure assessment.

There are various requirements for respiratory and other personal protective measures, which employers must provide at no cost. Respirators are to be used whenever the airborne levels of lead are at or above the 50-$\mu g/m^3$ PEL, and where engineering and work-practice controls are insufficient to reduce the levels below the PEL. OSHA has developed criteria for evaluating various types of respirators; these criteria are based on the "protection factor" (PF) of the respirator.

The standard requires employers to use engineering and work practice controls "to the extent feasible" to reduce exposures to levels at or below the PEL. Where utilization of such controls is insufficient, supplementary respiratory protection is used to reduce employee's exposures below the PEL.

"Feasible" engineering and work practice controls must be implemented, regardless of respirator usage, to reduce exposures to the PEL, or to the lowest level possible. Examples of engineering and work-practice controls include effective mechanical ventilation systems (particularly high-efficiency particulate air or "HEPA" systems), wet removal methods and enclosures. However, it is worth noting that no specific guidelines are provided in the standard defining "feasible."

Administrative controls involve rotating workers out of a contaminated area in order to reduce their TWA exposure. Note: While excluded from use for various other OSHA standards such as those for asbestos abatement, job rotation is acceptable as an administrative control type worker protection method under the new lead standard.

The standard requires that employers develop and implement a written compliance plan prior to beginning any job where the PEL limit will be exceeded. The plan must address a lengthy list of issues, including a description of lead-emitting activities, compliance methods, and subcontractor responsibilities.

The compliance program provides for frequent and regular jobsite inspections, and records of the monitoring and the compliance planning must be available to employees, their designated representatives, the National Institute of Occupational Safety and Health and OSHA upon request.

Visit the doctor

The medical surveillance provisions of the standard are part of OSHA's approach to minimization of lead related disease through early detection of medical effects associated with exposure. While employers are required to provide the medical examinations and procedures specified without cost to the employee at a reasonable time and place, the standard does not make participation by the employee mandatory. All medical surveillance provisions must be performed by, or under the supervision of, a licensed physician.

The medical surveillance occurs in two phases: an initial medical surveillance and an ongoing program. Initial surveillance is triggered by air lead exposures above the action level. Where testing indicates blood lead levels at or above 40 $\mu g/dL$, blood sampling and testing must be conducted at least every two months until two consecutive samples (at least 7 days apart) indicate levels below 40 ug/dL. Besides the 40-

μg/dL yardstick, another decision point is a blood level of 50 μg/dL or more. The employer must remove an employee from exposure to lead at or above the action level on each occasion that blood testing indicates the presence of lead above this level.

During the removal period, an employee whose blood lead level has not declined below 40 ug/dL is granted up to 18 months of medical-removal protection benefits. A return to the job is based on passing followup blood tests.

The standard requires the employer to provide information and training to all employees exposed to lead at or above the action level and to comply with OSHA Hazard Communication Standard (20 CRF 1926.59) and 29 CFR 1926-21. OSHA has also established other training requirements.

What to do now

The CPI has a variety of maintenance issues to deal with in complying with the OSHA lead standard. Attention should be paid first to steel structures, tanks and piping—all of which have traditionally been coated with lead-based paints—as well as general facility maintenance. CPI firms often use outside contractors for painting and paint maintenance, and proper relationships and responsibilities must be reestablished. In particular:

- ❑Use only individuals trained to deal with lead at the construction site (this is not only a requirement, but it is a prudent risk-management practice)
- ❑For outsourced maintenance, ensure that contracts, purchase orders, workplace practices and policies, as well as contractors themselves, are meeting the obligations of the new rule
- ❑For internal maintenance activities, begin testing for the presence of lead. These tests include X-ray fluorescence (XRF), paint-chip analysis, colormetric test strips or similar devices

While the standard does not designate a numerical criterion for leaded paint, a good rule of thumb would be that levels of lead at or above 1.0 μg/cm^2, as measured by XRF, or 0.5% (laboratory analysis of paint sample) will generally trigger exposure reduction actions. Nondetection of lead should generally rule out the need for exposure reduction. But the middle ground–between nondetectable and 1.0 μg/cm^2–will generally require monitoring to determine actual exposures. ■

Edited by Nicholas Basta

The author

Kirk Sweetland is a vice president and principal scientist at Environmental Science & Engineering, Inc. (8901 N. Industrial Rd., Peoria, IL 61615; tel: 309 692 4422). He has over 23 years' experience in the environmental field, and currently oversees his company's Industrial Hygiene efforts. Active in the American Soc. for Testing of Materials (ASTM) and other organizations on lead standards, he has worked previously on this issue in the foundry industry. He has a chemistry degree from Bradley University (Peoria, Ill.).

DISPOSE HAZARDOUS WASTES SAFELY

Inspector inserts coliwassa tube into drum to remove a sample

The route of wastes must be documented by means of a manfest

The penalties for not complying with federal regulations can be harsh

John O. Lucks, Safety-Kleen Corp.

Before 1986, companies that generated small quantities of hazardous waste (200 to 2,200 lb/month) were for the most part unaffected by U.S. environmental regulations. Laws pertained only to "high volume" generators. But things have changed. In 1986, the U.S. Resource Conservation and Recovery Act (RCRA) regulations for small quantity waste generators went into effect. Since then, companies have had to identify their wastes and are liable for them. They also have had to learn how to comply with the stacks of regulations dealing with these issues.

Complying with these regulations can be time consuming, confusing and costly. According to the U.S. Environmental Protection Agency (EPA), many companies still do not know the procedures. They are not sure whether the wastes they generate are considered hazardous, not aware of their liabilities, and do not know what types of disposal methods are best suited for their operations.

Moreover, some companies have not decided whether to get rid of their wastes themselves or hire a disposal firm. And, if they elect to use an outside company, they are uncertain as to what type should it be.

Not complying with RCRA regulations can mean extensive fines and penalties. Felony charges can be lodged against a person who, without a RCRA permit, knowingly transports, or arranges for the transportation of, hazardous wastes to another facility. Firms that treat, store or dispose of hazardous waste, or violate the conditions of a permit are also liable.

Penalties for such violations can range up to $50,000 for each day of violation or imprisonment for up to five years, or both. In light of the possible penalties, choosing acceptable and dependable methods for handling and disposing hazardous wastes becomes very important.

Most small-quantity waste generators do not have a department, or even a person, dedicated to the task of handling hazardous wastes. Their volumes often do not warrant even one full-time position. Therefore, the burden of compliance frequently falls on the operations or maintenance department.

A waste disposal company can lighten the burden of compliance for a small-quantity waste generator. It can assist with manifest paperwork, laboratory analyses and pickup scheduling, as well as with waste minimization programs.

Available alternatives

There are several types of firms that will remove hazardous wastes or contaminated fluids from your site: trans-

Originally published March 1990

porters, which pick up and haul away wastes; transfer and storage companies (also known as waste brokers), which collect wastes from several plants and combine them into shipments large enough to be taken to a large disposal firm; and treatment, storage and disposal firms, which ultimately dispose of, recycle or treat wastes.

Be warned, however, that you are responsible for your hazardous waste, regardless of which firm transports, stores or disposes of it. Therefore, you should exercise great care when selecting a disposal firm. When investigating such firms, thoroughly evaluate their credentials, facilities and practices.

Fractionation tower at this recycle center separates waste mixture

Is the firm certified?

Companies that remove, process and dispose of wastes must have an EPA identification number. Not a permit, this number is required for tracking and controlling companies involved in disposing of hazardous wastes.

Waste disposal companies are also required to obtain permits specifying the types of waste they are equipped to handle. In addition, most states require these companies to have a state permit number. Before contracting with a firm, check its permits, and get its EPA identification and federal and state permit numbers. If your waste is sent to another state or transferred to another company, be sure to obtain this information from the firm with which you are dealing.

Additional pointers

Other points you might inquire about:

- Is the company operating under a final permit, or has it only completed and submitted an application for it?
- Have there been, or are there now, actions against the company by state or federal regulatory agencies regarding its permits or approvals?
- Have there been, or are there now, actions against the company by a citizen's or another such private group?
- Does the company have appropriate insurance and a sufficient finances to cover clean up and accident damages, as

Blended with fuel, wastes can be burned to make cement, as in this facility

well as legal proceedings stemming from them?

Before contracting with a firm, find out exactly what services it provides. Is it simply a transporter of wastes, a broker, or does it offer treatment, storage and disposal? Can it handle all your wastes, or are you going to have to deal with several firms? Having one company handle all your wastes makes it easier to keep track of where they have been sent and how they have been processed.

Find out how the company stores, treats and disposes wastes. By the end of this year, land-filling will end as a disposal option for small-quantity waste generators. Afterwards, only incineration or recycling will be legal options. Of the two, recycling is preferred because it is less expensive and eliminates unwanted byproducts.

The facility must be adequate

If the disposal firm does not own and operate the facility that processes the wastes, determine what company does and investigate it thoroughly. Check the company through your trade association, or inspect its facility personally.

Although a general inspection will not reveal everything, it can disclose how a waste disposal firm handles its business. Check the general condition of the site. Is the equipment in good condition? Are operations well run? Is there evidence of prior leaks or spills? If there seems to be a large number of drums at the site, check of the label dates randomly to see if the drums have been at the site for a long time, possibly a sign of problems.

Examine the facility's operating, inspection and training records. Look at the manifest records to determine if they are up to date. Inspect the waste analysis data to see if incoming wastes have been routinely screened. Additionally, look for previous violations. Such records are available to the public through state EPA offices. Most waste treatment, storage and disposal firms have had some past violations, but persistent and serious violations can be a warning sign.

Before hiring a transporter, determine what services the firm will perform (e.g., loading drums, pumping tanks, labeling containers and filling out manifest and shipping documents).

Drums are checked to ensure that wastes will be shipped safely

Inspect its trucks to make sure they are in good working order. Other concerns: the general geology and hydrology at the disposal plant, and what emergency help you can expect from within the plant and the neighboring community.

Before signing a contract, ask about the disposal firm's personnel: their education, training, experience and abilities. Find out if the company has a training program. After hiring a firm, find out who will be the principal person to contact at each facility, along with phone numbers. This is important if questions or problems should arise.

Essential: a certified laboratory

By law, hazardous-waste disposal companies must know what materials they are handling. Therefore, a shipment of hazardous waste must be analyzed by a certified laboratory before it can be removed by the disposal company. Usually, the laboratory must be an independent one, but it may belong to the disposal company. This service, if available from the disposal company, is often faster and less expensive because such a laboratory typically specializes in analyzing wastes.

The laboratory should be certified as being capable of performing the required tests. Check with the state, or another certifying agency to see whether the laboratory has been approved. Also find out if the waste haulers and disposal companies in the area will accept the laboratory's results, provide guidance on how to collect, store and transport samples (if required), and supply sample containers. Don't forget to determine the turnaround time and cost of analyses.

According to the "cradle to grave" provisions of RCRA you remain ultimately responsible for disposing of the hazardous wastes that you generate. This includes anything that the hauling or disposal company may do. So, obviously, it is in your best interest to carefully select such a company. ■

The author

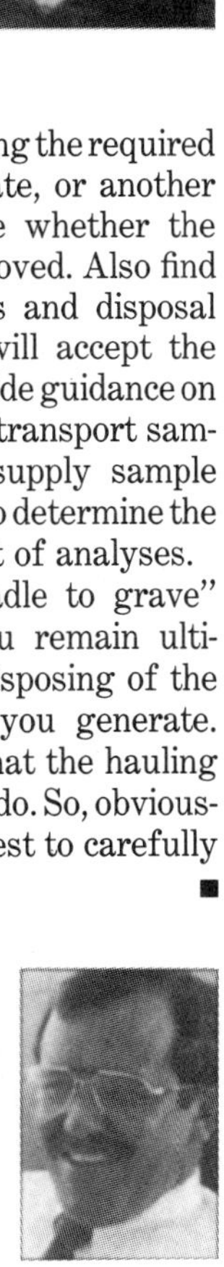

John O. Lucks is the Industrial Marketing Manager for Safety-Kleen Corp. (777 Big Timber Rd., Elgin, IL 60123; tel.: 708-697-8460). He attended Rose College in Midwest City, Oklahoma, where he received a degree in journalism. He has served as chairman of Elgin's solid waste alternatives task force, he is currently serving as the chairman of the Elgin United Way Youth Services Panel and as the city's Commissioner of Planning and Land Use.

HANDLE COMMERCIAL-GRIGNARD REAGENTS

Philip E. Rakita
CECA, Group Atochem
John F. Aultman, Laura Stapleton
Atochem North America

Grignard reagents are widely employed in the production of pharmaceuticals and their intermediates, flavor and fragrance chemicals, and polymerization catalysts. Many products can be formed by a coupling reaction between Grignard reagents and a variety of substrates.

Grignard reagents are formed as a result of a reaction between metallic magnesium and an organic halide (chloride, bromide, or iodide), in the presence of an ether solvent, and in the absence of water. The formation of Grignard reagents, and the many coupling reactions involving Grignard reagents, are vigorous and exothermic. Hence, it is imperative that laboratory, process, and pilot plant studies be conducted prior to full-scale production runs, to set the necessary process parameters.

Due to the extremely volatile nature of these reactive chemicals, clear guidelines are needed for the safe handling, production, and use of Grignard reagents. To avoid uncontrolled reaction, their manufacture and use must be carried out in inert atmospheres, primarily because of the volatility of the ether required to keep the reagent and the synthesized product in solution. Careful control of the feed rate, agitation, and heat removal is also required.

Ether is the problem

Ether provides the best solubility for Grignard reagents, thereby maximizing their activity. To reduce the risk associated with using volatile ether, other electron-donor solvents, such as amines, could be used as an alternative solvent.

However, since a hydrolysis step is usually required to isolate products formed by Grignard reactions, water soluble amines are not recommended. For the commercial use and manufacture of Grignard reagents, ethers and ether-hydrocarbon mixtures are the preferred solvents.

Historically, diethyl ether was the common solvent used in the preparation of Grignard reagents. Its use is limited by its extreme flammability and relatively low boiling point, which precludes its use in reactions with high working temperatures.

Tetrahydrofuran (THF) is a safer alternative solvent that has been widely used since the 1950s in the commercial-scale production and use of Grignard reagents. THF is a cyclic ether with greater solvent power, and a boiling point 60°F greater than that of diethyl ether. Occasionally, a mixture of THF and toluene is used, as toluene improves the solubility of some Grignards.

The 1912 Nobel prize for chemistry was given to French chemists Victor Grignard and Paul Sabatier for their development of Grignard reagents. Using the general formula SUBSTRATE + RMgX → PRODUCT, a variety of products can be synthesized using the Grignard reaction. This list (right) represents some of the substrate-product combinations available, and demonstrates the versatility of this reaction

As reactions involving Grignard reagents are exothermic, the associated heat must be controlled. Typically, reactions involving Grignard reagents are carried out in a jacketed, glass-lined-steel reactor (the lining providing protection against acidic conditions experienced during hydrolysis).

To control the heat of the Grignard reaction mass when THF solvent is being used, ambient cold water from a well or municipal supply may be used in the jacket. For reactions using diethyl ether as the solvent, on the other hand, chilled water or brine is required to keep the reaction mass below the ether's boiling point. Since Grignards are particularly reactive with water, liquid and gaseous hydrocarbon coolants may be employed instead, to eliminate the risk that could arise should an aqueous cooling system leak.

Regulatory compliance

Before a U.S. chemical manufacturer can scale up production of Grignard reagents from the lab to the plant, appropriate Federal, state, and local air permits must be obtained, or existing permits modified. Requirements and

SCALE SAFELY

GRIGNARD REAGENTS	
REACT WITH:	**TO YIELD:**
• Ortho esters	• Acetals
• Carbon dioxide • (2-keto) from Isonitriles	• Acids
• Aldehydes • Epoxides • Oxygen • Acid halides • (diols) from Lactones • Ketones	• Alcohols
• Isonitriles	• Aldehydes
• Amino ethers • Azides • Oxime sulfonates • (secondary) from Aldimines • (tertiary) from Iminium salts • (2-keto) from Isocyanates	• Amines
• Chloroformates	• Esters
• Acetals • Peresters • Cyclic acetals	• Ethers
• Alkyl halides • Sulfates	• Hydrocarbons
• Acid halides • Acids • Anhydrides • Amides • Carboxylic acids • Carbamoyl chlorides • Bromohydrins • N,N-Disubstituted carbamates • 2-Pyridine thiol esters • Esters • Iminium salts • Nitriles • Isonitriles • α, β-unsaturated ketones • Imidazolines • Ketenes • (β-hydroxy) from Isonitriles	• Ketones
• Sulfur	• Mercaptans
• Silylenol ethers • Vinyl halides • Vinylic sulfides, sulfones, or ethers	• Olefins
• Metal halides	• Organometallics
• Isothiocyanates	• Thioamides

Pressure or vacuum control, and inert-gas blanketing are required to control reactivity

approval periods will vary depending upon the jurisdiction.

For chemicals that do not appear on the U.S. Environmental Protection Agency's (EPA) list of "Chemical Substances in Commerce", it is necessary to submit a Premanufacture Notification (PMN) to the EPA prior to making a new material for commercial purposes. EPA approval is based on adequate documentation of planned manufacturing procedures, anticipated byproducts, health and environmental-effects test data, and procedures for safe use, handling and disposal.

When shipping Grignard reagents, one should review international transport regulations. U.S. import and export restrictions on Grignards are listed in the Toxic Substances Control Act (TSCA), Section 12. For export, restrictions may be imposed by the receiving country. Note that the majority of the regulatory considerations for the use of Grignard reagents will apply to the solvents used rather than the Grignards themselves.

Properties and reactivity

Grignard reagents have the general formula RMgX, where R is an alkyl, aryl, or other organic group, and X is a halide. Grignard reagent solutions react vigorously with water, forming the appropriate alkane or aromatic hydrocarbon:

$$RMgX + H_2O \rightarrow RH + Mg(OH)X$$

If the organic group R has four or fewer carbons, the alkane formed will be a flammable gas under normal conditions. Appropriate precautions are required to prevent the uncontrolled release of such a volatile product.

Grignard reagents will also react exothermically with oxygen upon hydrolysis to form alcohols:

$$2RMgX + O_2 \rightarrow RO_2MgX + RMgX \rightarrow 2ROMgX + 2H_2O \rightarrow 2ROH + Mg(OH)_2 + MgX_2$$

and with carbon dioxide followed by hydrolysis to form carboxylic acids:

$$RMgX + CO_2 \rightarrow RCO_2MgX + H_2O \rightarrow RCO_2H + Mg(OH)X$$

If the proper environment and reaction conditions are not maintained, unwanted byproducts, such as alcohols, carboxylic acids, carbon-carbon coupling products (R-R), and hydrocarbons (RH) will be formed. Such side reactions will have two effects:

- The amount of available Grignard reagent will be decreased, based on the stoichiometry of the desired reaction
- The removal of such byproduct impurities may be difficult

Packaging and transportation

Grignard reagents are supplied commercially in 55-gallon, 16-gage stainless steel tighthead drums (DOT-17C in the U.S.; the international equivalent is IATA

*Additional assistance on potential requirements under the Toxic Substances Control Act (TSCA) can be obtained through the EPA's Office of Industry Assistance (Washington, D.C.) at (202) 554-1404.

Section V

How to Guard Against Mishaps

PREVENT ACCIDENTS

NED SHAW

BEFORE THEY HAPPEN

Successful hazard and operability studies require planning, thoroughness and effective direction

Frank H. Curry
Curry Hydrocarbons

A hazard and operability analysis, or hazop, is a technique used by engineers to study and track the processes of their facility. The method helps workers to identify possible plant-related risks, typically those caused by equipment malfunctions or human error. It also helps determine the situations, or consequences, that can be caused by these malfunctions and errors.

The hazop is one of the most-popular methods used by engineers to fulfill the process hazard analysis requirements of OSHA's rule 29 CFR 1910.119 (Process Safety Management of Highly Hazardous Chemicals). This rule mandates that companies possessing a hazardous substance in excess of a certain amount and operating in the U.S. must develop and implement a process safety management system (PSM). Loss departments often require a series of hazops as part of the implementation of the PSM.

A hazop focuses on "deviations" from design intent, or undesired actions by machinery or workers. It is performed by a team of employees (described below) that studies a facility's procedures and equipment, specifically looking for deviations.

The team typically organizes its study by "process sections," or "items," such as piping, vessels or electrical systems. Items are groups of equipment with definite boundaries (e.g., a line between two vessels), or specific locations on P&IDs (e.g., individual reactors).

The team examines the probable consequences of each deviation, and decides whether these consequences pose a significant hazard. "Significance" covers a whole spectrum, from fatal to negligible. For example, a bucket of gasoline sitting next to a burner might cause a large explosion that could kill employees and destroy property. This consequence — such an explosion — would be considered very significant.

On the other hand, an off-spec ±0.1% pressure change in a piping system for a pressurized vessel is an example of an insignificant deviation that is not likely

Originally published June 1995

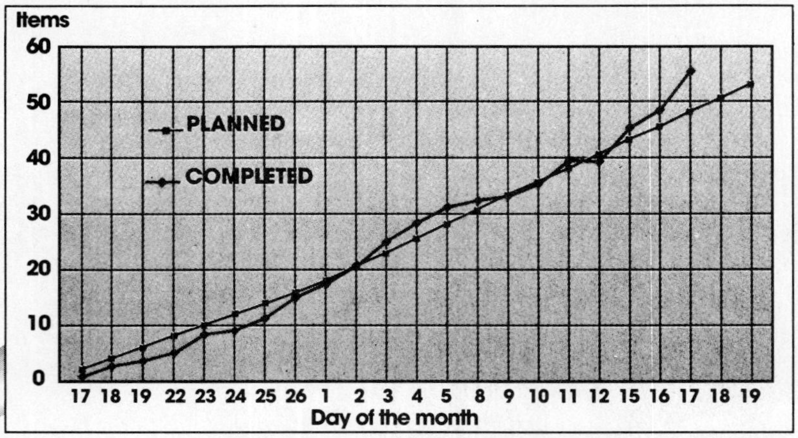

FIGURE. The actual progress of a hazop team, measured in terms of items completed, lags behind the scheduled rate at the beginning; however, by the end of the meetings, the team exceeds the planned rate and even finishes two days ahead of schedule

MEETING PROGRESS STATISTICS

	Section* 1	Section * 2	Unit
Items	32.0	23.0	55.0
Deviations	241.0	170.0	411.0
Days	13.0	8.0	21.0
Hours	58.5	36.0	94.5
Items per day	2.5	2.9	2.6
Deviations per hour	4.1	4.7	4.3

*"Section" refers to the two major parts of a refinery unit: Section 1 is the crude distillation unit, and Section 2 is the vacuum distillation unit

TABLE 1. Overall, the hazop team handled about 4.5 deviations per hour

to cause any consequence of concern.

After listing these consequences, the team studies the safety equipment and procedures that already exist to safeguard against the identified deviations. If these safety measures are inadequate to prevent an event with significant consequences, the team then recommends "actions" to eliminate the deviation or to mitigate its consequences (Table 2, p. 82).

For example, to eliminate concerns regarding overpressure in a heat exchanger, the team might suggest increasing the exchanger's design shell pressure to allow the shell to contain the results of a tube rupture. Or the team might recommend reducing pump shutoff pressure. In another instance, the team might recommend specifying double seals on a pump that handles a liquid with low-autoignition temperature, to reduce leakage concerns.

A successful hazop requires detailed information about the design and operation of the plant. It can be done during or after the detailed design stage.

A hazard and operability analysis study is significant undertaking. For example, a hazop study for a single large process unit can involve up to seven people for as long as twelve weeks. Meetings are typically 4.5 hours long and may take place as frequently as five days a week. Two important aspects of the hazop are discussed in following sections: the team members and meeting procedures.

The hazop team

A hazop is conducted by members of the company that owns the facility to be evaluated, sometimes with an outside contractor. The leader and other members must be chosen carefully.

For hazop meetings to be effective, a trained and experienced leader, who is sometimes called the facilitator, is required. This person can be from the client company, from the contractor or from some third-party organization. He or she need not be familiar with the plant, but should know the language and terminology used in the industry.

Some of the team members must have detailed knowledge of the process. Various disciplines should be represented on the team. These disciplines include engineering, operations, instrumentation and maintenance.

The actual mix of people will depend on the nature of the process and the focus of the study. For example, if the study is made during engineering, the engineering team will need strong representation. If the plant is in operation, then members of the company's operating staff will most likely predominate.

Since hazop meetings should not be scheduled to exceed five hours per day, it is common for the team members to continue to remain active in their regular jobs. However, alternates should be designated to cover for them while they are in the hazop meetings.

Logistics of hazop meetings

Keeping track of hazop meetings with a simple chart helps the team visualize its progress. The chart format also allows mid-course corrections and provides a basis for keeping management apprised of progress.

Preparing the baseline. A multiweek hazop can require hundreds of hours of labor. It is helpful to prepare a cumulative plot of the process sections studied (also called items or nodes) versus days scheduled for meetings (Figure). A single line with a slope of two to three items per day is typical as an initial baseline.

Plotting the results. At the end of each daily meeting, the hazop leader, or the person designated as "scribe," should record the number of items completed (rounding to the nearest whole number). It is useful to record daily progress with a spreadsheet program, which can automatically generate a progress-report graph (Figure).

This graph is useful to project and operating management. The graph can be distributed after a few days of meetings. It provides tangible evidence of team progress and lets management know when to expect the hazop recommendations, or "action items."

Making recommendations. Action items that the team recommends typically include suggestions for design changes, procedural changes or areas for further study (e.g., adding a redundant pressure alarm or reversing the sequence of two operating steps). These recommendations can be issued for planning purposes during or immediately after the meeting phase of the hazop study. If the hazop is part of a grassroots project or revamp, the team's findings will likely affect the design and implementation of the ongoing project. These progress reports will be appreciated by the project manager.

Documents, such as the completed progress-report graph (Figure), meeting

SAMPLE RECOMMENDATIONS FROM A HAZOP TEAM

Actions	Responsibility	Status*	Priority**	Target date
1. No action required. Safeguards are adequate; do not need to increase. (See: 1.1, 1.3, 1.4, ...8.1, 8.2)				
2. Check PSV-105[5] sizing to handle (a) fire case, (b) tube rupture on reboiler, (c) total loss of reflux to stripper, (d) loss of cooling to condenser EX-102, (e) instrument or controller failure, (f) instrument air failure, (g) power failure, etc.See: 1.1, 2.1, ...)	Tom Jones	Study	8	94/11/11
3. Modify TRC-103[6] on 3"-P-102[1] to TR-103[6] (See: 1.2, 1.5, 8.3)	Tom Jones	Incomplete	3	94/10/21
4. Add high-flow alarm to FT-101[3], which is independent of FRC-101[2] (See: 1.3, 8.3)	John Smith	Incomplete	6	94/12/7
5. Low low level swith from V-101[7] to trip FV-101[4] closed using solenoid (See: 1.3)	John Smith	Incomplete	6	94/12/7
6. Add check valve to 3"-P-102[1], close to stripper feed inlet (See: 1.3, 1.7)	Matt Lee	Incomplete	4	94/6/15
7. Provide low-flow alarm on FT-101[3] (See: 1.6)	John Smith	Incomplete	6	94/12/7
8. Consider need for on-line analyzer for column feed (See: 1.8)	Tom Jones	Study	3	94/12/21
9. Consider need for environmental monitors (See: 1.9, 4.9, 5.3, 6.6)	Daisy Patel	Study	6	94/11/11
10. Consider adding nitrogen pressure-balancing system to makeup in the event of sudden loss of steam on reboiler (See: 2.4, 2.5, 2.6, 4.5, 6.4, 7.4, 7.5, 8.4)	Tom Jones	Study	3	94/12/21

1 line number
2 flow control loop
3 flow transmitter
4 flow valve
5 relief valve
6 temperature control loop
7 upstream feed drum (not in this unit, but affects the section)

* These labels pertain to the status of the listed actions. "Incomplete" means that the change is under way, but not complete; "Study" infers that an engineering or other type of study is needed before the team can decide what to do. Other typical entries can be: "in progress," "cancelled," "in 1996 budget," and so forth.

** These numbers are based on a scale from 0 to 10, where 10 denotes highest priority. Some engineers prefer using a scale of 1 to 4.

DEVIATIONS FOR NODE 1*, Feed Lines 2"-P-101 and 102

1.1	High pressure
1.2	High temperature
1.3	High flow
1.4	Low pressure
1.5	Low temperature
1.6	Low or no flow
1.7	Reverse or misdirected flow
1.8	Contamination
1.9	Leak
1.10	Rupture

*Flow rate of 100,000 lb/h at 220°F and 90 psig, upstream of FV-101

TABLE 2. Each recommendation in this report excerpt applies to the deviations in parentheses. Action 1, which needs "no action," applies to deviations 1.1, 1.3, 1.4,... Note, however, that some deviations require actions; e.g., 1.1 appears under Action 2, and 1.3 falls under Actions 4 and 5. This is because some deviations have more than one cause; and though some causes do not require action, others do. These charts were built with DDM-Hazop software; for more information, circle number 625 on the Reader Service card

TABLE 2A (INSET). Several deviations are listed for each item, or "node" (only Node 1 is shown here). Note that each deviation may have multiple causes, each of which may require a different action to be taken (Table 2). For example, Deviation 1.1, high pressure in feed lines, can be caused by high upstream pressure due to high pressure in a feed drum, or by a leaking heat exchanger, or by a heat-exchanger rupture, or by a failed pressure valve

record and derived statistics (Table 1, p. 81) provide a convenient record for the hazop report. These also provide a basis for future hazop scheduling.

Starting slow, but steady. The team should spend few minutes each day discussing its progress against the baseline. It is usual for a beginning study to lag below the baseline for a few meetings (Figure). Typically, the group uses some time during initial meetings to identify and correct its deficiencies.

For instance, some team members who are unfamiliar with the hazop process will need some coaching. Or the team may find that an area of expertise is lacking and that it needs an additional team member, at least part time.

The group, which has not worked together before, must establish an efficient way of collaborating. The data available to the team may need supplementation. Or perhaps the spreadsheet program has a steep learning curve, and the scribe needs a few hours to get used to it.

Pointers for the leader

The leader must balance the demands of the schedule against product quality. A team with a clear target will remain motivated; the members need gentle, but firm, direction. Allowing members to take "all the time they want" can demotivate some team members.

In a well planned study, the leader can prepare deviations appropriate to each item before each meeting. This helps minimize meeting time.

However, the leader must by no means limit the team to the list. He or she should encourage others to add deviations to the list, prompting the team for additional deviations at the end of the discussion of each item. This covers possible simple omissions by the leader and ensures thorough analysis.

The use of software to facilitate hazop meetings is becoming common. The software should save the following data, as they are generated by the team: causes of deviations, consequences, safeguards and action items. This information should be accessible as needed, whether for the current project or a future one.

A few simple, practical pointers help improve efficiency in hazop meetings. For example, the leader should start the meetings on time, rather than wait for latecomers.

Providing good coffee and lunch at clearly designated times is also helpful. A productive schedule has meetings starting at 8 a.m., pausing at 11:45 for a half-hour lunch in the meeting room, and reconvening for another 45 minutes. This schedule provides plenty of meeting time and allows team members to get back to their regular duties by 1:00 p.m. ■

Author

Frank H. Curry, P.E., is president of Curry Hydrocarbons, Inc. (CHI; 55 Nugget Ave., Suite 202, Scarborough, ON M1S 3L1; Tel. 416-298-9563), an engineering company that provides process design, economic analysis and operating assistance to the refining, petrochemical and utilities industries. Curry has process design and plant operations experience with Texaco Canada Ltd. and Suncor Inc. He has a B.Sc. in chemical engineering from the Univ. of Durham (Durham, U.K.) and an M.A.Sc. in management sciences from the University of Waterloo (Waterloo, Ont.).

FORESTALL FAILURES BY NONDESTRUCTIVE TESTING

Helmut Thielsch, P.E., and **Florence M. Cone,**
Thielsch Engineering, Inc.

Nondestructive examination to spot defects is being increasingly applied to ensure the continued integrity of process equipment. This is particularly true in the chemical process industries (CPI) where the catastrophic failure of a component may endanger plant personnel, result in undesirable environmental consequences and cause costly business interruptions.

To eliminate such failures, there are numerous codes and standards governing the fabrication, inspection and erection of pressure vessels and piping systems. In the U.S., these include the American Soc. of Mechanical Engineers' (ASME) boiler and pressure vessel code and the various ASME and American National Standards Institute (ANSI) B31 codes for pressure piping (in particular, the B31.3 code for pressure piping in chemical plants and petroleum refineries).

The U.S. codes may also be used for pressure vessels fabricated in other countries. In addition, other national standards, such as AD Merkblatter, the German equivalent to Section VIII of the ASME Boiler and Pressure Vessel Code, and those of Britain and Japan may be applicable.

The various codes cover a wide range of CPI equipment, including fired and unfired pressure vessels, heat exchangers, pipes, boilers, reformers, and compressors and other rotating machinery operating under a variety of service conditions. For example, process equipment may be subject to sub-zero operating temperatures where embrittlement is a possibility, or it may run at elevated temperatures where failure by creep is a concern. A number of other operating conditions may also result in service-related deterioration, such as corrosion, erosion, or cracking or embrittlement (such as from thermal or mechanical fatigue, or shocks).

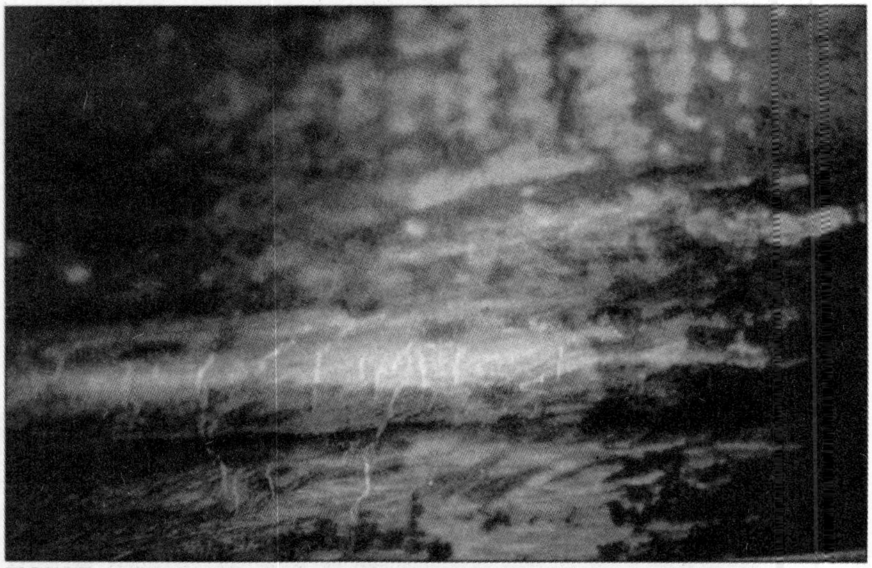

FIGURE 1. Wet fluorescent-magnetic-particle testing reveals cracking in a deaerator

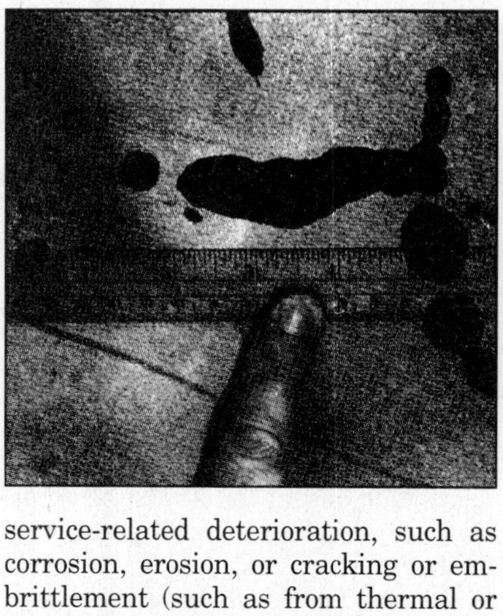

FIGURE 2 (below, left). Liquid-penetrant testing shows the presence of caustic stress-corrosion-cracking in this layered reactor

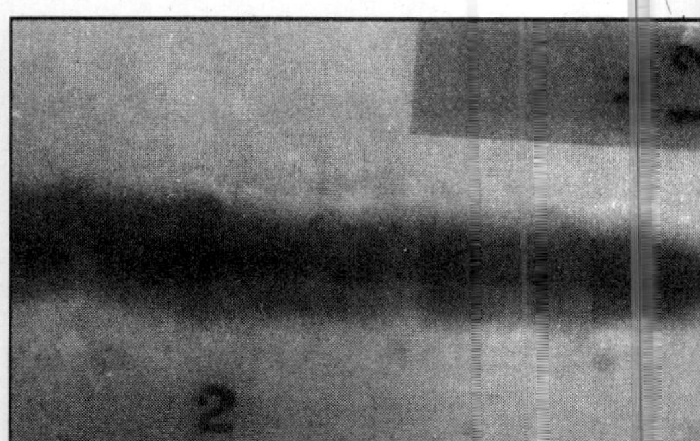

FIGURE 3. Defective pipe welds show a lack of fusion and incomplete penetration

These types of deterioration can generally be detected by nondestructive examination before they reach a dangerous level. For best results, however, the nondestructive examination techniques must be carefully selected and properly applied, and the results of the tests interpreted in a meaningful manner.

Several real-world examples of how

Originally published November 1994

to choose the right test methods, and of the best ways to interpret the results are given in this article. These case histories include ones where the right methods and interpretations saved millions of dollars in repair and shutdown costs.

On the other hand, the wrong test methods or misinterpretation of test results may result in shutting down a plant or, worse, in fixing nonexistent problems. Such misinterpreted "defects" may arise from inconsequential or noninjurious conditions, such as original manufacturing or fabricating defects, and superficial surface fissuring. Corrective action to fix these nonservice-related defects are unnecessary and in some cases, undesirable. However, all such defects should be documented and monitored, to ensure that they do not progressively worsen to cause potential failures.

Types of defects

If service-related deterioration is detected in the early stages, it may be possible to change the operating conditions to eliminate the source of the deterioration. At the least, early monitoring can reduce the rate at which any deterioration occurs. Early detection of service-related deterioration will also minimize any required repairs.

Nondestructive examination includes such conventional methods as

magnetic-particle, liquid-penetrant, ultrasonic and radiographic testing. Other techniques involve eddy-current and strain-gage testing, inspection of pipe internals using pipe crawlers and videoborescopes, and "replication" methods.

Although nondestructive testing encompasses a variety of techniques, these can be classified into two categories: surface and volumetric. As the names imply, surface examination will identify defects on the surface of components, while volumetric examination spots defects below the surface.

Common surface-examination techniques include magnetic-particle and liquid-penetrant testing. Volumetric examinations involve probing subsurface defects with radiographic equip-

The payoff is in improved safety and fewer process upsets

ment (using X-rays or radiation from isotopes), ultrasonic instruments, and eddy-current devices. In general, it is prudent to inspect a component using both surface- and volumetric-examination techniques, since this significantly increases the probability of detecting defects, corrosion or cracking that are likely to cause failure.

A magnetic outlook

Examination with magnetic-particle testing can successfully detect surface defects in ferromagnetic materials. The technique is most commonly applied to girth and seam welds in piping systems and pressure vessels.

Before testing, the surface of the component must be cleaned to remove oxide scale or other coatings and deposits that could interfere with the inspection. Sandblasting, power wire brushing or grinding is generally adequate to ensure a clean surface. Note that each of the cleaning techniques may result in some variations in the results.

If the inspection involves a weld, an electric yoke (or prod) is placed parallel to the weld and energized to introduce a magnetic field into the weld. The weld is then dusted with magnetic filings, which will be attracted to interruptions or discontinuities in the magnetic field. Since cracking and other linear weld defects produce interruptions in the magnetic field, the defects will be effectively outlined by the magnetic particles. The yoke is then placed at right angles to the weld, and the inspection is completed over the length of the weld. If the test shows linear "indications" – defects, cracking or corrosion – in the weld or the adjacent base metal, or alongside the edge of the weld, then a groove is carefully ground across the indications to determine their nature and to confirm their magnitude.

One variation of magnetic-particle testing is wet fluorescent-magnetic-particle testing. Here, a bath of fluorescent particles suspended in a light oil medium is sprayed onto the surface. Then, a magnetic field is introduced and the surface is inspected with an ultraviolet or "black" light. Wet fluorescent-magnetic particle testing is generally considered to be more sensitive than dry magnetic-powder testing.

The internals of deaerators, which are commonly found in CPI plants that generate their own power or process steam, are usually inspected on a regular basis using fluorescent-magnetic particles. The method is also suitable for checking the internals of ferritic-steel pressure vessels and tanks.

Typically, cracks in deaerators tend to occur in the heat-affected zone of the weld and may be oriented perpendicular or parallel to the weld (Figure 1). Cracking results primarily from residual welding stresses, and from water hammer and pressure cycling.

Extreme care must be taken in evaluating the results of magnetic-particle testing since certain noninjurious surface conditions bear a strong resemblance to progressive cracking. For instance, a surface lap in a vessel could easily be mistaken for progressive cracking. In such cases, exploratory grinding can confirm that the indication is a surface lap that had been introduced into the shell's plate during manufacture, and that the crack had not grown during the ensuing years of service.

Liquid penetrant testing

In contrast to testing with magnetic particles, examination using liquid penetrants is not limited to just ferromagnetic materials. Rather, it can detect surface defects in any type of nonporous material. However, liquid penetrant techniques are more labor intensive and time consuming, and so are applied where the material is non-magnetic or where examination by magnetic particles is not permissible.

Testing with liquid penetrants also requires surface preparation of the component. However, sandblasting should be avoided since the peening action of the abrasive grit may close up tight indications or cracks, rendering them undetectable.

Once the surface is prepared, the component is cleaned with a solvent to remove grease and other substances that can interfere with the resulting inspection. The surface is then sprayed with the penetrant, a visible red or fluorescent dye that is drawn by capillary action into any defects open to the surface. After an adequate time, the penetrant is wiped from the surface and a developer applied. The latter acts as a blotter, drawing any penetrant trapped in defects back to the surface, and providing a contrasting background for the penetrant.

Figure 2 is an example of typical cracking detected by liquid-penetrant testing. Initially, a visual examination of a multilayered urea reactor revealed that an overflow condition had occurred. As a result, the vessel was stripped of insulation and inspected by liquid penetrants, revealing stress-corrosion cracking due to caustic liquids.

Early detection of the crack by visual and liquid-penetrant examination enabled the vessel to be repaired onsite, rather than be dismantled and shipped to a fabricator for repairs. In fact, the savings in repair time and downtime was many millions of dollars.

As with magnetic particle testing, extreme care must be taken to differentiate between progressive deterioration, such as from caustic-stress-corrosion cracking, and less consequential conditions. As a case in point, a valve body was recently inspected by liquid-penetrant testing to confirm its suitability for service. The inspection revealed cracking along the inside diameter surface of the valve.

However, a careful engineering interpretation of this indication confirmed that it represented an original casting defect that had not progressed during service. The indication appeared essentially identical after an additional 60,000 h of operation, confirming the interpretation.

Radiographic examination involves the use of a radioactive source, such as iridium 192 or cobalt 60, in conjunction with X-ray film. In practice, the source is placed on one side of the component being examined and film is placed on the opposite side. The X-rays emitted by the source travel through the component and expose the film. Any variations in the thickness or in the density of the component will result in variations in the film exposure.

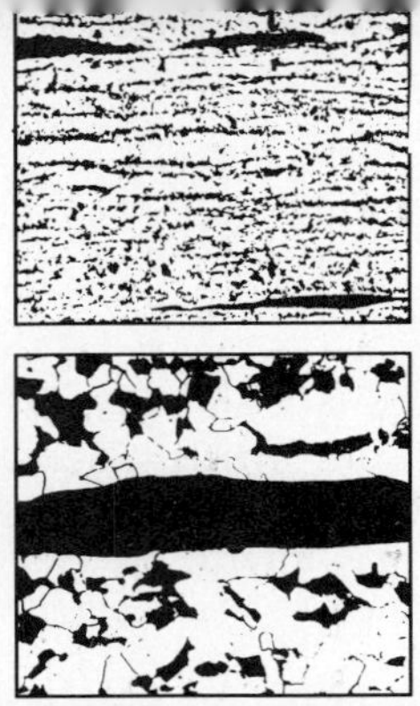

FIGURE 5. The photomicrographs of stringer-type laminations can be misinterpreted as cracking

FIGURE 4. A pipe that has been ruptured due to severe corrosion

Defects that form during casting, forging and extrusion, such as blow holes, shrinkage and cracks, reduce the actual thickness of metal that the X-rays have to pass through. This results in different exposure densities in the radiographic film. Since inadequate fusion, incomplete penetration, porosity and various other weld defects also effectively reduce the thickness, radiography can be used to detect and record the presence of subsurface weld defects in pipes or pressure vessels.

Radiography reveals

Radiographic examination is frequently required by various codes and standards governing the fabrication of pressure vessels. The extent of the examination depends on a vessel's pressure and temperature ratings, the corrosiveness or toxicity of the fluid, or the potential consequence of a failure.

Note that in certain cases where radiographic examination is required, it may not be the most suitable method – that is, it may not readily detect cracking conditions. Recently, for instance, a catastrophic failure of a stripper vessel occurred soon after the vessel had been placed in service. Subsequent metallurgical examination indicated that the weld that failed contained severe cracking that had been introduced during the vessel's fabrication. This weld had been inspected radiographically and had been found acceptable.

Such cases illustrate the importance of selecting the correct inspection method or combination of inspection methods. The failures also emphasize the necessity of using qualified inspection personnel with the necessary experience to select the most suitable methods.

Continuous advances in radiography have been made in the last 40 years. Radiographs of welds in piping or pressure vessels from the fifties and sixties may contain defects that were not detected or, if detected, were not considered sufficiently severe to be rejected.

However, if these welds are radiographed with modern equipment, some previously undetected defects could become apparent (Figure 3) In these cases, evaluation confirmed that the defects had not gradually deteriorated, and that crack propagation had not occurred. Actually, this gives rise to the eternal debate over whether the defect had developed over time or whether it had been there all along.

Ultrasonic examination is a highly refined technique in which a transducer is used to direct an ultrasonic beam into a component. The beam

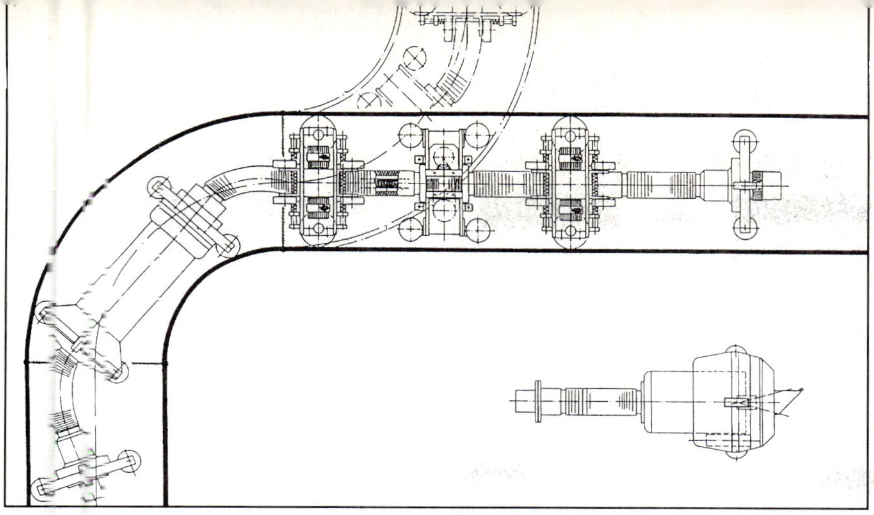

FIGURE 6. A schematic drawing of a pipe crawler with videoborescope for detecting cracks

passes through the component with little loss of energy; when it reaches the opposite side the beam is reflected back to the transducer.

The transducer can measure the energy of the beam and the time it takes to reach the opposite side. Thus, if the velocity characteristics of the material are known, ultrasonic examination can be used to determine the thickness of that material.

A sound test

Ultrasonic testing can also detect subsurface flaws. Specifically, such flaws act as an interface and reflect a part of the beam back to the transducer. The remaining portion of the beam passes through the flaw and reflects off the opposite side of the component. Thus, there will be two signals, or reflections, from a single beam.

Although ultrasonic testing is relatively simple in theory, it is quite complex in practice. The complications arise from the geometry of the component being tested, and in the various ways the results can be interpreted.

Ultrasonic examination is particularly important for detecting corrosion in piping, pressure vessels, tanks, heat exchangers and other equipment, both during normal operation and during shutdown periods.

Figure 4 illustrates the rupture of a pipe that corroded from water that had collected along the base of a piping system during shutdowns. Because the petroleum product flowing through the pipe contained relatively high sulfur levels, sulfuric acid formed. In time, the acidic water corroded the wall of the pipe to such an extent that it was unable to withstand the operating pressure and burst.

Although ultrasonic techniques are well established, the accuracy and repeatability depend on the calibration procedure, the inspection report and operator effectiveness. For example, an ultrasonic test carried out on a vessel was interpreted to represent a 3.5-in.-deep crack in the 5.25-in.-thick shell. Subsequent removal of a 4-in.-dia. plug from the vessel showed that the crack was due to intermittent stringer-type laminations in the shell's plate material (Figure 5).

A misinterpreted test can result in the fixing of a nonexistent problem

The calibration procedure involves both the instrument and the operator. The instrument calibration should be performed using a standard that is as similar as possible to the material and geometry of the part to be inspected. Whenever possible, a section of a similar part that has been removed from service should be used.

It is important for plant personnel or the inspection contractor to develop and maintain an inventory of these specific calibration and comparison standards for use in subsequent inspections. It is also desirable to record and store calibrations for retrieval during future inspections.

The "calibration" of operators involves informing them of the type of defects that may be encountered. Whenever possible, samples from failed parts should be provided. This will allow operators to develop signal-recognition skills.

The accuracy of inspection reports is extremely critical since they have to be compared with future ones. Whenever possible, all reportable reflectors should be digitally recorded and archived on computers to enable comparisons in the future.

Finally, the effectiveness of the equipment operator must never be overlooked. Even with state-of-the-art equipment, the inspection is only as good as the operator. The operator must be qualified and should have experience performing the same or very similar inspections. In addition to qualified operators, experienced engineering personnel should be available to provide meaningful evaluations of all indications.

Ultrasonic inspections using digitized values and signal analysis with personal computers can improve the accuracy of interpretations. Such inspections use a standard ultrasonic pulser-receiver with a high speed analog-to-digital converter. This allows the normal analog ultrasonic signal to be digitized and recorded on a computer disc. The transducer position is also recorded at the same time.

Once these data are recorded, several analytical tools can bet used to evaluate and characterize the signal. These include 3-D imaging, filtering, and frequency and power-spectrum analysis. These tools also provides a permanent record of the inspection results and calibration data – valuable for future comparison.

Testing with eddy currents

In an eddy current examination, an object is placed in a magnetic field produced by a conductor carrying an alternating current. The magnetic field introduces eddy currents in the test object. These eddy currents, in turn, produce an additional magnetic field in the vicinity of the test block. This secondary magnetic field is then compared with a magnetic field produced by a standard object of known characteristics and evaluated.

Eddy current testing is particularly useful for examining non-ferrous tubes

FIGURE 12. Closeup view of a groove in a reactor vessel shell

FIGURE 13 (below). Cracking in a weep hole

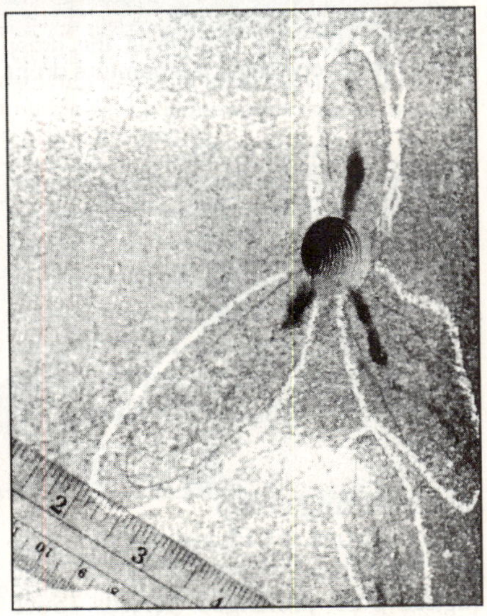

Examining weep holes

Pressure vessels that have multilayer shells with liners, or a single shell with a liner, and that are designed to Section VIII of the ASME Boiler and Pressure Vessel Code, generally are provided with weep holes. The purpose of weep holes is to allow detection as early as possible of any leak that develops in the vessel's liner.

Since inner liners may be made of stainless steel, nickel alloys, titanium or other materials, any leak that develops through these liner materials, either as a result of a crack or a defective weld, can result in corrosion of the carbon-steel or low-alloy-steel shells, or of multiple-layer shells. If this corrosion, which sometimes may be relatively rapid, is not detected and continues to occur, a major rupture of the vessel may occur.

A typical crack in the stainless-steel liner of a pressure vessel is shown in Figure 11. This crack had been detected by visual examination and confirmed by liquid penetrant examination of the inside of the liner during a shutdown. As the crack had progressed all the way through the liner, the carbon-steel shell gradually corroded. Fortunately, since the liner cracking had been detected in the early stages, significant corrosion of the shell had not occurred, and the vessel was easily repaired by welding the liner.

Figure 12 shows cracking in a 10,000-psi pressure vessel for processing acetic acid. The cracking had occurred through a $^1/_2$-in.-thick stainless-steel liner after about two years of operation. This resulted in corrosion extending deeply into a 2-in.-thick shell plate that was surrounded by an additional 8 in. of thick layered steel bands intended to place the shell under compression. The corrosion produced grooves and penetrations about $1^1/_2$ in. deep into the 2-in.-thick shell. The leakage was detected by an inspection of the weep holes.

Subsequent evaluation had confirmed that repairs by welding were feasible. The repairs took approximately one month to accomplish. The replacement time of the vessel if repairs had not been possible would have involved nearly two years.

Another urea plant involving a pressure vessel of multilayer construction with cracks at a weep-hole location is shown in Figure 13. This cracking had been the result of stress corrosion cracking of the carbon-steel-shell layer by the process liquid on the inside of the vessel.

Thus, during inspections of vessels, including ones with layers or internal liners, an inspection of all the weep holes is very important. Inspections should be performed by experienced engineering personnel and should include:

- Visual examination
- Moisture checking
- Soap-bubble testing
- Detailed weep-hole examinations with lighting sources
- Borescope examinations, if the diameter of the weep holes is sufficiently large to permit a borescope to be inserted
- Deposits collection, using tubular bore scrapers. This permits inspectors to determine whether the deposits are the result of external atmospheres or of corrosion from leaks inside the vessel. The equipment should also be examined to determined the locations of the deposits
- Deposits analysis
- Deposits evaluation, to determine whether the source of the deposits is external or internal

If weepage is noted, the equipment should be shut down and opened up for an internal inspection. It should be noted that weep holes in multilayered vessels may involve continuous holes or staggered holes, with the staggering varying from one layered vessel plate to the next.

In some instances, it may be necessary to drill a special hole into the shell to verify leak conditions at suspected areas. This drilling must be performed very carefully and must not extend into the inner shell. ■

Edited by Jayadev Choudhury

The authors

Helmut Thielsch is president of Thielsch Engineering, Inc. (195 Frances Ave., Cranston, RI 02910–2211; Tel: 401-467-6454; Fax; 401-467-2398). He was formerly vice president of research, development and engineering at ITT Grinnell Corp., where he was employed from 1954 to 1984. Mr. Thielsch is the author of the book "Defects and Failures in Pressure Vessels and Piping," extensively used as a reference book of material failures. He has also authored 150 professional and engineering articles covering various aspects of failure analysis, piping and pressure vessel materials, quality assurance and testing. He is a Fellow of the ASM, ASME, ASNT and AWS.

Florence Cone is a senior metallurgical and welding engineer at Thielsch Engineering, Inc. Her assignments include metallurgical examinations and failure analyses of equipment in CPI plants. She has investigated failures in turbines, pumps, valves, shaft bearings, fans and commercial and industrial vehicles. Other projects she has worked include assessing damage by fires to equipment. Ms. Cone received a B.S. in metallurgical engineering from Carnegie Mellon University. She is a member of ASM and the Technical Assn. of the Pulp and Paper Industry.

COPING WITH ELECTROSTATIC HAZARDS

Graham Hearn
Wolfson Electrostatics

In the winter of 1982, the manager of a packaging plant, aware of the potential hazard from static electricity, installed grounding rods inside the plant's 12-metric-ton fiber-reinforced-plastic (FRP) silos. Two weeks later, during filling, one of the silos underwent a violent explosion resulting in extensive damage. Personnel working in the area were shaken up but, fortunately, otherwise unhurt.

The cause of the explosion was an electrostatic spark igniting a dust cloud in the silo's headspace above the bulk product. The newly installed rods did not extend fully to the base of the silo, but ended about a meter short. The bulk product was a relatively good electrical conductor, even though it was a granular organic material. During pneumatic transfer into the silo, it had developed a high electrostatic charge. And due to the insulating nature of the FRP tank, the charge on the granules could not dissipate, and accumulated on the growing mass of material until it reached the tip of the grounding rod. The subsequent electrostatic spark ignited airborne fines and caused the explosion.

These types of accidents are not uncommon in industries where powdered, granular or pelleted materials are handled. Liquids are also at risk. In cases where flammable materials are present, the results can be disastrous. Many engineers in the chemical process industries (CPI) are perplexed by electrostatic hazards, yet their causes and effects are not too difficult to understand. Most significantly, it is possible to drastically reduce their dangers by following some practical guidelines.

Reducing risk

In most CPI processes, flammable atmospheres can be conveniently divided into three groups: gases and vapors; liquid mists or aerosols; and dust clouds. In all these cases, the atmosphere becomes a fire risk only when it is exposed to, or mixed with, air. There is a "flammability range" associated with each type of flammable material, defined by the volume percentage of the material in air. There are many references that give exact data on these ranges (*1, 2*).

For many common industrial solvents, this range lies between 1 — 15%. The lower percentage is usually called the "lower explosive limit" (LEL); the higher is the "upper explosive limit" (UEL). There are, however, some notable exceptions, such as acetylene (1.5 to essentially 100%) and carbon disulfide

Originally published November 1991

(1 — 60%). In addition, increased temperatures and pressures above atmospheric can significantly increase the flammability range.

In many industrial situations, the containment of reactions involving flammable solvents and gases (as either ingredients or reaction products) quickly establishes and maintains an overrich vapor-air mixture for combustion within the reactor vessel. Examples where flammable atmospheres are likely to be created, however, during:

- The beginning or end of a process when solvents are piped in and out of vessels
- The pneumatic addition of solid constituents to vessels containing flammable solvents
- Inspection or the manual addition of constituents when the vessel is opened to atmosphere
- In and around vents or leak points

Another technical term connected with flammable mixtures is the *flashpoint,* which is defined as the lowest temperature at which a substance in an open vessel will create a momentary flash of fire when a flame is brought near it (at a somewhat higher temperature, it becomes a self-sustaining fire).

Danger: aerosols

From the point of view of static electricity, solvents and volatile liquids below their flashpoint do not generally constitute a hazard, since a static spark will never contain enough energy to raise the bulk temperature high enough to sustain a fire. Fine mists or aerosols, however, can be ignited by static discharges even though the liquid is below its flashpoint. Mists of kerosene, for example, can be ignited 20°C or more below its flashpoint of 38°C.

High-pressure leaks and the agitation of low-viscosity liquids are two examples of how such atmospheres could be produced. Hydraulic oil, although not normally considered an ignition risk, has been cited as the cause of more than one serious industrial fire where leaks have occurred.

A large number of industrial processes involve handling powders or dusts at some stage. Generally, although fires can occur in powder layers or in bulk powder in storage vessels, there will only be an explosion if the particles are dispersed in a cloud. The particles comprising the cloud must be separated enough for air to have free access to their surface, but they must also be close enough for one heated particle to affect the next.

This would imply that there is a UEL and LEL for powders, based on density, just as there is a UEL and LEL for vapors based on concentration. An LEL for powders can in fact be determined fairly easily, although in practice, the UEL is more difficult to quantify. The LEL depends on the material and its particle size; generally speaking, for powders that are fine enough to form a cloud, the concentration is above the LEL when visibility through the cloud is limited to a few meters.

In order to gauge the degree of hazard associated with flammable materials, a number of tests and observations must be performed. If a material will ignite and propagate a flame, a minimum ignition energy (MIE) measurement will quantify its sensitivity to electrostatic discharge. The MIE of a material can be defined as the energy of the least energetic spark capable of igniting and propagating flame through that material. The value of MIE obtained for any one material will depend on a number of factors including concentration. In the case of mists or dust clouds, the particle size will greatly influence MIE, with finer particles having a lower MIE.

Flame initiation

In general, gases and vapors are more sensitive to ignition than dusts are, although the energy required to initiate flame through a dust cloud can, for some materials, be as low as 1 mJ — well below the energy in some electrostatic sparks. Assuming the presence of a flammable atmosphere, a hazard will exist whenever there is the possibility of ignition from an electrostatic spark discharge.

It is difficult, if not impossible, to predict the amount of static charge a process may generate. This is mainly because charge generation by friction is a surface phenomenon and therefore any surface contamination or oxidation will have a significant effect.

The most commonly encountered sources of incendiary electrostatic discharge are isolated conductors such as sections of ungrounded equipment constructed from metal. These objects can acquire high-potential charge due to friction, charge induction or conduction. Another source is rotating machinery (such as centrifuges) that build up a charge because they are improperly grounded.

The energy of the discharge can be expressed in terms of the voltage obtained by the object, and its capacitance. Table 1 provides typical capacitances of common industrial items.

Insulating materials, which include most industrial powders and some liquids, will also acquire electrostatic charges. In these cases, the discharges from the surface are of limited energy since the charge is relatively immobile due to the high resistivity of the material. The reported theoretical maximum spark discharge energy from a plain insulating surface is 4 mJ, which is capable of igniting many solvent vapors and gases.

Special consideration should be given to thin plastic layers or linings that incorporate a grounded metal backing,

Knowing the electrostatic properties of products and equipment aids in preventing fires and explosions

such as a pipe or vessel lined with polytetrafluoroethylene (PTFE) or other plastic. Due to the large effective capacitance of this configuration large surface charge densities can be achieved, giving rise to highly energetic, lightning-like sparks known as propagating brush discharges.

The final source of electrostatic sparks in many CPI installations is from individual workers. In this case, there are a number of possible charging mechanisms. Operations such as the pouring, sieving or vacuum transfer of granular or powdered materials are notorious for generating electrostatic charges. Personnel in the vicinity of these operations can themselves become charged with 10,000 V or more. The hazard from charged personnel is particularly serious because not only is the human body an isolated conductor capable of producing an incendiary spark but it is also mobile, and can therefore carry dangerous potentials into hazardous areas.

Armed with this basic understanding of what constitutes an electrostatic hazard, the process or safety engineer is now ready to examine a plant to uncover potential problems. The figure shows a decision tree for evaluating the plant on a systematic basis. It is important as a first step in the investigation to identify the presence of flammable media. This is not always as straightforward as it might seem, since complex chemical processes can generate small quantities of hazardous materials whose hazardous state may only exist for a short period of time.

The spark energies of even small objects are sufficient to ignite hydrocarbon vapors

Conductor	Capacitance in picoFarads	Spark energy in milliJoules at 20kV
Small metal items (spoons, scoops and such)	10 — 20	2 — 4
Containers <0.5m³ volume	10 — 100	2 — 20
Containers >0.5m³ volume	50 — 300	10 — 60
Human body	200 — 300	40 — 60
Reactor vessels and other larger equipment	100 — 1,000	20 — 200

At the beginning of a blending operation, for example, the action of an agitator can generate a fine mist of solvent in the head space of a vessel that could be ignited by a static spark. This could occur despite the solvent itself being well below its flashpoint.

Once any possible flammable atmospheres have been identified, they must be characterized in terms of their sensitivity to ignition. A large amount of data exists for solvents and dusts, and

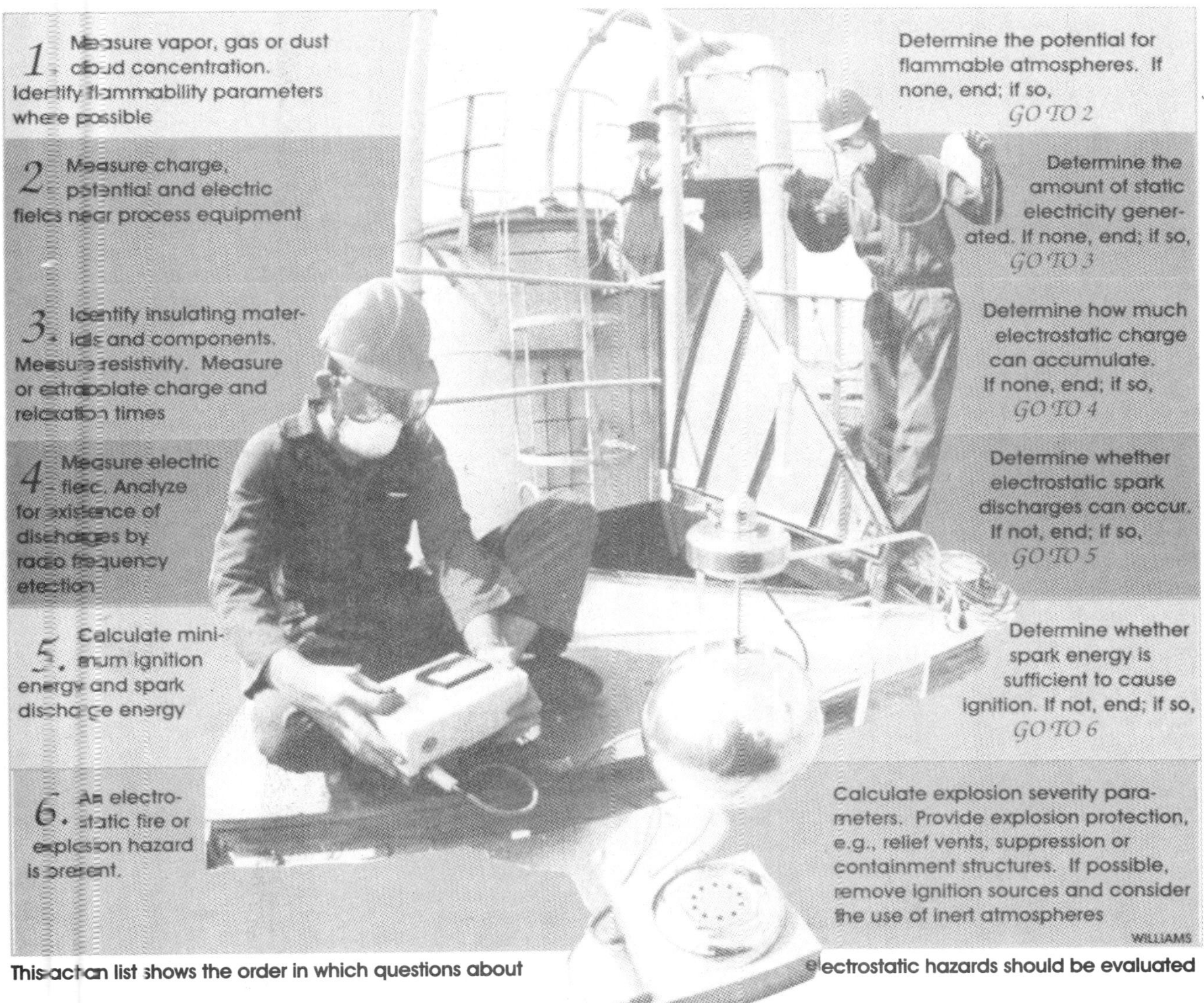

This action list shows the order in which questions about electrostatic hazards should be evaluated

new products can be tested to determine their flammability parameters.

For a spark to occur, electrostatic charge must first be generated. Electrostatic charging is inevitable in many industrial processes, but it can successfully be detected and monitored. Threshold values for the maximum electric field, net charge and charging currents exist for safe operation of many basic CPI processes. Care must be taken during measurement, however, since even the presence of monitoring equipment itself can influence the true effects. Hiring a consulting service may prove beneficial.

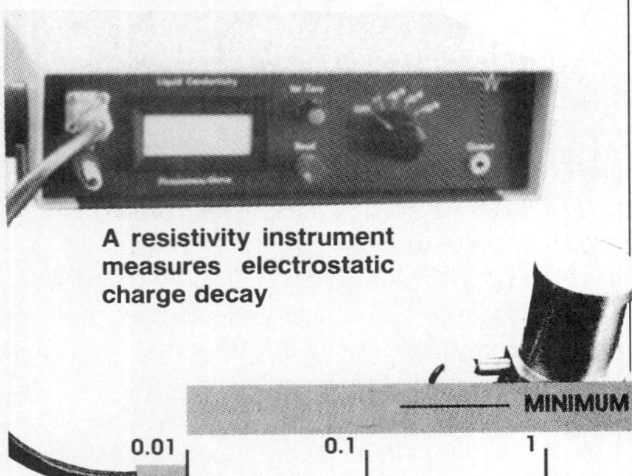

A resistivity instrument measures electrostatic charge decay

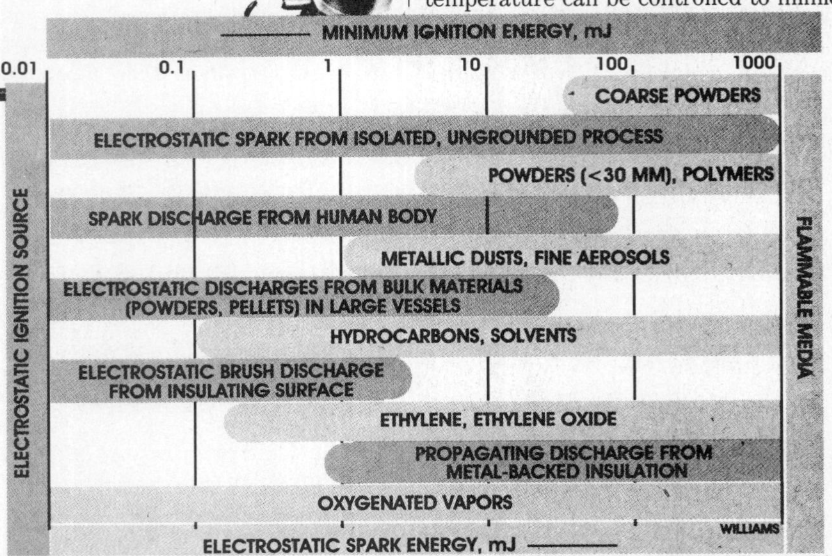

By plotting spark energies (from left to right) along with ignition energies for flammable media (right to left), it becomes clear that many such combinations represent an electrostatic hazard

Electrical insulators

For electrostatic charge to accumulate, some components of the process must be relatively electrically insulating. Seawater running through metal pipework, for example, would hardly be likely to cause a hazard. Most industrial powders are electrically insulating, as are some organic solvents, such as aliphatic or aromatic compounds. The increased use of plastic or composite pipes and linings contribute to a greater degree of hazard. Operations conducted in glass-lined reactor vessels can be a risk as well.

Other factors useful for assessing electrostatic hazards are the resistivity and rate of charge decay of the material of interest. Generally, materials exhibiting resistivity values of less than 10^{10} Ohm-meters will dissipate charge readily and therefore present little risk. These parameters can be measured at the plant or in the laboratory where humidity and temperature can be controlled to mimic worst-case conditions. Commercial instrumentation now exists for making these measurements in the field.

It has already been mentioned that sparks from isolated conducting objects such as a metallic piece of equipment can be highly incendiary and, in fact, this is the origin of most industrial accidents caused by static electricity. Straightforward grounding can normally be applied in most situations once trouble spots have been identified.

Mobile conductors such as transportation devices (cars, conveyors, and so forth) can be maintained at near-zero potential by the use of conductive flooring or conductive wheels. Workers can use conductive (i.e., "antistatic") footwear. A maximum resistance to ground of 10^8 Ohms is generally recommended in situations where the MIE of flammable materials present is below 100 mJ. It is also possible to use industrial-scale ionizers to neutralize highly charged insulating materials, but the effectiveness of this measure is variable.

If, after all these considerations, the risk of an electrostatic discharge cannot be ruled out, then must ask whether the discharge can cause ignition. A rough indication can be obtained by measuring or estimating the maximum possible discharge energy. If this is near or above the MIE of the the surrounding atmosphere, a hazard exists. (In practice, the igniting power of electrostatic discharges is related not only to their energy content but to other parameters such as the polarity of the discharging surface; these factors will also influence the probability of ignition.) At this point, one has to consider the more expensive options of nitrogen blanketing or the provision of explosion relief, suppression or containment.

Finally, it should be appreciated that electrostatic hazards are governed by the statistical probability that all the conditions necessary for charge generation, energetic discharge, ignition and flame propagation are present at exactly the "right" time for an explosion to occur. For this reason, the occurrence of these incidents is relatively rare. Unfortunately, even a single incident in many thousands of operation cycles can have dire consequences. ■

Edited by Nicholas Basta

References

1. Bretherick, L., *Handbook of Reactive Chemical Hazards, 3rd ed.*, (Butterworths, London, 1985)
2. National Fire Protection Assn., *Properties of Flammable Liquids*, (Quincy, Mass., 1969)

The author

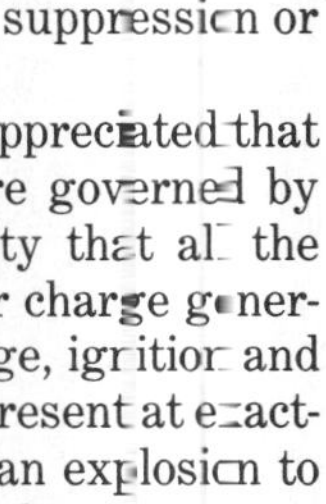

Graham Hearn is a senior consultant engineer with Wolfson Electrostatics Unit based at Southampton University in England. Wolfson Electrostatics (Department of Electrical Engineering, University of Southampton, SO9 5NH; tel: [0] 703 552266; fax: [0] 703 593015) provides consulting services to industry worldwide on hazards prevention and protection. Hearn holds an honors degree in physics, and is a member of both the Institute of Electrical Engineers and the Institute of Chemical Engineers in the United Kingdom.

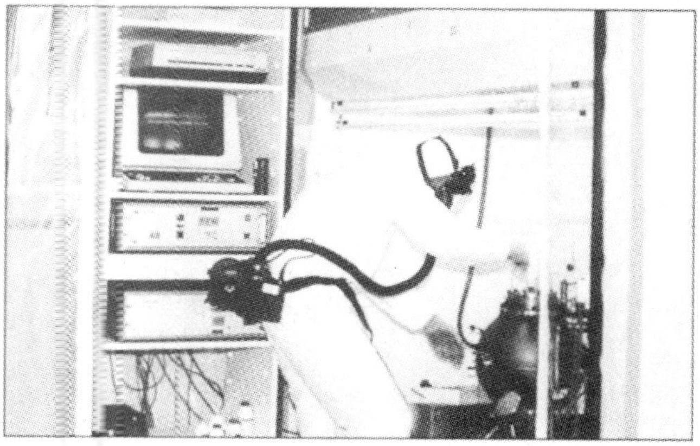

KEEP DUST-CLOUD EXPLOSIONS AT BAY

Vahid Ebadat
Chilworth Technology, Inc.

FIGURE 1. The minimum ignition temperature required to cause an explosion is determined visually by exposing the dust cloud to various temperatures
FIGURE 2. The severity of dust-cloud explosions is checked in this 20-L spherical bomb

Assess the hazards of process dusts in your plant

Last year, the U.S. Occupational Safety and Health Administration (OSHA; Washington, D.C.) — as part of its Process Safety Management rule [1] — required companies to compile detailed data on the hazards posed by the process materials and intermediates they handle. One category of the information required covers the thermal behavior and explosion properties of bulk powders and dusts.

Over the years, as the production and use of powders have grown, so have the number of explosions and fires associated with them. These incidents range from small fires — resulting in loss of production — when bulk powders decompose at high temperatures, to devastating ones — resulting in considerable loss of life and property — when dust clouds ignite suddenly.

The need to generate data on dust clouds is critical, since nearly every sector of the chemical process industries (CPI) handles powders at some stage of production. Among the industries with the most flammable powders are those processing chemicals, pharmaceuticals, plastics, resins, dyes, pigments, foods and metals [2]. Examples include grain, starch, polyethylene, carbon, coal, sulfur, iron and copper.

Further, since the information will be employed in so many critical applications, a systematic approach is needed to compile the data and ensure its accuracy. A step-by-step approach will also help plant personnel identify potential hazards, and develop various options for promoting safety.

OSHA expects the compiled data to serve as a resource for performing process hazards analyses, developing training programs and operating procedures, and conducting plant prestartup reviews. The information can also be used by contractors to monitor plant employees, local emergency-preparedness planners to keep track of process hazards, government regulators to oversee a plant's compliance with safety rules, and insurance agents to assess accident premiums.

TABLE 1. The frequency of occurrence of various ignition sources that resulted in dust explosions in the U.K. (1979–1984)

Ignition source	Number of incidents	Percent of total
Mechanical failure; friction	56	18
Overheating and spontaneous heating	51	17
Flaming materials	44	15
Tramp metal	21	7
Welding; cutting	20	7
Static electricity	11	4
Electrical	10	3
Other	7	2
Unknown	83	27

Source: British Materials Handling Board, Dept. of Trade and Industry

The explosion potential

In order for a dust-cloud explosion to take place, the dust and the atmosphere in which it is suspended should have the following characteristics:

- The dust must be combustible
- The oxygen content of the atmosphere in which the dust is dispersed must be sufficient to sustain combustion
- The dust concentration must be above the minimum explosible concentration
- The particle size distribution must be capable of supporting combustion
- There must be an ignition source of sufficient energy to initiate the explosion

Besides dust clouds, the handling of bulk powders also creates the potential for explosions or fires, or both. For example, a large number of organic and

Originally published December 1993

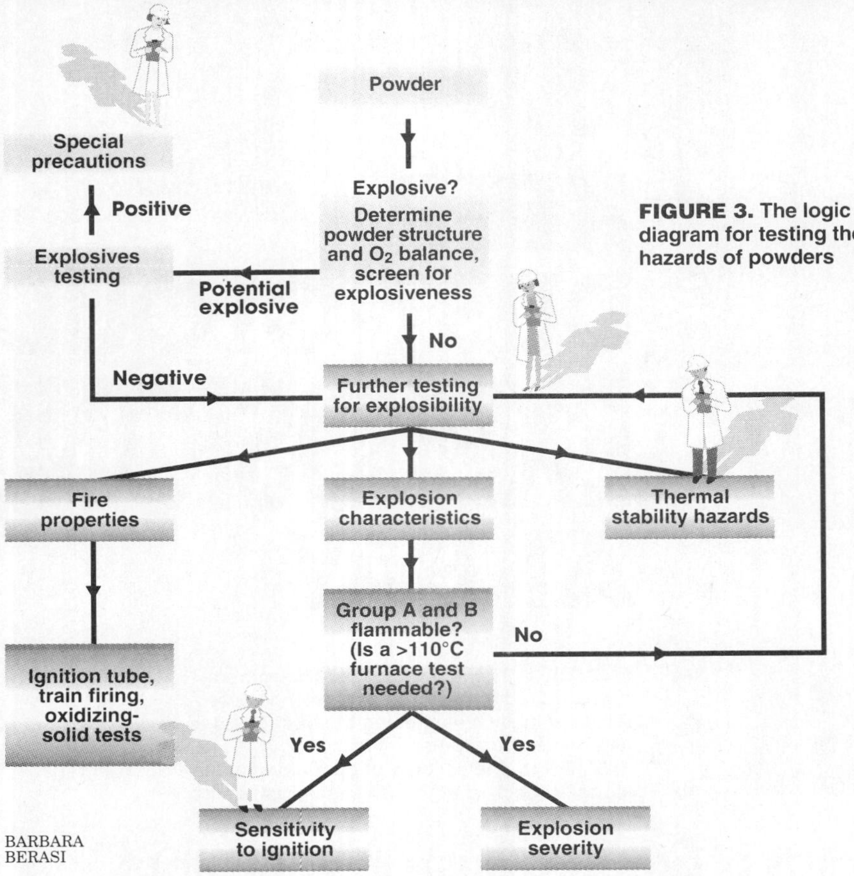

FIGURE 3. The logic diagram for testing the hazards of powders

BARBARA BERASI

some inorganic materials decompose on heating, generating exothermic reactions that degrade product quality and result in fires.

Thus, it is important to determine the maximum temperature to which the product can be exposed before it undergoes a chemical reaction or decomposes. In many cases, a material does not have a unique or sharply defined decomposition temperature, since the latter depends on such factors as air movement over or through the powder, oxygen concentration, the duration of exposure to a particular temperature, shape or geometry of the holding vessel, particle size and bulk density of the powder, and product contamination.

Ignition sources

To cause an explosion of a dust with the right properties (combustibility, dispersibility, size and concentration) requires an ignition source with the appropriate amount of energy. However, opinions vary as to the relative importance of different ignition sources in various powder-handling industries.

In general, the following sources of ignition may have sufficient energy to initiate a dust explosion: flames; sparks from welding or cutting operations; friction or impact sparks; electric sparks; electrostatic discharges; and hot surfaces. Table 1 shows the frequency of occurrence of different types of ignition sources for a number of reported dust explosions in the U.K. between 1979 and 1984.

Dust layers and deposits can be ignited by sparks, incandescent particles and other high-temperature sources. An important feature of dust deposits is that they may also be subject to smoldering combustion. This can occur by a process of self-heating when the temperature of the dust layer or bulk material is raised to a level at which the heat liberated by the exothermic reaction is sufficient to exceed the heat losses. Eventually, the smoldering material may burst into flame, especially when disturbed.

An initial screening test is necessary to identify and isolate powders that may have deflagration or detonation properties. Figure 1 shows a logic diagram for dust explosion testing.

The tests for explosibility involve consideration of the chemical composition and the oxygen balance of the product molecules. If the chemical screening indicates that the material cannot produce adequate quantities of heat by oxidation, then dust explosions can be excluded and further testing is unnecessary. (Note: When handling explosive materials, care should be taken to observe the precautions specific to the explosives industry.)

The first stage of a hazard assessment is to determine whether a dust cloud will explode when dispersed and exposed to an ignition source, such as electrostatic sparks or heated surfaces and environments. This is done by visual observation in an *Explosion Classification Test* (ECT), in which a suspended dust's propensity for propagating flames or generating pressures determines whether or not the material is capable of initiating and sustaining an explosion.

Dust-cloud ignition

Under the ECT, Group A defines dusts that burn and propagate flames, while Group B are dusts that do not ignite. However, since some Group B powders are combustible at higher temperatures, it is important to carry the ECT for Group B powders at a temperature of 1,000°C. This will demonstrate whether the dust cloud will explode if subjected to operating temperatures above 110°C.

Next, one needs to investigate whether a dust cloud will self-ignite when it contacts a hot surface, such as the metal casing of electrical equipment (Figure 1). This test, carried out in a furnace whose temperature can be adjusted up to 1,000°C, determines the *Minimum Dust Cloud Ignition Temperature,* also known as the auto-ignition temperature.

The third test is the *Minimum Ignition Energy* test, which measures the smallest amount of energy that is required to initiate a dust-cloud explosion. These data are used to determine whether electrostatic discharges are capable of igniting the dust cloud.

Once the information on the dust cloud's flammability is obtained, the next stage involves measuring the *Dust Explosion Violence*. The data obtained from this test will be required for the design of explosion-protection measures, such as relief venting, and suppression or containment of explosions.

In this test, a dust-air mixture is introduced into a 20-L sphere and ignited by chemical ignitors (Figure 2). The resulting explosion pressure vs. time curve yields values of the maximum pressure, P_{max}, and the maximum rate

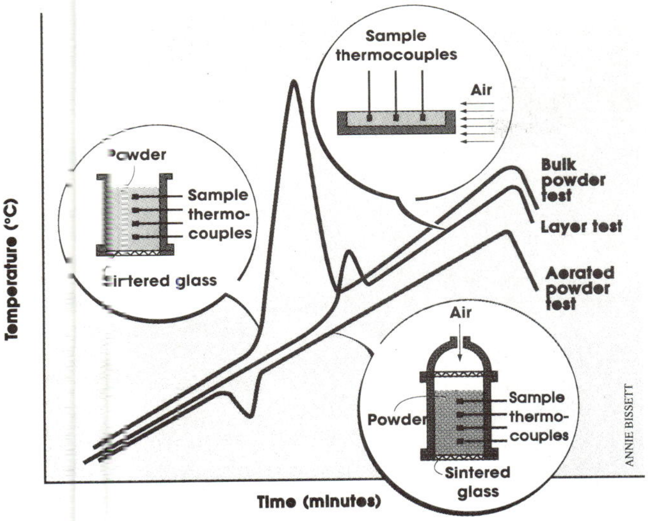

FIGURE 4. Various types of cells are used for testing the explosibility of bulk powders. These include: a diffusion cell for testing the hazards of bulked powders (top left), a cell for checking the hazards of air flowing above powder layers (top right), and an aerated cell for testing dust hazards when air flows through the powder (bottom)

of pressure rise, $(dp/dt)_{max}$, gives an indication of the relative severity of the explosion.

Bulk-material testing

Just as with dust clouds, a reliable assessment of the thermal instability of dust layers and deposits and of bulk materials subjected to heat can be made with small-scale laboratory testing. For example, the situations in which bulk powders are subjected to elevated temperatures, such as those occurring at the bottom of some dryers, silos or bag houses, can be simulated by the *Bulked Powder Test*.

In this test, the amount of air available to the powder is limited to natural diffusion via a sintered-glass disk at the bottom of the test cell (Figure 4). The powder is then heated isothermally at various temperatures, or ramped up, typically at a rate of 0.5°C/min. Depending on the temperature at which the material undergoes exothermic activity (or explodes or decomposes), a maximum temperature for the largescale operation can then be recommended.

In many industrial situations, however, air is forced or induced through a bulk powder, such as at the end of a fluid-bed drying cycle when the fan is switched off but an induced flow of air may be present. To assess the probability of a bulk powder to decompose under such conditions requires the use of the *Aerated Powder Test*, to determine the minimum ignition temperature for the onset of exotherm activity.

Figure 4 shows the temperature and time relationship of a powder tested in an aerated cell, in which air is blown through a cell filled with bulked powder. Here, the powder undergoes endothermic activity at 148°C.

Similarly, a *Powder Layer Test*, in which air flows above a dust layer is required to determine the minimum ignition temperature of a powder layer exposed to a flow of heated air. This test simulates tray drying and other operations where heated air flows over the product, but not through it, as in the Aerated Power Test. Such conditions also arise on ledges and walls of dryers. Figure 4 shows the exothermic activity of a powder subjected to the layer test.

As can be seen, a dust explosion can be prevented by: eliminating the dust cloud; operating below the minimum explosible concentration; eliminating all possible ignition sources; and, operating in inert atmospheres (such as by inerting vessels with nitrogen). However, process safety should be based on the elimination of potential ignition sources only as a first step, rather than as the ultimate basis for plant safety.■

Edited by Jayadev Chowdhury

References

1. Kuryla, M. and Yahay, S., New safety rules add to plant managers' worries, *Chem. Eng.*, June 1992, pp. 153–160.
2. Eckhoff, R.K., "Dust Explosions in the Process Industries," Butterworth Heinemann (Stoneham, Mass.), 1991.

The author

Vahid Ebadat is the technical manager of Chilworth Technology, Inc. (Princeton Corporate Plaza, 11 Deer Park, Suite 204, Monmouth Junction, NJ 08852-9623; Tel: 908-274-0900), a firm of consulting engineers and scientists specializing in assessing the hazards posed by process fires and explosions. The company provides testing services for determining the potential for dust explosions, and consulting on explosion prevention and protection.

Table 5: PLASTICS

Material	Median particle size, μm	Minimum explosive concentration g/m³	P_{max}, bar ga	$(dP/dt)_{max}$, bar/s	K_{St}, bar-m/s	Dust Hazard Class
Polyacrylamide	10	250	5.9	12	12	1
Polyacrylonitrile	25		8.5	121	121	1
Polyethylene (low-pressure process)	<10	30	8.0	156	156	1
Epoxy resin	26	30	7.9	129	129	1
Melamine resin	18	125	10.2	110	110	1
Melamine, molded (wood flour and mineral-filled phenol-formaldehyde)	15	60	7.5	41	41	1
Melamine, molded (phenol-cellulose)	12	60	10.0	127	127	1
(poly) methyl acrylate	21	30	9.4	269	269	2
(poly) methyl acrylate emulsion polymer	18	30	10.1	202	202	2
Phenolic resin	<10	15	9.3	129	129	1
Polypropylene	25	30	8.4	101	101	1
Terpene-phenol resin	10	13	8.7	143	143	1
Urea-formal-dehyde cellulose, molded	13	60	10.2	136	136	1
(poly) vinyl acetate-ethylene copolymer	32	30	8.6	119	119	1
Polyvinyl alcohol	26	60	8.9	128	128	1
Polyvinyl butyral	65	30	8.9	147	147	1
Polyvinyl chloride	107	200	7.6	46	46	1
(poly) vinyl chloride-vinyl acetylene emulsion copolymer	35	60	8.2	95	95	1
(poly) vinyl chloride-ethylene-vinyl acetylene suspension copolymer	60	60	8.3	98	98	1

tain the structural integrity of the collector at the chosen P_{red}.

According to NFPA, external wind loads upon an explosion vent are to be taken into account when determining the actual P_{stat} value at which the vent will open. This requirement is essentially ignored by most people, because no standard method for determining the external wind-load effect on an explosion vent has been developed. If the collector is located outdoors, it is best to locate an explosion vent on the prevailing leeward side.

Vent ducting

As noted earlier, it is sometimes necessary to vent the explosion panels through a duct to the exterior of a building. Such ducting directs the flame front, the hot gases and the pressure away from personnel and adjacent equipment. It also lessens or eliminates the threat of a secondary explosion caused by dust on the floor, and of indoor structural members being dispersed and ignited by the deflagration gases.

According to NFPA's criteria, the cross-section of the duct must be greater than the cross-sectional area of the vent. Furthermore, the duct must be as straight as possible.

The addition of vent ducting requires the dust collector (and the duct itself) to withstand a higher internal pressure. The NFPA-68 document includes a graph* that shows the increase in P_{red} if a vent duct is applied. The graph deals separately with ducts that are less than, and ones that are more than, 3 m long.

The duct length makes a big difference. Consider, for instance, the case when P_{stat} is 0.05 bar and P_{red} without ducting is 0.10 bar (both of these being commonly employed values). If a duct of less than 3 m is added, the graph stipulates a rise in P_{red} to 0.415 bar—but if the duct is instead longer than 3 m, P_{red} rises all the way to 1.17 bar.

This strongly suggests that if ducting is required, the engineer should strive for a layout that keeps the duct length under 3 m. Even this length usually requires heavy reinforcing of the collector shell. If staying under 3 m is difficult or impossible, one may wish to refer to a publication available from the U.K.'s Institution of Chemical Engineers, entitled "Guide to Dust Explosion Prevention and Protection / Part 3 - Venting of Weak Explosions and the Effect of Vent Ducts;" it covers situations requiring ducts over 3 m long, with or without bends.

Don't overlook thrust

In addition to the effect of pressure on the dust collector walls, NFPA-68 states that the magnitude and duration of the mechanical thrust forces generated by the deflagration gases must be calculated, and taken into account during the design of the dust collector's supporting structure. Formulas for these two quantities can be found in the NFPA-68 document.

NFPA allows one to assume that the reactive force occurs in a plane that passes through the geometric center of the dust collector. On the other hand, the structural designer cannot take it for granted that if two vents are 180 deg apart, the reactive forces cancel each other out; it is in fact quite likely that they will not open simultaneously.

And keep in mind…

The explosion-vent calculations and decisions discussed here are not affected by the presence or absence of the various other safety features with which a dust collector is likely to be outfitted, especially if the system is handling an unusually explosive dust. These features include: explosionproof electrical enclosures and motors, static-resistant drive belts, non-sparking fans on the

*It appears in Figure 5-4(b) on p. 68-18 of the document.

"clean" air side of the process flow, nonsparking dampers, grounded filter bags for dissipation of static charges, fire-suppression systems, contact closures indicating an open explosion vent, and automatic isolation valves on the collector inlet duct to prevent backflow of deflagration gas and flame.

Compliance with NFPA criteria does not end after the vented collector has been installed. Maintenance requirements involve basic common-sense responsibilities such as making sure that the vents are not damaged, obstructed or corroded, and, of course, immediately correcting any abnormalities noted.

Records of maintenance inspections are to be maintained. In addition to checking the actual vents, the author suggests that the inspection include an internal and external survey of the collector shell, vent ducting, and auxiliary equipment, such as rotary valves.

Finally, bear in mind that although NFPA-68 is the most widely accepted method for sizing explosion vents, other methods do exist. NFPA-68 is, in the author's opinion, the most conservative of the methods applied in the U.S. industry. Even so, whenever a potential explosion-hazard condition exists, be sure to contact the insurance engineer for the facility where the equipment is to be installed, to determine if explosion venting is required and, if so, how this venting should be sized and selected. ■

Edited by Nicholas P. Chopey

Caution: This article should not be used as the sole basis on which to design a vent system. Before any venting arrangement is selected, the reader should become thoroughly versed in NFPA-68.—*The author*

The author

Gregory J. Black is an applications engineer in Fuller-Kovako Corp.'s Air Pollution Control Group (2158 Avenue C, Bethlehem, PA 18017; Tel. 215 264-6893). He is also the Emergency Management Coordinator for Northampton Borough, Pa. Before joining Fuller-Kovako, he worked for the Powder Processing and Air Pollution Control groups of Hosokawa Micron. Holder of a B.S. in Mechanical Engineering from Fairleigh Dickinson U., he is a New Jersey Engineer-in-Training (E.I.T.) and an Associate Member of the American Soc. of Mechanical Engineers.

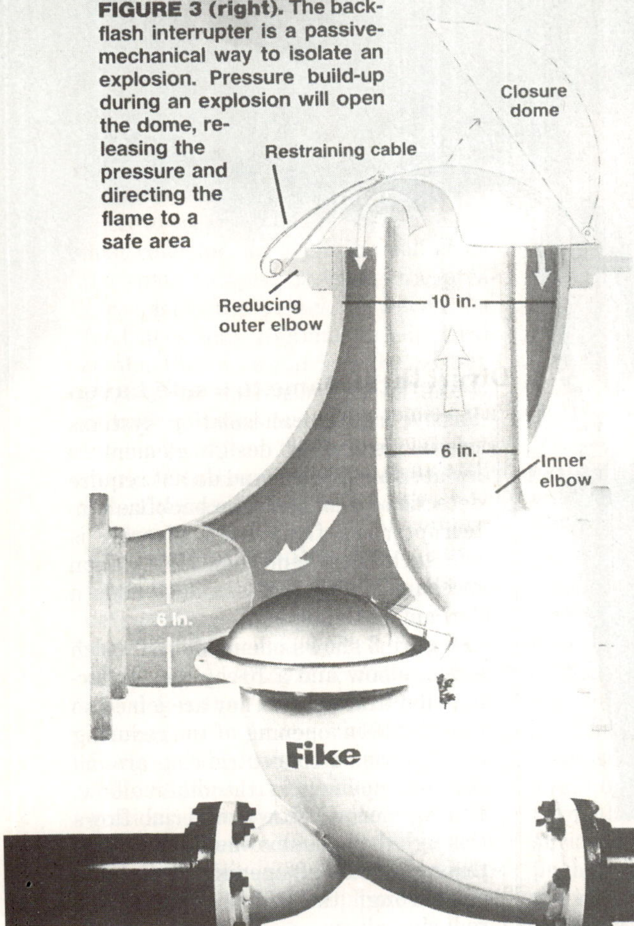

FIGURE 3 (right). The backflash interrupter is a passive-mechanical way to isolate an explosion. Pressure build-up during an explosion will open the dome, releasing the pressure and directing the flame to a safe area

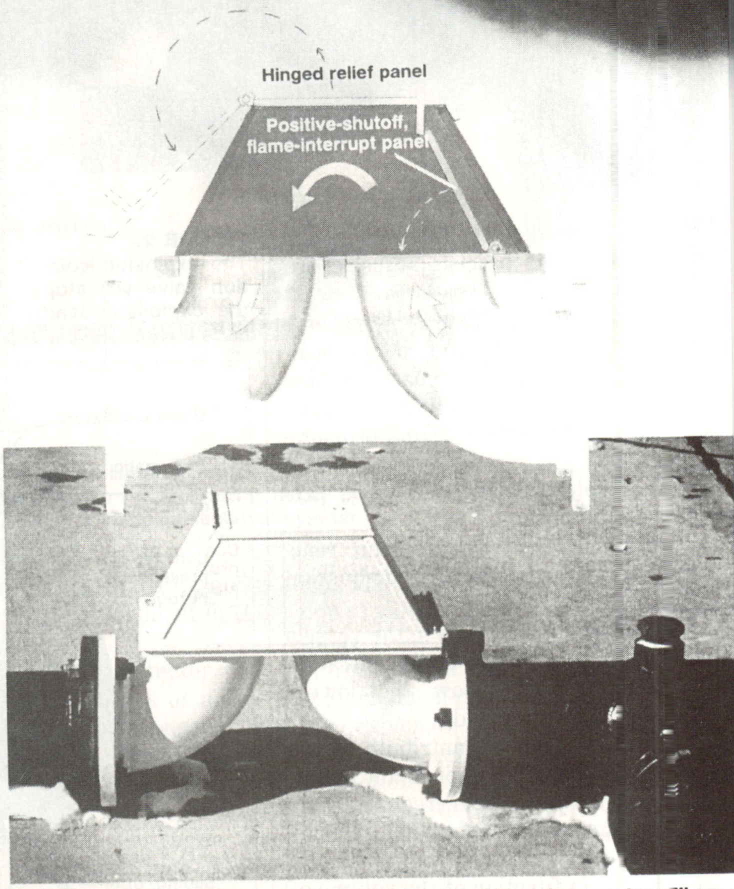

FIGURE 4 (above). The door on this type of backflash interrupter will open during an explosion. As it does, it releases a positive-shutoff panel inside the device, which cuts off the flow of combustible material

flow. In the event of an explosion, the pressure buildup opens the dome, and propagation of the explosion is halted, as the flame is released into an open area. Clearly, this area should be devoid of equipment or personnel, and should be constructed of fire-resistant material. Further, if the product in question is too hazardous for release to the atmosphere, this is not a suitable method of explosion isolation; in such cases, explosion-isolation valves or chemical-suppressant systems should be used.

Figure 4 shows a different version of the backflash interrupter, in which two 6-in. elbows are connected by a housing, which has a relief panel at the top and a positive-shutoff "flame interrupt" panel inside. During normal operation, the relief panel is closed and the flame-interupt panel remains open, allowing uninterrupted flow through the device. In the event of an explosion, the rapid pressure buildup opens the relief panel, which in turn trips the flame-interrupt panel into a closed position, to create a physical barrier to isolate the explosion.

The passive-mechanical backflash interrupter is relatively new compared to explosion-isolation valves and chemical-isolation systems. Available in sizes ranging from 2 to 12 in. in dia., they provide a simple, low-cost alternative to the other, electrical response systems. ■

Edited by Suzanne Shelley

The author

Kris Chatrathi is director of product development and testing for Fike Metal Products (704 S. 10th St., P.O. Box 610, Blue Springs, MO 64013, Tel.: 816-229-3405) a div. of Fike Corp. He earned his B.S., M.S. and Ph.D. degrees in chemical engineering from the University of Arkansas. Chatrathi is a member of the American Institute of Chemical Engineers (AIChE), and serves on the American Soc. of Testing and Materials (ASTM) committee on the hazard potential of chemicals.

References

1. Bartknecht, W., "Explosions: Course Prevention, Protection," Springer-Verlag, New York, 1981.
2. Field, P., "Dust Explosions," Elsevier, New York, 1982.
3. Chatrathi, K., and R. DeGood, Explosion isolation systems used in conjunction with explosion vents, Powder and Bulk Solids Conference Proceedings, Chicago, pp. 114-127, 1989.

BRACE YOUR PLANT AGAINST A DUST EXPLOSION

Thomas Godbey, DCE, Inc.

Throughout the chemical process industries (CPI), substances such as plastics, fertilizers, fuels and pharmaceuticals, are routinely dried to powder or dust. When suspended in air or other oxidants, these dusts are capable of producing a dangerous and costly explosion.

While it is difficult to precisely define the explosion risk created during the handling of potentially explosive materials the risk is ever present. Explosions should be anticipated, and preventive measures must be taken in the design and selection of bulk-solids-handling equipment, such as dust collectors, conveyors, bins, grain elevators, and size-reduction equipment, such as grinders, crushers and pulverizers.

Selection of the correct explosion protection cannot be left to chance. The risks include the possible loss of life, damage to capital equipment, loss of production and increased insurance premiums. Many of the hazards can be effectively avoided by thoroughly testing the dusts to ascertain their explosion potential, and taking the appropriate protective measures.

The potential explosion hazard involved in dust handling has prompted testing, federal regulations, and insurance considerations. The sheer power of dust explosions was dramatically demonstrated by the destruction of several large grain silos in the U.S. in 1977. Although much has been done to minimize risk, such incidents are reminders of the devastation and loss of life that any explosion can inflict inside an industrial plant.

To regulate dust handling in the CPI, the U.S. Occupational Safety and Health Administration (OSHA) requires the presence of protective systems, including hooding, ducting and dust-collection devices. Most major dust-collection vendors provide such systems.

Suspended powders, volatile gases and just one spark: a deadly combination

There are three pieces to the explosion puzzle: a combustible dust, air or another oxidant, and a source of ignition. During operation, a cloud of finely divided particles is held in suspension in the vessel of a dust collector — making the perfect explosion environment, if other critical components (such as volatile gases) were to be introduced.

If a source of ignition initiates the combustion of a dust cloud, the gases in the cloud will rapidly expand, due to heat developed during combustion. If a dust-collector vessel constricts this gas expansion, a rapid pressure build-up inside the collector casing will cause a violent explosion.

When the flame speed in a explosion is less than the speed of sound at the appropriate pressure, the explosion is called a *deflagration*. When the flame speed is greater than the speed of sound, the result is a *detonation*. The majority of industrial explosions are deflagrations.

Some explode, some don't

Many commonly handled dusts have been tested to determine the degree of explosion hazard that each poses, and the results are published by the National Fire Protection Assn. (NFPA 68,

Originally published April 1991

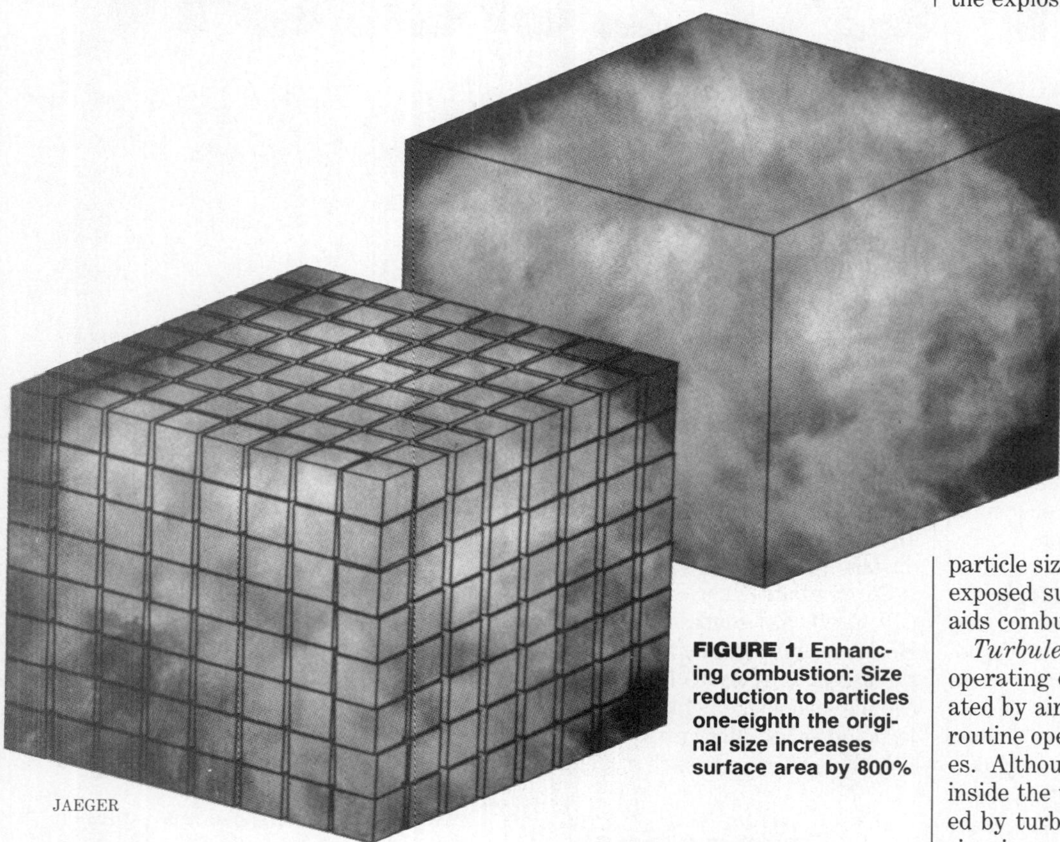

FIGURE 1. Enhancing combustion: Size reduction to particles one-eighth the original size increases surface area by 800%

"Venting of Deflagrations," 1988 Ed.). The explosion potential of a dust confined within a vessel can be characterized by maximum explosion pressure (P_{max}) and the maximum rate of pressure rise (dp/dt_{max}).

If the explosion potential of a particular dust is unknown, a sample may be submitted to a commercial testing laboratory for characterization. To classify a dust in terms of the risk it poses, the substance is subjected to a source of ignition and described as a dust that either does or does not propagate flame.

If the dust does propagate flame, further tests are conducted to determine the following parameters: the minimum ignition temperature, the minimum concentration of dust (by weight) required for an explosion, and the minimum ignition energy in joules.

Also measured are the maximum permissible oxygen concentration (oxygen deprivation will inhibit ignition of the cloud), the maximum explosion pressure and the rate of pressure rise. Pressures as high as 1,035 kN/m^2 (or 150 lb/in^2) and rates of pressure rise as high as 140,000 kN/m^2/s (or 20,000 lb/in^2/s) have been measured in dust-explosion tests.

In addition to the physical characteristics obtained from testing the dust itself, a number of other factors affect the explosion potential of a dust cloud:

Source of ignition — As intuition would suggest, a large ignition source causes more-rapid combustion than a smaller one. A typical small source may be a spark or contact with a hot surface. A typical large ignition source may be a flaming dust cloud passing through the ducting.

Particle size — While the maximum pressure of a dust cloud is relatively unaffected by the size of the individual particles, the rate of pressure rise increases significantly as particle size decreases. Additionally, as particle size is decreased, the amount of exposed surface area increases, which aids combustion (Figure 1).

Turbulence — Even under normal operating conditions, turbulence is created by air flow a dust collector during routine operation and cleaning processes. Although the maximum pressure inside the vessel is only slightly affected by turbulence, the rate of pressure rise increases significantly as turbulence increases.

Moisture — The presence of moisture in the dust particles raises the minimum temperature required to ignite the dust cloud. The presence of moisture in the gas surrounding the dust, though, has little effect that entrained moisture is quickly atomized after an explosion begins.

Hybrid mixtures — These are mixtures of flammable gases with dusts. Some dusts are not flammable in air, but become flammable with the addition of even a small amount of flamma-

STEPHEN WALSH

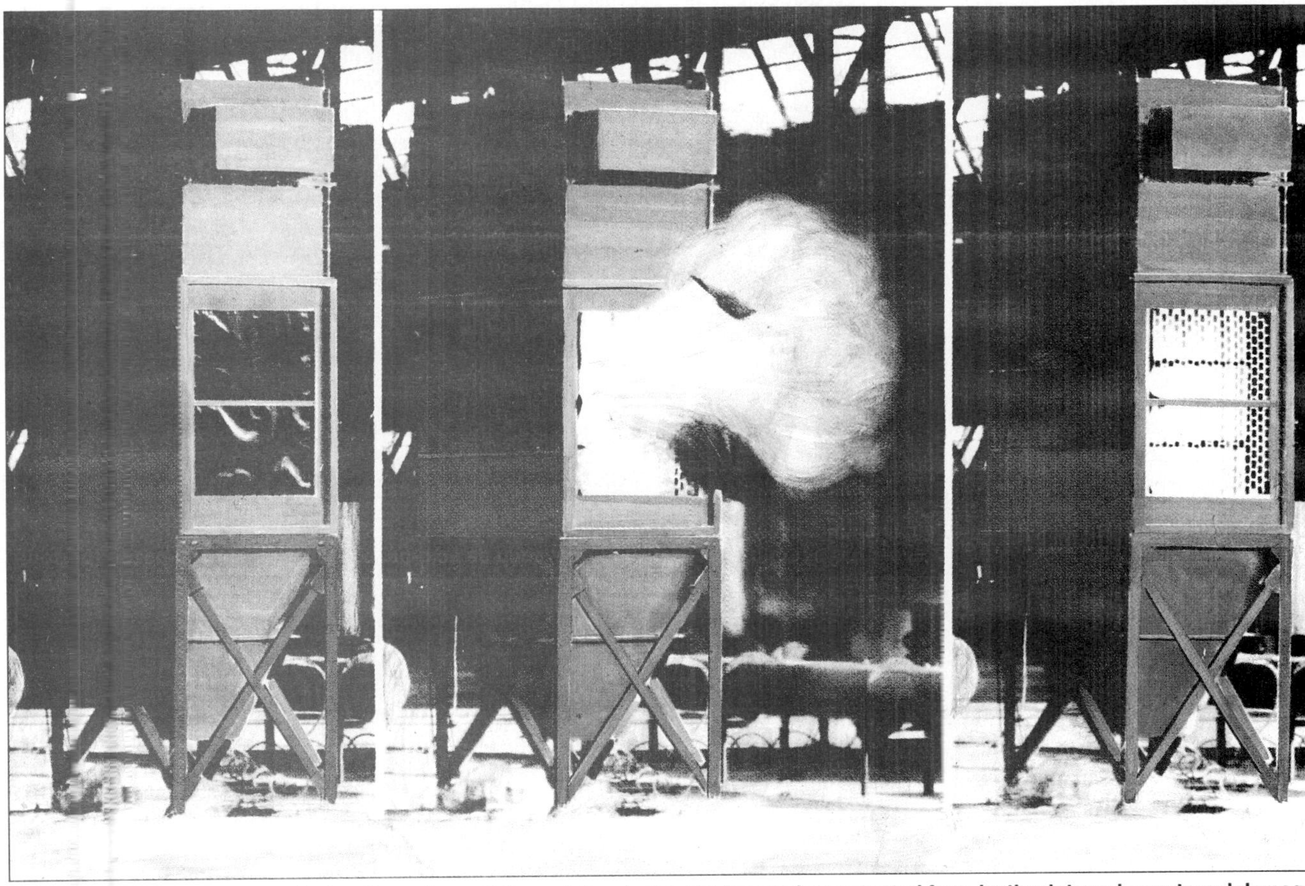

FIGURE 2 By venting an explosion through a relief panel, the dust collector can be protected from lasting internal or external damage

ble gas. The gas increases the risk of explosion, and considerably reduces the minimum ignition energy required to initiate an explosion.

A match to a tinderbox

Sources of possible ignition, present in every process plant, must be identified and controlled to minimize the risk of a dust-cloud explosion. Flames and smoldering particles can arise from dryers, grinders, furnaces, kilns, ovens and mechanical handling equipment, such as conveyors, or simply from bad housekeeping.

Friction on equipment, such a bulk chemical grinders, creates hot surfaces, which can ignite a dust cloud. Spontaneous combustion is also a risk when certain materials are contained in bulk.

Finally, the movement of air in dust-collector ducts creates static electricity. The static discharge from dusts can reach an energy level up to 50 millijoules (mJ). If the minimum ignition energy required to ignite a particular dust cloud is below that level, such static electricity may be all that is needed to ignite the cloud. Some dusts can be ignited by an energy source of only a few millijoules.

Defuse an explosive situation

Awareness of the risks involved in working with dusts and powders in a process environment is of paramount importance. A prevention program aims to minimize each of the risks. In addition to preventive measures, a number of steps can be taken to reduce the destructive effects of an explosion:

Limit static buildup — To eliminate one possible ignition source, specific dust-collector components should be grounded, to continuously discharge static electricity as it builds up. The development of "epitropic" filter fabric is a recent innovation that helps to disperse the build-up of static charges in a dust collector.

Epitropic filters are made from conductive fibers that have been impregnated with carbon. The filters must be connected to an effective grounding point on the outside of the dust collector. The fabric's resistance is less than 1×10^8 ohms. According to the NFPA (77, "Static Electricity," 1988 Ed.), a resistance of 1×10^{10} ohms or less will provide an adequate leakage path to bleed away static build-up in most applications.

Inerting — The addition of an inert gas to replace oxygen in a dust collector can prevent an explosion, ensuring that the minimum oxygen content required for ignition is never reached. In open-circuit dust-collection systems (those that bring in and exhaust air), however, inerting is often not economical, as the open system means a constant loss of expensive inerting gases.

Explosion suppression — This protective method requires the early detection of an explosion, usually within the first 10 milliseconds. Once ignition is detected, an explosion-suppression device injects a pressurized chemical suppressant into the collector vessel, to displace the oxygen and impede combustion.

Such suppression systems can be operated in conjunction with rapid-acting isolating valves that are often used in both the inlet and the outlet headers of a dust collector. Such isolation valves are often used when toxic dusts are being handled. The valves prevent an explosion from sending toxic dusts into the ducting leading into and out of the collector.

Explosion containment — There are specialized dust collectors on the market that have been designed to withstand the maximum pressure generated during an explosion. Commercially available dust collectors, however, are often designed to withstand only 7–14 kN/m^2 (1–2 psi); this is not sufficient to contain an explosion in progress.

Explosion relief — This method of alleviating rapidly building pressures is commonly used with dust collectors. As pressure increases quickly leading up to an explosion, a relief vent opens to allow the rapidly expanding gases to escape, effectively reducing the pressure build-up. Relief vents should be located on the "dirty" (unfiltered) side of any filter, as the filter media itself acts as a barrier to the expanding gases.

The necessary area for such a relief vent is a function of the vessel volume, vessel strength, and the maximum pressure that the vent closure can withstand, as well as the rate of pressure rise characteristic of the dust in question.

The most comprehensive design basis for venting dust deflagrations is contained in VDI Richtilinie 3673*. The work is based on data obtained from an extensive test of four dusts in four different-size vessels. These test results are presented in a series of nomographs now published by the NFPA (68, "Venting of Deflagrations," 1988 Ed.).

Pressure-venting options

Two main types of explosion vents are available: explosion-relief doors and bursting-panel relief vents. Explosion-relief doors should be lightweight and retained by springs, by magnets, or by gravity alone. Flexible "pop out" panels also may be included in this category.

In most cases, the restraining pressure of the door is listed as the stated relief pressure. This alone is not always a true representation of relief pressure.

* Pressure Release of Dust Explosion, VDI Richtilinie 3673, Verein Deutscher Ingenieure - Kommission Reinhaltung der Luft, Dusseldorf, 1979 and 1983, VDI Verlag GmbH, Dusseldorf

FIGURE 3. Rapid-fire test results: 180 milliseconds from beginning to end

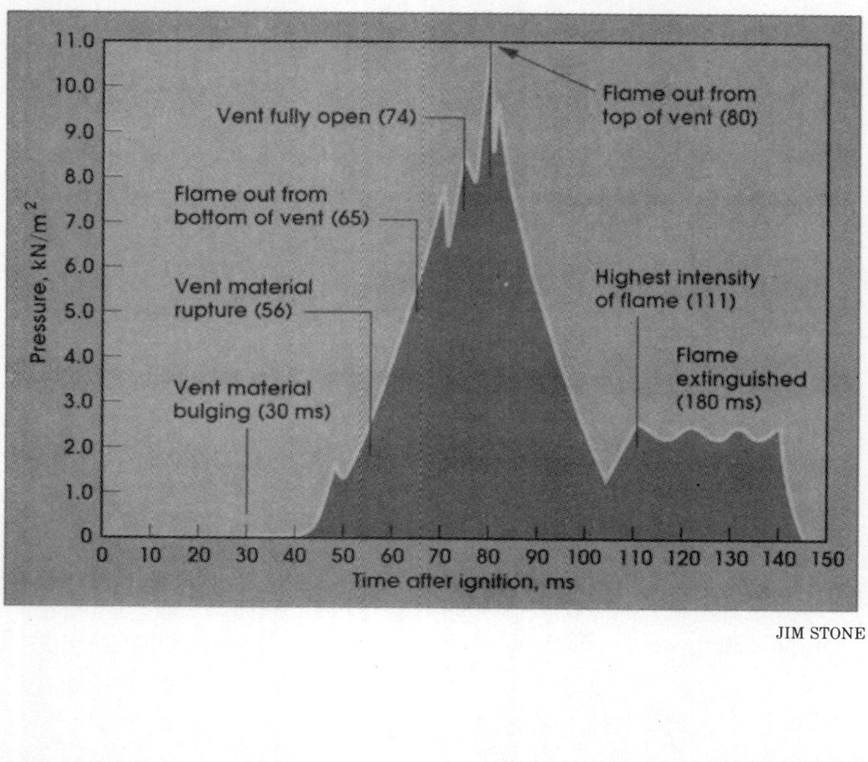

JIM STONE

In a recent series of tests carried out under actual explosion conditions, the inertia of the door was found to be an additional factor. Thus, the force needed to blow out these doors is often greater than first assumed. The total weight of the door assembly, including any insulation and permanently mounted hardware should be as low as is practical, and in no case should it exceed 3.5 lbs/ft^2 (NFPA 68, "Venting of Deflagrations," 1988 Ed.).

One disadvantage of using an explosion-relief door is the considerable increase in force that may be added by corrosion or freezing between the door, the casing and the restraint interfaces. The use of hinges as a means of restraint should be avoided, as the high negative pressures following an explosion could result in the door swinging closed and causing an implosion of the filter casing.

Often held in place by a non-corrosive magnetic strip, the door is maintained in a vertical plane to eliminate water traps that could develop when used outside. Wire ropes are often used to prevent the door from flying off during an explosion.

In addition to explosion-relief doors, bursting-panel relief vents are also used to alleviate pressure buildup inside a vessel, to prevent an explosion. In this method, a membrane of known bursting pressure is used instead of a door over the vent opening.

There are many membrane materials available. Comprehensive testing is required to select the materials with the optimum bursting characteristics

For dust collectors, the membrane must be: weatherproof; sufficiently resilient to withstand the normal working pressures experienced within the collector; capable of withstanding abrasion and chemical attack from the dust being handled; and conductive to prevent the accumulation of a static charge.

Figure 2 shows a violent explosion in progress, rupturing a rear-mounted bursting panel on a dust collector. The pressures recorded throughout this explosion test on the dust collector's rear-mounted bursting panel are shown in Figure 3.

Ideally, the best location for an ex-

FIGURE 4. A spectacular aspirin-powder explosion through a top-mounted relief vent. Typically, the pressures recorded during these instances are higher than during the explosion of a rear-mounted pressure-relief vent, because of turbulence created by the filter bags in the dust collector

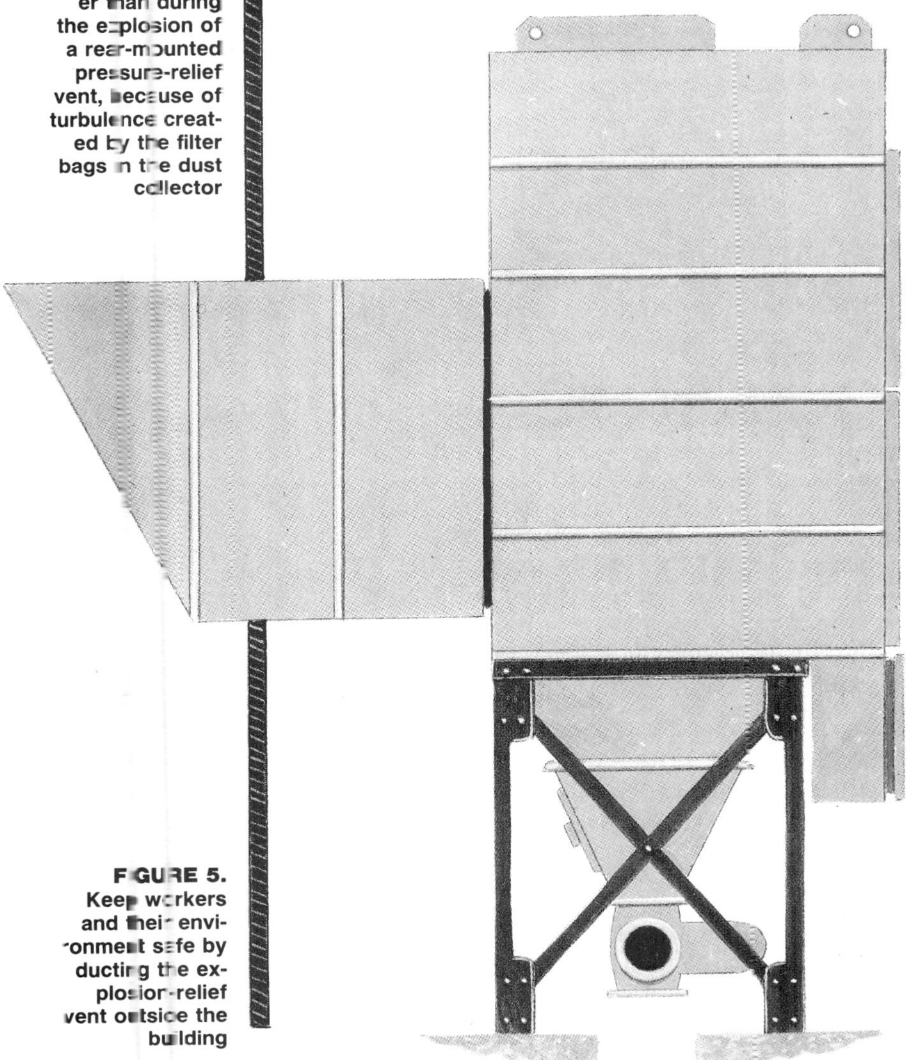

FIGURE 5. Keep workers and their environment safe by ducting the explosion-relief vent outside the building

STEPHEN WALSH

plosion-relief vent would be as close as possible to the source of ignition; however, predicting the ignition source is nearly impossible.

Realistically, the vent should be positioned where it is least affected by internal fittings, which could increase turbulence if the flame front has to pass through them. The vent may be on the side or the top of the equipment (Figure 4).

The best position for an explosion-relief vent is found to be adjacent to the hopper. Access to this position, however, may be precluded by the factory layout.

Extreme caution is advised when using explosion-relief vents. Any dust collector equipped with explosion-relief vents should be sited outside of the facility, if possible. Otherwise, an explosion inside the collector would cause the relief doors or panels to blow out into the plant, possibly injuring personnel, damaging the facility, or polluting the air.

When factory layout prevents putting the whole unit outside, the relief vent should be ducted away, to a safe place outside the building. To prevent blockage during an explosion, the ducting from an explosion-relief vent should be as short as possible (preferably no longer than 10 ft) and with as few directional changes as possible. Both the length and the shape of the ducting will inhibit its effectiveness.

The relief duct must be designed to withstand the same pressure buildup as the dust collector. Figure 5 shows a rear explosion-relief vent that has been ducted to the outside of the building. Many options are available for weather protection for outside relief ducting. Again, the chosen method must offer the least resistance to the explosion path.

Edited by Suzanne Shelley

The author

Thomas Godbey, Sr. is the technical manager at DCE, Inc. (11301 Electron Dr., Jeffersontown, KY, 40299, Tel.: 502-267-0707). Godbey holds a B.S. degree in agricultural engineering from the University of Kentucky (Lexington, KY), and has more than 17 years of experience in the air pollution control industry as an application and product engineer. He has published several articles on basic industrial ventilation, static grounding and explosion venting, and has lectured extensively in the U.S. and Canada. ■

FIRE WALLS

THE LAST LINE OF DEFENSE

Know what they are, how they work and when to use them

Peter J. Gore Willse and Anson C. Smith
Industrial Risk Insurers

Fire walls are necessary for two reasons. First, they contain fires and limit fire damage. Second, fire walls are typically required by national, state and local safety codes to protect employees and property under certain circumstances.

A fire wall is a structure, usually freestanding, that protects an interior area in a building. In most cases, a fire wall is entirely contained within a building. In this article, four types of walls are defined. These are differentiated by how they are built and by their rated ability to resist fire.

There are no hard and fast requirements governing the thickness and materials used in the walls. They are primarily categorized on how long and how well they are able to withstand fire. Even the most robust of walls, however, must be properly maintained to provide adequate performance.

Fire walls should be built with designs and materials that have met the requirements of ASTM standard E 119 (Standard Method of Fire Testing of Building Construction and Materials). ASTM E 119 provides specific instructions for fire-endurance tests, such as placing a fire wall sample in a furnace and heating it to a particular temperature for a specified time. Characteristics to be measured include the length of time for which the wall remains structurally intact, as well as its ability to limit the amount of heat passing through it.

A fire wall and a standard fire wall (one with no openings) are designed to remain standing when the structures around them collapse. To withstand the expansion that accompanies heat, they are usually thicker than walls with lower fire ratings. If they are of considerable height and length, they may be buttressed by crosswalls or pilasters.

Fire barriers and fire partitions subdivide areas and can be attached to, or supported by, structural columns. Although these have lower ratings than fire walls, they, too, must be built according to designs tested by nationally recognized testing laboratories. Some fire walls, which border areas containing explosive materials, must resist the

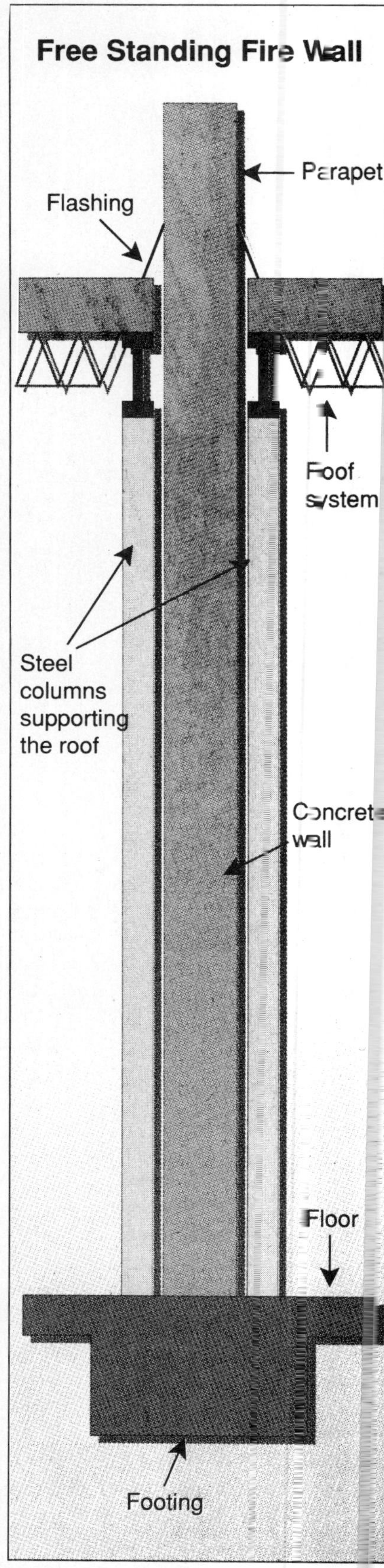

FIGURE 1. A freestanding fire wall, which generally extends above the roof to form a parapet, does not have a structural function. It supports only its own weight and is tied to the roof solely by the roof flashing

Originally published July 1993

overpressure and missiles that can result from the explosion of these materials. These walls require complex engineering designs and are beyond the scope of this article.

Most fire walls are built to designs tested and listed by testing laboratories. These designs, however, only recognize fire rating (i.e., how long a wall can withstand fire). There are other design factors that should be taken into account. Fire walls should evaluated for the following criteria:

Ability to withstand fire. The wall is rated according to ASTM E 119 for the length of time for which it endures fire.

Stability under applied design loadings. These should include dead loadings — e.g., from heavy equipment or other material hanging from the wall — as well as live loadings, such as wind, snow, and earthquake forces.

Other forces that could affect the wall. These include collapse of the roof, building contents (such as elevated vessels or racks) or adjoining structures; or explosion of the contents of the area, such as pressure vessels or flammable material.

Effect of thermal expansion on the building's structural steel and on the wall itself. The former depends upon the length of the joists and the height of the steel columns; the latter, upon the construction material, height and width of the wall.

Standard fire wall

The standard fire wall is a freestanding masonry wall with no openings (Figure 1). It has at least a 4-h, fire-resistance rating. The standard fire wall is designed to contain a fire in the fire's area of origin, even after other firefighting efforts, such as sprinklers, fail. In addition, it must:

- Be reinforced, supported or designed to withstand damage from the fall, collapse or expansion of stored items
- Resist fracture, penetration and fragmentation that can be caused by a fire
- Extend from exterior wall to exterior wall, or from one standard fire wall to another
- Extend vertically and continuously through all storeys of the building and through the roof to form a 3-ft (0.9-m) parapet over the highest point in the roof or of any structure within 25 ft (7.6 m) of the fire wall
- Have expansion or control joints, which protect a wall against buckling when rising temperatures cause the wall to expand. Control joints, which are normally narrower than expansion joints, may be spaced at 100- to 150-ft (30–46 m) intervals in masonry walls
- Have a minimum 3- to 6-ft- (0.9 to 1.8-m-) long wingwall, a 30-ft- (9.2-m-) long endwall, or a 3-ft- (0.9 m-) long extension wall, as needed (Figure 2). Endwalls and wingwalls* are blank masonry structures (i.e., having no openings) that prevent fire from passing around or through fire walls. These are joined perpendicularly to the fire wall's ends, forming part of the building's exterior walls.

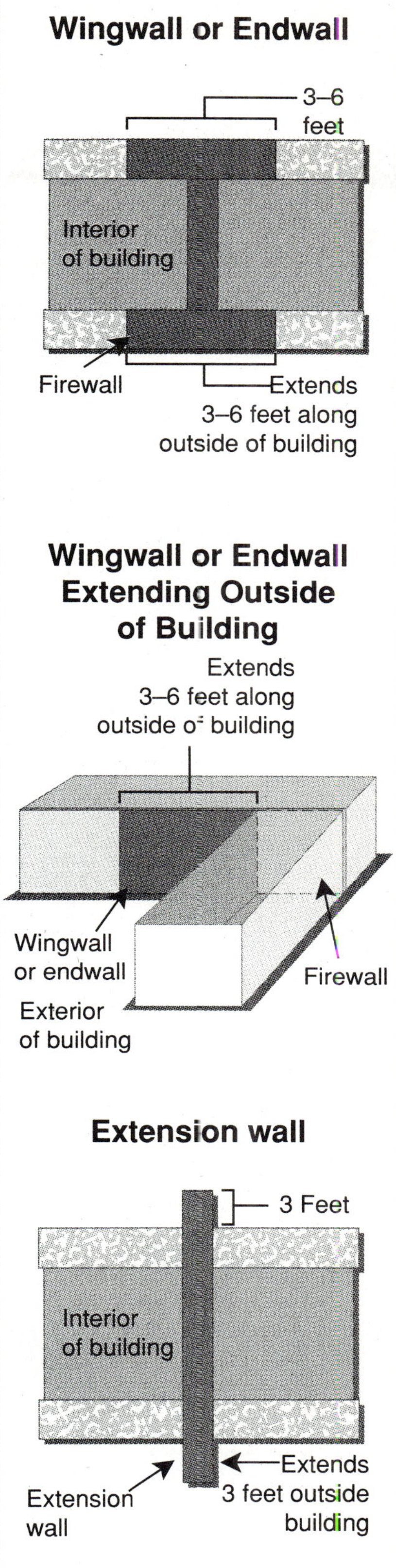

REBECCA PERRY

FIGURE 2. Wingwalls and endwalls meet an interior fire wall perpendicularly. Where the external walls of two buildings meet, an endwall extends past an interior fire wall at the juncture of the "L" (middle). Extension walls (bottom) extend on each side of an interior fire wall

Where two buildings meet perpendicularly to form an L-shape, an endwall is appended to an interior fire wall to extend 30 ft (9.2 m) beyond the vertex of the L (Figure 2, middle). It should be constructed of blank masonry, or other fire-rated material.

Extension walls can substitute for endwalls or wingwalls (Figure 2, bottom). These extend, in line with the fire wall, 3 ft (0.9 m) beyond the building's exterior walls.

A fire wall is similar to a standard fire wall, but has protected openings and a 3- to 4-h fire-resistance rating. This classification also includes tied walls and double one-way walls, which will be discussed in this section.

There are several types of fire walls and standard fire walls. These include:

Freestanding fire walls. These have no ties to the building other than roof flashing, which is a thin piece of sheet metal that horizontally joins the wall to the roof. This type of wall is usually made of brick, concrete, or concrete masonry blocks and is connected to its footing with reinforcing bars.

The wall is usually installed in an expansion break in the building framing. It must rely on its own strength to overcome forces such as expansion caused by temperature differences inside and outside the wall, or collapse of building components or stock. Reinforcing rods or pilasters

*The terms endwalls and wingwalls are generally used interchangeably in the industry.

are installed in the wall to give it the required strength.

Clearance between the wall and the structural framework (i.e., roof, structure and columns) must be maintained to allow for the deflection of the wall and expansion of the steel.

Tied fire walls. These are masonry or other rated constructions that encase the building's steel framing, or that are tied to the steel framing and coated with a fireproofing material (Figure 3). These walls are held in place by the steel framework.

To maintain the walls' stability, the roof steel on either side must be strong enough to resist forces caused by steel collapsing on a fire-exposed side. Bolts are usually attached to the roof beams and passed through the wall.

Double one-way fire walls. This type comprises two one-way walls, which are walls that are rated as fire walls, but have exposed structural steel on one side. The one-way walls are placed back-to-back, with a minimum 7-in. (178 mm) separation between them. Each wall must have at least a 3-h, fire-resistance rating.

Double fire walls are most commonly built when a fire wall is required between an existing structure and a new building. An existing wall secured to the existing building frame can be upgraded to the required fire-resistance rating. A new, rated fire wall can be raised next to it and secured to the new building frame. If a fire destroys one wall, the wall supported by the other side should remain standing.

Fire barriers

Although they can have lower fire-resistance ratings than do fire walls, fire barriers must also be built according to designs tested by nationally recognized laboratories. Fire barriers subdivide floors and can be attached to or supported by structural members.

Fire barriers, which have a 2- to 3-h fire-resistance rating, are usually nonload-bearing walls that extend from the floor to the roof, and are supported by columns. All supporting structures, such as roofs, columns or floors, should be as noncombustible or fire resistant as the barriers they support.

Fire barriers do not require parapets or endwalls, and are generally not freestanding. They restrict the initial flow of heat within the area of origin, and limit the operation of sprinklers outside the fire zone. They also provide employees with sufficient time to coordinate and begin manual firefighting efforts from adjacent areas.

Fire barriers are most effective when heat and smoke vents are provided and sprinklers are operable. A fire barrier supplements a sprinkler system. For example, a sprinkler system might be adequate for the area it protects. However, during a fire one or two sprinkler heads may not work properly due to lack of maintenance or a mechanical problem. In this case, the fire barrier supplies protection by containing the fire.

If there is a possibility of an intense fire on either side of a barrier, a double or freestanding barrier can be constructed between a building's structural columns. A double barrier is similar to a double, one-way wall, but comprises two, one-way fire barriers rather than two fire walls.

The one-way fire barrier is the simplest of fire barriers. It is designed to withstand a fire on one side only. It is effective for separating an area with low-combustible loading (i.e., an area whose contents will generate relatively little heat during a fire) from a higher-hazard one. It is tied into the building's structural steel or other framework on the low-hazard side.

During a fire, steel can begin to weaken at about 1,000°F (540°C) and cause the roof to collapse. As the roof collapses, it starts pulling on the wall. For this reason, a single, tied fire wall is relatively ineffective. However, it can be useful to protect a building from fires in adjacent buildings.

A one-way fire barrier should be a nonload-bearing wall:

- Installed in line with a row of a building's structural columns, and having through-barrier ties to building framing on both sides
- Installed to follow a line of structural columns so it can benefit from the columns' vertical strength and resist twisting forces as much as possible. All columns and trusses parallel to and within 1 ft (0.3 m) of the barrier should have the same fire-resistance rating as the barrier

Where two buildings are joined with

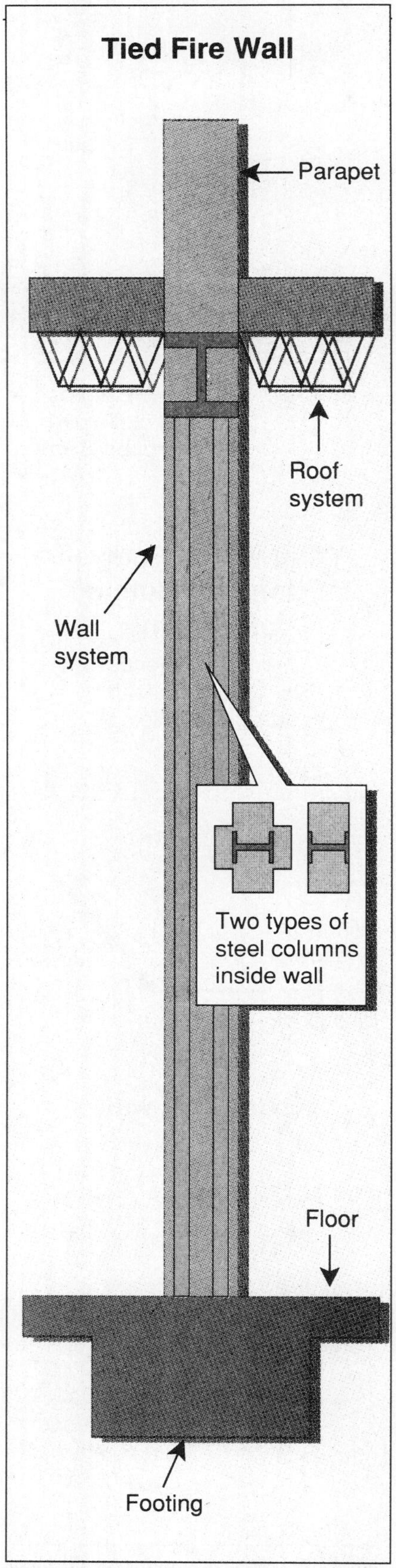

FIGURE 3. A tied fire wall, which is structural, supports its roof. Its construction can contain one of two types of steel columns (inset); the walls are coated with a fireproofing material

a barrier, each side of the building must be balanced for stability. That is, the structural steel must be made to support equal loads on both sides. This must also be done in cases where a building is separated into two sections by a barrier. In the barrier, ties to building framing in the plane of the barrier should be placed at the same level or levels on both sides.

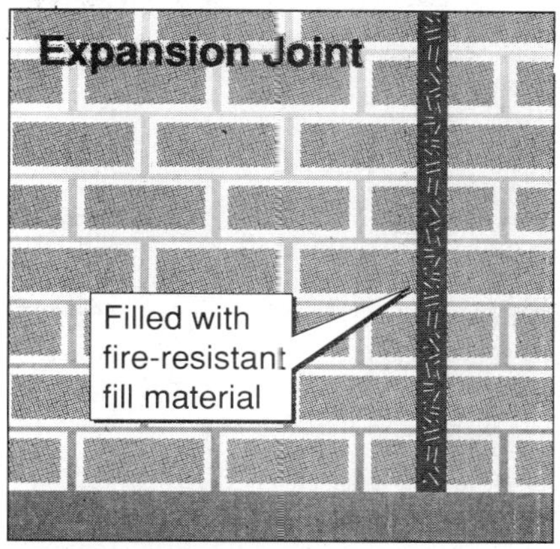

FIGURE 4. Expansion joints allow walls to expand at high temperatures

Fire partitions

Fire partitions subdivide areas and can be attached to or supported by structural members. The least protective walls in this article, fire partitions extend to the ceiling only and are constructed of less-fire-resistive materials than fire barriers. However, they too must be built according to designs certified by testing laboratories.

The terms barrier and partition are generally used interchangeably in the industry. However, there is a difference between the two structures. Fire partitions have a 1- to 2-h fire-resistance rating. In all other respects, they are similar to fire barriers.

Opening protection

The biggest cause of fire wall failure is unprotected or improperly protected openings, such as doorways, conveyor openings and cable penetrations. Such openings have allowed the spread of some of the most destructive industrial fires in history.

To maintain the integrity of the wall and to keep fires from spreading, the number and size of openings in a fire wall should be minimal. Openings should not constitute more than 25% of the area of any fire wall and should have the proper protection.

All openings in fire walls and fire barriers should be protected with equipment listed by testing laboratories. Opening-protection components include automatic fire doors, shutters and dampers; or through-penetration, fire-stop assemblies. They should have fire-resistance ratings equal to or greater than the walls in which they are installed (Table). Manually operated doors, windows, shutters, and water curtains are not viable means of protecting openings in fire walls or fire barriers.

Fire door assemblies should be listed by a nationally recognized laboratory. The listing should be based on testing in accordance with codes ASTM E 152, UL10B or NFPA 252 (all are titled Standard Methods of Fire Tests for Door Assemblies). These tests, similar to testing methods for fire walls, involve placing a sample of a fire door in a furnace and rating its endurance.

These tests are not usually performed on doors larger than 120 sq. ft (11 m^2). Therefore, if an oversized door is required, the one selected should be certified and labeled as an "oversized door."

Fire doors should be simple, direct acting, and reliable. When selecting a design, be sure that

- Self-releasing features are incorporated in the design of the track, chain, supports and other components
- Sprinkler discharge will not impede operation of any fusible link on the fire door. A fusible link, which consists of two pieces of metal soldered together, is typically installed on a fire door, holding the door open. When the ambient temperature in the area reaches a certain level, the solder melts, allowing the door to close
- If a conveyor belt carries objects through an opening protected by a fire door, counterweights, springs and other components must be designed with enough strength to push material on conveyor belts out of the way, allowing the fire door to close

Fire doors are rated as 4-, 3-, 1½- and 1-h, as well as 45, 30 and 20 minutes. The doors are also rated at three temperature levels that indicate the maximum temperature transmitted to the unexposed side after 30 minutes: up to 250°F (121°C), up to 450°F (250°C), and up to 650°F (361°C). In other words, the lower the temperature rating, the more protection the door offers, since it conveys less heat than a door with a higher rating.

A fire door with a temperature level

RATINGS OF STANDARD FIRE WALLS, WALLS, BARRIERS AND PARTITIONS

TYPE OF CONSTRUCTION	RATING	CONFIGURATION
Standard fire wall	4-hour minimum blank with no openings	Parapet extends above roof with wingwalls, endwalls or extensions
Fire wall	3 to 4 hour with protected openings	Parapet extends above roof with wingwalls, endwalls or extensions
Fire barrier	2 to 3 hour with protected openings	Wall extends from floor to beneath roof deck
Fire partition	1 to 2 hour with protected openings	Wall extends from floor to ceiling

Source: Industrial Risk Insurers

Table. Any wall with less than a 1-h fire resistance is unacceptable as fire cut-off.

higher than 250°F (121°C) may allow sufficient heat from one side to pass through to ignite combustibles on the other side. This is why combustibles placed too near a fire door sometimes ignite from a fire on the other side of the door.

Types of fire doors include: swinging, horizontally sliding, vertically sliding and rolling. Rolling doors are metal doors that unroll downward to close. A door should be chosen depending upon how it will be used and what space clearances it will have.

The biggest cause of fire wall failure is unprotected openings

Hardware is a critical part of the fire door. A door that does not operate properly in an emergency is as useless as no door at all.

Booklets of listings issued by testing laboratories provide recommendations of specific hardware to use with types of doors. The hardware should be installed according to manufacturers' specifications.

Fire doors used for egress must be designed and installed in accordance with the requirements of NFPA 101-1991, Life Safety Code. Single-personnel doors (i.e., those allowing one person at a time to pass) should not be larger than 3 ft x 7 ft (0.9 m x 2 m). Glass is allowed in a ½-hour-rated fire door.

Fire doors should be provided in all openings in fire walls and fire barriers as outlined in the Table. Barriers (i.e., bumpers or guards, such as metal posts) should be installed to prevent vehicular traffic from damaging the doors and their mechanisms.

Closing mechanisms

Fire doors are arranged to close automatically via links with certain devices. The most common such device is the heat-sensitive fusible link. A fusible link is installed over a door opening and at the ceiling on both sides of the wall. When activated, the link releases a latch, which releases the door or the counterweights that trigger the operation of the door.

Another device is the fire detector. Fire detectors are located on each side of the wall, either over the opening or at the ceiling. When activated, the detector releases the door or a set of weights that triggers the operation of the door.

A third device is a fire-suppression system. This can be a sprinkler system, water-flow alarm, carbon dioxide system or foam system. When the system is activated, it automatically releases the fire door.

Through-penetration protection

In fire walls or fire barriers, "through penetrations" that allow passage of cables, ducts, pipes and conduits also present problems. Thus, they should be sealed with noncombustible material having a fire-resistance rating equal to that of the wall.

If an opening is no longer required, it should be permanently sealed. Be sure to include through penetrations when calculating whether openings in a fire wall occupy 25% of its area.

Ducts should be made of noncombustible materials and have automatically activated fire dampers where they pass through the fire wall or fire barrier. A fire damper is a mechanism, placed in a duct, that closes off the duct when the temperature reaches a certain level.

Improperly designed ducts can be problematic. For example, during a fire, the roof can collapse, exerting pressure on a duct. If the duct is not designed to break free from a fire wall, it can exert excessive pressure on the wall, enlarging its through-penetration opening.

To avoid this scenario, ducts should pass around a wall, as should piping, conduits, cables and cable trays. If any of these components must pass through the wall, they should do so as near the floor as possible, preferably as near as possible to the wall's intersection with other walls.

Equipment passing through a wall more than 3 ft (0.9 m) from the floor should be designed to break away during a fire. Slip joints can be installed to allow ducts and cable trays to detach from the wall. A slip joint is a plastic sleeve that fits into an opening in a wall; the two ends of the duct or cable tray meet inside the sleeve.

Information about the type of fire-control walls for various occupancies can be found in the appropriate National Fire Protection Assn. (NFPA) codes, as well as in IRInformation, a publication of Industrial Risk Insurers (Hartford, Conn.). NFPA is currently drafting a document on how to build fire walls and barriers. The document is available to the public in draft form (readers can obtain a copy by calling NFPA headquarters in Quincy, Mass. at 1-800-344-3555 and asking for Proposed NFPA 221). The document is still open to public comment. ■

Edited by Irene Kim

The authors

Anson C. Smith is publications editor at Industrial Risk Insurers (IRI; P.O. Box 5010, 85 Woodland St., Hartford CT 06102-5010; Tel. 203-520-6016, fax 203-549-5780). He is responsible for writing and producing The Sentinel, IRI's quarterly magazine (international circulation: 64,000). He oversees production of IRI's employee magazine and handles a variety of marketing and public-relations projects. He has a bachelor's degree from Boston University (Mass.) and has completed master's degree work at Rhode Island College (Providence, R.I.).

Peter J. Gore Willse is a research consultant in IRI's National Research Department. He conducts research and writes for the IRInformation manual, and instructs employees in the areas of building construction, warehousing, combustion controls and special hazards. He has worked for IRI since 1979, as a field engineer and district supervisor in Florida, Germany, New Jersey and Connecticut. He has a B.S. in civil enginering from the University of Rhode Island. He is a member of the International Assn. of Arson Investigators, National Fire Protection Assn., American Soc. of Civil Engineers and Soc. of Fire Protection Engineers.

HOW TO FOIL THE CURSE OF CORROSIVE GASES

Richard Paciej
The BCC Group Inc.

This panel has been scarred by corrosion products in acid-gas service MEGAN REID

Understanding the quirks of various corrosion processes is important in selecting the right material of construction for components, such as piping, valves, and storage cylinders used in handling corrosive gases. The objective, of course, is ultimately to achieve a corrosion-free gas containment and distribution system.

Materials exposed to corrosive gases, such as H_2S and H_2Se, can corrode by direct reaction with the gas under relatively dry conditions. Under moist conditions, this corrosion is often accelerated. Progressive damages aside, the reaction products may be detrimental to critical manufacturing processes, such as semiconductor fabrication because of particles and gas-purity problems.

Underestimating the harmful effects of corrosion may invite premature equipment-failures, and heighten prod-

Originally published May 1991

Simple tests may provide the data needed to avoid equipment failure

uct contamination. But overestimating it may also be expensive due to overdesigning of systems. Also bear in mind that materials of construction, surface treatments, moisture content of the gas, and system purging all affect the overall corrosion-resistance of corrosive-gas handling systems.

Moisture + reactive gases

Many of the halogen-containing gases, such as HCl and WF_6, are not very corrosive when dry. However, when these gases react with moisture, their corrosivity increases significantly. For example, while 0.1 ppm moisture in HCl gas does not attack 316L stainless steel (SS), 200 ppm moisture can significantly corrode the SS's surface [1].

The reaction products of moisture with most corrosive gases are either acidic or alkaline (see Table 1). As shown, some of the reaction products contain HF acid or HCl acid, which are most corrosive to stainless steels.

Corrosion attack can be accelerated by the reaction of dry gases with moisture adhering to component surfaces. Moisture can reside in crevices or pits on machined metal surfaces, particles, hygroscopic corrosion products, lubricants, and polymeric materials. For example, inadequate purging of moisture from a diaphragm valve exposed to HCl gas can induce formation of large amount of corrosion products on the seat holder, and cause small-scale corrosion on the spring (Figure 1). Purging the valve with a dry inert gas, and reducing dead space (that is, unventilated regions) and surface area exposed to the corrosive gas are necessary to minimize corrosion.

Moisture is present on the surface of metals in three different forms. Water can be chemisorbed, physisorbed, or can bond to existing layers of water already held on the surface by even weaker bonds.

The condensed water can be removed by purging the system with a dry inert gas. However, the physisorbed and chemisorbed moisture must be removed by heat. For example, moisture residue may remain lodged on a ceramic filter surface even after purging with dry N_2 gas at 60°C for over two hours. Baking at temperatures of 260°F, and higher, is required to remove the remaining moisture.

FIGURE 1. Moisture condensation has induced local corrosion on the seat holder of a diaphragm valve

Moisture is also trapped in plating or passive films (of metal oxide or hydroxide, for example) fabricated by means of wet chemical processes. This can be removed only by purging or "degassing" at elevated temperatures during component fabrication.

Many faces of corrosion

The types of corrosion that form because of the presence of moisture in gas systems are quite varied. The major categories include uniform, galvanic, crevice, selective leaching, fretting, intergranular, or pitting corrosion.

Uniform corrosion is the general attack of a surface by the gas, such as HCl acid, or the gas and moisture reaction products. It is characterized by a reaction that proceeds uniformly over the entire exposed surface. This type of corrosion is the easiest to protect against. For example, nickel coatings can be used to prevent the uniform corrosion of steel exposed to liquefied HCl.

Galvanic corrosion is associated with a reaction resulting from the electrical coupling of dissimilar metals. The electrolytic connection between the dissimilar metals is made by a thin layer of moisture, or corrosion products containing moisture.

A number of techniques can be employed to minimize galvanic corrosion. For example, dissimilar metals can be insulated from each other using a nonconductor, or the materials can be chosen to minimize the electrical potential between them. Table 2 presents a galvanic series of various metals in sea water. Similar tables can be constructed for other environments. Typically, materials positioned near each other in the galvanic series should be used together to minimize galvanic corrosion.

Precipitates or impurities in metals and the metal matrix may act as a galvanic couple resulting in the corrosion of the least corrosion-resistant material. The rate of corrosion will depend on the relative activity and amount of each phase. For example, impurities in stainless steel containing sulfur or phosphorus increase the metal's overall corrosion rate as the impurity content increases. This is why 316L SS, which contains low impurities, is recommended for corrosive-gas systems.

Crevice corrosion is a form of localized corrosion occurring at locations where easy access to the bulk environment is prevented, such as at mating surfaces of metals or assemblies of metal and nonmetals. Moisture and gas can become entrapped in crevices and react to form a corrosive solution. Crevice corrosion can occur in valves at diaphragm-seat holder connections, and valve-cylinder contact areas.

Selective leaching or dealloying is a discriminatory process affecting one or more components of a solid solution alloy. A classical example is the dealloying of brass (CuZn alloy). Zinc is selectively corroded from the alloy because its corrosion resistance is lower than that of copper. Other examples are the dealloying of nickel from Cu-Ni alloys, iron from Fe-C (steel), and aluminum from Cu-Al alloys.

Fretting corrosion is the deterioration at the interface of two contacting surfaces under load, accelerated by relative motion between them that is of sufficient magnitude to produce slip. This type of corrosion can occur, for example, in valves where there is a metal-to-metal or metal-to-polymer rubbing, while the metal component is corroding away.

Intergranular corrosion is the preferential corrosion at or along the grain boundary of a metal or alloy. The Cr_2O_3

GAS	REACTION PRODUCTS
Hydrogen sulfide (H_2S)	Slightly acidic solution
Methyl chloride (CH_3Cl)	HCl solution
Monomethylamine (CH_3NH_2)	Alkaline solution
Phosphorous oxychloride ($POCl_3$)	HCl and acids of phosphorus (vigorous reaction)
Phosphorus pentafluoride (PF_5)	HF and phosphoric acids
Phosphorous tribromide (PBr_3)	Dense HBr fumes (violent hydrolysis)
Phosphorous trichloride (PCl_3)	HCl and phosphoric acid (violent reaction)
Phosphorous trifluoride (PF_3)	HF acid (slow reaction)
Silicon tetrachloride ($SiCl_4$)	HCl and clear gelatinous siloxane (hydrolysis)
Silicon tetrafluoride (SiF_4)	HF and silicic acids
Sulfur dioxide (SO_2)	H_2SO_3 acid solution
Trichlorosilane ($SiCl_3H$)	HCl acid (vigorous reaction)
Trimethylamine ($(CH_3)_3N$)	Alkaline solution
Tungsten hexafluoride (WF_6)	HF and tungsten oxyfluorides (violent reaction)
Xenon difluoride (XeF_2)	HF acid solution
Ammonia (NH_3)	Alkaline solution
Arsenic pentafluoride (AsF_5)	Toxic fluorides similar to HF (hydrolysis)
Arsenic trifluoride (AsF_3)	Toxic fluorides (vigorous hydrolysis)
Boron trichloride (BCl_3)	$B(OH)_3$, H_3BO_3 and HCl
Boron trifluoride (BF_3)	$(BF_3OH)^-$ and HF acids
Boron-11 trifluoride ($B^{11}F_3$)	HF acid
Carbon dioxide (CO_2)	Carbonic acid
Chlorine (Cl_2)	HCl and HOCl acids
Diborane (B_2H_6)	Boric acid (H_3BO_3), $B(OH)_3$ and H_2 gas
Dichlorosilane (SiH_2Cl_2)	HCl acid and mixture of polymeric siloxanes
Dimethylamine ($(CH_3)_2NH$)	Alkaline solution
Ethyl chloride (C_2H_5Cl)	Chloride and phosgene vapors (hydrolysis)
Fluorine (F_2)	O_3 and HF acid solution
Germanium tetrachloride ($GeCl_4$)	HCl acid fumes (vigorous hydrolysis)
Germanium tetrafluoride (GeF_4)	HF acid (hydrolysis)
Hydrogen halides (HX)	Acidic solution and fumes
Hydrogen selenide (H_2Se)	Slightly acidic solution

TABLE 1. Products of corrosive gas reactions with water

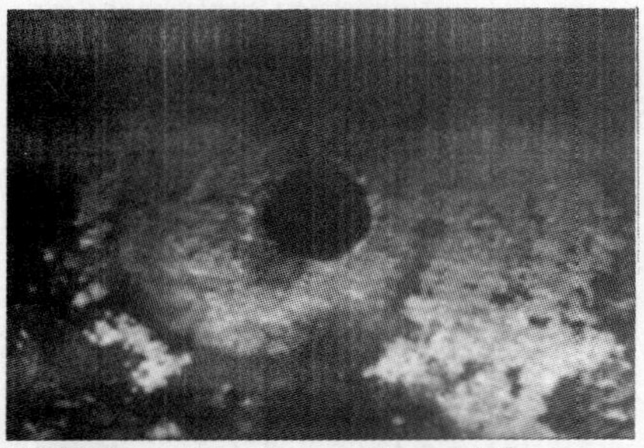

FIGURE 2. The large pit and HCl corrosion products on the inner surface of a tube have resulted from its exposure to HCl gas

MEGAN REID

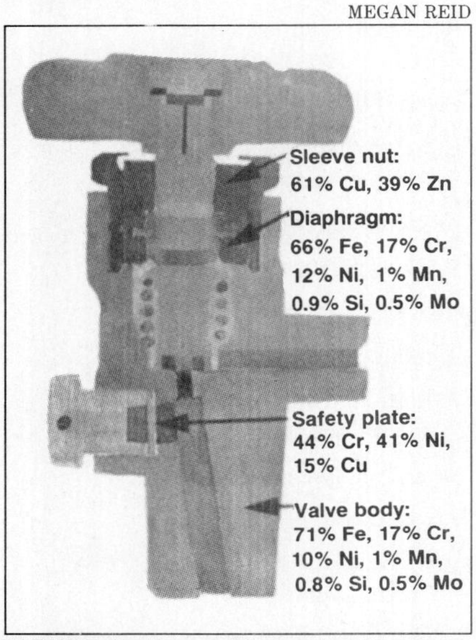

FIGURE 3. Energy Dispersive Spectroscopy (EDS) allows analysis of a valve's material composition for comparison with the listed materials of construction

The Cr_2O_3 oxide film of stainless steel contains imperfections especially at the grain boundaries, and is attacked by chlorides. The particles in the grain boundary act as a cathode and the precipitate-free zone acts as the anode, which is selectively corroded. Intergranular corrosion may occur in the weld-heat affected zones of stainless steels when non-L grades are used.

Pitting corrosion is a localized phenomenon that can penetrate deep into the metal, and is prevalent in passive alloys such as stainless steel and aluminum alloys. Figure 2 shows an actual pit in a tube from a failed HCl gas system. The molybdenum in 316L SS is beneficial in minimizing this form of corrosion.

The environmental factors contributing to pit formation are film-reducing ions (for example, Cl^-), O_2, temperature, and gravity. The metallurgical factors contributing to pit formation are nucleation of pits at heterogeneities such as inclusions, grain boundaries, and dislocations. Because of these heterogeneities, a passive film (of metal oxide or hydroxide) is always less effective than one would expect. High-purity stainless steels with minimum sulfur, oxygen, and carbon contents tend to be more resistant to pitting corrosion.

Materials of construction

Systematic evaluation of the materials of construction is necessary to ensure proper materials compatibility. Passiv-

ACTIVE END
Magnesium
Magnesium alloys
Zinc
Aluminum
Cadmium
Mild steel
Wrought iron
Cast iron
13% cadmium (Type 410) stainless steel, without passive (oxide or hydroxide) layer
50-50 lead-tin solder
18% Cr-8% Ni (Type 304) stainless steel, without passive layer
18% Cr-8% Ni-3% Mo (Type 316) stainless steel, without passive layer
Lead
Tin
Muntz metal (61% Cu-39% Zn)
Manganese bronze
Naval brass
Nickel, without passive layer
Yellow brass
Red brass
Copper
Silicon bronze
70% Cu-30% Ni
Nickel, with passive layer
70% Ni-30% Cu
18% Cr-8% Ni-3% Mo (Type 316) stainless steel, with passive layer
NOBLE END

TABLE 2. Galvanic series in sea water

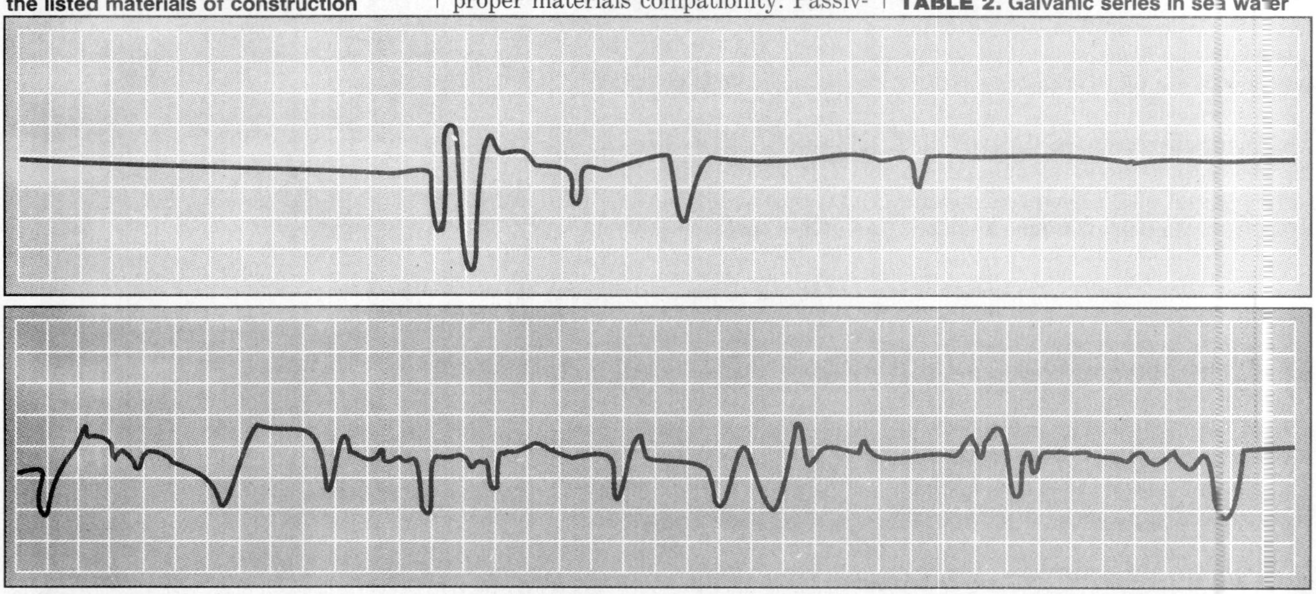

FIGURE 4. Traces from a profilometer show the effect of cleaning methods, such as acid-based (top) and caustic-based (bottom) washing, on the surface roughness of an aluminum-alloy cylinder

TABLE 3. Stability tests of aluminum cylinders exposed to various gas mixtures indicate that after one, two and four months of exposure, there is no degradation of gas products

ALUMINUM ALLOY	GAS MIXTURE
6351	20% H_2S + 20 ppm H_2O
6351	1% SO_2/N_2 + 20 ppm H_2O
6351	5% NO/N_2 + 20 ppm H_2O
6351	NH_3 + 20 ppm H_2O
6061	20% H_2S + 20 ppm H_2O
6061	1% SO_2/N_2 + 20 ppm H_2O
6061	5% NO/N_2 + 20 ppm H_2O
6061	NH_3 + 20 ppm H_2O

ation treatments (via formation of a thin, protective film, usually of metal oxide or hydroxide, on the surface) and contamination from fabrication should also be evaluated. The evaluation should include the critical components and lubricants, which may not be identified as such in the bill of materials of construction from the vendor.

Rapid techniques for such evaluations include Energy Dispersive Spectroscopy (EDS) for metal, Fourier Transform Infra-Red Spectroscopy (FTIR) for polymers and lubricants, and X-ray Photoelectron Spectroscopy (XPS for passive coatings and surface contaminants. Figure 3 shows an example of an evaluation of the materials of construction of a valve.

The surface-finish methods used, whether mill finish or electropolish (EP), greatly affect the corrosion resistance of gas systems. Minimizing surface area is useful both for curtailing corrosion as well as reducing the number of particles that can absorb or adsorb moisture and gases, and subsequently contribute to contamination of the gas. Rough surfaces hold more moisture, generate more particles, and usually corrode at a higher rate when exposed to corrosive gases. A profilometer is commonly used to measure the roughness of metallic surfaces.

The surface roughness of aluminum cylinders can be significantly reduced by properly treating the cylinder with clearing agents following fabrication, as shown in Figure 4. These surfaces can also be characterized by Scanning Electron Microscopy (SEM).

Materials compatibility

In some cases, reliable corrosion data are not available in the literature, and it is impractical to perform field tests to determine a component's material compatibility with the gas. Thus, simulated tests in a specially constructed setting must be performed.

The author carried out accelerated corrosion tests on pneumatic diaphragm valves, made of stainless steels, to evaluate the effects of HCl gas on valve life. For these units exposed to humid air for four hours a day, corrosion did not affect their performance because they all exceeded the manufacturer's specified life of 5,000 cycles of opening and closing.

However, there was an increase of corrosion products on the valve surfaces as the exposure time (to HCl gas and humid air) increased. Adsorption or condensation of some residual moisture on

FIGURE 5. Fretting corrosion at the tip of a stem from a packed valve (left) is the result of exposure to BF_3 gas

the valve's machined surfaces may account for the valve-surface corrosion.

Dry HCl gas is compatible with stainless steel, with little or no corrosion expected. Similar tests on valves with electropolished surfaces showed only a thin layer of corrosion product for all exposure conditions, demonstrating the significant benefits of electropolishing. Visual examination of the effect of direct exposure of components over time is often enough to resolve the question of materials compatibility.

Often, corrosion processes can be detected by gas analysis. Analyzers can sense an increase in reaction products from corrosion, such as H_2O, N_2, or H_2S. The author conducted a number of tests on 6061 and 6351 aluminum alloy (AA) gas cylinders with various gas mixtures containing moisture. After one, two and four months of exposure, no degrada-

FIGURE 6. Valve spring, seat and seat holder exposed to a mixture of H_2 and N_2 remain corrosion free (top), while valve components exposed to HCl gas are severely corroded (bottom)

tion of gas product has been detected (Table 3). XPS and SEM analyses also indicated there has been no reaction with the surfaces of either alloy, and therefore, the gases were considered compatible with the two aluminum alloys.

Electrochemical corrosion tests, such as potentiostatic polarization and polarization resistance, are accelerated procedures that give relative information on the behavior of various materials with solutions. These solutions have the chemical makeup that is similar to that produced by a corrosive gas in water.

American Society of Testing and Materials (ASTM) standards G3-74, G5-87, and G-59-76 describe the accepted procedures for electrochemical testing [2]. A typical test setup is made up of a galvanostat and programmer (to supply the requisite voltage and current), a working or test coupon (metal under study), a counter electrode, a saturated calomel reference electrode, and an electrolyte.

FIGURE 7. Moisture can react with HCl gas to corrode a leak in the valve

FIGURE 8. Direct reaction of valve components with H_2S results in progressive (from left to right) corrosion of Al-Si-bronze packed valves in a 10%-H_2S gas

Failure analysis

Device failures offer the most relevant data for materials compatibility. In addition, they offer an opportunity to understand corrosion processes as they occur in the field because of the presence of moisture. The following are three case studies for BF_3, HCl, and H_2S gases.

Case study 1: Type 303 SS diaphragm and packed valves were installed in systems handling BF_3. Engineering literature indicates that the recommended materials for *dry* BF_3 gas are Al-Fe-bronze alloy, SS, copper, nickel, Ni(70%)-Cu(30%) alloy, aluminum and the more-noble metals. On the other hand, copper or Ni(70%)-Cu(30%) alloy, and the more-noble metals would be recommended if the intent was to handle *wet* BF_3 gas. The known reaction products are aqueous HF acid and paste-type corrosion products.

During the handling of BF_3 gas, corrosion products built up on the internal surfaces of valve components, blocking the gas ports (Figure 5). This prompted the new recommendations: Change valve material from stainless steel to Ni(70%)-Cu(30%) alloy if handling of moisture-laden gas is anticipated; avoid exposing components to moisture; bake and vacuum treat components to purge moisture; and inspect the units for internal corrosion.

Case study 2: Targeted for analysis were Type 316 SS diaphragm valves and piping exposed to HCl gas. Recommended materials from materials compatibility studies for *dry* HCl gas are SS, copper, nickel, Ni(70%)-Cu(30%) alloy, and the more-noble materials. If the intent was to handle *wet* HCl gas, the corresponding recommendations would be nickel, silicon-bronze, tantalum, zirconium, and the more-noble metals. Aqueous HCl acid and the corrosion products are expected to make up the corrosive environment when the gas reacts with moisture contamination on surfaces.

It was found that a panel had failed due to internal surface pitting of the pipes and corrosion at fittings. Prior to failure, also corroded were the gaskets and a fitting seat (Figure 6), allowing moisture to enter the tubing. Corrosion products collected on the seat and spring of the valves whereas a similar valve in a H_2-N_2 gas service in a panel next to the HCl-service panel was free from corrosion products (Figure 7). Therefore, recommended solutions are: Avoid exposing components to moisture by proper equipment-maintenance; use baking and vacuum treatments to purge moisture; use a higher-grade HCl gas; install a moisture-absorbing filter in the gas line.

Case study 3: Type 304 stainless steel diaphragm valves were brought in contact with a gas stream of 148 ppm H_2S in N_2. The usual recommended materials are Al-Si-bronze, SS, and iron. Here the corrosive environment results from direct reaction of platinum-clad copper with H_2S gas.

The platinum cladding cracked due to overtorqueing of the safety plug, exposing the copper to the corrosive gas. The result: failed disk, and gas leaking through the safety plug. The recommendations are: Replace safety disk with a compatible material, such as 316 SS; avoid overtorqueing the safety plug to avoid cracking of the platinum cladding.

Moisture: an arch accomplice

The lessons to be learned are that machine components are susceptible to corrosive attack both when the gas reacts with moisture to form a corrosive solution, and in some cases by direct attack of the gas (Figure 8), as illustrated in Case Study 3. Electropolishing significantly improves corrosion resistance of components in corrosive-gas systems.

Component corrosion-resistance can be improved by reducing alloy impurities. Evaluation for compatibility with the bill of materials of construction, elimination of surface contamination, and proper gas-system conditioning (via purging, vacuum treatment, and so on) are important factors in minimizing corrosion. Additional materials-compatibility information can be obtained by reviewing the literature, performing accelerated corrosion studies, or evaluating and learning from field studies.

Edited by Gulam Samdani

Acknowledgments

I would like to thank Mr. Fred Kies, Mr. Paul Damiani, and Ms. Leslie Link from The BOC Group Technical Center, and Mr. Jack Wert, AIRCO, Murray Hill, N.J., for their assistance on various phases of projects described in this article.

References

1. Hardwick, S. J., Corrosion of Type 316L SS in Anhydrous Hydrogen Chloride, Electrochemical Society Spring Meeting, Abstract No. 19, May, 1989.
2. ASTM Standards, G3-74, Conventions Applicable To Electrochemical Measurements in Corrosion Testing; G5-87, Standard Reference Test Method for Making Potentiostatic and Potentiodynamic Anodic Polarization Measurements; and G59-78, Conducting Potentiodynamic Polarization Resistance Measurements, *Annual Book of ASTM Standards,* Section 3.02, Wear and Erosion; Metals Corrosion, ASTM, Philadephia, PA 19103.

The author

Richard Paciej is a section manager at the Group Technical Center, The BOC Group, 100 Mountain Ave, Murray Hill, NJ 07974 USA; tel: (908) 771-6393. He works in the areas of materials selection and performance, and ultrapure gas distribution systems. He came to the firm in 1988 with four years of experience as a materials-engineer at the U. S. Naval Air Development Center (Warminster, Pa.), where he performed research in corrosion and materials science. He is a member of National Assn. of Corrosion Engineers, American Soc. for Metals, American Soc. for Nondestructive Testing, and Electrochemical Soc. He received his B.S. from Wilkes College (Wilkes-Barre, Pa.), and M.S. and Ph.D. from the Univ. of Virginia, all in materials science and engineering. ■

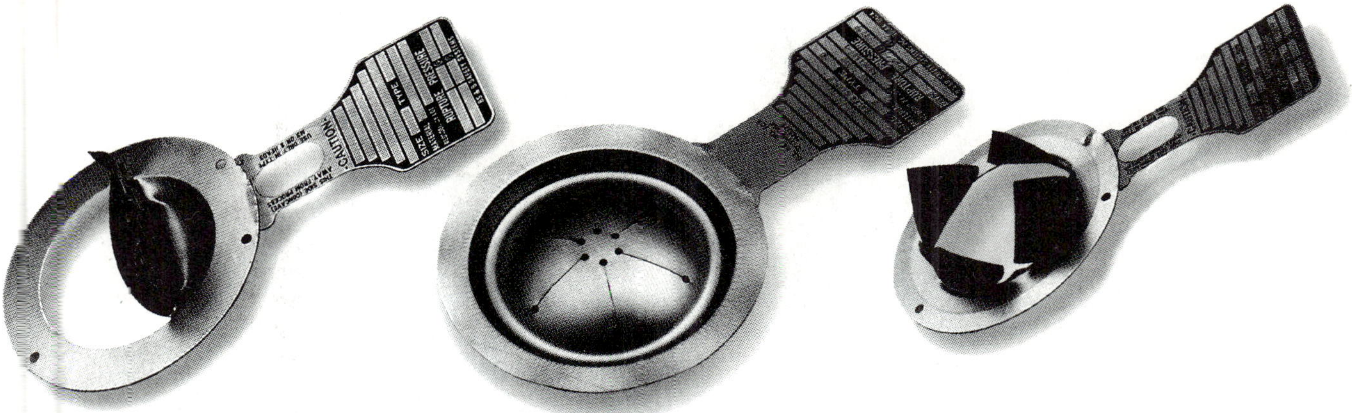

KEEPING SCORE IN RUPTURE DISKS

New designs ensure that there's one for every use

Geof Brazier
BS&B Safety Systems

At first glance it may seem that the world of rupture disks isn't one bursting with innovation — after all, some of the first designs of more than 60 years ago are still in use today. But take a closer look: Rupture disk technology is hardly standing still.

Over the years, the disks have evolved from simple calibrated foils to highly-specialized devices that employ advanced designs and manufacturing techniques to ensure safety in critical applications. The new disks generally feature high accuracy, zero leakage, long service life, and non-fragmentation (the ability not to splinter in action).

Recent products use special metals and alloys to withstand corrosive process conditions and offer a much wider range of operating conditions than those of early devices. Manufacturers are also introducing rupture disks that are intrinsically fail-safe — i.e., the units burst at or below their rated burst pressure even if they are damaged during installation or are improperly installed.

A complementary technology

The early history of rupture disks is tied hand-in-hand with the emerging petrochemical industries in the U.S. The disks found an initial niche as an economic alternative to safety relief valves, because rupture disks could vent gases more quickly than valves in applications prone to sudden pressure rises.

Beginning in the 1980s, the use of rupture disks took off dramatically, mainly in response to heightened environmental and safety concerns. One reason is that both the American Soc. of Mechanical Engineers Pressure Vessel Code and the International Standards Organization's ISO 6718 standard permit the use of a rupture disk at the inlet of a safety relief valve.

In this scheme, the rupture disk is the first to open in the event of system overpressure; the vented process fluid then contacts the safety relief valve, which releases the fluid if the pressure is excessive. Once the pressure drops to normal levels, the valve reseats itself and, thus, reseals the system.

There are five advantages in using a rupture disks in conjunction with a safety valve. The rupture disk:

- Acts as a leakproof seal to eliminate process leaks to the atmosphere
- Extends valve life by isolating the valve from corrosive materials
- Permits the use of less-expensive valve trim
- Minimizes product buildup on the valve's mechanical components, and so extends the period between overhauls
- Seals off the valve from the process stream and so permits in-place valve testing

The 'conventional' family

In classifying rupture disks, some of the earliest designs, evolving between 1930 and 1960, are considered "conventional" devices. This family consists of three basic types: tension-loaded solid metal disks, tension-loaded composite metal disks, and carbon disks.

Today, vendors offer these three types of rupture disks in a wide range of configurations. The devices comes in diameters ranging from ⅛ in. to 60 in. The disks are offered with differential pressure ratings as low as 1 psi, to ones that go up to 150,000 psi and above. On the negative side, these rupture disks generally have a limited service life, which can be up to about 10 times lower than those of the latest designs. Also, they fragment at the time of rupture, rendering them unsuitable for use in combination with safety relief valves.

Tension-loaded solid metal disk. As illustrated in Figure 2A, the solid-metal disk is essentially a domed metal membrane whose burst pressure is controlled by the disk's thickness and nominal diameter, and the disk material's mechanical properties, such as tensile strength. During the disk's manufacture, it is subject to a forming pressure that exceeds the recommended service

Figure 1 (above). The current crop of rupture disks are leakproof, very accurate and feature long operating lives

Originally published September 1993

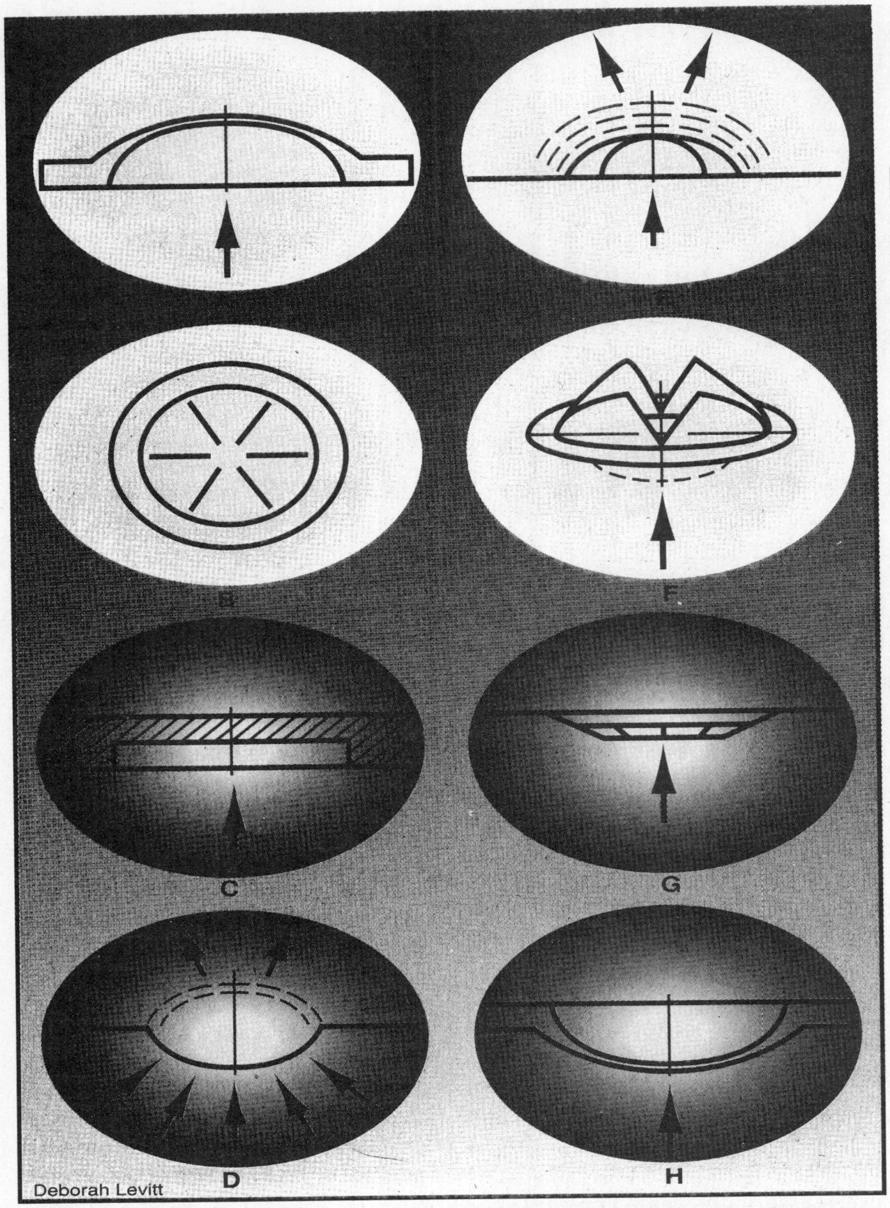

Figure 2. The burst pressure of a tension-loaded, solid metal disk (A) is set by the material's thickness and mechanical properties, and the disk diameter; in the composite metal disk (B), burst pressure is modified by a pattern of slots. Carbon disks (C) are suited for corrosive applications, while reverse-buckling disks (D) are used at the inlets of safety relief valves. Scored disks are made by marking a disk with a sharp knife, followed by pressurizing from one side (E); a scored reverse-buckling disk does not fragment upon bursting (F). New disk designs include the truncated conical disks (G), and the "universal" disk (H), which is scored along the circumference

pressure limit of 70% of the disk's burst rating. This forming action is a key quality-control step, since any faulty material used in making the disk membrane is likely to fail when subjected to this pressure.

Tension-loaded composite metal disk. This type of composite slotted disk meets the need for rupture disks that can operate at burst pressures lower than those of solid metal disks. In the composite metal disk, the burst pressure is still controlled by metal's thickness, dimensions and mechanical properties, but the disk itself is modified by a pattern of slots that converge at its center (Figure 2B). The closer the slots converge, the lower the burst rating for a given material.

The composite metal disk requires a seal to be placed on the process side to provide leak tightness. The seals are usually fabricated from plastics.

Carbon disks. The third member of the conventional family is the graphite disk (Figure 2C). The latter's forte is in applications operating at very low differential pressures (typically below 15 psi) and requiring chemical resistance — a combination of requirements that solid-metal and composite-metal disks cannot easily meet.

The intrinsic "weakness" of carbon makes it ideal for fabricating rupture disks that can burst at low pressures. Carbon's chemical inertness is also a plus; however, in order to overcome the material's natural porosity, the carbon disks are impregnated with a resin.

Reversing the trend

The reverse buckling disk made its appearance in the 1960s as a "non-fragmenting" rupture disk that can be used at the inlets of safety relief valves (Figure 2D). The rupture disk employs the compressive strength of the material to regulate burst pressure — i.e., process pressure is exerted on the convex side of the disk (resulting in compressive pressure), rather than the concave side (which causes tensile pressure).

The disk itself is a hemispherical structure that buckles at pressures that are determined by the radius of the disk's dome, and the thickness of material and its modulus of elasticity. Since the modulus of elasticity can be fixed for a given disk material, the burst pressure can be calculated by knowing the dome's thickness and radius.

Reverse buckling disks are available in sizes ranging from ¼- to 60-in. in diameter. They can operate at up to 90% of the rated burst pressure and are suited for use under cyclic pressure conditions, such as successive vacuum and positive pressure events.

In operation, the domed structure of a reverse buckling disk snaps through to form a mirror image — but without actually bursting. The actual cutting of the dome is achieved by a knife blade that is installed on the disk holder.

The use of a knife blade does have some limitations. For example, if the disk is installed under uneven bolt torques, the dome can become unstable even under normal low-pressure conditions. The lower pressure, in turn, causes the dome to buckle at slower speeds than it would at its rated — and higher — burst pressure. As a result, the force of the dome on the cutting knife is less than that needed to cut open the metal membrane. Also, the knife blades themselves can become blunt or corroded after several ruptures, curtailing the ability of the disk to open correctly.

Many of the limitations of conventional and reverse buckling disks can be overcome by a carefully controlled process known as in scoring. In the technique, a flat disk is scored (or marked) — typically with a cross-shaped imprint — using a hardened steel knife. Next, the scored disk is pressurized from one side, producing a domed structure. The dome is then carefully heat treated, producing a rupture disk that is capable of resisting continuous pressure cycling very close to the rated burst pressure (Figure 2E).

As with conventional disks, scored disks are installed with process pressure exerting on the concave side of the membrane. This subjects the disks to tensile loading. The burst pressure is controlled by a combination of the disk's standard characteristics — thickness, dimensions and tensile properties — together with the influence of the score line, which creates a point of weakness at the disk's center.

The scored disks generally open without fragmentation, and can be utilized in applications where the working pressure conditions are relatively static. They also offer slightly longer operating lives, compared with solid-metal, slotted-metal and carbon disks.

When reverse buckling disks are scored, the membrane's burst pressure is determined primarily by the shape of the dome. However, the scored disks don't require a knife for opening, since on reversing, the metal splits evenly along score lines to give a full, unobstructed opening (Figure 2F).

Recent innovations

Rupture disks have traditionally been limited only to applications in which a gas pocket is available to transmit sufficient energy to ensure a clean rupture of the disk. This limitation has been recently overcome by a reverse buckling disk that is capable of operating under full liquid service.

The reverse buckling liquid-service disk achieves this with a circular scoring design, rather than with a typical cross-pattern. Further, the outlet side of the disk incorporates a structural support that ensures that available energy is fully transmitted to the score line, resulting in a complete opening.

Another innovation in reverse buckling devices is the use of truncated conical disks that burst at differential pressures as low as 8 psi, and result in full opening diameters of up to 2 in. at rupture (Figure 2G). This disk, a frustrum reverse-buckling device, is of particular interest in the biotechnology, pharmaceutical and food-processing industries, where sterile, stainless-steel construction is necessary for protecting low-pressure equipment.

In 1991, a circular-scored reverse buckling disk took its first steps as a "universal" rupture disk — one that could be used in all applications (Figure 2H). In this design, the disk is scored along the circumference — rather than in a cross-pattern. This scoring pattern yields a clearly defined opening when the disk undergoes rupture.

The circular-scored disk can withstand full vacuum and repeated pressure cycling, and does not fragment on rupture. It can be used in liquid, gas or two-phase flow conditions, and is rated for operation at up to 90% of rated burst pressure (which ranges from 30 to 750 psig). It is available in diameters of 1- to 8-in.

An additional feature of the circular-scored disk is its fail-safe design. For example, if the disk is damaged or installed in its holder in the wrong direction, it will still rupture at or below the required rating. ■

Edited by Jayadev Chowdhury

The author

Geof Brazier is the International Market Development Manager for BS&B Safety Systems (7455 East 46th St., Tulsa, OK 74145; Tel: 918-622-5950). He has worked in the rupture disk industry in the U.S. and Europe for more than 15 years. He was educated at the University of Bristol in England as a physicist.

The modular system was installed quickly, expediting recovery from a serious accident

TEMPORARY ROADWAYS SPEED ACCIDENT CLEANUP

Cleanup after an explosion destroyed two polyethylene units at the Phillips 66 Co.'s Houston Chemical Complex last October (*Chem. Eng.*, Nov. 1989, p. 17) was a monumental task. But the job of dismantling, transporting and disposing of the wreckage was made easier by a temporary roadway system developed and patented by Uni-Mat International, Inc. (Houston).

Modules simplify construction

Uni-Mat's temporary roadway is modular and is based on 8-ft by 14-ft prefabricated mat sections made of oak lumber. The basic module consists of a layer of four mats placed upside down and locked in place by a single mat placed solid-side up in the center of the four base mats. The pad is then extended as far as necessary by adding more such modules, and finished on the sides and ends by adding 4-ft by 14-ft mats to the top layer. With this arrangement, the load on one top mat (112 ft^2) is dispersed over an area of 448 ft^2 of bottom mats.

Because the mats measure 8 ft by 14 ft, they can be shipped by truck, rail, barge or ship. They can be stacked, making them easy to load and unload with a forklift. Each mat is 4 in. thick and weighs about 1,200 lb.

At the Phillips site, Uni-Mat provided a 40-ft × 300-ft crossing over railroad tracks that run alongside the plant (behind the truck in the photo on p. 150). A 40-ft × 1,000-ft roadway extended from the end of the crossing and into an open field; this section had several "fingers" off to each side and a 100-ft × 150-ft turnaround area at the end. Another section, also 40-ft × 1,000 ft, ran alongside the tracks (perpendicular to the

A top mat is placed over the intersection of four upside-down mats to lock the sections into place

Temporary roadways provided access for the heavy equipment that removed plant debris and an area for storing the debris prior to its final removal

NOT JUST FOR ACCIDENTS

The Uni-Mat system has numerous other applications, including chemical-process-industries (CPI) plant construction and expansion. For example, during the construction of a polypropylene unit at a large chemical complex, a major bottleneck was the transportion of eight sets of reactor legs through the plant. Each leg was 115 ft long and weighed 41,000 lb; with the tractor assembly, the truck and its load were 160 ft long and weighed over 109,000 lb. The existing roadways through the plant had several sharp right-angle turns that could not be negotiated with this long, rigid load.

The obvious alternatives for moving the reactors into the plant were to build additional permanent roads or remove existing plant facilities. The latter would have required shutting down ongoing processes; both alternatives were expensive and time consuming.

Instead, Uni-Mat installed 250-ft-long roadways — with widened entrance approaches — that ran diagonally between two perpendicular sections of pavement, thus eliminating the 90-deg. turns. The roadways were installed in 4 h per section, and removed in the same amount of time. Each reactor leg was transported from the ship, into the plant, across the Uni-Mat road and placed on the staging area in just 6 h. When the project was finished, the roadway was removed.

Other projects where the mats have been successfully used include temporary docks, roads for hauling heavy loads (1,200 tons has been the maximum so far), and general construction projects where access across soft ground or sensitive areas is required.

roadway into the field); at one end, it connected with the main plant-access road (not shown, but off the left edge of the photo).

The roadways provided a means for heavy machinery to remove the plant debris. The 40-ft widths allowed several trucks carrying debris to travel at the same time, and also permitted backing up and dumping. The mats were installed in a pattern such that all of the roadways and the turnaround area were interlocked. Uni-Mat installed 1,600 mats, a total of 90,000 ft^2, in two days, using two 6,000-lb-capacity forklifts and two- to three-person crews.

During the cleanup, over 400 loads of debris (which has since been removed) were transported to the holding area. The dump site would not have been accessible without the Uni-Mat system. Having it nearby eliminated longer truck trips and allowed more work to be performed more quickly and with fewer vehicles and personnel. — *Uni-Mat International, Inc., Houston, Tex.* **461** ■

Section VI

Safe Operation and Maintenance of Equipment

HIGH PRESSURE OPERATIONS

PART 1

UPKEEP OF EQUIPMENT

Robert Andrews and Paul Weber,
Condux, Inc.

Safe and effective operation of high-pressure equipment in the chemical process industries has never been so important. Plant operators today contend with the burden of governmental regulations, the impact of legal liability, and the need for efficient, high-quality production in the face of intense global competition. They need to assure themselves that their procedures are both legally and technologically sound.

Procedures are changing as a result of the issuance of new U.S. regulatory standards, such as the Occupational Safety and Health Administration's codes for process safety management, or the Responsible Care program administered by the Chemical Mfrs. Assn. (Washington, D.C.). Part 2 of this report deals with these regulations and programs in more detail.

New regulations and safety recommendations call for changes in the operating procedures at many plants. There are effective ways of making these changes, and at the same time ensuring that process equipment for high-pressure operation, such as safety relief valves or pressure vessels, is properly maintained.

As operating pressure rises, so does the importance of maintenance and repair

High pressure basics

Many basic operations in the chemical process industries (CPI) run at high pressures. Steam-generating units run from a few hundred to a few thousand psi. Many chemical syntheses or polymerization reactions run in excess of 5,000 psi, and some above 10,000 psi.

For the purposes of this article, "high pressure" is considered to be in the 5,000- to 10,000-psi range, where a considerable number of process operations take place. Fired units generally run at lower pressures, especially when generating steam. Conversely, there are also some polymerization reactions run as high as 20,000 psi or more. Many of the operational procedures described here apply to both the higher and lower pressure ranges.

Only a pressure vessel that has been properly designed can be maintained well. High-pressure vessels generally have rounded shapes, to minimize the existence of corners and sharp bends where pressure could affect the material's integrity most easily. The use of finite-element analysis (FEA) programs on computers has become widely used in designing such vessels. These programs are able to calculate the stresses and pressure points in many complex types of vessel geometries.

Operators of high-pressure systems are concerned with the proper control of the reactions taking place in the vessels, and with the integrity of the as-designed piece of equipment. Corrosion

Grayloc Products

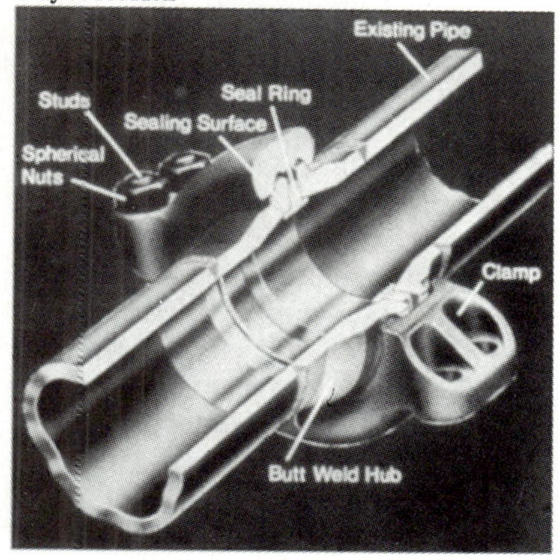

is a constant worry. Each of the entry points of a vessel — inlets, outlets, portals for analytical equipment, manways and the like — bear close monitoring. In addition, the equipment attached to a vessel, such as relief valves, rupture discs, probes and pipe flanges, must receive scheduled inspection.

FEA programs can also be useful for determining thermal stresses on a vessel as it operates. Stresses caused by vibration are an additional concern. Where fatigue is present, such as near reciprocating compressors, design to minimize stress-related cracking.

For many installations, the operating personnel are protected from the process by a barricade, which can be either concrete or steel. Barricades are essential in pilot-plant operations where there is uncertainty of the reactions under study. Hydrostatic testing of large vessels is another instance where a barricaded area is often desirable.

Know your fabrication shop

Because of the volume of specialized work and repairs that a large facility continually requires, many plant operators will set up a fabrication shop to handle this work. High-pressure equipment requires much more sophisticated machining, inspection and fabrication work. In some fabrication shops, a separate precision machine shop is needed. Such a shop is clean, with conditioned air, and with machine tools having the required precision. This technique is used for pressure assemblies in compressors subject to fatigue. Special welding facilities may also be necessary for the unusual alloys or types of welds to be completed.

Originally published September 1993

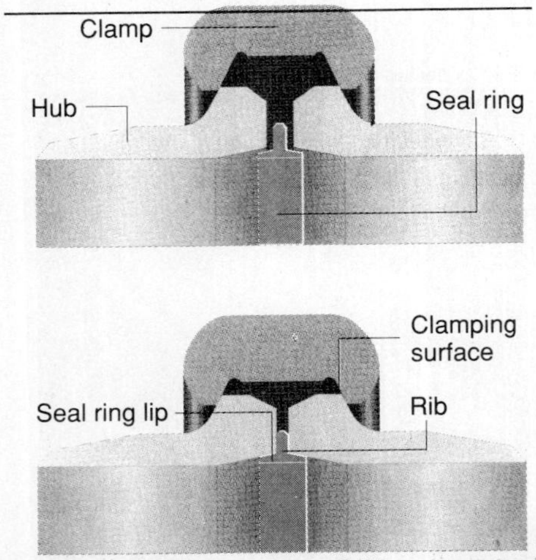

Figure 2. The top view of this connector shows its configuration before the pipe is pressurized; the bottom shows the self-reinforcing after pressurization

FIGURE 3. This 'Safety Selector Valve' has an internal valving mechanism that allows the user to isolate a safety relief valve for servicing, without shutting down a process

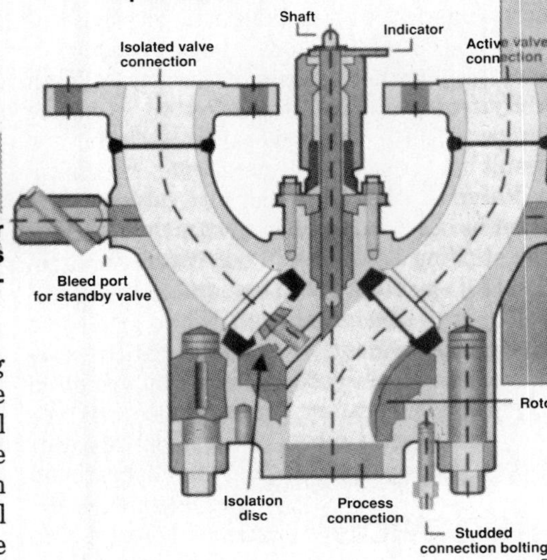

Anderson, Greenwood & Co.

Such operations as shrink-fitting members to improve fatigue resistance may be performed. Another useful technique is autofrettage, where the bore of equipment is subjected to an elevated pressure to result in local yielding at the bore. On release of the pressure, the yielded bore is put into compression and will thereby better resist fatigue loading.

Much of the equipment for high pressure operations is fabricated from high-strength alloy steels. It is imperative to consult with metallurgists to ensure that adequate specs cover the materials. Such factors as impact, notch sensitivity, corrosion and erosion resistance — in addition to the standard physical properties — are important. Note, too, that many of the high-strength alloy steels are not weldable; other, weldable alloys must be substituted.

It is very important that the maintenance organization making repairs utilizes materials identical to those specified by the designer. The repair shop must have a system to verify that materials received for the repair are the ones that were ordered. The vendor should clearly identify the material and send papers with the material certifying its metallurgy. In critical applications an additional metallurgical examination should be conducted by the repair shop to verify that the material is correct.

Only currently qualified welders should be allowed to make welded repairs on high pressure equipment. A detailed welding procedure must be written and followed. It should specify how each pass should be checked, what final inspection technique is to be used, and the procedure for stress relieving.

Cleanliness is imperative, as contamination of sealing surfaces can result in process leaks that only can be corrected by complete replacement of the sealing unit.

Some of the special considerations for high-pressure work are:

- ***Hydrogen.*** This gas increases in temperature as it expands. This can result in fire at leaks in a system
- ***Valve design.*** High-pressure valves must be designed so that adjustment of the stuffing box nut will not result in forceful ejection of the stem assembly
- ***Rupture disc discharge.*** When discharged suddenly, pressure-relief valves as well as rupture discs can generate large vibratory forces that are transmitted into process piping. A good design practice is to eliminate bends in the piping from these devices in favor of straight runs. Adequate support is necessary. It is also important to size rupture discs of different blowing pressures such that a holder for a low-pressure disc doesn't receive a higher-rated one
- ***Piping supports.*** Some installations are subject to significant levels of vibration or to thermal expansion and will require suitable support to accommodate the resulting stresses
- ***Safety relief valve fittings.*** The scheduled maintenance of relief valves is often inconvenient because production must stop. For this reason, one sometimes encounters installations with fittings to blank off the safety valve, and allow production to continue while the valve is serviced. This runs completely counter to the rationale for installing safety valves, and should not be tolerated. Similarly, no design should allow the flow intended for a safety valve to be manifolded. There are better devices to use, such as Figure 3 on p. 102
- ***Large vessel closures.*** Because of the high pressure load, these closures can require large-diameter bolts. To give adequate tightening, it may be necessary to provide for hydraulic tensioners to stress the bolts
- ***Moving parts.*** Mechanical linkages are frequently required in valves, agitated reactors and other components. Alignment is all important. Adequate cooling and special lubrication may be required for the sophisticated agitator shaft seals commonly used today

Inspection of fabricated equipment and piping is important to verify that all specifications are met. Checking for cracks can be performed visually, or with nondestructive testing such as ultrasonic, magnetic-particle, chemical-dye and X-ray analyzers. Machined parts should be checked for dimension, finish, and fit-up. Finished systems are hydrostatically tested to verify pressure integrity. In special cases, this testing may require use of strain gages secured to certain equipment pieces that may be of new or nonstandard design.

Maintenance philosophy

The purpose of the maintenance program for a high-pressure process is to assure that the integrity of the system is maintained, and that requirements of Sec. (j) of OSHA 29CFR 1910.119 are met (Part 2, p. 100). Here, it is assumed that the equipment was originally designed in accordance with the appropriate standards and codes as outlined in the previous section.

Experience shows that aside from professional operation of the process, quality maintenance of process equipment is the most important factor in achieving safe operation of the plant. The goal of the maintenance program is to maintain the equipment in a condition which continues to meet or exceed design standards and codes. All inspection, testing and repair work is aimed at this goal. The maintenance organization should be familiar with applicable national or local codes.

At pressures of 1,200 psi and higher, quality maintenance procedures and practices become more critical and deviations from them cannot be tolerated. Double and triple checking is mandatory. Machining tolerances and finishes must be religiously observed.

If the original compressor, valve or pump was made in a precision shop, then the repair must also be made in a precision shop. As with fabrication, mentioned earlier, substitute materials of construction cannot be permitted without a vigorous engineering and metallurgical investigation and approval by qualified personnel. Welding personnel, procedures and practices must be qualified before use. Care must be exercised during reassembly to assure no damage to replacement components. Do not deviate from design criteria. If a question arises about a procedure or practice, it must be answered before the work proceeds.

Design files

During installation and subsequent maintenance, opening and updating a *design file* for each safety relief valve, rupture disk and other critical piece of high-pressure equipment is a vital task. The file should include the design information for the device. This information is required as a basis for future maintenance work and incident investigation.

The establishment of these files is sometimes overlooked in the process of constructing or starting up a new plant and the information can get lost. Establishment of the file at a later date is difficult and time consuming if the design data and calculations have been lost and have to be done again.

For example, consider the steps to take when a safety relief valve is brought into service. During process design, the relief capacity needed to protect the system or vessel is calculated. The calculation determines the size of valve, set pressure, relieving capacity, orifice size and spring size. Usually the spring size is indicated by a spring number. This information is supplied by the manufacturer and should be used to check that the valve, as specified, has been received. The original calculations should become part of the design file.

The manufacturer's papers should also go, immediately, to the design file. These are important documents and will be used in the future to assure correct repair. The valve file should be a hard copy file. Valve data should not be altered and a hard copy file is the best insurance that this does not happen.

As part of normal, routine operations in commissioning the safety relief valve, it should receive a unique identification number (which is recorded), and a form should be attached to the valve whenever it is removed from service. Each safety relief valve should receive a unique number that is recorded in the design file. The number should be affixed permanently to the valve. This ensures that the correct design information can be retrieved from the file. The file can also be used to track the life history of the valve.

A healthy pressure vessel

It is important that pressure vessels be maintained at design and manufacturing specifications. Specific attention must be paid to gasket and flange surfaces of tubing fittings or access openings, because these are vulnerable to wear or degradation. All mating surfaces should be protected by soft covers such as wood or plastic when they are temporarily stored during maintenance inspection, cleaning and repair.

The frequency of maintenance depends on the type of process and can vary from one to five years. Processes that are corrosive or erosive often require the higher frequencies. A design file should be opened and maintained for the vessel and its structural components.

The normal maintenance of vessels consists of cleaning, inspection and testing. Visual inspection of the vessel is a preferred first step. If the inspection shows localized pitting or cracking, then more sophisticated metallurgical tests such as chemical dye, ultrasonic and X-ray analysis should be performed to determine the type and extent of repair required. A hydrostatic test should be performed to assure pressure integrity of the vessel. Hydrostatic tests do not ensure against pinhole leaks but do assure the vessel's capability to contain the process.

Welding by code

Welding repair on high pressure vessels and piping is one of the most critical types of repair. It must be done with qualified welders who not only have passed code tests but who have been practicing their skill. A welder who has passed the code tests, but who has not done any code welding for a long period of time, is not qualified unless he or she can pass the code test again. Shops doing code welding on pressure vessels need to have a solid training program to qualify their welders. They should have qualified procedures written for each welder to follow.

All pressure welds should be X-rayed. On critical vessels, inspection should be 100% X-ray. It is also recommended that welders stamp their work with an identifying mark that may help in determining the cause of any subsequent failure.

Vessels repairs must be made under the code under which the vessel was manufactured. Depending on the size, welded repairs may need to be stress relieved. Any defects found in the welds should be ground out and repaired. When bolted joints are made up, all nuts should be torqued to the value necessary to properly load the gasket. If an impact wrench is used for this

Shops doing code welding need to have qualified, written procedures to follow

purpose, it should be tested at least semiannually to assure the it puts out the correct torque. There are hydraulic tensioners on the market that stress the studs, giving greater assurance of proper joint makeup.

During the repressurizing of the process, each joint should be checked for leaks. If leaks are found, the joint should be tightened. If the leak cannot be stopped, the process should be depressurized, cleared, and the joint replaced. If the flanges are within specification a new gasket may be all that is necessary.

The pipes and tubes attached to the pressure vessel usually require a similar level of exactitude when performing maintenance and repair. Check that the correct material of construction and wall thickness are being used, according to design specifications for the job. Make sure that qualified welders, using qualified procedures, are on the job. Test joints for leaks.

Tubing joint makeup is just as critical as vessel joint makeup — and maybe more so, because some joints may not be accessible. Surface finish must be proper for the gasket that will be used. The gaskets, such as sprial-wound, copper, ring or delta, specified by design must be used. Substitutes should be avoided unless an engineering study of the joint design is conducted. The complete joint design includes flange, gasket and bolt loading. The bolt loading is critical, and the bolt should be stretched or torqued per design.

There are a variety of fittings that have been specifically designed for high pressure. They offer such modifications as collars that surround pipe flanges, so that the fitting is self-energized (Figure 2, p. 100). Here, too, manufacturers directions must be followed closely for proper performance.

Relieving pressure safely

A safety relief valve is used to protect both personnel and equipment. It is designed to safely relieve system pressure that rises above a given pressure. A strict, high-quality maintenance program for them is mandatory. The program should be a joint effort among engineering, operations and maintenance groups. Each group has a critical part to play and if one does not complete its responsibility, the whole program will be ineffective.

The maintenance program should consider both frequency and procedures for repairs. Normal frequency for maintenance inspection and repair for a safety relief valve in a clean process that does not form deposits is one year. For dirty processes that form deposits in piping and equipment, the inspection and repair should be more frequent.

On a new plant, the frequency of inspection and repair should be once a year, unless the best judgment of engineering and operations personnel de-

FIGURE 4. The popping pressure of a safety relief valve is determined on a test stand. If the valve's performance does not meet specification, further testing may be necessary

Calder Testers, Inc.

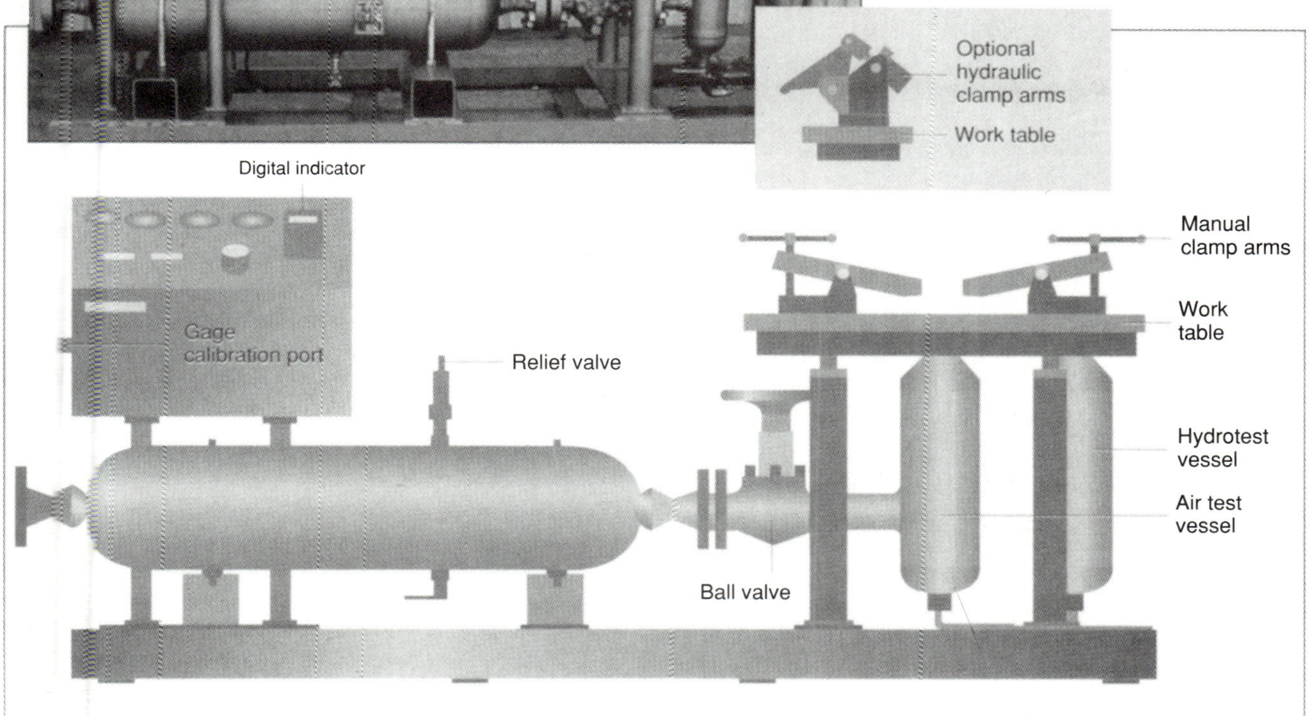

Figure 5. A safety-relief-valve test stand enables plant maintenance personnel to check valve performance regularly. The stand can be equipped with a variety of clamps or holding plates to handle most types of valves. Instrumentation records 'popping' pressure and other performance variables

mands greater frequency. One year should be the maximum until experience is gained. Changes in frequency should not be made without a thorough engineering review.

If it is necessary to continue operating the process beyond the time a valve is scheduled for maintenance, there should be an engineering committee headed by a plant staff member who will review the request and have the authority to authorize or not authorize the extension. Any request for extension should be accompanied by a:

- Compelling reason
- Procedure for safe operation during the extension
- Date for the inspection and test

Repair document

Maintenance procedures are very critical to a good safety relief valve program. The basis for the procedures is excellence in the documentation. A rigid system for identification and documentation should be established, in addition to the design file, and should cover:

- Unique identification of each valve and its location
- Steps for field removal, shop testing and shop repair
- Steps for reinstallation
- Records for each valve

The location of each safety relief valve should be identified and recorded in the file. This identification forms a double check that the correct valve is removed when maintenance is scheduled.

When the safety relief valve is being serviced, a form should be issued that will remain with the valve from the time it is removed until the time it is reinstalled. The form should be in a waterproof plastic envelope attached to the valve with a stainless steel wire. There should be no exceptions to this rule.

The form should contain the valve identification, its location, size, make, orifice size and spring number. An operator and a mechanic should jointly check the valve identification on the form against the valve to assure that the correct valve is to be removed. Both should sign that this is the case. When the valve is removed, the inlet and discharged piping should be inspected for foreign material. Any obstruction found should be noted on the form.

In the U.S., many state or local authorities may require that the repair shop have a "VR" stamp which shows the shop has been certified by a national board as qualified for safety relief valve repairs. The plant should check to see if this is a requirement. Many plants that repair their own valves have a VR stamp, even though it may not be required, because it gives added assurance that the work is done properly.

When the valve arrives in the shop, it should be inspected visually so that any foreign material in the inlet or discharge can be removed. If the valve is clean, it should be put on a test stand (Figures 4 and 5) before it is dismantled, and pressurized to determine the pressure at which it pops. Record the popping pressure on the form. If it is more than 5% different from the set pressure, an engineering analysis should be performed to determine the reason. A different frequency of repair or design change may be required to assure continued safe operation.

The valve should then be dismantled, cleaned and inspected. Any problems should be noted and written on the form. Assemble the valve, making sure that the spring number and orifice size correspond to the values on the form. If they are different, an engineering representative from the process area should resolve the discrepancy to ensure that the proper parts are used in the assembly. Changes in valve components should never be made without an engineering analysis of all the factors involved in properly relieving the process.

The relieving pressure should be reset and checked on the test stand. Set the pop pressure to the value contained on the form. If the valve operates properly, the shop mechanic should sign the form indicating the work is complete as specified on the form.

Experience shows that flange protectors should be installed before sending the valve to the field for installation. If this is not done flanges can be damaged, resulting in leaks when the process is brought up to pressure.

When the valve is received from the shop, a mechanic and an operator should inspect inlet and discharge openings for cleanliness and check the valve papers to make sure the information is correct. The papers should be completely filled out and each section signed by the mechanic doing the work. The operator should witness the installation and both should sign the form. The completed form should be placed in the hardcopy safety relief valve file as a permanent record.

Some special considerations apply to safety relief valve installations. In the first place, discharges from them should never be manifolded. If a valve on a system is being removed for maintenance and a second valve on the manifold releases pressure, the mechanic removing the first valve can be severely injured.

Similarly, the inlet of a single safety relief valve should never be valved-off or blanked while the system it is protecting is being operated. It is tempting to think that a system can be safely operated by close operator attention while the safety relief valve is removed for maintenance. This is an economic judgment, not a safety judgment, and as Murphy's law states, "If something can happen, it will happen." Even with close attention, reactions take place so quickly that manual corrections to the process cannot be quick enough.

If it is desired to operate a process and not shut it down to maintain the safety relief valve, there should be two identical valves installed each of which is capable of handling the full system discharge. These two safety relief valves can each be mounted on the outlet sides of a three-way valve, such as the setup shown in Figure 3. These valves are designed so that the system can be switched from one safety relief valve to the other. In the center position both valves are open to the system, and both cannot be closed at the same time so that the system is always protected.

Rupture discs

The primary purpose of rupture discs is to protect personnel and equipment from a rupture that would result in a catastrophic failure and injury. There is a secondary purpose — to protect the inlet of a safety relief valve when the valve is installed in a dirty process. (See the article on p. 155 for more details.)

Sometimes the process can form deposits on the inlet of a safety relief valve and it will not relieve at its set pressure. To prevent this, a rupture disc is installed ahead of the inlet to the valve. This leaves a void space between the rupture disc and the valve. A pressure gage or pressure sensor is installed to measure the pressure in this void. This will detect whether or not the rupture disc has failed and is not protecting the safety relief valve. The operator should check daily to determine if a pressure is detected.

Rupture discs should normally be changed during regularly scheduled turnarounds. The installations should be inspected for a build up of deposits both on the inlet and discharge sides. Based on these inspections a different frequency of change may be necessary. The deposits may affect the pressure at which the rupture disc fails.

When a rupture disc is replaced, a form should be initiated by the operating group. This form should show the location of the disc, its identification and its relieving pressure. The mechanic uses this information to get the proper replacement disc. Together the

Fike

FIGURE 6. The rupture disc, in the past an accepted way of preventing vessel overpressure, now provides protection for safety relief valves as well

operator and mechanic check the information on the new disc to assure it is the proper one. The actual replacement of the disc is not complicated but there are two cautions:

- Inspect the disc to assure there are no nicks or scratches. These can cause the disc to fail at lower than design pressure
- If the rupture disc is separate from the holder and not a machined reverse-buckling type, inspect the holder to ensure that it is smooth. Sharp edges and imperfections on the holder can lower the relieving pressure of the disc

The rupture disc form should be signed by the operator and mechanic when the installation is complete and then placed in the hard copy area file as a permanent record. Rupture discs are made to precise tolerances and should be stored where they cannot be damaged. Care should be taken to see that discs of different relieving capacity are identified and separated in the storage location.

Respect interlocks

Safety interlocks installed in the process are there to detect any process malfunction that can be missed inadvertently by an operator and to shut down the process. It is possible that these interlocks themselves can malfunction and occasionally trip the process even though there is no problem with the process. This can be an aggravation that sometimes leads a well-intentioned employee to bypass an interlock until a repair can be scheduled at a more convenient time instead of shutting down the process and immediately replacing the interlock.

Interlocks should never be bypassed unless each instance is authorized, in writing, by a member of the plant staff. The risks involved are great should a process upset occur when the interlock is bypassed.

Dealing with machine parts

Maintenance of high-pressure equipment with moving parts such as compressors, pumps and seals is also critical. Repair and assembly require high quality workmanship which at least duplicates that of the original manufacturer. Close attention must be paid to attainment of design tolerances and surface

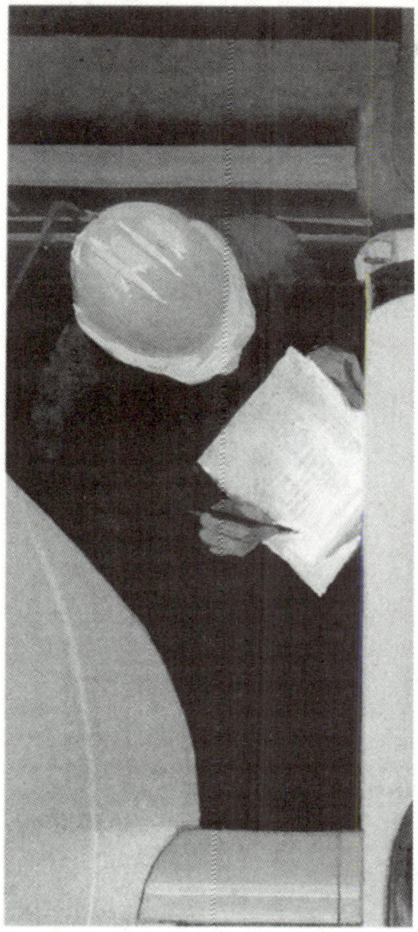

FIGURE 7. Accountability should be spread throughout a plant's staff

finishes. Although this principle applies at all pressures, it is critical at pressures of 5,000 psi and above. Tiny imperfections or scratches can form the escape path for high energy process molecules.

Equally important to surface finish is cleanliness. A good finish can be ruined if dirt contaminates the surface during handling and assembly. As mentioned earlier, a clean room is needed for some fabrication of original equipment, such as seals and valves. If this is necessary for original manufacture it is also necessary for repair and reassembly. If a repair shop does not use a clean room for that work it cannot guarantee the quality required.

Proper surface hardness must be reproduced on repair parts. For instance, reciprocating compressors usually require tough steel for compressor rods to withstand the cycling of stresses with a surface hardening, such as nitriding, in the packing area to reduce wear and maximize running time.

Alignment and balance become more critical as pressures increase. Alignment techniques using self-aligning coupling at ordinary pressures cannot be tolerated at the high pressures. Techniques such as optical or laser alignment are mandatory for high-speed, high-pressure centrifugal compressor trains. ■

Edited by Nicholas Basta

The authors

Robert K. Andrews is principal maintenance consultant at Condux, Inc., (201 Possum Park Rd., Suite 9, Newark, DE 19711; tel: 302 738 9200). Condux, a consulting firm established in 1985, provides engineering and safety services throughout the chemical process industries. Many of its principals worked formerly at Du Pont Co.

Mr. Andrews received a B.S.E. degree in mechanical engineering from Swarthmore College, and did graduate studies in engineering at the University of Buffalo and Lamar University. He is a registered professional engineer with experience in management and consulting for chemical, refinery and nuclear facilities in the U.S. and South America. He has also conducted continuing education courses in multiple engineering disciplines.

Richard W. Prugh is principal process safety consultant at Condux. He is a registered engineer and certified safety professional with experience in chemical process safety, analysis, explosion and toxic-vapor hazards analysis, and all phases of risk and safety management. He has a B.S.Ch.E. from the Massachusetts Institute of Technology, and an M.S. degree from Stevens Institute of Technology. Among his publications are articles for the *Encyclopedia of Polymer Science and Technology*, the *Encyclopedia of Chemical Technology*, and for AIChE's International Symposium on Loss Prevention and the Center for Chemical Process Safety.

Howard A. Smith, principal consultant on safety and health at Condux, received a B.S. degree in chemistry and mathematics from St. John's University, and an M.S. degree in analytical chemistry from the University of Colorado. He is a member of the American Chemical Soc. and the American Industrial Hygiene Assn. His experience includes safety and occupational health, epidemiology, toxicity analysis, auditing and training. He has published articles in the *J. of Occupational Medicine* and *Analytical Chemistry,* among others.

Paul R. Weber is a consultant in high pressure technology for Condux. Educated at the University of Wisconsin (B.S., mechanical engineering) and the University of Delaware, he is member of the Academy of Lifelong Learning at the Univ. of Delaware and serves as a technical interpreter at the Hagley Museum. He has published in *Mechanical Engineering,* and is a recognized specialist in the operation, maintenance and safety of high-pressure vessels and systems.

Donald G. Windsor, principal consultant for safety services and occupational health at Condux, is a certified safety professional with extensive experience in corporate safety programming, distribution and handling of hazardous materials, risk management and the development of loss prevention guidelines. He has a B.S. in chemistry from Dickinson College, and is recipient of the National Safety Council's highest award, Distinguished Service to Safety. He is the author of *The Art of Safety,* and has had articles published by the National Safety Council.

PART 2

MANAGING SAFETY

Richard Prugh, Howard Smith and Donald Windsor,
Condux, Inc.

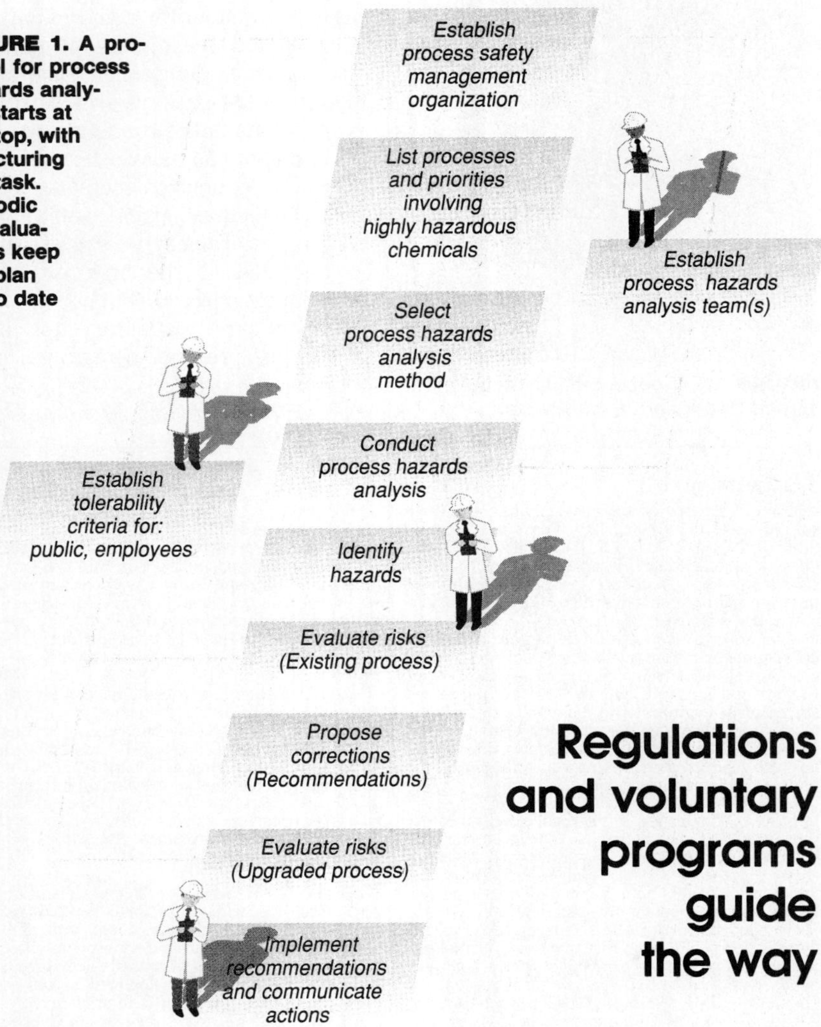

FIGURE 1. A protocol for process hazards analysis starts at the top, with structuring the task. Periodic reevaluations keep the plan up to date

BARBARA BERASI

Regulations and voluntary programs guide the way

Accidents are caused; they just don't happen. Numerous articles have been written by knowledgeable authors on what really causes accidents. There have been many heated disagreements on the statistics or percentages of mishaps that are "people" caused, "equipment" caused and so on. The numbers are not of primary importance. One way or another, plant operations must accommodate the actions of workers, and workers must adjust their habits to conform to a plant's safety requirements.

Managing safety in high-pressure equipment is not inherently different from managing safety in other types of operations. The fact remains, though, that the stakes may be considerably higher. Upsets of high-pressure processes frequently cause releases of hazardous materials or explosions.

A safety philosophy

A basic stance for all design and operational approaches should be that all accidents are preventable. Someone will inevitably refer to the bolt of lightning that strikes and causes a problem. Given that that may happen, and not often, the other 99.99% of incidents are not "acts of God."

Frequently, one feels that the emphasis on people suggests the operator has erred, pushed the wrong button, added the wrong ingredient or failed to follow other operating instructions. There are many other avenues of human error, including improper design, improper maintenance, using the wrong materials, inadequate training and poor operating instructions.

One might consider the three "ins" regarding safety:

- Design it in
- Build it in
- Keep it in

Besides these "ins," human factors are an ever-present element. Employees

Originally published September 1993

now assume greater responsibility regarding safety, for themselves and their co-workers. All involved workers, from designers to operators, share in the safety responsibility. All must accept that any accident (and therefore injuries) can be prevented. These shared responsibilities are more important in today's era, when many organizations have have changed their management methods. No longer are supervisors and managers constantly present.

Rules for the road

The safe design of high-pressure systems has been worked at for decades. Many precise design principles have been established, such as the ASME Code of Fired Pressure Vessels, Sec. 1, of Unfired Pressure Vessels, Sec. 8, or CGA Pamphlet 1 of the Compressed Gas Assn.

There are also many newly updated procedures for maintaining and repairing high-pressure process equipment. The Clean Air Act Amendments of 1990 contain two sections of importance to this subject: Sec. 112(r), Prevention of Accidental Releases, and Sec. 304, Chemical Process Safety Management. Sec. 304 leads to Process Safety Management of Highly Hazardous Chemicals, the title for a regulation developed by the U.S. Occupational Safety and Health Administration and published as 29 CFR 1910.119.

Sec. 112(r) created an expanded role for the U.S. Environmental Protection Agency (EPA) in the management of chemical process safety. It is seen as the next logical step in EPA's efforts to inform communities of nearby toxic chemical hazards through Title III of the Superfund Amendments and Reauthorization Act of 1986.

To implement the statutory mandate, EPA has developed two regulatory initiatives addressing accidental chemical releases. One is to create a list of regulated substances, and to issue a final regulation that identifies at least 100 of them and then define minimum threshold quantities that would trigger accidental-release prevention provisions.

The second approach is to manage risks by developing accidental-release prevention programs. The required management plan is initiated when a regulated substance is present at a facility in quantities greater than a threshold amount (based on the chemical list mentioned above). Once a plant is determined to fall under these requirements, a hazard assessment, a release-prevention program, and an emergency-response program must all be developed. EPA is required to finalize these rules by Nov. 15, 1993. Manufacturing facilities will have up to three years from that date to comply.

Safety in the workplace

OSHA's Standard 29 CFR 1910.119, Process Safety Management of Highly Hazardous Chemicals, directly impacts the operations of high-pressure systems in which any of its 128 listed chemicals are handled. Documentation, training and employee involvement are an integral part of the standard.

Among the elements of this rule that are of greatest significance to high-pressure operations is Section (d), which involves a review of the equipment, materials of construction, piping and instrumentation diagrams, relief devices, safety systems and other components of the manufacturing process. Section (e) addresses process hazard analysis (PHA), including thorough studies of the potential hazards of a production process. Details of PHA programs will be described below.

Section (j) addresses the mechanical integrity of process equipment in great detail, including maintenance procedures. It is the regulatory basis for many of the safety functions described in Part 1.

Other notable OSHA regulations for high-pressure operations are:

- OSHA 1910.146, which specifies procedures for opening process equipment or piping, and "confined-space entry" of workers into process vessels
- OSHA 1910.147 which specifies procedures for lockouts or tagouts of process equipment, as well as for opening this equipment

Responsible Care

In 1988, the Chemical Manufacturers Assn. (CMA; Washington, D.C.) established the Responsible Care program, aimed at improving the industry health, safety and environmental-quality performance. It also seeks to enhance communication and understanding toward the public's concerns about products and plant operations. The essence of the program is contained in ten principles and six management practice codes that convert philosophy into actions.

One of the guidelines that is more relevant for high-pressure safety specifies that CMA members should "operate plants and facilities in a manner that protects the environment and the health and safety of employees and the public." Another relevant guideline is to recognize and respond to community concerns about chemicals and plant operations.

There are several management-practices codes that turn this guideline into working procedures. For high-pressure equipment operators, the two most relevant ones are process safety and employee health and safety:

The Process Safety Code has a goal of preventing fires, explosions and accidental releases. It requires plant managers to:

- Have an ongoing safety program that measures performance, conducts audits and initiates corrective actions
- Conduct safety reviews on all new and modified facilities during design and startup
- Perform, and document, maintenance and inspection programs to ensure facility soundness
- Develop and put into place layers of protection to prevent the escalation of a

'What if' analysis cannot be relied on for identifying *unrecognized* hazards

single failure into a catastrophic event

- Train all employees in safe work practices
- Share relevant information with industry, government and the community to help prevent such incidents

The Employee Health and Safety Code has the goal of promoting the health and safety (H&S) of workers or people visiting company sites. It requires plant managers to:

- Develop occupational H&S programs and review their effectiveness
- Permit employees to participate in the development, implementation and review of H&S programs
- Select contractors on the basis of their own H&S programs, and to periodically review their performance
- Develop methods for determining trends in H&S data, and to identify areas for improvement
- Identify and assess potential H&S hazards
- Promote safety through equipment and job-task safety reviews
- Investigate worksite illnesses, injuries and accidents, and take corrective actions to prevent recurrence
- Develop communications and training programs for H&S information

There are relatively few specific requirements in the CMA program. Perhaps the most important is that process hazards analysis be included in the funding and schedule for new projects.

Plotting safe procedures

Both OSHA and CMA guidelines highlight the importance of process hazards analysis as an engineering tool for evaluating safety. OSHA 1910.119(e2) makes reference to a variety of methods to perform this analysis. Some are more rigorous than others, but care must be taken in using any of the techniques.

The so-called "what if" and checklist methods, recommended in subsections (e2i, e2ii and e2iii) usually cannot be relied on for identifying *unrecognized* hazards. Teams using these methods may fail to delve into the chemistry and process-control systems of processes with which they are have become familiar. Experience and insight are required to identify hazard scenarios that are not readily apparent.

It is also worth noting here that the OSHA standard requires that some attempt be made to evaluate (i.e., quantify) the hazards of a process-control failure. Although it may be easiest to quantify worst-case consequences, the greatest potential risks — particularly to employees — may arise from smaller, more-frequent "loss of containment" incidents.

The "failure modes and effects analysis" (e2v) may be the easiest to use, because it begins by placing all process components on a list, and then identifying possible modes of failure (such as open, shut or stuck). A disadvantage of this method is that including humans as "components." Identifying human failure modes is difficult at best. If control of a process is primarily manual, another analytical method should be used.

The Hazop (Hazard and Operability) method (e2iv) has recently become a favorite, because its use can lead to identification of human as well as component failures. This method starts with the definition of the design intention, in terms of the required ranges of temperature, pressure, flow and so forth. "Guide words" are then applied to each of the process variables, to identify possible deviations from the design intention.

The next step is to analyze the causes and consequences of deviations, and to determine the effectiveness of protective systems. Recommendations are developed for scenarios where the protective systems are considered to be inadequate.

The final methodology listed by OSHA is fault tree analysis, which is useful for determining, quantitatively, the presently unknown frequency of a particular undesired event. Starting from this "top event" (such as a one-ton release of toxic vapor), the analyst identifies its direct and immediate causes. Each of the causes then becomes an event that must be similarly investigated. This deductive reasoning is used to develop a "tree," downward toward basic causes, for which occurrence frequencies are available.

Criteria are being developed for the

The greatest potential risks arise from smaller, more-frequent 'loss of containment' incidents

tolerability or acceptability of a top event occurrence. Such criteria are needed where employees or the public could be injured by exposure to such an event.

Assign accountability

To ensure the safe and optimum performance of high-pressure systems, as with most other CPI operations, an organized management system must be in place, understood and followed enthusiastically. A clear, concise written policy addressing management's principles and commitment on safety, health and environmental responsibility is the starting point.

Systems should be in place, recognized by all and functioning, to address safety and health planning, procedures, priorities and support. Accountability must be discussed, agreed on and communicated to everyone. Experience shows that the subject of accountability cannot be overemphasized.

Responsibility and accountability should be spread throughout the entire organization and everyone should understand and accept the responsibilities assigned to them. People should feel personally accountable for safe performance, and goals and objectives to meet and maintain safe operations should be part of individual performance appraisals.

This commitment is backed up by another management system, that of performing field audits, incident reports and investigations. Operating rules should be reivewed and updated periodically as well. Audits should not focus exclusively on equipment, but also on the people responsible for maintaining safety.

Engineers are objective-driven. Objectives are valuable. They force us to evaluate our capabilities, decide what we can accomplish and how it can be brought about. In order to be effective, however, objectives must be communicated and then acted upon. How many times have we found, when clearing out a file, last year's objectives, untouched? One way of setting objectives is to involve the whole organization; not the whole plant or department but a smaller group working together. And this builds "buy in."

Working together as a team, the group can meet and exceed objectives. And the responsibility is placed where it belongs, on those executing the project, whether it is in the design, construction or operation phase. As with all good management systems, responsibility goes hand in hand with accountability. This means that when objectives are not met, the consequences are negative. Performance should be based, in part, on whether safety objectives are met. This is a method that can rapidly communicate to all whether management is truly concerned with safety.

It is imperative, in pursuing a goal of no accidents, to include a preventive approach. It is much better to keep an explosion from happening than to investigate it after the fact. Accepting that there may be an incident, however, one must have in place a system to properly investigate, determine causality and define preventive measures. Here again, the system must force action on preventive recommendations, else such reports will be relegated to a file for "safe" storage.

Now, a word of caution. Having determined the proper corrective course, but not implementing it, may place some persons in a precarious legal position. Should a similar event recur, the courts may find that having such knowledge and taking no action constitutes criminal behavior. Consult your legal advisor, or better yet, take the corrective action! ■

PRESSURE VESSELS DESIGN OPER

K. P. Singh, Holtec International

An enclosure devised to sustain a pressure differential between its inside and outside is termed a "pressure vessel." Usually, the pressure on the inside is higher than that on the outside, although the opposite is not uncommon. In the chemical process industries, the fluid on the inside is often not only at a high pressure, but also hot, flammable, noxious, or otherwise dangerous. Because of such hazards, close attention must be paid to leak tightness in the design and operation of pressure vessels.

It was not so much leaking joints, however, but the alarming human toll at the turn of the last century that spurred the development of codes pertaining to the design, inspection and operation of pressure vessels. Boiler explosions, a common occurrence in U. S. industries in the 1860s, had become a national issue by 1910. This led to the development and publication of the ASME Boiler Code, which sought to standardize the design, manufacture and testing of boilers.

Most states in the U. S. rapidly adopted the ASME Code as a statutory document. Zealous enforcement dramatically reduced the incidence of boiler explosions. Today, pressure-vessel explosions are rare, and almost always caused by unmonitored corrosion or incorrect operation. The Code has become the sourcebook, the authority on the correctness of pressure vessel design and manufacturing. It has also evolved into a voluminous document.

Published in 1989, the latest issue of the Boiler and Pressure Vessel Code of the American Society of Mechanical Engineers contains eleven sections. A 1986 report in *Chemical Engineering* by Stanley Yokel of MGT, Inc. provides a comprehensive account of the workings of regulatory bodies and committees, and of the Code [1]. By explaining the underlying principles of pressure-vessel design and their relationship to the Code, this report complements the earlier one.

FIGURE 1. Two basic parts of vessels are plate- and shell-type elements

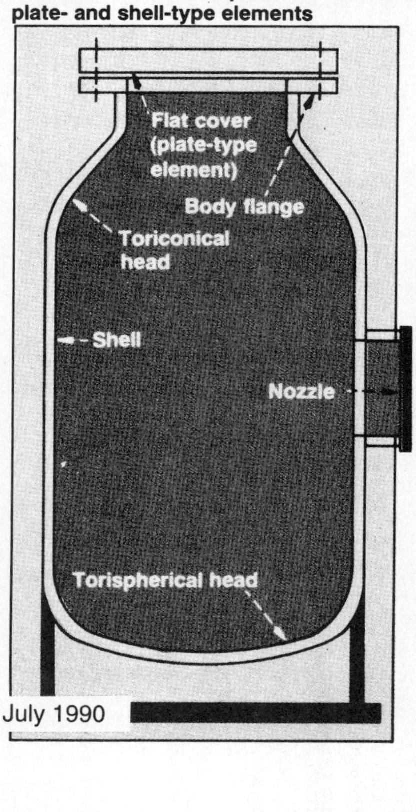

Vessel parts

The basic geometries involved in a pressure vessel, no matter how complex in the appearance, can be divided into two categories: those with plate-type and those with shell-type elements. The flat cover in Figure 1 is a plate-type element. It resists pressure by flexing (i.e., bending). A shell-type element, on the other hand, resists pressure by a "membrane action," which is similar to how a balloon contains pressure.

Figure 2a shows a solid circular plate supported at its outer edge and subjected to a lateral pressure p. Under pressure, the loaded face becomes concave and tensile stress develops on its convex face (Figure 2b). The stress across the thickness of the plate varies from tensile at the convex face to compressive at the concave face.

Classical plate-bending theory is based on the assumption that the stress level varies linearly as in Figure 2c. The two surfaces of the plate are stressed most, and the middle of the plate is not stressed at all. This shows that the plate-type configuration does not make optimum use of a material's strength.

Shell-type elements are generally cylindrical, spherical, ellipsoidal, torispherical or toriconical (Figure 1). A pressure vessel may be a combination of a plate-type element and one or more of the shell-type elements. Figure 3 shows a pressure vessel consisting of a cylindrical main shell, two hemispherical headers, and several nozzle connections, which are essentially cylindrical shells having flat covers. In general, the shell-type element is the preferred form for constructing a pressure vessel because it requires

Drawings by RON CARBONI

Pressure vessels come in a wide variety of shapes and sizes, but their design is still based on a small number of key concepts

considerably less metal thickness than the plate-type.

Figure 4a shows the end view of a cylindrical shell under internal pressure p. The Figure 4b cutaway illustrates the pressure resisting action. The value of the stress, which is uniform across the thickness of the shell, can be computed by a direct force balance. If the stress is σ, the resisting force is represented by $2L\sigma t$, and the applied force by $(L)(2R)(p)$. Equating these two forces yields the Lame equation, perhaps the most widely recognized formula in pressure-vessel design:

$$2L\sigma t = 2LRp, \text{ or } \sigma = pR/t \quad (1)$$

In vessel design, the vessel radius and the allowable stress (σ) are usually specified. To calculate the vessel's wall thickness, the engineer, therefore, uses Lame's formula in the form:

$$t = pR/\sigma \quad (2)$$

Plate vs. shell shape

The following problem reveals the relative effectiveness of plate and shell shapes. What is the ratio between the required thickness of a circular plate and a shell shape, both of radius R, of identical material, and subject to the same pressure p?

Calculate shell thickness with Equation 2. The equivalent formula for a circular plate is more complicated [2]:

$$t_p = \{[3(3 + v)/8][p/\sigma]\}^{1/2} \times R \quad (3)$$

Here, t_p = required plate thickness, and v = Poisson's ratio (equal to 0.3 for nearly all metals).

For $v = 0.3$:

$$t_p = 1.1R \times (p/\sigma)^{1/2} \quad (4)$$

To calculate the thickness ratio, combine Equations (2) and (4):

$$t_p/t = 1.11(\sigma/p)^{1/2} \quad (5)$$

According to Equation (5), a plate for a carbon-steel vessel (ASME SA 516 Grade 70, allowable stress = 17,500 psi) designed for 100 psig will be 14.7 times thicker [i.e., 1.11 (17,500/100)$^{1/2}$] than a shell of identical diameter. This clearly indicates the economy of material to be realized from using shell-type shapes rather than plate shapes for pressure vessels.

Stress classification is central to all pressure-vessel and heat-exchanger design codes. It is closely linked to weld-joint categories and weld-type classifications, which form the bases of code design work [2].

Although the stress field developed in a pressure vessel is complicated, one can usually find two principal stress directions, each of which can be represented at a point as the sum of (1) an average or "membrane" component, (2) a linearly varying or "bending" moment, and (3) a remainder or "peak" component.

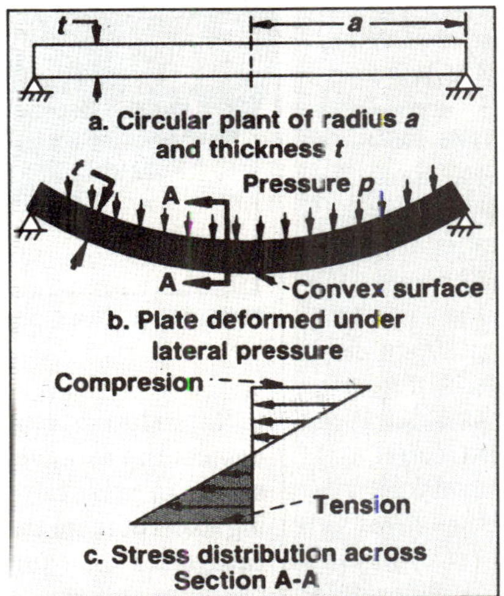

FIGURE 2. A plate does not take full advantage of a material's strength

Shell-type pressure parts, such as heat exchanger shells, channels, heads and nozzles resist pressure primarily by membrane action. Most design rules in the Code involve computing and prescribing limits on pressure-induced membrane stresses in shell-type pressure parts, and bending stresses in plate-type pressure parts. These stresses are formally referred to as "primary membrane" and "primary bending" stresses. The word "primary" signifies that the particular stress originates from a primary loading (such as internal pressure) and that the stress is not localized. A temperature gradient, on the other hand, is considered a secondary load, because the stress produced by it simply accommodates a difference in free thermal expansion.

Primary loads, such as pressure, also produce membrane and bending stresses that are very localized. The stresses produced in a shell by the pressure induced flexing of an integrally-welded flat cover are known to attenuate rapidly from the junction between shell and cover. Rapidly decaying stresses of this

ATION

MAINTENANCE

A short glossary of key terms in the design of pressure vessels

Design pressure is the gage pressure at the top of the vessel. It and the corresponding metal temperature are stamped on the nameplate. Because a pressure vessel may be connected to other vessels (including heat exchangers), more than one design pressure may be required to define the loading. The Code encourages setting design pressures higher than normal operating pressures with margins that allow for probable pressure surges.

When units are connected, the maximum difference in pressure between adjacent chambers should be considered in setting the design pressure. Although not explicitly stated in the Code, design pressures can differ in different parts of the same pressure chamber of a hybrid pressure vessel, such as a heat exchanger. For example, if one of the design pressures on the shellside of an exchanger is absolute vacuum ($p = -p_a$), and the normal operating pressure (with a suitable margin) on the tubeside is positive (p_t, for example), the design pressure for such parts as the channel cover and flange will be p_t but $p_t + p_a$ for the tubesheet and tubes.

Design temperature is the expected mean metal temperature corresponding to the design pressure. It can differ for different pressure parts if operating conditions ensure a defined temperature variation. For example, the design temperature for the inlet of a recuperator or fluegas cooler may be much higher than the outlet.

Maximum Allowable Working Pressure (MAWP) is the maximum pressure expected for the specified operating temperature. It is figured separately for each part of a pressure chamber. The computed stresses (calculated by means of applicable Code equations) from MAWP and any other likely loadings should not exceed the Code allowed stress values. Any metal thickness specified as corrosion allowance is not to be include in these computations.

The calculated MAWP for each pressure part is adjusted by the amount of the static head of the contained fluid to arrive a MAWP at the top of each pressure chamber (see Part 2 of this report). The overall MAWP is the smallest of the adjusted MAWPs for all of the pressure chamber parts. It can be determined for more than one operating temperature.

Allowable stress for materials in tension at discrete design temperatures are listed in Code tables. Allowable values in compression, which depend on the "slenderness" of the pressure part, are presented in a series of charts based on the ratios of the characteristic dimensions of the pressure parts.

Joint efficiency defines the quality of a welded joint, which depends on such factors as joint preparation, type of welding process, and the control of the welding process parameters. Welds are commonly examined radiographically. For each type of joint and the extent of the examination, the Code prescribes a joint efficiency, *(E)* which is to be multiplied with the allowable stress (σ) to yield an effective allowable stress (i.e., *SE*).

TABLE 2. ASME Code equations for designing cylindrical shells

Member	Minimum thickness, t	Maximum pressure, p	Code reference	Limitation
Thin shell (in terms of inside radius) Longitudinal joint	$\frac{pR}{SE-0.6p}$	$\frac{SEt}{R+0.6t}$	UG-29	$p \leq 0.385SE$ $t \leq 0.5R$
Circumferential joint*	$\frac{pR}{2SE+0.4p}$	$\frac{2SEt}{R-0.4t}$	UG-29	$p \leq 1.25SE$ $t \leq 0.5R$
Thin shell (in terms of outside ratius R_O)	$\frac{pR_O}{SE+0.4p}$	$\frac{SEt}{R_O-0.4t}$		$p \leq 0.385SE$ $t \leq 0.5R$
Thick shell, longitudinal joints (inside radius)	$R(z^{1/2}-1)$	$SE\frac{(z-1)}{(z-1)}$	Appendix I	$p > 0.385SE$ $t > 0.5R$
(outside radius)†	$R_O\frac{(z^{1/2}-1)}{z^{1/2}}$			
Circumferential (inside radius)	$R(z^{1/2}-1)$	$SE(z-1)$	Appendix I	$p > 1.25SE$ $t > 0.5R$
(outside radius)**	$R_O\frac{(z^{1/2})-1)}{z^{1/2}}$			

*Equations based on stress across circumferential joints govern only if the efficiency of the circumferential joint (Category B)is less than one half of the longitudinal joint
†For outside radius in this case: $z = (R+t)/R = (R_O)^2/(R^2) = (R_O)^2/(R_O-t)^2$
**For outside radius in this case: $z = (p/SE + 1)$

on the gasket deformation for sealing (Figure 9e). The gaskets are usually a ring of elastomer (e.g., rubber) or a hollow metallic tube.

Flange facing finishes

Critical to the sealing worthiness of a flanged joint is the type and texture of the surface finish of the flanges. Finish type and quality are designated by five surface types: rounded-nose spiral, spiral serrated, concentric serrated, smooth, and cold water.

Rounded nose spiral, or stock—This finish is a continuous spiral groove. For nominal sizes 12 in., and smaller, it can be generated by a 1/16-in.-radius round-nose tool at a feedrate of 1/32-in. per revolution. For sizes larger than 12 in., the tool nose-radius is increased to 1/8-in. and the feed rate to 3/64 in. This finish is mostly used with fiber, rubber and other organic gaskets. It is often called the "phonograph" finish.

Spiral serrated—The continuous spiral groove on this face is generated by a 90-deg.-included-angle V-tool. For all flange sizes, the groove is 1/64-in. deep and the feedrate is 1/32-in.

Concentric serrated—This facing is the same as the spiral serrated, except for the groove being concentric circles.

Smooth—This finish has no visible tool markings. It is dressed up with an abrasive, such as emery paper.

Cold water—This name is derived from a surface whose appearance is as mirror-like as a quiet body of water. It is produced by a wide tool operating at a high speed. It is seldom used because it is expensive and fragile.

Gasket selection

The gasket is the heart of a bolted joint. Usually made from an elastomeric-plastic material that is softer than the flange face, the gasket deforms under bolt pressure to fill up the irregularities of the flange face. In this condition, it is considered to be seated.

The building up of pressure in a vessel separates the flanges imperceptibly at the gasket, which must be sufficiently resilient to maintain enough residual surface pressure to keep the joint from leaking. Gaskets are made from many types of materials. The ASME Code list some common ones. The proper choice of material depends on operating conditions, the mechanical attributes of the flanged joint and on metallurgical considerations. Some factors to consider when selecting gasket material are pressure, temperature, corrosion and cyclic operating conditions.

For a particular diameter, a pressure vessel's operating pressure determines the magnitude of the hydrostatic load that the bolts must resist. Therefore, the bolt load must at least exceed the hydrostatic load. Before the vessel is pressurized, the gasket bears the entire bolt load. Therefore, it must be strong enough to withstand the bolt load without being crushed or extruded. This rules out such soft materials as organic fibers for high-pressure service.

For other than metallic gaskets, the operating temperature of the fluid contained by a joint must be less than a permissible temperature. Such activities as oxidation, galvanic corrosion and dezincification must be considered when selecting metallic gaskets. Fluctuations in pressure result in repeated loading and unloading of the gasket. Spiral-wound and hollow tubular gaskets withstand such a condition quite well.

Flanges for very high pressure

Bolted joints are widely used with pressure vessels in low- to medium-pressure service. Because the thickness of a flat cover increases directly in proportion to the product of shell diameter and the square root of the pressure, bolted-flange closures for large-diameter vessels at very high pressures could become heavy and unwieldy.

In addition, the number and diameter of the bolts required for the necessary preload may exceed what can be arranged on the minimum circumferential pitch (set to accommodate a box wrench) on the bolt circle. This may necessitate having two bolt circles, greater moment arms and, thus, a still heavier flange. Therefore, when the pressure/diameter ratio is such that the thickness of the shell exceeds 1½ in., the engineer should consider a "boltless" flange.

Boltless flanges can be divided into two categories: axially locked and pressure actuated. The first is infrequently used in the chemical process industries. The second is available in a number of styles.

The pressure-actuated joint is truly

Governing Code standards for the mechanical design of pressure vessels

A pressure vessel's operating conditions determine which parts of the ASME Boiler & Pressure Vessel Code govern its design. Section VIII, Div. 1 of the Code applies if the design pressure exceeds 15 psig or is full vacuum. This coverage is limited to a maximum pressure of 3,000 psig, above which additional design considerations are recommended. Section VIII, Div. 2, being analysis oriented, does not prescribe pressure limits. It allows thinner walls at the expense of additional design analysis, material inspections and testing during fabrication.

Section III followed the advent of commercial nuclear power. Here, pressure vessels and heat exchangers for nuclear power plants are divided into three classes (or grades). Three subsections (NB, NC and ND) cover the design and construction of Class 1, 2 and 3 equipment, respectively. Class 1, the most stringent, requires design by stress analysis. Class 2 permits design by analysis (Paragraph NC-3200) or by formula (Paragraph NC-3300). Class 3 prescribes design by formula, and is equivalent of Section VIII Div. 1. The nuclear engineer evaluates the safety function of each pressure vessel and heat exchanger, and designs the equipment according to the appropriate Code class.

The design of supports for Section VIII, Div. 1 vessels are not prescribed in the ASME Code. Section III does include subsection NF, which prescribes the design of supports for Class 1, 2 and 3 heat exchangers and pressure vessels.

The standards of the Tubular Exchangers Manufacturing Assn. (TEMA) supplement the ASME Code on the design of heat-exchangers. The TEMA Standards are actually a collection of TEMA-R, TEMA-C and TEMA-B [*4*]. TEMA-B has been promulgated as an American national standard (ANSI B-78).

The Heat Exchanger Institute also has published standards for heat exchangers [*5,6*]. These deal with design aspects not covered by the ASME Code, such as the design of tubesheets, baffles, tie rods and spacers.

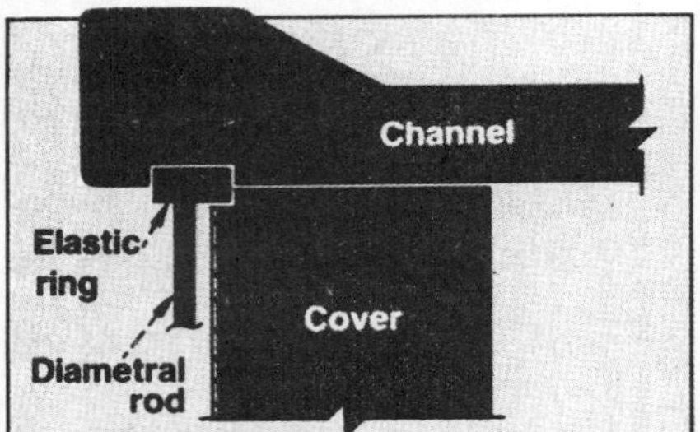

a. Rectangular seal-ring closure

FIGURE 10. Fitting a seal ring into a boltless flange needs a special arrangement

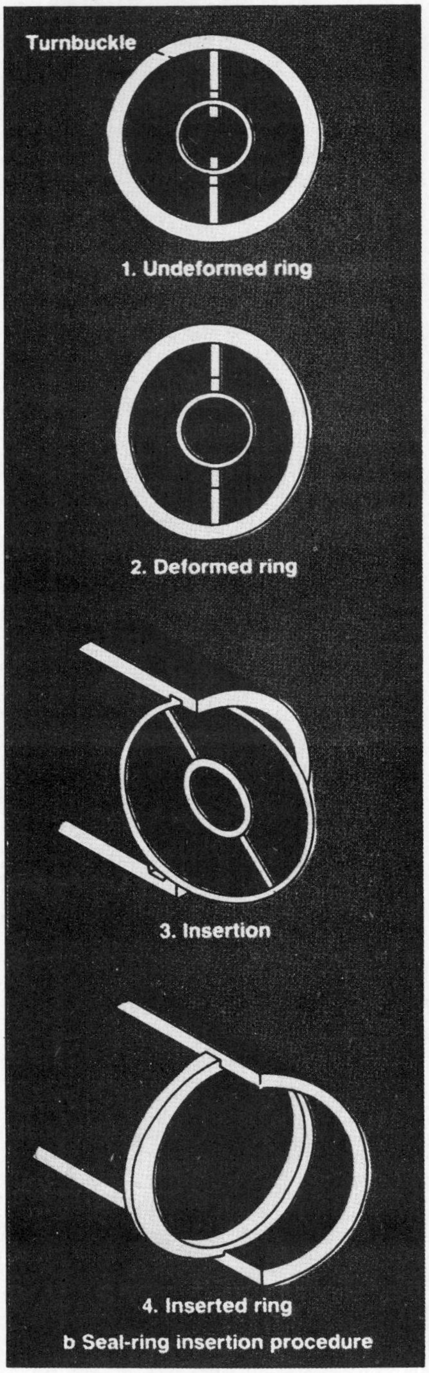

b Seal-ring insertion procedure

boltless. This class of closures is based on the "unsupported area" principle (Figure 10a). An endless elastic ring is forced into a circumferential groove. For this, the ring is equipped with a diametral rod having a turnbuckle. By means of the turnbuckle, the ring is deformed into an oval. The initial deformation is carefully controlled so that the ring is not permanently set. The ring is inserted into the groove as shown in Figure 10b. Such rings are made of a machinable material that can withstand the axial trust of a pressure load without failing. The compressive bearing stress, σ, on the seal ring can be calculated from:

$$\sigma = D^2 p/4(D' - W)W \tag{7}$$

Here, D' is the cover diameter. Selecting the appropriate value of W (see Figure 10a) sets the most desirable value of σ. This closure has been used with pressure as high as 6,000 psi.

The wedge seal-ring closure consists of five principal parts: a shell, which has a machined circumferential slot near its end; a cover, which resists the pressure loading the same way that a laterally-loaded flat circular plate does; a retainer shoe, which is a segmented split ring made to fit the circumferential slot in the shell; a solid follower ring; and a five-sided seal ring (Figure 11a).

The retainer shoe cannot be installed into the shell slot unless it is made in segments, the number of which should be minimized to limit the reduction in ring strength (Figure 11b). The retainer shoe should fit into the slot with minimal clearance longitudinally (less than 1/16 in.). It should provide adequate bearing surface for the follower ring.

The solid follower ring should fit into the inside diameter of the shell barrel with minimal radial clearance (1/32 in.). The dimensions of the seal ring are important design parameters that can be varied to obtain the optimum ring cross-section for particular operating conditions.

The cover is machined so that the seal ring's inclined face can ride on the machined surface. Internal pressure pushes the seal ring up the inclined machined surface of the cover. The ring material should be softer than the shell and cover materials and be metallurgically compatible with them. Materials that tend to bind or to gall each other under pressure should be avoided. ■

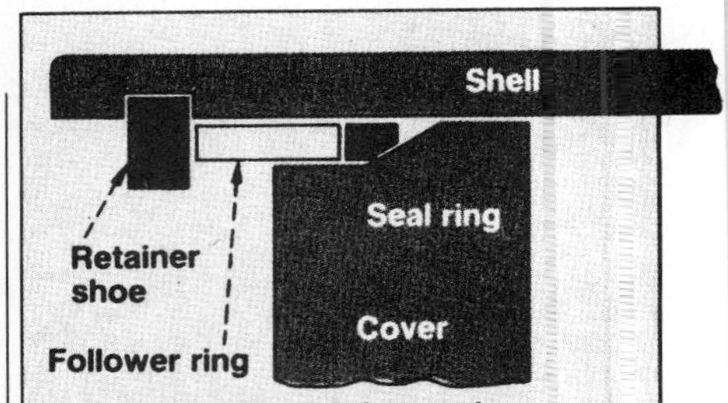

a. Wedge seal-ring enclosure

FIGURE 11. Only a segmented seal ring can be installed in the wedge closure

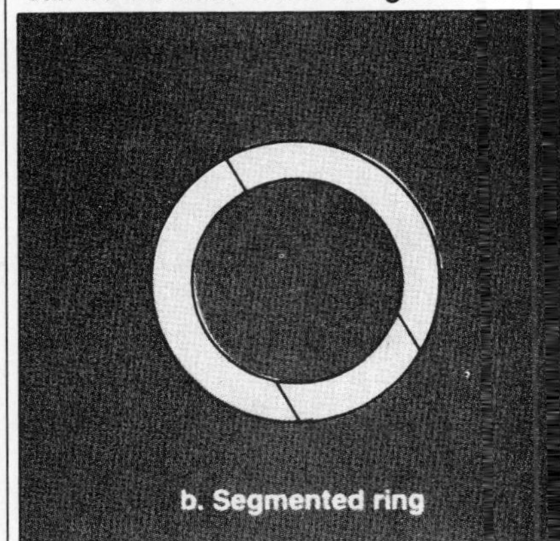

b. Segmented ring

References

1. Yokell, S., Understanding the Pressure Vessel Codes, *Chem. Eng.*, McGraw-Hill Inc., May 2, 1986, pp.75-85.
2. Singh, K. P., and Solar, A. I., "Mechanical Design of Heat Exchangers and Pressure Vessel Components," 1st Ed., Arcturus Publishers, Cherry Hill, N.J., 1984.
3. ASME Boiler and Pressure Vessel Code, American Soc. of Mechanical Engineers, New York, N.Y., 1989.
4. Standards of Tubular Exchanger Manufacturers Assn., 7th ed., TEMA Inc., Tarrytown, N.Y., 1988.
5. Standards for Power Plant Heat Exchangers, 1st ed., Heat Exchange Institute, Cleveland, Ohio, 1980.
6. Standards for Surface Condensers, 8th ed., Heat Exchange Institute, Cleveland, Ohio, 1984.
7. Burgreen, D., "Pressure Vessel Analysis," Arcturus Publishers, Cherry Hill, N.J.
8. Pressure Vessel Handbook, 6th ed., edited by E. F. Megyesy, Pressure Vessel Handbook Publishing Co., Tulsa, Okla., 1983.
9. Harvey, J. F., "Pressure Component Construction," Von Nostrand, N.Y. 1980.

The author

K. P. Singh is president of Holtec International (Cherry Hill, NJ 08003; tel.: 609-424-0999). The recipient of a Ph.D. degree in mechanical engineering from the University of Pennsylvania, he has more than 50 papers published, and is the principal author of the book, "Mechanical Design of Heat Exchangers and Pressure Vessel Components," (Arcturus Publishers) and a patented co-inventor. A Fellow of ASME and an adjunct professor of mechanical engineering at the University of Pennsylvania, he has served as the chairman of TEMA Technical Subcommittee and on ASME's Pressure Vessel and Piping Committee.

PRESSURE VESSELS TOTAL DESIGN SPECIFICATION PART 2

Glenn D. Nasman, Badger Engineers, Inc.

Process engineers must consider more than the usual parameters when issuing specifications for the mechanical design of pressure vessels. To come up with vessels that perform satisfactorily in difficult services, they must go beyond designating vessel and nozzle sizes, design pressures and temperatures, and corrosion allowances.

For instance, the design pressure is normally that at the top of a vessel, and it usually coincides with the relief-valve setting. In addition to specifying this pressure, the process engineer must advise the vessel designer of the maximum level of the liquid, and the liquid's maximum specific gravity. This information will enable the designer to take into account the static head at the bottom of the vessel. This is critical because, even with a liquid having a specific gravity of one, the pressure at the bottom increases almost five lbs/in.^{2} for every 10 ft of liquid height.

Short-term conditions

The ASME Pressure Vessel Code requires that all pressure vessels be designed for "the most severe condition of coincident pressure and temperature that will be expected during normal operation" [*1*]

There can be no deviation from this requirement, even if the severe condition is short-term and occurs only occasionally.

If the condition governs the vessel design, the process engineer should inform the designer of it because it could significantly influence the material selection. For example, suppose that the temperature of a vessel that normally operates at 650°F could rise to 850°F during an upset. Because the vessel would be shut down immediately, the condition would be short-term.

A designer who selects a material based on the 850°F design temperature without knowing the frequency and duration of operations at this temperature will very likely choose a low-alloy steel. Carbon steel, the most common material chosen for noncorrosive environments between –20°F and 800°F, weakens with long exposure to temperatures higher than 785°F. This is known as graphitization, a cumulative process whose progress depends on how long the material remains at a high temperature. If the high temperature last for only a short time, a vessel of carbon steel would be a reasonable and economical choice.

When a vacuum is possible

A process engineer will sometimes specify that a vessel be designed for full vacuum even though it is not intended for vacuum service. He or she may do this out of concern that, for instance, a vacuum might be pulled during a steamout. The process engineer should tell the design engineer whether or not the vessel actually will actually operate under vacuum when full vacuum is specified, because this will affect the design.

A major concern about vacuum operation is flange leakage. Many gaskets for pressure service are much

SUSAN COHEN

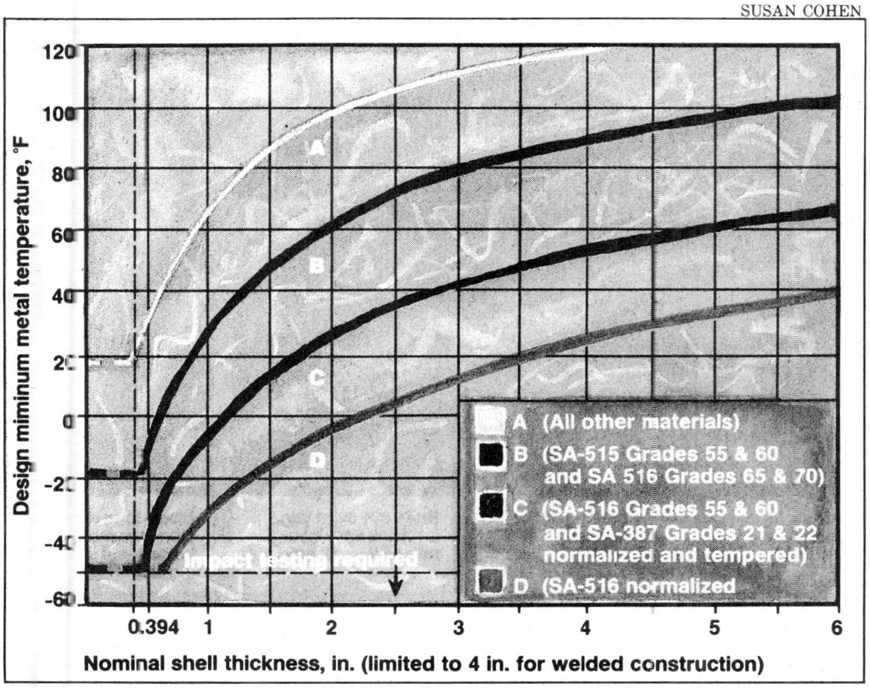

FIGURE 1. These curves help determine whether a vessel must be impact-tested

Originally published July 1990

The process engineer must communicate all the factors that could influence the mechanical design

less effective under vacuum. Because sheet gaskets tend to be sucked in, a designer will often upgrade to a spiral-wound gasket. Leakage is such a concern at large nozzles that engineers commonly specify stub-end connections for all nozzles larger than 24-in. diameter, and have connecting piping welded directly to nozzles, to eliminate flanges and, thus, potential leak sources.

Another concern is the method of analyzing the external nozzle loads on a vessel in vacuum service. If the internal pressure will be atmospheric or positive, the effect of external loads transmitted into the vessel wall by a nozzle or a clip are usually analyzed according to the P. P. Bijlaard procedure outlined in Bulletin No. 107 of the Welding Research Council (New York, N.Y.). This method cannot be applied to vessels in vacuum service because the deflection due to external loads in the radial direction further distorts the roundness of the vessel shell, increasing the likelihood of buckling. Consequently, the designer must analyze external loads by means of another method, such as finite-element-analysis or the calculation method outlined in Code Case N-284 of the ASME Nuclear Code. Still another option is to stiffen the shell with external ribs to ensure that its wall will not buckle.

Low-temperature service

Section VIII, Division 1 of the Pressure Vessel Code, as revised in 1987, recommends that the vessel designer consider a design minimum metal temperature when selecting a material of construction or deciding whether or not a material should be impact tested. Paragraph UG-20(b) of the Code states that this "consideration shall include the lowest operating temperature, operational upsets, atmospheric temperature and any other sources of cooling."

The designer can obtain the lowest ambient temperature from climatic data for a particular site, but the process engineer must inform him or her if other factors can lower the vessel temperature below the minimum ambient. After the minimum design temperature has been established, the designer can decide if the combination of this temperature, the metal thickness and the induced stress in it will require whether the vessel wall should be impact tested.

The simplified impact-test exemption curve shown in Figure 1 includes some common carbon and low-alloy materials used in the fabrication of pressure vessels. Vessels need not be impact tested at minimum metal temperatures during startup, shutdown and depressurization if the vessel pressure is controlled so that the ratio of induced stress in tension to allowable tensile stress is reduced in accordance with the Code. It is the process engineer who must advise the designer whether or not the pressure can be controlled during such such operations as startups and shutdowns.

If a vessel's contents could kill

The process engineer must also inform the designer if the contents of a vessel could be lethal. The Code defines lethal substances as "poisonous gases or liquids of such a nature that a very small amount of the gas or of the vapor of the liquid mixed or unmixed with air is dangerous to life when inhaled." Some examples of such substances are hydrocyanic acid, carbonyl chloride, cyanogen, mustard gas and xylyl bromide.

The Code requires that all such vessels be stamped identifiably and that the manufacturer comply with the applicable requirements. Advised that the vessel will contain a lethal fluid, the designer will require that all the butt welded joints be radiographed, that carbon- or low-alloy-steel vessels be heat treated after welding, and that flanges be weld-necks or lap-joints with forged stub-ends.

Hydrogen service

Vessels containing hydrogen are subject to damage from the excessive absorption of hydrogen by the metal. A vessel is considered to be in hydrogen service if the partial pressure of the hydrogen exceeds 87 psia (600 kPa absolute) and the vessel is hotter than 300°F, or the partial pressure exceeds 145 psia (1,000 kPa absolute) regardless of the vessel temperature.

At moderate temperatures, hydrogen damage occurs when corrosion on the inside metal surface generates hydrogen atoms. These diffuse into the metal, rather than combine into hydrogen molecules. Many of the atoms diffuse right through the vessel and form hydrogen molecules on the outside surface of the vessel. However, ydrogen atoms that diffuse into a void in the metal will combine into molecular hydrogen. Because hydrogen molecules cannot diffuse, the concentration and pressure of the hydrogen gas in the void may increase until the metal blisters and ruptures. Killed steels are much more resistant to hydrogen blistering because they contain far fewer voids.

High temperatures and pressures promote hydrogen attack. In this form of damage, hydrogen atoms diffuse into the metal, where they react with iron carbides to form methane, usually at grain boundaries. Because the methane cannot diffuse out of the metal, the concentration and pressure of the methane in the collection points will increase until the metal cracks or blisters. In the later stages of hydrogen attack, the metal's tensile strength, hardness and ductility have deteriorated substantially because the steel has been completely decarburized throughout its cross-section.

How fast the hydrogen permeates the steel depends on its partial pressure and temperature. Therefore, the process engineer must inform the designer of the partial pressure when issuing the specification. Guidelines for selecting materials for vessels in hydrogen service are presented in the Nelson diagram of API Publication 941 [*2*].

The Nelson diagram, which is based on service experience, indicates the temperature and pressure limits below which a particular material *has not* failed. No implication is intended that a material *will not* fail at conditions below the line, only that no such failures have been reported. If additional failures should be reported, the curves would be shifted downward. Because the diagram is updated periodically to reflect new failures, only the latest revision of API Publication 941 should be consulted. For illustrative purposes, a simplified Nelson diagram is shown in Figure 2.

In general, the Nelson diagram indicates that the higher the percentage of chromium in a low alloy steel, the greater is the steel's resistance to hydrogen attack at elevated temperatures. Chromium, a carbide stabilizer, ties up the carbon, lowering the amount of carbon available for forming methane.

Austenitic stainless steels are immune to hydrogen attack because their chromium content is high. In addition to selecting materials according to the Nelson diagram, the vessel designer should consider the following when designing a vessel that will be in hydrogen service:

1. Use killed carbon steel when carbon steel is acceptable

2. Avoid slip-on flanges because they provide a void in which molecular hydrogen and methane can accumulate

3. Eliminate voids by specifying full-penetration welds for all internal and external attachments

4. Provide vent holes for all internal and external reinforcing pads

5. Specify a gasket material and flange finish that will not allow hydrogen to leak

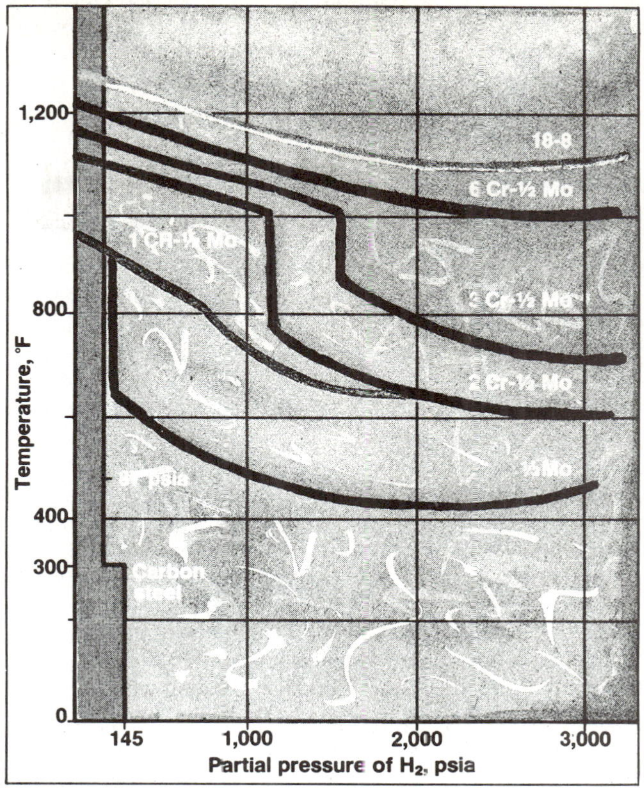

FIGURE 2. A Nelson diagram helps the engineer to make the best choice of materials for vessels containing hydrogen [*2*]

There are eight forms of corrosion that process engineers should consider when specifying pressure vessels

Caustic containment

Caustic soda (sodium hydroxide) is the most common alkaline material in the chemical process industries. At elevated temperatures, carbon steel is subject to caustic cracking (sometimes called caustic embrittlement) and rapid corrosion. The process engineer must specify the highest metal temperature and caustic concentration expected.

Below certain temperatures, steel will not crack in caustic solutions regardless of the level of residual stresses created in fabrication (i.e., forming or welding). Above these temperatures and depending on the caustic concentration, a vessel should be heat treated after being welded (Figure 3). The treatment reduces residual static stresses that contribute to stress corrosion cracking.

Above certain higher temperatures, carbon steel corrodes at an excessive rate, and nickel and nickel alloys become the preferred material of construction. Nickel can be used for practically all concentrations and temperatures of caustic soda.

Corrosion's many faces

Corrosion can be classified into eight forms on the basis of the appearance of the corroded metal [3]. The process engineer should consider these when specifying pressure vessels:

- Uniform (or general) attack
- Galvanic (bi-metal)
- Crevice
- Pitting
- Intergranular
- Selective leaching
- Erosion
- Stress

Attack corrosion is the most common. It spreads uniformly over an entire surface. The metal wall becomes progressively thinner and finally fails. If the rate of corrosion is known, (from, for example, operating experience, laboratory tests or publications), the process engineer can specify a corrosion allowance to cover the life of the vessel.

Galvanic corrosion occurs most often when two dissimilar metals are in contact in a corrosive or conductive solution. The contact increases the corrosion rate of the more-anodic member of the couple.

If the process engineer must select combinations of metals, he or she should select those that are closest together in the galvanic series for seawater, even though the galvanic series for the actual process medium could yield more reliable results. Threaded joints of materials far apart in the series should be avoided. Galvanic corrosion does not occur when the metals are at temperatures high enough to keep them dry because there is no electrolyte to carry the current between them.

Crevice corrosion occurs in such shielded areas as cracks and joints. To function as a corrosion site, a crevice must be wide enough to let liquid enter it, yet narrow enough for the liquid to remain stagnant. Typical locations for crevice corrosion are behind welded metal attachments that have not been welded all around, underneath nuts and washers, between metal parts that have been clamped or bolted together and on flange faces that are in contact with fibrous gaskets that draw moisture (as a wick does).

Metals that depend on oxide films or passive layers for corrosion resistance are particularly susceptible to crevice corrosion. If the process engineer is aware that a metal is susceptible to crevice corrosion in a particular service, he or she should advise the vessel designer to take appropriate precautions. These precautions include specifying that all internals be attached by full-penetration welds or continuous fillet welds (eliminating internal bolting), that all sharp corners and stagnant areas be avoided to promote drainage, and that gaskets be solid and nonabsorbant (such as those made of polytetrafluoroethylene).

Pitting is a localized form of corrosion that not only makes pits but sometimes even holes in metals. Pitting resembles crevice corrosion, except that it does not need an existing crevice because it makes its own. Most pits start and grow downward from the top of horizontal surfaces. They also develop on vertical surfaces but rarely on the underside of horizontal surfaces. Once started, conditions in a pit become progressively more favorable for further corrosion.

Stagnant conditions promote pitting. Austenitic stainless steels are more susceptible to pitting corrosion than any other group of metals. Most pitting is associated with halide ions, particularly the chlorides, the bromides and the hypochlorites. The process engineer should, of course, specify materials that will not be pitted by the fluid in a vessel.

Intergranular corrosion occurs when there is localized attack at, and adjacent to, grain boundaries. There is little corrosion within the grains. The most common example of this type of corrosion is that found in the sensitized heat-affected zones next to the welds of austenitic stainless-steel vessels.

Adding chromium to ordinary steel makes steel more resistant to many substances. However, this benefit is lost if the chromium combines with carbon at the grain boundaries during welding. A band on either side of a weld will become "sensitized," a condition that makes the area vulnerable to intergranular attack by some substances.

Intergranular corrosion near welds does not occur in all environments because austenitic stainless steels are often used where the full corrosion resistance of the alloy is not required (e.g., the material was selected for sanitary or high-temperature service). All the metal in a welded vessel should be the

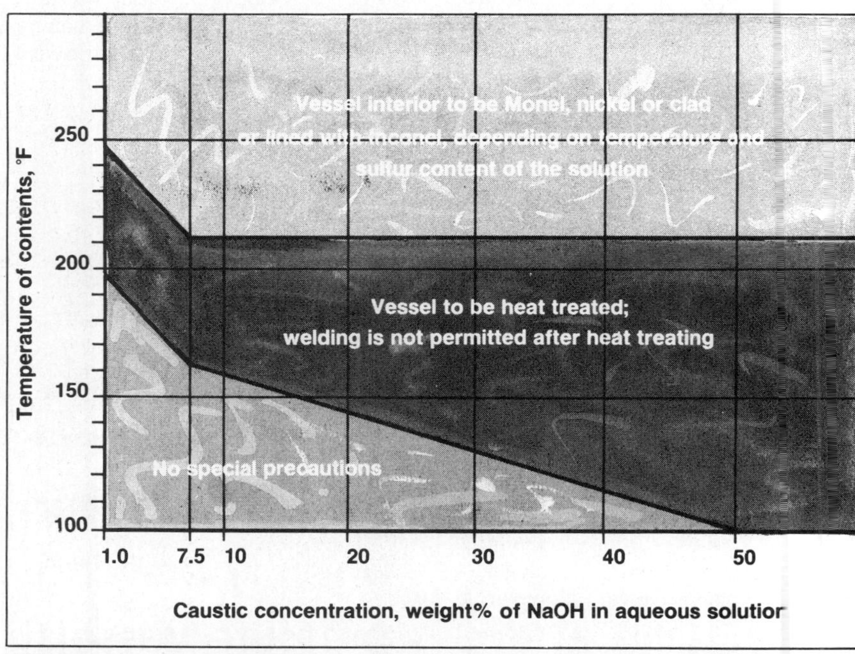

FIGURE 3. Caustic concentration and shell temperature influence metal heat treatment

To avoid or lessen stress corrosion, eliminate crevices at welds and heat treat metals that have been welded

most corrosion resistant, if the fluid is severely corrosive. In such cases, the process engineer should specify a stainless steel that will minimize chromium depletion in the heat-affected zones of welds. The stainless steel should contain stabilizing elements, such as columbium (Type 347) or titanium (Type 321), or be one that has a low carbon content (Type 304L or Type 316L).

Selective leaching occurs when corrosion removes an element from a solid alloy. Some examples of selective leaching include the removal of zinc from brass, aluminum from aluminum bronzes and the graphitic corrosion of gray cast iron. Of course, a metal susceptible to selective corrosion should not be specified for such service.

Erosion corrosion is the accelerated metal loss caused by the relative movement between a corrosive fluid and a metal surface. Metal is removed as dissolved ions and solid corrosion products are swept away from the metal surface by a rapidly moving process stream, which creates a grooved pattern. The process engineer should be aware that erosion can occur at nozzles and inside pipes (especially at elbows), at orifices, baffles and on the inside of vessels having agitators or spray nozzles.

A metal having a protective film that is hard and dense and that forms rapidly will resist erosion better than one that has a film that is brittle and easily worn away. Some design features that reduce erosion damage include wear plates at impingement spots, large pipe diameters (to reduce fluid velocity), thick nozzle necks, inlet flow directed to the center of a vessel rather than to its wall, and hard-facings or weld overlays on surfaces.

Stress corrosion cracking results from the combination of stresses and corrosion. The metal surface is mostly unattacked except for fine cracks through it. The cracks, which appear to be randomly oriented, can vary from a single vein to complex branching. The cracks eventually propagate through the vessel wall, creating leaks. Among common examples are carbon steels exposed to, for example, sodium hydroxide or wet hydrogen sulfide, and austenitic stainless steels exposed to chlorides, hydroxides or sulfur compounds.

Whether or not cracking occurs depends on the degree of stress, the concentration of the corrosive substance and the temperature. Bulk concentration alone does not necessarily cause the cracking. More critical is a rise in concentration in crevices or on surfaces that are alternately wet and dry.

The process engineer should advise the designer if the vessel's environment could promote stress corrosion cracking. Measures that avoid or alleviate it include eliminating crevices at welds and following welding with heat treatment. ■

References

1. "ASME Boiler and Pressure Vessel Code," American Soc. of Mechanical Engineers, New York, 1989.
2. "API Publication 941, Steels for Hydrogen Service at Elevated Temperatures and Pressures in Petroleum Refineries and Petrochemical Plants, 4th ed., 1990.
3. Fontana, M. G., "The Eight Forms of Corrosion," Process Industries Corrosion, National Assn. of Corrosion Engineers, Houston, Tex, 1975.

The author

Glenn D. Nasman is a principal vessel engineer for Badger Engineers, Inc. (One Broadway, Cambridge, MA 02142; tel. 617-494-700). He received a B.S. degree in mechanical engineering from Northeastern University and is a member of the American Soc. of Mechanical Engineers.

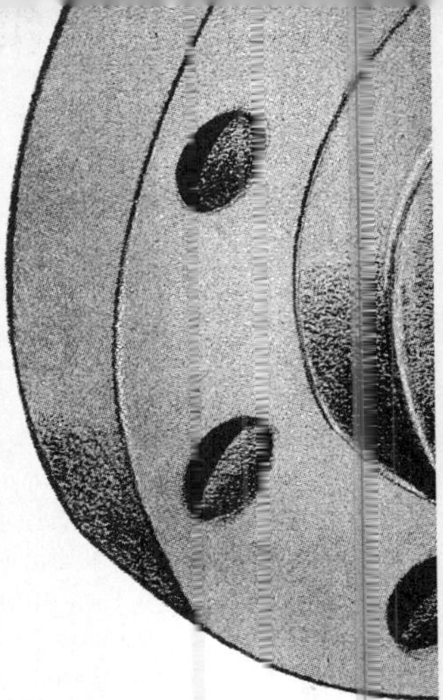

Michael Ridey
McCanna, Inc.

Ball-valve seals can blow out if body-cavity pressures reach excessive levels

RELIEVING

Ball valves of the "soft-seated" type offer significant performance advantages over other traditional valve styles. One of the foremost is the ability to provide reliable, tight shutoff. However, tight-shutoff ball valves will trap fluid in the body cavity between the seats when the valve is closed. When the valve is then subjected to a temperature increase — often merely the warming of a cold line to ambient temperature — a potential hazard is created.

Simply stated, fluid trapped in the body cavity tries to increase in volume as it's warmed, but can't. The result is a rise in body-cavity pressure (Fig. 1). Depending on the fluid's thermal expansion characteristics and the amount of temperature increase, body-cavity pressure can rapidly build to a dangerous level. Excessive cavity pressure can not only blow out a valve seat or a body-joint seal, but can also rupture the valve body. When the fluid in the valve is toxic, flammable or explosive, the results can be devastating.

Ball valve manufacturers offer various features to counteract this problem in services where it is anticipated. Expansion chambers and pressure-relief valves can be connected to the valve's body cavity to handle excessive cavity pressure, but these solutions are not very cost effective and increase system complexity.

The two methods more commonly used by ball valve manufacturers to control cavity pressure are internal (or upstream) venting and self-relieving seats. We will compare these two methods, and review the benefits and shortcomings of each.

Controlling cavity pressure

In the upstream-venting design, a small hole is drilled through one side of the ball perpendicular to and into the ball's flow passage, connecting the flow passage and body cavity to the upstream side of the valve when the ball is in the closed position. This configuration bypasses the upstream seat, and provides a continuous vent path for cavity pressure (Fig. 2).

Top-entry floating-ball valve designs allow the vent hole to be located in the valve body behind the upstream seat. This provides continuous venting of the cavity to the upstream side, and eliminates the potential problem of assembling a valve with a vented ball facing the wrong direction.

In the self-relieving-seat design, the seats act as internal "relief valves" to open a vent path from the valve body cavity to the line, spontaneously relieving body-cavity overpressure (Fig. 3). Self-relieving seats are intended to act as "normal" valve seats, unless cavity pressure increases beyond a predetermined level. At this point the seats act to relieve the overpressure. The mechanism by which self-relieving seats operate varies, depending on the type of ball valve in which they are used.

In trunnion-mounted ball valves, the ball is supported by trunnion bearings that fix its location in the body. To seal the valve, the seats move against the ball, aided by a mechanical device (typically a spring) behind the seat, and by line pressure. Floating-ball valves work in the opposite way; the seats are fixed in position, and the ball moves towards one seat (following line pressure) to seal.

Trunnion-mounted valves are "upstream seating," so called because when the valve is closed, the higher, upstream line-pressure acting on the upstream seat forces it against the ball, helping to close the valve tightly. Floating-ball valves are "downstream seating," since the upstream line pressure forces the ball against the downstream seat when the valve is closed to tighten the seal.

Trunnion-mounted ball valves are inherently "self relieving." In a cavity overpressure condition, the pressure forces acting on the cavity side of the seat overbalance the pressure and mechanical forces holding the seat against the ball, causing the seat to move away slightly, venting the overpressure. Relief occurs towards the lower line-pressure side of the valve, following the path of least resistance. With both sides at equal pressure, relief is towards the side with the lower mechanical-seat load.

Soft-seated floating-ball valves are *not* inherently self-relieving. The capability must be designed into the seat, and is generally accomplished by

Originally published December 1991

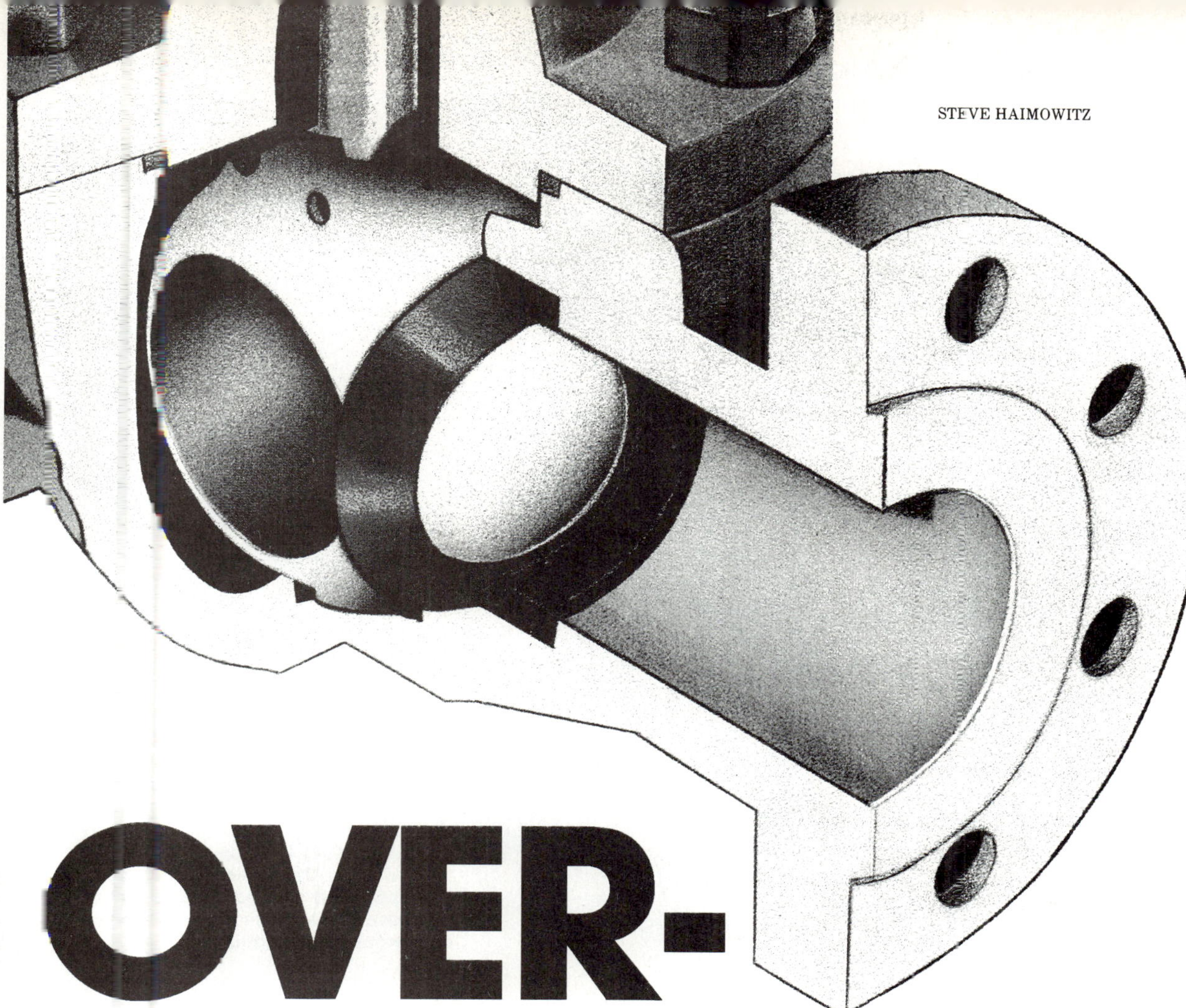

STEVE HAIMOWITZ

OVER-PRESSURE

permitting a limited amount of seat flexure. When cavity pressure increases to a predetermined level, the seat flexes away from the ball, allowing pressure to escape. These seats are typically polytetrafluoroethylene-based material, which despite its other outstanding properties is relatively inelastic. The amount of flexure and where it occurs must be closely controlled to prevent permanent distortion of the seat when relieving.

The basic concept is that of a simple cantilever beam having one end fixed and the other free; force applied along the beam causes the free end to deflect. The deflection and stress in response to pressure load must permit sufficient flexure for relief without overstressing the material. Machining a slight angle on the seating surface in the body, or removing some material from the back of the seat, creates space for the seat to flex, as well as limiting it.

Floating-ball valves with self-relieving seats are intended to relieve pressure towards the higher pressure (upstream) side of the valve. Line pressure loads on the downstream seat increase its flexural resistance, while the relatively lightly loaded upstream seat is flexed away from the ball by pressure forces on its cavity side, venting the overpressure to the line upstream. When both sides of the valve are at equal pressure, relief is toward the side with the weaker seat.

Pros and cons

Another feature of double-seated floating-ball valves is their ability to control flow in either direction. Self-relieving seats become attractive from the standpoint that the valve's bidirectional sealing capability remains intact.

Upstream-vented valves are unidirectional, providing tight shutoff in one flow direction only. This gives self-relieving seats another slight advantage over upstream venting, in that valves with self-relieving seats may be installed without regard to flow direction, while upstream-vented valves will not function properly if installed with the vent facing downstream.

Spontaneous cavity relief towards the low-pressure side presents a potential hazard in trunnion-mounted ball valves. A valve closed to allow maintenance to be performed downstream will relieve cavity overpressure in that direction, which could expose workers to hazardous fluids, release toxic fluids into the atmosphere, or damage equipment. For this reason, trunnion-mounted valves are generally fitted with a body vent (a notched pipe plug or simple vent valve) to bleed-off cavity pressure, nullifying the self-relieving feature.

Similarly, a closed ball valve (either

OPEN AND SHUT CASES

It's worth noting that ball valves can trap fluid in the body cavity when open as well as when closed. The trapping mechanism is the same, but the results can be quite different, depending on the circumstances.

When the line is flowing it is unlikely that the trapped fluid will cause an overpressure condition, since it will essentially remain at line temperature. In a shutdown line, self-relieving seats would function the same as if the valve were closed, and an upstream-vented ball would vent into the ball flow passage, rather than upstream. The effects of this are dependent on the fluid released, as discussed elsewhere in this article.

This becomes a problem in applications where self-relieving seats or upstream venting would not normally be considered. A system that gets flushed or steamed out, even only once at startup, can trap the flushing fluid (or steam condensate) in the cavity when the valve is opened. If the fluid is water, for example, and the line normally operates at 400°F, the trapped water turns to steam and resulting pressures can blow out a soft valve-seat, making the valve inoperable or unable to shut off the flow. Fortunately, this is easily remedied by providing a small vent hole in the ball-stem slot, to equalize body-cavity and line pressures.

This feature is automatically provided in some valve designs, but is optional in others. It should be a standard requirement of ball valve users; it doesn't hurt anything by being there, but its absence can cause unexpected difficulties.

To put this in perspective, consider this scenario: A 4-in., full-port floating ball valve with self-relieving seats in an outdoor line carrying dry, liquid chlorine at 80 psi at 500°F is inadvertently left closed when the line is drained and purged for maintenance. As a result, about 1 liter of chlorine is trapped in the body cavity. Piping on the downstream side of the valve is removed for repair.

It's a hot sunny day, and before long the sun heats up the valve to 150°F, increasing cavity pressure to roughly 300 psi. The seats are designed to relieve between 150 and 400 psig, and one seat does at 250 psi. The chlorine immediately vaporizes, forming about 460 L of gas. The blast of chlorine gas, if it doesn't kill anyone outright, is enough for a usually fatal concentration in air of 1,000 ppm in an area 25 feet on each side.

Which seat spontaneously relieved? The upstream seat into a system thought free of chlorine, or the downstream seat into the work area? The question can't be answered until it happens, which unfortunately is too late. □

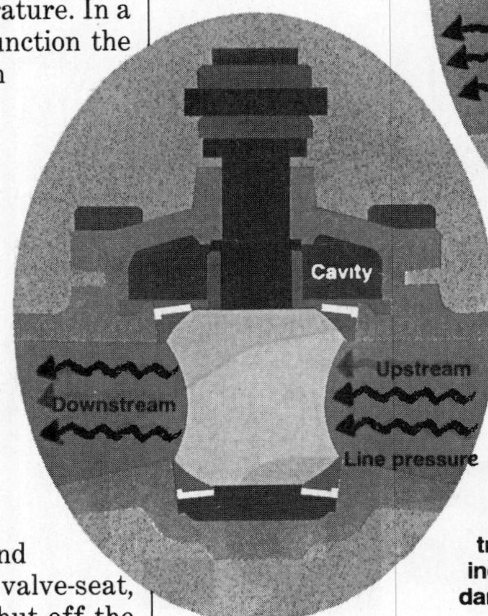

FIGURE 1. When the ball valve is completely open, fluid flows right through (left). As the valve closes, fluid enters the body cavity through the ball port (top). With the valve closed, fluid is trapped in the body cavity; as the valve warms up, the trapped fluid tries to expand and can increase pressures in the body cavity to dangerous levels (right)

trunnion mounted or floating ball) with self-relieving seats in a unpressurized line will relieve towards the weaker seat. This is made more dangerous by the fact that the relief direction is completely unpredictable. Unless cavity pressure is bled off, this may result in a hazardous or toxic discharge into a work area. Even if a means to bleed cavity pressure is provided, a toxic fluid can't be bled into the atmosphere, so the bleed-off would have to be contained as well.

FIGURE 2. In upstream venting, a small hole connecting the body cavity to the upstream side of the valve relieves cavity overpressures

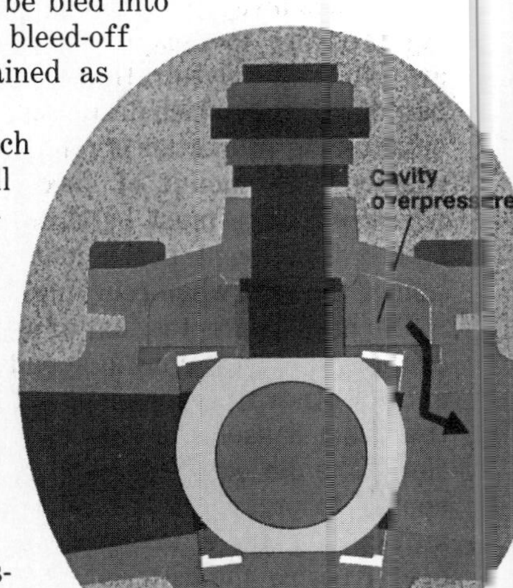

The pressure at which self-relieving seats will function is not very predictable due to variations in material properties and manufacturing. Seat-relief pressure will vary from valve to valve and will seldom, if ever, occur at the actual design value. Seat relief is more likely to occur within a range of pressures, which may span several hundred psi. Relief

pressure is also sensitive to line temperature, the fluid being handled, and seat wear. Self-relieving seats are usually limited in choice of material, and the available seat materials may not be suitable to application requirements. All of these factors create uncertainty about the performance of self-relieving seats.

Good applications for self-relieving seats in floating ball valves are actually rather limited. Services involving slurries or precipitating solids, solidifying or self-polymerizing fluids, or generally "dirty" systems are not good applications. This is because solid materials contained in the fluid stream can build up in the space behind the seat, interfering with flexure and, thus, pressure relief. In fact, installing floating-ball valves with self-relieving seats in such applications creates problems where previously there were none.

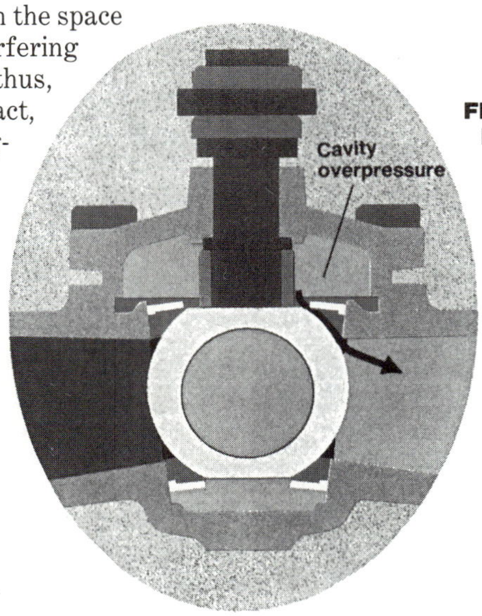

FIGURE 3. When body-cavity pressures rise above preset limits, self-relieving seats flex away from the ball and open a vent path to the upstream line

The issue: safety

Prevention of cavity overpressure in ball valves is a safety issue. The overriding consideration in selecting valves for such applications should be reliability of the technique used to avoid cavity overpressure. By far the most reliable means for ensuring positive-cavity-pressure relief is the simple upstream vent. Since this method continuously vents the cavity to the upstream line, cavity pressure cannot exceed line pressure, which can be safely controlled by system relief valves. Further, the direction of cavity relief is a constant; there will be no unpleasant surprises for maintenance personnel downstream of an upstream-vented ball valve.

However, all designs involve tradeoffs, and upstream venting is no exception. Bidirectional sealing must be given up when upstream venting is used, but may be no great loss if line flow is never reversed. However, upstream venting is only a tradeoff of a performance feature; it is not a compromise of plant safety.

Ball valve manufacturers can supply valves with either method of cavity relief, depending on the application. They can be consulted for recommendations on specific applications, and will provide sound advice.

Recommendations

When evaluating ball valves for applications where fluid trapped in the body cavity can cause a hazardous condition to develop, the following should be kept in mind:

1. *Options for bidirectional shutoff.* Review the operational requirements for the valve carefully. If bidirectional shutoff is a necessity, self-relieving seats must be considered. However, the potential problems that can occur with self-relieving seats must weigh heavily in any decision to use them.

Keeping in mind that safety principles call for elimination of potential hazards, use self-relieving seats only if you believe the benefits of bidirectional shutoff outweigh the potential risks. If there are doubts, however, do not use them.

2. *Upstream venting for overpressure.* Remember that upstream venting is simple, positive, predictable and continuous. Cavity pressure cannot exceed line pressure when this method is used, so an overpressure condition will not occur in the first place.

Don't be mislead by claims about the virtues of self-relieving, seated, ball valves. If it sounds too good to be true, it probably is.

3. *Bidirectional shutoff plus upstream venting.* If you like the upstream-vent concept but still need bidirectional shutoff, contact a ball valve manufacturer who specializes in severe services and difficult applications. There, experienced ball-valve engineers can evaluate your application and possibly recommend alternative solutions. That is the business they are in; take advantage of their experience and expertise.

4. *Safety always comes first.* Take no unnecessary risks when selecting ball valves for severe services, or any service where control of cavity pressure is required. Don't wait for the U.S. Occupational Safety and Health Administration, the Environmental Protection Agency, or the courts to tell you what should have been done. ■

Edited by Jayadev Chowdhury

The author

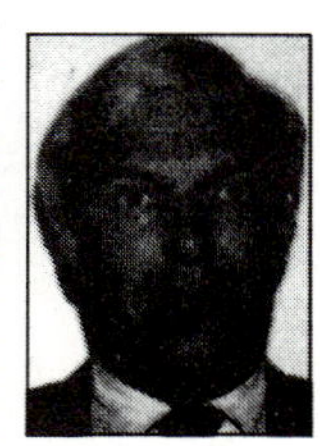

Michael Ridey is Manager of Proposal Engineering at McCanna, Inc., 400 Maple Ave., Carpentersville, IL 60110. He holds a Bachelor of Science degree in Mechanical Design from Bradley University. He began his career as a design engineer at Newport News Shipbuilding, and has been with McCanna for the past 18 years, holding various managerial positions in Product Design Engineering, Marketing, and Valve Applications Engineering. He is co-editor of the *Flow Manual for Quarter-Turn Valves*, published by McCanna.

PUMPING DIFFICULT FLUIDS

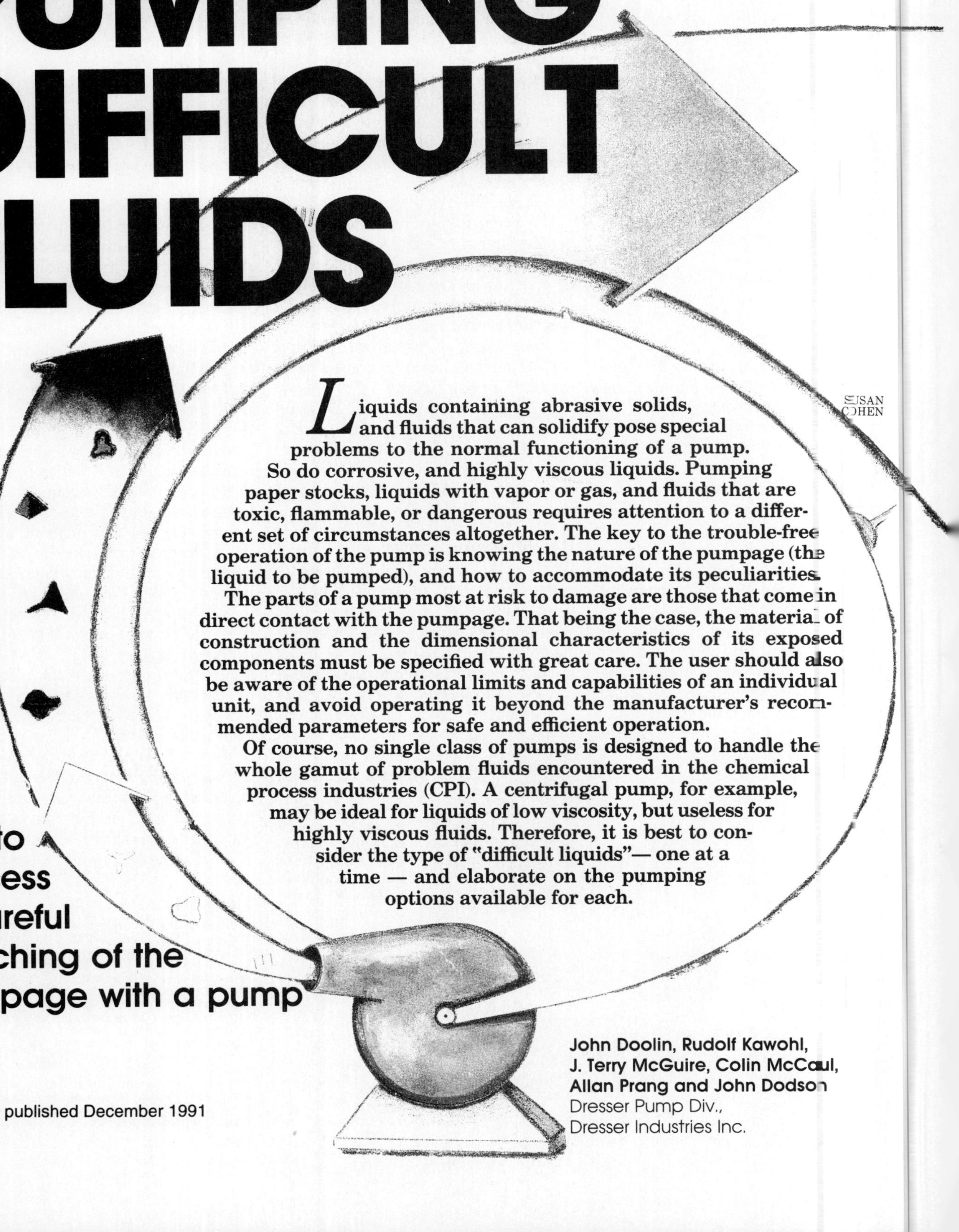

SUSAN COHEN

Key to success is careful matching of the pumpage with a pump

Liquids containing abrasive solids, and fluids that can solidify pose special problems to the normal functioning of a pump. So do corrosive, and highly viscous liquids. Pumping paper stocks, liquids with vapor or gas, and fluids that are toxic, flammable, or dangerous requires attention to a different set of circumstances altogether. The key to the trouble-free operation of the pump is knowing the nature of the pumpage (the liquid to be pumped), and how to accommodate its peculiarities.

The parts of a pump most at risk to damage are those that come in direct contact with the pumpage. That being the case, the material of construction and the dimensional characteristics of its exposed components must be specified with great care. The user should also be aware of the operational limits and capabilities of an individual unit, and avoid operating it beyond the manufacturer's recommended parameters for safe and efficient operation.

Of course, no single class of pumps is designed to handle the whole gamut of problem fluids encountered in the chemical process industries (CPI). A centrifugal pump, for example, may be ideal for liquids of low viscosity, but useless for highly viscous fluids. Therefore, it is best to consider the type of "difficult liquids"— one at a time — and elaborate on the pumping options available for each.

John Doolin, Rudolf Kawohl, J. Terry McGuire, Colin McCaul, Allan Prang and John Dodson
Dresser Pump Div.,
Dresser Industries Inc.

Originally published December 1991

LIQUIDS CONTAINING ABRASIVES

The abrasive solids suspended in a high-velocity liquid stream can cause severe wear on the internals of a pumping unit. Examples of such abrasive streams include slurries found in the mining industry, and water with flyash captured in fluegas desulfurization processes. The parts most at risk to wear in a centrifugal pump handling abrasives are the impeller and the interior of the casing. To provide abrasion resistance, the use of hard materials of construction, or wear-resistant liners, is recommended.

STEPHEN
WALSH

Hard-metal construction

The materials commonly favored for their resistance to abrasion (Table 1) feature superior hardness and mechanical strength. All parts exposed to abrasive pumpage are either made thicker, or protected by wear plates. Besides being cost-effective, they must also meet the requirement of being corrosion resistant.

The shapes of the impeller and the casing are designed to maximize their hydraulic performance. More often than not, the optimal performance is a compromise between the highest-possible efficiency and what can be reliably achieved given the abrasive properties of the pumpage.

Of course, the higher the rotational speed of the pump rotor, the greater will be the abrasive impact. But too slow a rotor is often not economical from an operational point of view. Striking a balance between economics and the reliability of performance calls for past experience in each specific application.

The classic style of the hydraulically sound design of a casing, along with its impeller, is not necessarily good for handling highly abrasive liquids. Ideally suited for pumping clean or mildly abrasive liquids, the impeller imparts sufficient energy to the pumpage by the action of its vanes. Only by varying the choice of the material of construction for the impeller and the casing can one adapt this standard design for pumping moderately abrasive fluid mixtures

However, because of limitations in the fabrication techniques used for their construction from hard-metal alloys, there are less-complex hydraulic shapes available for abrasion-resistant pumps. The casing form (Figure 1; bottom) is a compromise between a volute and a circle. The volute tongue is retracted to minimize the likelihood of its premature damage from abrasion. To steer clear of the abrasive solids, the impeller hub is recessed. In addition, the impeller passages are larger than what would normally be required to meet the hydraulic requirements.

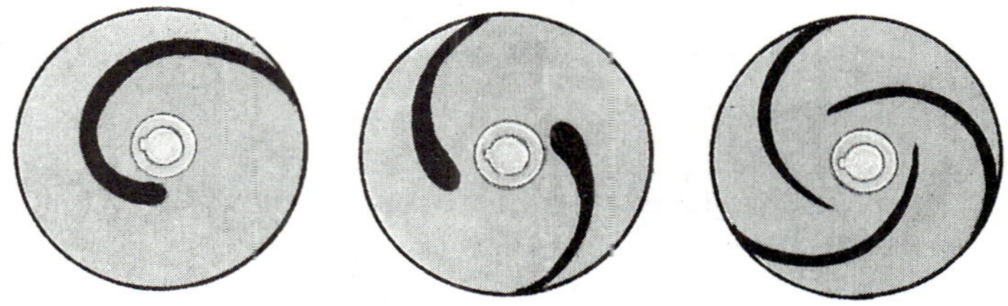

FIGURE 1. The hard-metal design provides relatively large impeller passages to steer clear of abrasive solids (below). The number of vanes can be varied to accommodate unusually abrasive fluids (above)

The shapes and sizes of vanes can vary widely. A sample of the variety of vanes designed for specific tasks is shown in Figure 1 above. Typically, an impeller is equipped with two or three vanes, but certain large pumps are often built with four-vane impellers. On the other extreme, there are even single-vane impellers designed to handle particularly difficult pumping jobs.

The hard-metal design allows relatively easy adjustment of the shaft in the axial direction. However, any change in the impeller diameter (by trimming, for example) to match pressure-head requirements is not practical. As a result, such pumps are usually pulley driven. By adjusting the ratio of the diameter of the driver and the follower pulleys, one can modify the impeller speed. An alternative to this, of course, is to use a motor equipped with a variable-frequency drive.

The use of hard-metal alloys also leads to some other design specifications, such as very rigid shafts; strong bearings to match; and shaft sealing by means of mechanical seals, stuffing-box packings with clean liquid injection, or back vanes that produce a "dynamic sealing" effect . The hardness of the metal alloys ensures protection of the casing interior from abrasive damage.

The impeller is threaded to the shaft, and all parts susceptible to wear are constructed from hard-metal alloys. The hydraulic form of the impeller is kept very simple.

Depending on the nature of the abrasive solids, and the operational requirements, the user needs to specify the appropriate pumping unit on a case-by-case basis. To avoid unnecessary complications, keep in mind the following suggestions:

- Limit the developed pressure head to about 300 ft
- Install pumping units in series to handle pressure heads higher than 300 ft
- To minimize wear and power loss, select the pump size and the impeller speed so that its best efficiency point

COMPOSITION	ABRASION RESISTANCE	CORROSION RESISTANCE	HARDNESS BRINELL NUMBER
Machinable			
Grey cast iron	Poor	Poor	180 - 350
Bainite cast iron (1 - 4.5% Ni; 0.5 - 1% Mo)	Good	Fair	250 - 350
Ferritic cast iron with chrome (28 - 34% Cr; Mo)	Good	Good	250 - 380
Chrome-alloyed cast iron (24 - 30% Cr; Ni and Mo)	Good	Good	240 - 360
Not machinable after hardness treatment			
Martensitic cast iron (3 - 4.5% Ni; 0.5 - 1.2% Cr)	Good	Poor	350 - 450
White cast iron with chrome (11 - 28% Cr; Ni, Mo and Cu)	Very good	Good	450 - 650
White cast iron alloy (2.5 - 11% Cr; 3 - 6% Ni)	Excellent	Good	450 - 650
Chromium steel (13% Cr; 1.8% C)	Very good	Fair	400 - 650

TABLE 1. Hard-metal construction of pumps to handle liquids containing abrasives can offer corrosion resistance as well

(BEP) comes closest to the required flow

- Choose a unit larger than what would be required to solely meet the operational specifications, and run it at a slower rotating speed, to obviate excessive wear of parts

Rubber lining

Because of their ability to absorb the shock energy of abrasive particles, rubber materials provide an interesting solution to abrasion of parts exposed to the pumpage. In particular, when the abrasive particles are small, rubber-lining of the parts provides effective resistance to wear.

Typically, the external metal casing is made of cast iron. The impeller skeleton is constructed from welded steel, or ductile iron. Natural rubber, neoprene, nitrile, and chlorosulfonated polyethylene are the materials of choice for lining the metal casing, or coating the molded impeller. The hardness for natural rubber ranges from 38 to 42 Shore Durometer (SD) while that for synthetic elastomers varies from 50 to 52 SD.

The form of the impeller and the volute is quite simple (Figure 2). The impeller sides and the vanes are very thick by design. There are generally three or four vanes of a simple, curved shape. Despite these simplicities the efficiency of these rubber-lined units is reasonably high. To obtain the best performance, bear in mind the following recommendations:

- Limit impeller peripheral speed to 100 ft/s, thus limiting the developed head to 145 ft
- Use the units in series to develop pressure heads higher than 145 ft
- Allow handling of solid particles of sizes up to 0.1 in. if they are sharp; otherwise, allow particles up to 0.2 in.
- Because trimming of the impeller diameter to match the required pressure head is not practical, adjust impeller speed with pulleys
- Choose the speed of impeller rotation and the size of pump so that the work is carried out at or near the best-efficiency point supplied by the manufacturer
- Do not use rubber-lined units when the inlet pressure is less than the atmospheric pressure, to avoid the danger of detaching the liners
- Limit temperatures from 150°F to 225°F, depending on the rubber composition

Recessed impeller

Instead of vanes, a recessed impeller consists of a disc equipped with ribs (Figure 3). The recessed device generates a forced vortex while transmitting its energy to the pumped liquid. Because the volute passage is widely open, there is a gentle transmission of energy. As a result, a recessed-impeller pump is commonly used for handling liquids that contain gas, fibrous materials, hard particles (as large as one-third the size of the volute), such as sand or stone, or even fragile materials, such as food products.

However, the transmission of energy is not as efficient as that done by the classical impeller. As a rule, one should not expect more than 70% of the efficiency of a conventional impeller. The selection criteria suggested above for hard-metal pumps are also valuable for

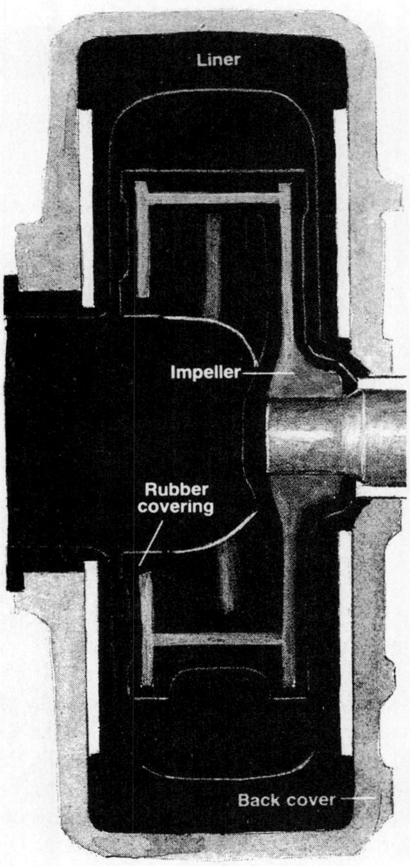

FIGURE 2. When the abrasive particles are small, rubber lining of the casing and the impeller provides effective resistance to wear

the recessed-impeller units. One additional point: Because of their lower efficiency, use of the recessed units is limited to pumping applications below 250 hp.

Single screw

Designed to give positive displacement, a single-screw pump consists of a rotor, which is a screw thread in form, with a circular section. The rotor turns within a stator of the same form, but is equipped with an oblong section (Figure 4).

One of the main advantages of the single-screw pump is that it does not require valves. Furthermore, the construction is simple, although it has complex kinematics. For example, to create a volume between the stator and the rotor, the latter executes a planetary movement. The rotor turns on its own axis C_r (Figure 4) while it rotates around the stator's axis C_{st}. This translates into an alternately linear movement of the area S_r of the rotor along its axis.

To accommodate the complicated motions inside, the rotor geometry can be quite complex. The points of contact P_1 and P_2 continually change their position leading to friction with the stator. The flow direction changes with the direction of rotation. Generally, the flow direction is chosen to ensure that the suction pressure is on the sealing side.

The shaft sealing is accomplished by use of either packing or a mechanical seal. It is recommended that the rotor be made from coated or hardened stainless steel, and the stator from extrudable plastics with hardnesses ranging from 65 to 68 SD. The materials for the pump casing and rotor are chosen for their resistance to chemical aggressiveness.

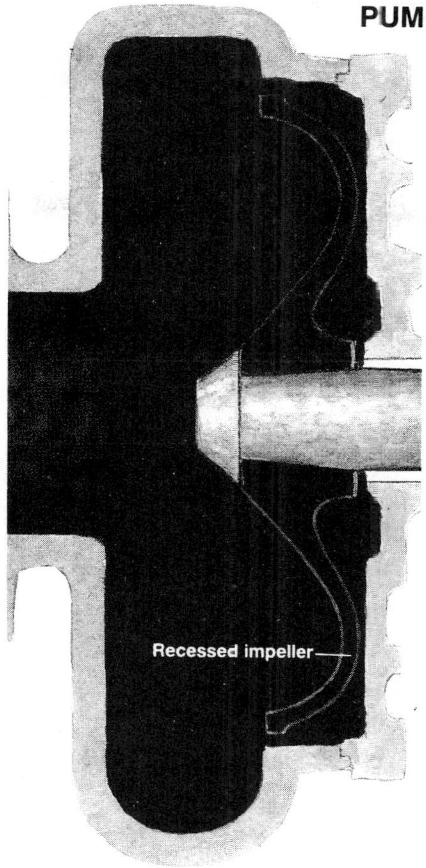

FIGURE 3. If the pumpage contains fibrous materials or hard particles, a recessed impeller equipped with ribs is best to handle the job

TABLE 2. The higher the number of stages in a single-screw pump assembly, the greater the differential pressure between the discharge and inlet

NUMBER OF STAGES	PRESSURE DIFFERENCE, psi FOR FLUIDS THAT ARE		
	LIGHTLY ABRASIVE	MODERATELY ABRASIVE	STRONGLY ABRASIVE
1	70 – 110	60 – 85	45 – 70
2	145 – 190	115 – 145	85 – 115
4	290 – 360	230 – 290	175 – 215

During selection of a particular unit, remember the capabilities, as well as limitations, of this class of pumps:

- Use the recommended number of stages (Table 2) required to develop the needed ranges of differential pressures
- Expect to be able to deliver flowrates up to 2,000 gpm
- For slurries, keep the rotor speed roughly between 150 and 500 rpm
- Maintain the net positive suction head (NPSH) at about 5 ft
- Do not allow particles of dimension greater than about one-third the size of the entrance area between the stator and the rotor

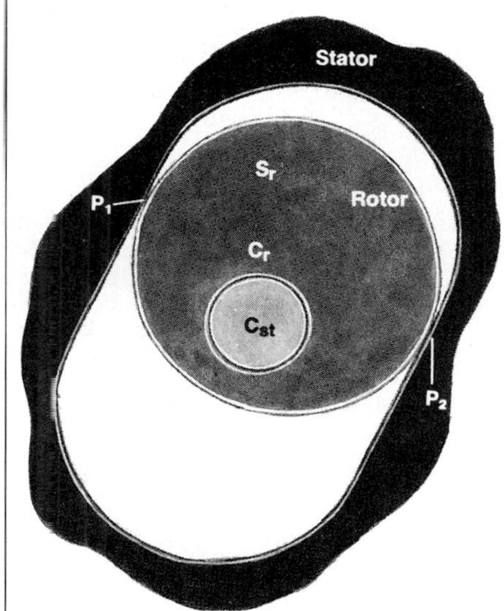

FIGURE 4. A single-screw pump (below) can handle slurry flowrates up to 2,000 gal/min. To create a volume for the pumpage, the rotor and stator undergo complex kinematics (above)

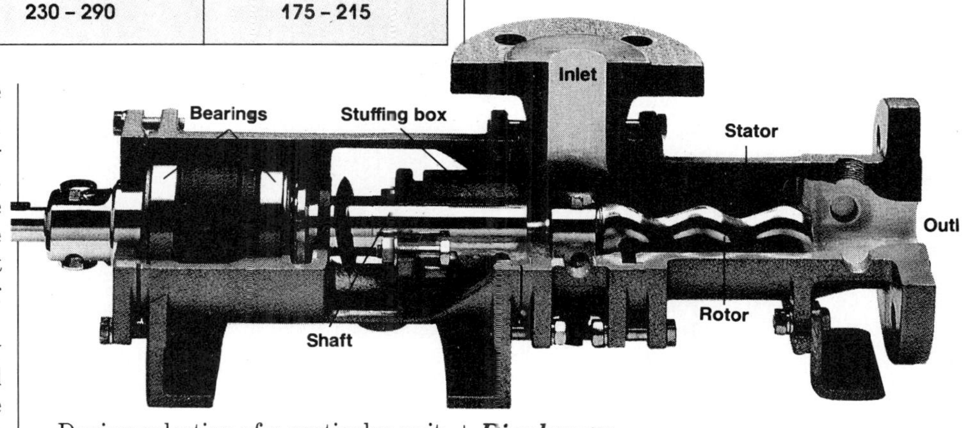

Diaphragm

Billed as a typical reciprocating device of very simple construction, a diaphragm pump uses a flat, flexible diaphragm that is hermetically sealed in a housing. The pumped liquid stays on one side of the diaphragm while a moving fluid on the other side imparts its pressure via the diaphragm. Clearly, the discharge pressure of the pumpage cannot be greater than that of the motive fluid.

A typical direct-acting diaphragm pump (Figure 5) employs compressed air as the motive fluid. It uses a duplex construction with a linkage between the two diaphragms. The energy of the motive fluid is generated in a separate compressor section. Unlike all other reciprocating pumps, the diaphragm pump does not use seals. However, the valves used for controlling the inflows and outflows of fluid streams represent the weak points in a diaphragm pump.

The materials of choice for the pump casing include metals and nonmetals that offer the required level of resistance to corrosion and abrasion. Reinforced synthetic rubbers, such as neoprene, butyl and fluoroelastomers, or tetrafluoroethylene fluorocarbon polymers are commonly used for the diaphragm, while various elastomers and elastomer-coated metals are popular for the valves.

During selection of a diaphragm pump, consider the following:

- The user can vary the pumping capacity of a direct-acting unit by regulating the motive fluid
- There is no danger in dry running
- The capacity can be up to 750 gpm
- Differential pressures up to 125 psig can be achieved with a direct-acting diaphragm pump, although pressures up to 5,000 psi are possible with "power diaphragm" units activated by liquid from a reciprocating power pump

Warren Rupp Inc.

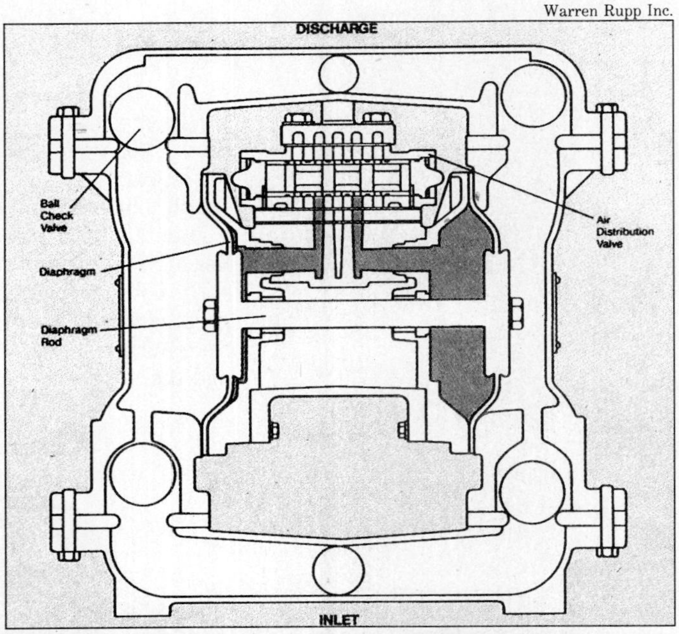

FIGURE 5. Unlike all other reciprocating pumps, the diaphragm pump (left) does not use seals. The duplex construction (below) uses a linkage between the two diaphragms

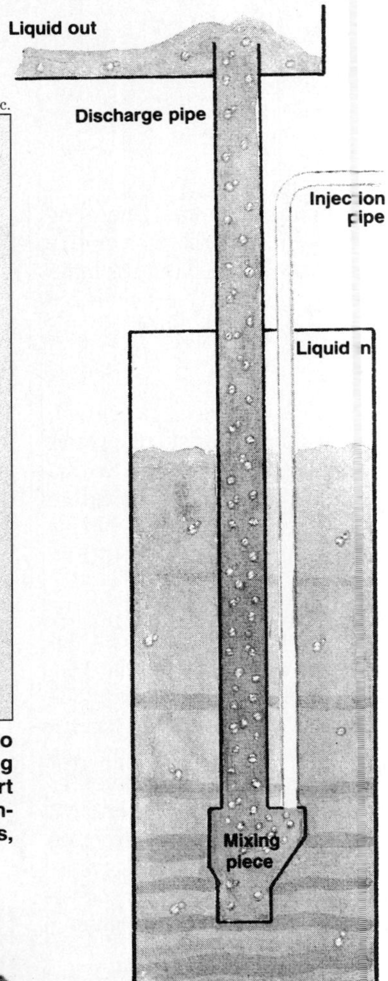

FIGURE 6. Featuring no moving parts, this pumping device (right) uses air or inert gas for handling liquids containing abrasive sands, clinkers and cinders

Air or inert gas

The ensemble (Figure 6) consists of a discharge pipe, an injection pipe, and a mixing piece. Using air or a neutral gas under pressure, this pumping device allows an extremely simple construction, without any moving pieces. The unit can be used for handling liquids with abrasive contents, such as sand, clinker and cinders.

The compressed air or gas is injected through the injection pipe. The injected air mixes with the liquid to be pumped. Although the mixture is really nonhomogeneous, it is quite common to consider it to be a homogeneous liquid of lighter density. The density difference between the interior and the exterior of the discharge pipe causes a 'lift' for the air-liquid mixture, inducing a pumping action.

The extreme simplicity of the components allow their easy fabrication. They can be constructed from virtually any material, metallic or not. For example, steel (alloyed or not), cast iron, rubber-coated steel, and plastics can be used as materials of construction.

A few pointers to keep in mind during selection of the air pump are as follows:

- The dimension of the particles that can be pumped are limited to half the size of the discharge pipe section
- At atmospheric pressure, 2 to 3 ft^3 of air can raise 1 ft^3 of water up to 50 ft, while 4 to 5 ft^3 of air is capable of raising 1 ft^3 of water up to 200 ft
- Efficiency (defined as the ratio of energy of hydraulic gain to the energy of consumed air) is about 40%. Of course, this does not include the efficiency of the compressor. □

LIQUIDS THAT SOLIDIFY

A circumstance to be avoided during handling of a liquid is its solidification inside a pump. Besides causing damage to the internal components, solidified products can completely block the flowpath.

The nature of solidification varies with the liquid, so it is necessary to classify the types before discussing how to avoid the associated problems. In pumping practice, there are four circumstances that can lead to pumpage solidification. In other words, the liquid being pumped is most likely to solidify if:

- The melting point of the pumped product is above ambient temperature
- The liquid crystallizes on contact with the atmosphere
- The pumpage "plates out" a metallic deposit under high pressure
- The liquid is being pumped in a polymerized state

High melting point

The most common liquids in this class are heavy hydrocarbons. Most applications involve pumping such liquids at temperatures well above their melting point. If the process is continuous, there is no need to add heat to the pump because the residence time is likely to be too short for solidification to occur.

Pumping at temperatures close to the melting point, or batch pumping generally dictates that heat be added to the pump to avoid any tendency of the liquid to thicken and solidify. This is usually accomplished by incorporating jackets around the pump (Figure 7), and providing the necessary heat by circulating hot water, steam, or some other convenient heated fluid from a plant's distribution network.

In cases where the pump is not, or cannot be, equipped with jackets, but the unit needs to have the heat added to ensure that the pumped product remains liquid, various forms of external tracing are applied. Some of them are bonded to the exterior surfaces of the pump with heat-transfer cement to improve their effectiveness.

There are situations when it is not convenient, or even feasible, to add heat to the pump. In such cases, it is necessary to completely evacuate the unit once pumping ceases. To do this quickly, the unit must have adequate provisions for draining and flushing. Designs intended for such services usually have the casing arranged in a manner that it can be easily opened to allow complete drainage.

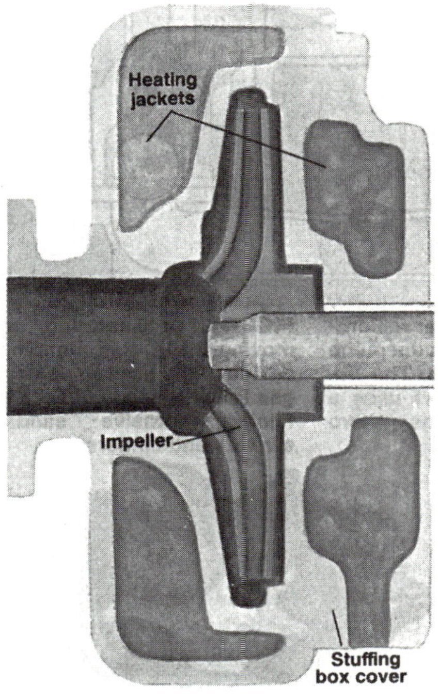

FIGURE 7. The heating jacket provides enough heat to keep the pumpage temperature well above its melting point

Fighting crystallization

The greatest concern with liquids that crystallize on contact with the atmosphere is impairment of the pump's shaft seal. In units equipped with axial-face (mechanical) seals, crystalline deposits can restrict movement of the flexibly mounted face, thereby increasing leakage. To make matters worse, the crystals can cause rubbing contact between the shaft and the stationary seal-face. The result is fracture of the seal face. Pumps equipped with packed-box seals also suffer severe erosion of the sleeve, and a high rate of packing wear.

There are two generally accepted means of avoiding difficulty with crystallization at the seal. The simplest is to continually remove the crystal deposits with what is called a seal "quench" mechanism (Figure 8), in which an inert fluid is supplied from an external source, and then usually piped to a drain.

A more elaborate approach is to inject the seal housing with an inert liquid to keep the pumped liquid away from the seal. With a single mechanical seal, the flow required may be quite high, and may even pose problems of product dilution. Resorting to a double mechanical seal avoids the dilution, but adds considerable complexity to the seal.

What if the liquid plates out?

Some liquids, such as solutions of metal salts, tend to deposit or plate out under the influence of high pressure. If such liquids are sealed with mechanical seals, the high pressure that may exist between the seal faces may be sufficient to cause plating while the pump is stationary. Should this happen, the seal faces will be bonded together, and the seal will be destroyed when the pump is next started.

The simplest solution is not to use a mechanical seal at all. In applications where a mechanical seal is mandatory, a double seal has to be used.

Avoid polymerization

In many processes, the liquids being pumped polymerize, meaning that they undergo a chemical reaction while being pumped. Whether this represents a

FIGURE 8. Crystal deposits formed at the seal can be washed away with an inert fluid

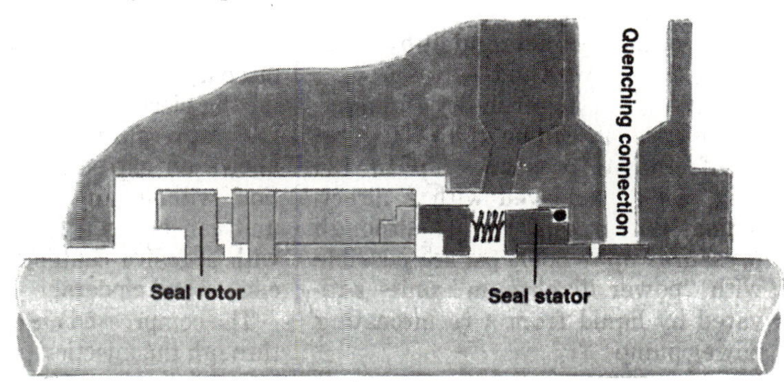

problem depends on the rate of reaction, the point in time at which polymerization was initiated, and the physical nature of the polymerized state.

Frequently, there is no notable change in the liquid before it passes through the pump. In other cases, the liquid contains dispersed solids that can clog the flowpath, or bind to the interior of the pump. Open impellers offer some immunity to the binding, provided the solids concentration in the pumpage is not unusually high.

Another approach favored by some users is to flush the regions of the pump where binding is likely to occur. The liquid used for such flushing will, of course, mix with the product, and so must be compatible with it.

In yet other cases, the liquid may solidify in the pump if it is left standing full for an extended period of time. Draining the unit is the solution if there is a likelihood that this may occur. □

HIGHLY CORROSIVE LIQUIDS

In the manufacture of strong acids, such as HCl, H_2SO_4, H_3PO_4 and so on, the engineer has to select appropriate materials of construction for the pump to handle the corrosive liquids. Proper selection requires detailed knowledge of the fluid characteristics, including pH, temperature, aeration, and the presence of contaminants.

Indeed, the level of contaminants can significantly alter the severity of corrosive attack. For example, trace quantities of cupric salts in 30 mol % H_2SO_4 will drastically reduce the rate of corrosion in Type 316 stainless steel. The same contaminant, however, will induce rapid failure in some nickel-based alloys handling HCl.

Corrosion rates are usually based on test data for wrought materials. Experience has shown that the cast analog of many wrought materials can have erratic, and often inferior, corrosion resistance. Because pumps use predominantly cast components, this can lead to premature failure. For example, Type 20 alloy is an austenitic stainless steel that was originally developed for H_2SO_4 service. Its cast version, CN7M, has often shown unsatisfactory corrosion resistance in both H_2SO_4 and HNO_3 environments.

Recent developments in foundry technology, such as the introduction of argon-oxygen decarburization, allow the production of steel castings that have very low contaminant levels. Indications are that castings produced in this manner have improved corrosion resistance, similar to that of the wrought product.

Corrosion-resistant alloys

Stainless steels are the most commonly specified materials to handle corrosive liquids. Among them, the austenitic alloys, such as Type 316L, are most frequently used. Containing 18% chromium, 8% nickel and 3% molybdenum, Type 316L offers good corrosion resistance over a wide pH range. Furthermore, the molybdenum addition provides a measure of resistance to pitting and crevice corrosion in fluids containing chlorides. The alloy also has excellent resistance to a wide range of neutral, alkaline, and acid salt solutions (other than halogen compounds), organic acids, weak bases, and strong bases at moderate temperatures and concentrations.

Performance in inorganic acids depends on the type of acid and, in some cases, on the concentration and temperature. A primary limitation of Type 316L and all 300-series austenitics, is their susceptibility to stress-corrosion cracking in solutions containing chlorides, and caustics, at elevated temperatures. They are also subject to pitting and crevice attack in chlorides.

The rate of attack is a function of pH, chloride level, and fluid velocity. The susceptibility of Type 316L to stress-corrosion cracking is related to the presence of 8% nickel in the alloy. Higher nickel levels reduce the likelihood of cracking.

An alternative to Type 316L is often the Type 20 alloys, which combine immunity to stress-corrosion cracking with superior resistance to hot H_2SO_4, and other corrosive chemicals. These alloys are fully austenitic, and contain about 30% nickel. Cast versions, known as CN7M and CN7MS, are widely used in pumping applications that are considered beyond the capabilities of Type 316L. Bear in mind, however, that Type 20 alloys are not particularly resistant to localized corrosion by chlorides.

A new class of austenitic stainless steels, containing 6% molybdenum, has been developed to handle corrosive environments for which neither Type 316L nor Type 20 alloys are particularly well suited. Known commercially by the designations AL6XN, 254 sMo, and 1925hMo, these alloys have increased molybdenum for maximum resistance to pitting.

This characteristic is needed in applications involving exposure to acidic fluids containing oxidizing agents. Papermill bleaching operations, for example, have frequently upgraded from Type 316 or 317 stainless steel to the 6%-molybdenum alloys, in order to handle low-pH waters containing ClO_2 and other potent oxidants.

Duplex stainless steels, which have a microstructure of 50% austenite and 50% ferrite, are often an economic alternative to austenitic alloys. The duplexes share with the Type 20 alloys excellent resistance to stress-corrosion cracking. Corrosion resistance in most environments is equal or superior to that of Type 316L.

Duplexes are also superior to Type 316 in terms of resistance to chloride pitting. They have considerably better mechanical properties. Duplex-steel pumps are handling corrosive fluids in a variety of applications, including flue-gas desulfurization, and seawater injection for enhanced oil recovery.

Nickel-based alloys, especially Type 625 and C276, should be considered when pumping highly corrosive acid-chloride solutions. Both contain high levels of chromium and molybdenum for resistance to strong oxidizers, such as ferric and cupric chlorides, chlorine, formic and acetic acids, acetic anhydride, seawater and brine solutions. The C276 alloy is frequently used to handle sodium hypochlorite solutions.

For specialized pumping applications, there are other metals and alloys For example, nickel is suited for pumping caustic soda at all concentrations and temperatures. It is also used in a variety of other industrial applications involving exposure to strong alkalis. Monel is often used in the manufacture and purification of hydrofluoric acid.

An outstanding characteristic of titanium is its resistance to chlorine and its compounds. It is frequently used to handle chlorites, hypochlorites, chlorates, perchlorates, and similar chemicals. Titanium is also resistant to nitric and chromic acids. Moreover, high-performance pumps handling seawater and brines have been built in titanium.

Organic resins

Plastic pumps, either solid or lined, are in some instances a viable alternative to alloy construction. These units usually have a stainless-steel shaft. They offer excellent chemical resistance to a variety of fluids. Solid epoxy-pumps, for example, resist hydrochloric, phosphoric and sulfuric acids, ammonium, ferric and zinc chlorides, and mixed acids and alkalis over a wide pH range. They are often much less costly than alloy alternatives.

A wide range of other organic resins, including polypropylene, polyphenylene oxide, polytetrafluoroethylene, and polyvinylidene fluoride are also being used. Nonmetallic pumps are generally limited to relatively low-temperature and low-pressure applications, although some of the newer high-performance resins, such as polyamide-imide, can tolerate continuous exposure to 500°F.

Concerns about the structural integrity of solid nonmetallic pumps often restrict their use to noncritical applications. Lined units do not have this limitation, but could suffer rapid deterioration in the event of a defect in the lining, allowing the pumpage to come in contact with the metal backing. For these reasons, nonmetallic pumps are only used to a limited extent in applications involving exposure to highly corrosive liquids. □

LIQUIDS WITH VAPOR OR GAS

Two-phase-flow pumping applications include situations where undissolved vapors or gases are being carried by the pumpage. One example is biological-fluid processing, such as the fermentation process used in yeast production. In this application, the process liquid is circulated from the bottom to the top of an aerator that injects large amounts of air into the process liquid. The circulating pump may be expected to handle as much as 50% air by volume.

Oil production from wells often require the pumping of crude oil that contains large amounts of natural gas mixed with the oil. In many cases, the inlet pressures for the pump range from 100 to 200 bars.

Other applications may be those where the inlet pressure is below atmospheric pressure. As a consequence, air can leak into the system, resulting in gas-liquid mixtures that must be handled by the pump. Even small amounts of air can cause problems because the air expands substantially under low pressure to increase its volume, particularly at the inlet of the pump impeller.

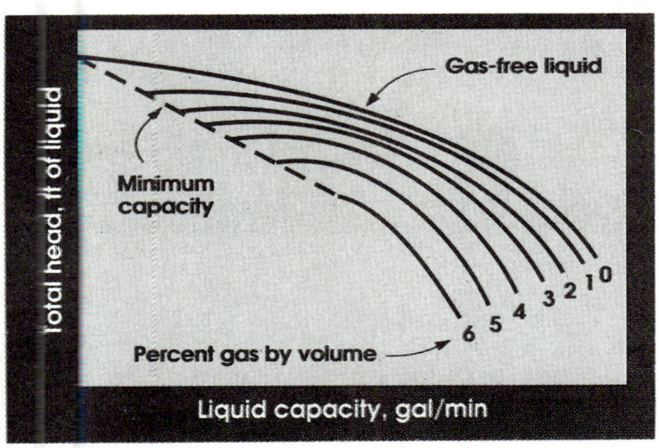

FIGURE 9. The reductions in capacity and head result from the gas blockage of the flow, and reduced specific gravity of the pumpage, respectively

JIM STONE

Effect of gas on performance

The most dramatic effect of gas or vapor on centrifugal pump performance is the complete blocking of the impeller inlet as the pump becomes "airbound." When this happens, the impeller acts as a centrifuge, and tends to separate the heavier liquid from the gas that builds up at the impeller inlet. At low flowrates, the liquid flow cannot even carry the air through the impeller, and the gas bubble grows until it completely fills the impeller inlet. The result is complete cessation of liquid flow.

Even when small amounts of gas are carried through the impeller, the liquid capacity and pump discharge pressure are reduced (Figure 9). This reduction is the result of the blockage of the flow by the gas, and a reduction in developed head due to the reduced specific gravity of the pumped mixture.

It can be seen from Figure 9 that even with small percentages of air, the unit stops pumping due to accumulated air in the impeller when operating near the zero-flow condition of the pump. High velocities at higher flowrates can carry with it higher percentages of gas. Therefore, when gas entrainment is a potential problem, pumps should be operated at or beyond the BEP specified by the manufacturer.

Inducers or inlet boosters (Figure 10) are devices designed to benefit the functioning of the impeller in that they increase the fluid pressure before the mixture enters the pump. This increase in pressure reduces the volume of the air, thereby reducing its negative effect on the impeller performance.

Laboratory tests have shown that pumps with higher specific speed (that is, units that produce high flow and low head) are affected less by the presence of gas than those with low specific head (low flow and high head). In some cases, it may be helpful to use a high-specific-speed booster pump in series with a low-specific-speed pumping unit in order to minimize the effect of the gas.

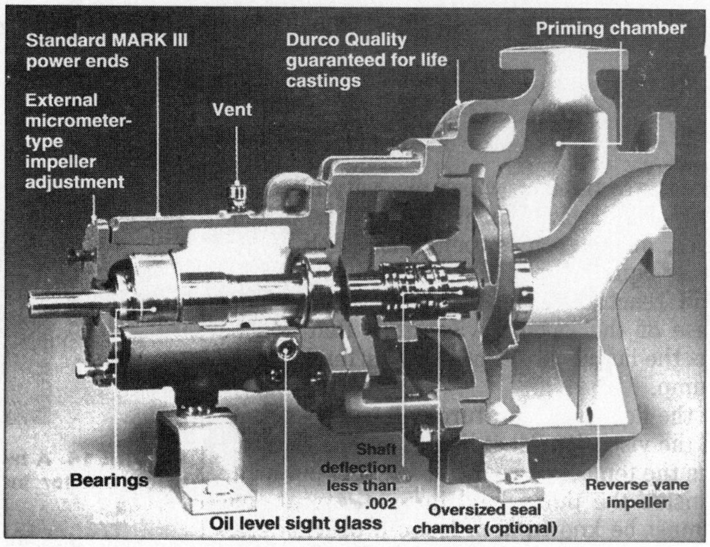

The Duriron Co.

FIGURE 11. A self-priming pump evacuates all the entrained air from the inlet pipe

Another helpful action is to provide a gas vent at the pump inlet. The suction pipe should be sized about twice as large as the flange at the pump inlet in order to keep inlet velocities low. A vent connection should be located at the top of the pipe, close to the pump so that gas can escape back to the source.

If the pump takes suction from a closed tank, it may be possible to pressurize the inlet, thereby reducing the volume of entrained gas, or turn some vapors back to liquid. Where vapor is the primary problem, subcooling of the inlet pipe will be helpful. This will also tend to turn vapor back to liquid, and thus reduce the volume of free vapor that must be handled by the pump.

Special designs

A patented design allows entrained gas to be vented through the back of the impeller shroud, and eventually back to the fluid source. The back of the impeller is also designed in a way that prevents the vented gas from mixing with the main flow stream.

Self-priming pumps (Figure 11) can handle higher amounts of gas, especially at low flow. These units will not become airbound because, in order to make them self-priming, they are designed to handle 100% air by evacuating the air from the inlet pipe. However, they will suffer the same reduction in capacity and head as a conventional centrifugal pump.

Rotary pumps, especially the screw-type units, can also handle large amounts of gas because they use positive displacement, and therefore, are not subject to the problems caused by the complex fluid dynamics associated with the centrifugal pumping action. The single-screw pump is particularly suitable in that it can handle low-viscosity liquids. Many other rotary units, such as gear and vane pumps, depend on the viscosity of the liquid being pumped to lubricate the moving parts. □

STEPHEN WALSH

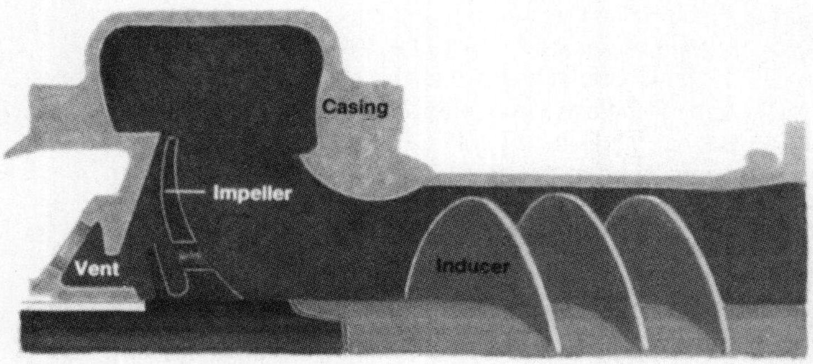

FIGURE 10. Inducers increase the fluid pressure at the inlet to improve pump performance

HIGHLY VISCOUS LIQUIDS

The absolute viscosity of a fluid is the ratio of the shear stress (force applied divided by the area upon which force is acting) to the shear rate (velocity of the moved liquid layer divided by the distance between the layer upon which the force is applied and the static layer). This quantity is a measure of the resistance to movement between adjacent layers of fluid, or the resistance of the fluid to flow. Its reciprocal is called fluidity. Examples of unusually viscous liquids include hot asphalt, lubricating oil, plastic polymers, and food products, such as molasses.

Highly viscous fluids present a unique challenge to the pump user because of the many different variables involved. But knowing some basic physical and engineering properties can facilitate the selection of the right pumping equipment.

If the viscosity is constant at a given temperature, regardless of the shear rate, the liquid is called Newtonian. When the shear stress is plotted against the shear rate, all Newtonian liquids represent a straight line (Figure 12). Most common oils exhibit this property, and therefore, are Newtonian.

However, in many liquids, the viscosity varies with the shear rate. Thixotropic liquids are those in which viscosity decreases as the shear rate becomes larger. Many emulsions, plastisols, rubber cements, and ketchup fall into this category.

JIM STONE

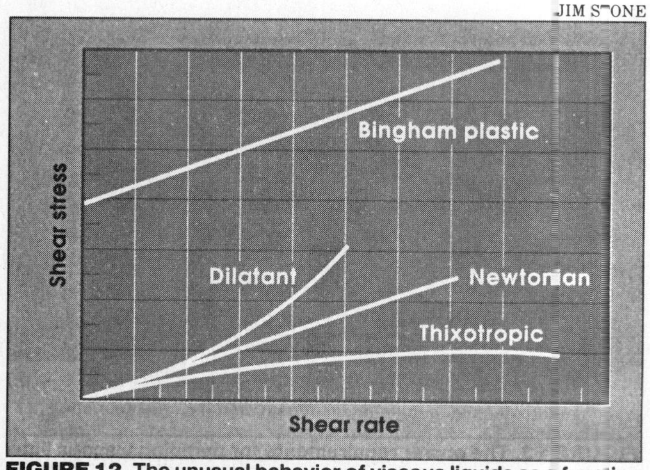

FIGURE 12. The unusual behavior of viscous liquids as a function of shear rate must be kept in mind during selection of a pump

Dilatant liquids are those in which viscosity increases with the shear rate. Examples of such fluids include mineral slurries, and starch suspensions. Each of the different types of liquids, broadly known as non-Newtonian liquids, will react differently during the pumping process because of the varying viscosities that will be encountered as the liquids are exposed to different shear rates in the pump.

In addition, the liquid temperature will have a significant effect on the viscosity. Most liquids exhibit decreasing viscosity as the temperature increases. The pumpage temperature inside the pump, and its viscosity at that temperature, must be known in order to properly determine the overall performance.

Viscosity also plays an important part in determining the condition of service that the pump must meet. For example, the higher resistance to flow (characteristic of unusually viscous liquids) implies a higher system resistance that must be overcome for pumping to take place. This usually requires larger-diameter piping for suction and discharge.

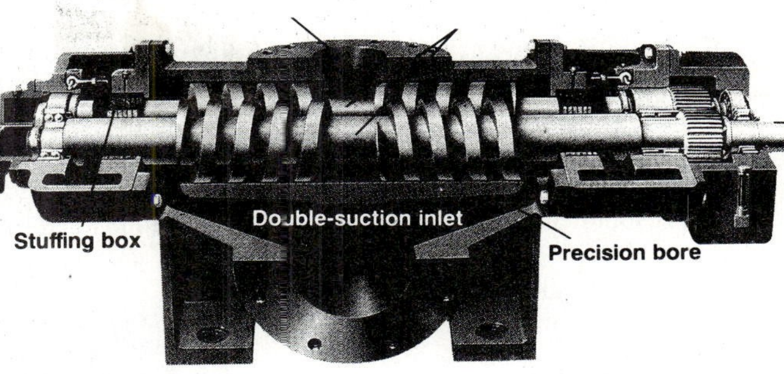

FIGURE 14. A two-screw rotary pump generates positive displacement for the pumpage by rotation of the intermeshing screws

JIM STONE

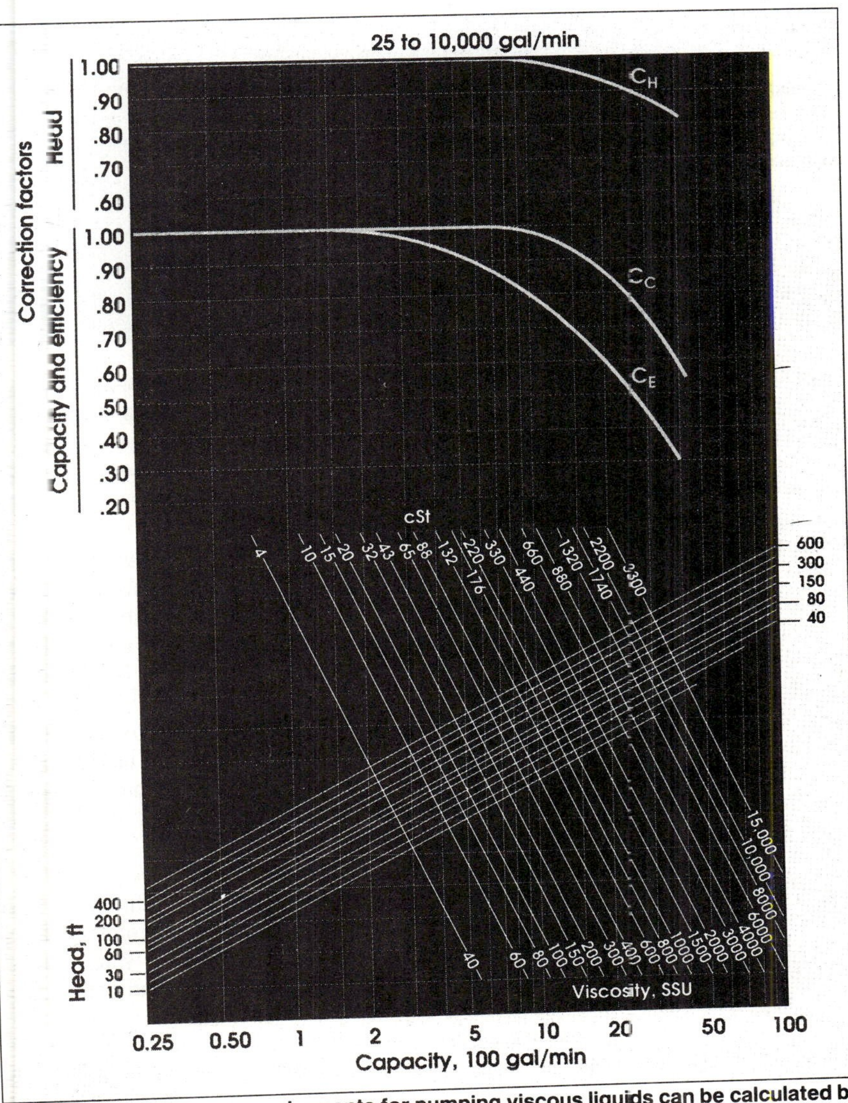

FIGURE 13. The power requirements for pumping viscous liquids can be calculated by using water as a standard for comparison, and then introducing the appropriate correction factors'

Centrifugal action

For liquids with viscosity up to 3,300 cP (15,000 SSU), centrifugal pumps usually can do the job quite well. However, proper corrections need to be made in evaluating the overall performance of the unit. With Newtonian fluids, the viscosity is constant with shear rate, and therefore, is relatively constant in all areas of the pump.

However, with non-Newtonian liquids, the viscosity will vary in different areas of the pumping device due to the different shear rates. Analysis has shown that the shear rates and viscosities in the volute casing will generally dictate the performance of a standard single-stage centrifugal unit.

The highly viscous liquid causes a drop in the head generated by the pump, thereby requiring a higher power input compared to that for pumping a less-viscous liquid, such as water. Correction factors (Figure 13) have been developed for calculating the pump performance with a variety of viscous liquids. Applied to the head, power requirement, and efficiency, these factors (originally published in the Hydraulic Institute Standards) pertain to a number of single-stage centrifugal pumps with discharge sizes varying from 2- to 8-in dia.

The head, capacity and efficiency ratios can be calculated from the chart at various capacities, and a predicted performance-curve established for a given viscous liquid. The driver for the pump must be sized correctly to ensure there is sufficient power available.

certain applications, however, the hazardous nature of the liquid makes sealing so difficult, or zero leakage so important that it becomes necessary to resort to some form of sealless pumping unit.

Among the hermetically sealed pumps canned-motor and magnetic-drive units are gaining in popularity. For a cost comparison of these sealless as well as mechanically sealed pumps refer to "Figuring the Relative Cost of Pumps," *CE*, August, pp. 129 – 136. The strengths and weaknesses of magnetic-drive units are discussed in "Sealless Magnetic-Drive Pumps — Stronger Than Ever," *CE*, August, 1990, pp. 112 – 117, while the operational dos and don'ts are indicat-

Although a single mechanical seal is quite effective and reliable in a large number of sealing applications, it always allows a small leakage to the atmosphere across the face providing lubrication. In hazardous applications, therefore a single seal is not usually appropriate.

The most common solution is to put two seals back-to-back, and provide a nonhazardous barrier fluid between the two seals (Figure 16). This barrier fluid must be compatible with the pumped product. Also, it must be kept at a pressure of at least 15 psi above the pump pressure at the inboard seal.

Today, there are a wide variety of seal designs in the market, enabling the user to pump fluids at pressures over

information on mechanical sealing options, refer to "Sealing Off Fugitive Emissions," *CE*, May, pp. 189 – 198.

Condition monitoring

All applications involving hazardous liquids must have some form of monitoring device, either to simply warn of a problem or to indicate the advisability of the complete shutdown of a pumping system. For canned-motor, eddy-current- and magnetic-drive pumps, for example, there are three main areas that need continuous monitoring:

- Inadequate flow of the cooling medium to the containment shell, and insufficient flow of lubricants to the bearings, that may result from blockage to internal passageways, or seizure of impeller

HAZMAT STORAGE

Changes in building codes coupled with the increasing cost of compliance have made the task of building a hazardous-materials storage facility ever more complicated. A detailed step-by-step approach to design, coupled with sound engineering judgement, is needed to produce an economical and safe facility that satisfies building requirements.

Several major building codes are used throughout the U.S., including the Uniform Building Code (UBC), the Southern Building Code, and the BOCA (Building Officials & Code Administrators International) Basic Building Code. In addition, there are codes or standards that deal with other aspects of construction, such as fire prevention,

CONTENTS OF A TYPICAL BUILDING CODE

ADMINISTRATIVE
Title, scope and general application
Organization and enforcement
Permits and inspection

DEFINITIONS AND ABBREVIATIONS

REQUIREMENTS BASED ON OCCUPANCY
Classification by use or occupancy
General requirements for all occupancies
Requirements for assembly buildings, commercial and industrial buildings, educational buildings, hazardous buildings, hospitals, and dwellings

REQUIREMENTS BASED ON TYPES OF CONSTRUCTION
Classification of all buildings by

complexities of the process and in achieving the most desirable design.

1 Determine the hazard classes and the quantities of the materials to be stored. Typical hazard classes include: combustibles, flammables, oxidizers, toxics, poisons and so forth. Any changes that are anticipated in the list of materials to be stored should be taken into consideration. When upgrading existing facilities, an onsite inspection and evaluation can be helpful in establishing the materials and the quantities that are involved. Inventory turnover rate also should be taken into consideration.

2 Establish the type of material handling and storage systems to

Positive displacement by rotation
In rotary pumps, the pumping action is created by displacement of a given volume of liquid through the pump by means of intermeshing screws, gears, lobes, or some other positive-displacement action between the rotor(s) and the stator. For example, a two-screw rotary unit (Figure 14) can be used for pumping a range of highly viscous liquids. The various velocities and clearances in these pumps result in different shear rates and subsequent variations in viscosity when handling non-Newtonian liquids.

In contrast with centrifugal pumps, the rotary units normally operate bet-

include handling of liquids with viscosities up to 500 million SSU. Increased clearances between components inside, lower-than-normal operating speeds and special arrangements for the inlet — all improve pump performance with highly viscous liquids. □

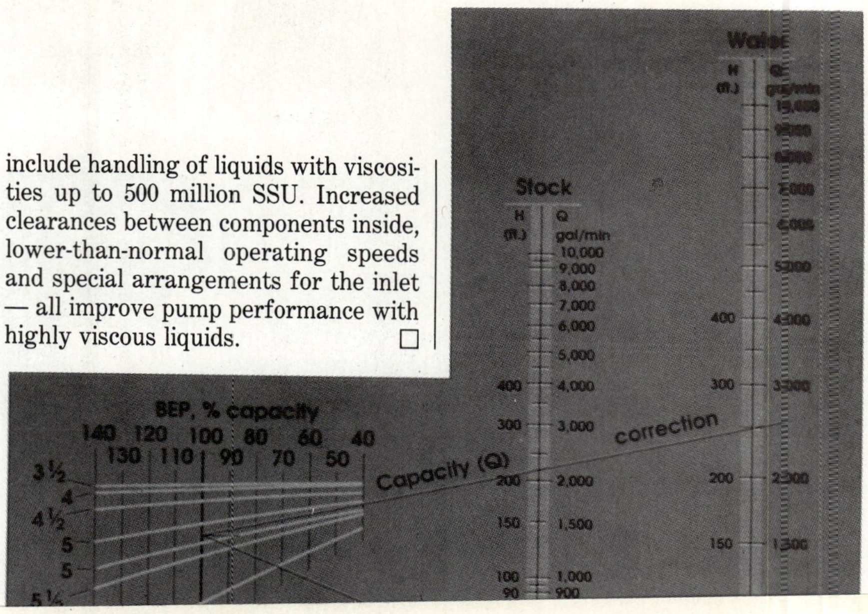

FIGURE 16. Leakage from a double seal is minimized by maintaining the barrier fluid at a pressure at least 15 psi above the pumpage pressure

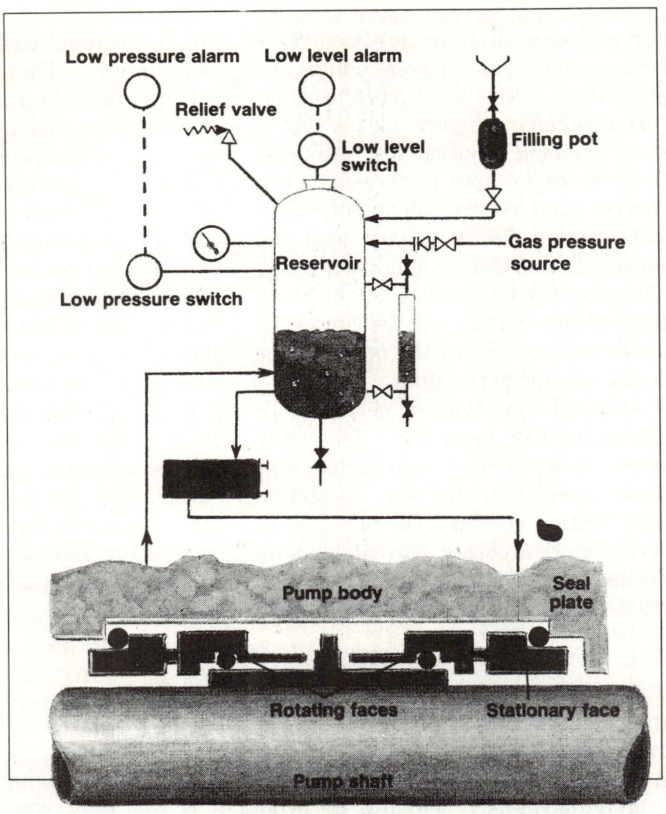

• Empty suction, or a closed suction valve that may cause the pump to run

uct from the containment shell

The monitoring of anomalies can be

discharge, level switches in the suction, and amperage-detection devices sensing change in power input. However, none of these will pinpoint the effect of internal passageways being blocked.

The most effective device to meet all the requirements of the first two symptoms listed above is a temperature sensor on the outside diameter of the liquid-containment shell. When positioned as close to the outer rotor as possible, and set to disengage and turn off the motor as soon as the temperature rises above working conditions, the sensor can effectively avert the dangerous consequences of a malfunction.

Leakage of the pumped product can be detected using either capacitance or conductivity probes inserted into the bottom of the housing around the outer rotor, or stator windings. The progressive wear in bearings and out-of-balance placement of shafts can be monitored by detecting unusual vibration at the earliest possible stage.

Leakage of the barrier fluid into the pumped product, or to the atmosphere denotes a problem, too. This is normally detected using a pressure switch in the barrier-fluid circulation

4 Compute the occupant load — the total weight of materials to be stored. Engineers also should establish whether — based on the hazard classes of the materials involved — area separations will be necessary. If dealing with different occupancies, it may be best to satisfy the specific requirements of the most-restrictive classification.

5 Establish the type of construction required based on occupancy. Most codes designate construction types as noncombustible, combustible, fire-resistant and nonfire-resistant.

For facilities that require area separations, a minimum wall rating of 1-4 hours is normally needed. Definitions of those ratings are specified in the applicable code.

6 Determine the location of the facility on the site. The proximity of the storage building to property lines and other structures along with the hazard classes of the materials involved will help establish the necessary exterior wall ratings.

7 Calculate the allowable floor area — the maximum approved space based on occupancy classification and construction type. Increases may be allowed based on the location of the facility on the site.

If the desired floor area exceeds that which is allowed, the limit can be increased by choosing a more fire-resistant construction type or creating two or more separate areas by the use of separation walls.

8 Determine the height and number of stories. Again, the occupancy classification and construction type will establish the limit. Generally, a hazardous-materials storage facility is limited to one-story structures without basements.

9 Review the proposed building layout against code requirements for exits, stairways and corri-

DON'T FORGET THE DETAILS

When designing a hazardous-materials storage building, good engineering judgment is the guiding principle. There are several details to keep in mind:

Containment — Size impoundments for the largest spill or fire hazard possible. Then design sumps, floor recesses, slopes, curbing and refuge sidewalks accordingly.

Chemical resistance — Check the resistance of paints, coatings, floorings, mechanical and electrical equipment and racks and shelves used against the chemicals to be stored. Make sure that manufacturers' installation or application recommendations are followed.

Explosions — If any of the chemicals to be stored present a fire or explosion hazard when exposed to heat or flame, the walls, floor, roof and doors need to be properly constructed.

Exits — In hazardous areas, secondary exits usually are required for areas greater than 200 ft². Verify that storage-rack layouts enable adequate egress.

Ventilation — Continuous mechanical ventilation is usually required in hazardous materials storage facilities at an exchange rate of 1 cfm/ft² or greater.

Electrical design — Power and lighting must be designed for hazardous occupancies according to the applicable code. Emergency power may be required in some circumstances.

Material storage — Many buildings designed for the storage of hazardous and flammable materials use racks for high bay storage. Rack layout, clearances and structural details must be checked to ensure that they can properly handle the various types of pallets and materials weights. Fork lifts need to be compatible with the rack system. Possible risks must be assessed. Hazard analysis techniques — including fault-tree, energy-trace and barrier analyses — may be useful in determining the best storage system.

Emergency equipment — Showers and eye-wash stations must be located as specified in the applicable code. It may also be required that personnel in the structure wear full-face respirators in case of spills or leaks.

Security — When the materials stored include explosives, poisons or alcohols, the facility must be secured against unauthorized access.

Environmental — Storage for spill-cleanup equipment and supplies should be provided for.

The costs for building hazmat storage facilities are high. A detailed step-by-step approach to design can go a long way in helping engineers keep the facilities within budget. However, cost must be tempered with good engineering judgement if the facility is to serve its function over the long haul. □

The authors

Waldemar Seton is president, founder and principal chemical engineer of SJO Consulting Engineers, Inc. (1500 S.W. 12th, Portland, OR 97201; tel: 503-226-3921). His four-decade career includes responsibilities in project management, construction scheduling, cost estimation and process engineering for large industrial facilities. A graduate of Stanford University, he is a registered professional engineer in Oregon, California and Idaho.

Thomas Harris is an architect-project manager for SJO. Since joining SJO, he has been involved in design and project management on commercial and industrial projects. He holds a bachelor's degree in Architecture from the University of Colorado and is a registered professional architect in the states of Oregon, Washington and Colorado.

Concern in the U.S. for the safety and welfare of workers engaged in the handling and storage of hazardous materials has led to strong Occupational Safety and Health Administration (OSHA) regulations in this area. The rules are becoming stricter and more specific on almost a monthly basis, and the fines for infractions have been increased sevenfold in the last year. For the economic viability of their companies, as well as ensuring safe practices in the plant, engineers as well as operators must keep abreast of rules, regulation and guidelines. The challenge: Maximizing compliance and efficiency, while minimizing costs.

Beginning this year, fines for serious violations of federal health and safety standards have jumped to $7,000 from $1,000 in 1990. Willful and repeat violators will be penalized $70,000 per act. So, it is important for engineers to find economical ways to comply with regulations governing storage of hazardous materials.

To create safer working environments, there is a regulatory trend to isolate storage areas from habitable space, and to

STORING HAZARDOUS WASTES SAFELY

Gary Van Valkenburgh
Unified Safety Corp.

Free-standing structures provide security, mobility and economy

Originally published September 1991

may touch off a series of explosions at | to prevent leaks and overfills from | corrosion by keeping unprotected tank bottoms from contacting soil di-

* *See,* Out of sight, not out of mind, *CE*, January 1991, pp. 30–35

Originally published July 1992

			public way
Factory fabricated	275 or less 276–750 751–12,000 12,001–30,000 30,001–50,000	5 10 15 20 30	5 5 5 5 10
Field erected	50,001–100,000 100,001–500,000 500,001–1,000,000 1,000,001–2,000,000 2,000,001–3,000,000 3,000,001 or more	50 80 100 135 165 175	15 25 35 45 55 60

proper inerting system (as outlined in NFPA 69, "Standard on Explosion Prevention Systems"), these distances can be cut in half. If fire protection for properties adjacent to the liquid storage is not available, then these distances must be doubled. This table was adapted from Tables 2.1 and 2.6 of NFPA 30

THAT COVERS YOUR TANKS

• ***The Oil Pollution Act of 1990 (OPA)***
OPA makes the onshore tank owner (or operator) liable for removal costs and damages if oil or other hazardous substances are discharged to navigable waters and adjoining shorelines. OPA specifies the type of damages the owner must pay, and establishes a federal fund for damages caused by delinquent owners.

Enforced by the U.S. Coast Guard, OPA places the burden on owners and operators to respond to discharges (spills, leaks, pumping or dumping) by underwriting the cost of clean-up operations, personnel and equipment. Monetary damages may be assessed, should natural resources be destroyed.

Designed solely to protect the nation's waters, OPA does not address soil or groundwater contamination unless the surface water and groundwater are hydrogeologically connected. An estimated 80% of ASTs are impacted by the OPA regulations.

• ***The National Pollution Discharge Elimination System***
This set of regulations (Section 405 of the Water Quality Act) expands the Act under EPA and relates to ASTs indirectly, since it regulates the discharge of pollutants into U.S. surface waters.

The impact for AST owners is that runoff collected in the diked areas of ASTs cannot be discharged directly into a storm sewer unless certain water-quality standards have been met. To comply, tank owners are often forced to use separators or other forms of water treatment prior to dis-

that vapors must be recovered as tanks are loaded.

Title III identifies certain hazardous air-toxic pollutants (189 chemicals in all) that must be controlled. The petroleum industry's biggest concern is with the control of benzene vapors commonly liberated from most petroleum fuels.

• ***Comprehensive Environmental Response, Compensation and Liability Act (CERCLA)***
CERCLA establishes liability for each of the polluters at a given site. In the Superfund Reauthorization Act portion of CERCLA, Title II targets owners of aboveground tanks containing fluids other than petroleum, and requires that they report spills of hazardous wastes.

Title II also mandates that responsible parties provide liability and cost coverage for clean-up, and establishes clean-up requirements. It also specifies that public health officials and other agencies be guaranteed access to certain chemical information, so that emergency-planning measures can be assured. Specific language exists on whether chemicals exceed threshold planning quantities.

• ***Occupational Safety and Health Administration (OSHA) standards***
OSHA's national standards (29 CFR, 1910, 106) for the safe handling, storage and use of flammable and combustible liquids are largely based on National Fire Protection Association (NFPA) requirements for flammable and combustible liquids. OSHA regulations apply specifically to liquids with flashpoints below 200°F. OSHA is currently proposing new

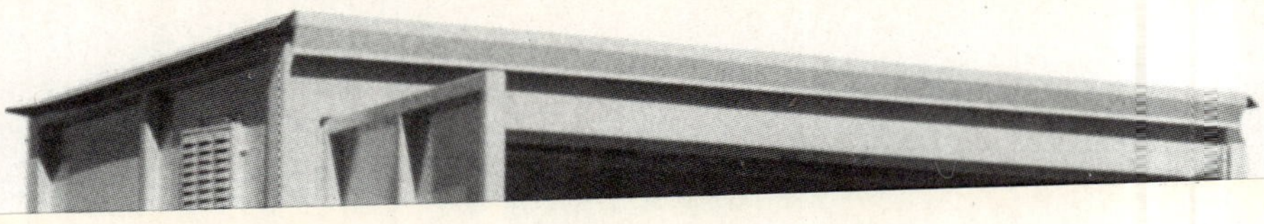

rectly. Many newer ASTs include factory attached steel-containment dikes, which serve the same purpose.

Leak detection While no specific regulations exist yet for outfitting ASTs with leak-detection instrumentation, tank owners are installing such equipment as a preventive measure. The field of options — which includes vacuum or pressure monitoring, electronic gauging, and the use of optics and sniffers — is evolving rapidly, and is too broad to detail here. Many double-walled ASTs use a sensing device within the interstitial space, to detect any releases between the walls.

Operating limitations ASTs require certain design considerations that are not typically needed with USTs. For

STI

Steel tanks with a tight-wrapped, double-walled construction enjoy increased protection against leakage. For additional protection, leak-detection devices and thermal insulation can be sandwiched between the walls

instance, compared to ASTs, buried tanks experience a much narrower range of temperature fluctuation (typically, temperatures fluctuate by no more than 10°F over the entire year). This relative temperature stability makes it is easier to control the formation of vapors and the desired viscosity of fluids contained in USTs.

Conversely, exposed aboveground tanks are subject to much-more-variable temperature conditions. In cold climates, many tanks (and connective piping) carrying heavy fuel oils are outfitted with steam tracing to maintain the desired viscosity to promote product flow. In warmer climates, excessive vapor formation from chemicals and petroleum in ASTs may call

THE REGULATORY QUILT

While leaks from underground tanks often seep slowly and silently into soil or aquifers, the concern about ASTs in that an overfill or tank rupture will allow large quantities of product to escape into a navigable waterway or drinking water source, causing an immediate — and potentially devastating — pollution incident.

To protect the nation's groundwater and soils, the pressure is on for U.S. lawmakers to establish a comprehensive regulatory program for aboveground storage tanks that would be similar to that now in place for underground tanks. Unlike the comprehensive technical standards mandated by EPA for underground storage tanks in 1988 (40 CFR 280), no single set of regulations exists for aboveground storage tanks.

Several bills to regulate ASTs have been introduced in the U.S. Congress in recent years. Proposed legislation would specify design and construction criteria; set operations and maintenance standards; require financial responsibility; and dictate cleanup requirements. The most recent bill (S. 1761), reintroduced in September 1991, would establish leak detection, corrosion protection, structural integrity and secondary containment for ASTs.

The following are the existing technical specifications that should be implemented in conjunction with local environmental, zoning and fire codes.

• ***Spill Prevention, Control & Countermeasures Program*** EPA's Office of Solid Waste & Emergency Response is currently working to modify the AST requirements that are now codified as the Spill Prevention, Control and Countermeasures (SPCC) Program (40 CFR 112). To date, EPA's success with inspection and enforcement of SPCC has been stymied because the perception of ...

would require either annual testing of buried piping to demonstrate system integrity, or monthly monitoring of buried piping for leaks. Such requirements would involve shutting down product flow and conducting water or air tests to lock for leaks. Tank integrity testing is also recommended.

To bolster the facility-inspection program, Phase I also seeks to enforce the collection and standardization of inventory information from existing facilities. The inventory question would require the creation of a national AST registration database, which would list the number of tanks, and the individual and aggregate capacities stored at each site. The Phase I proposal was published in the Federal Register, October 22, 1991 for public comment; EPA expects to finalize the revisions by late 1992.

Phase II upgrades were recommended by a special SPCC task force following the collapse of a 4-million-gallon AST in January 1988), which spilled huge volumes of diesel fuel into the Monongahela River near Pittsburgh. Phase II would require operators to develop site-specific emergency contingency plans, and to demonstrate preparedness by maintaining access to equipment and personnel necessary to contain a leak.

The proposed changes would force operators to prepare and implement an emergency-contingency plan *prior to* beginning operations; currently, operators have six months to prepare — and 12 months to implement — the plan. The proposed changes would also require annual discharge-response training for all site employees. Phase II changes, which are expected to create a serious financial burden for tank owners, are expected to be released for public comment by the end of 1992.

CARLA MAGAZINO

Backward from effect to cause

The first step — identifying the problem — will usually be simpler if column performance has suddenly deteriorated than if capacity or efficiency declined gradually. Except when related to auxiliary equipment or instrumentation, most distillation problems arise from flooding (impeded liquid flowpath), excessive pressure drop (impeded vapor flowpath), or inefficiency (insufficient liquid and vapor contact). Of course, these problems do not usually occur independently.

After attempts to stabilize column operation, usually at reduced throughput, the troubleshooter next begins a detailed evaluation of the available information. This involves interviewing operators, reviewing previous problems and their solutions, collecting recorded data, making diagnostic tests, and calculating heat and material balances.

In some cases, the symptoms will strongly indicate a certain problem and cause. At other times, the troubleshooter must consider a series of causes and look for data that support or eliminate

through the column (Figure 1). Usually uncontrolled, flooding can stabilize, as when it reaches an area of extra-large tray spacing and sufficient head develops for the liquid to flow down without flooding the next tray.

A column's maximum hydraulic capacity is the highest throughput at which it will operate without flooding (Figure 2a). This capacity is a function of the vapor and liquid rates, column geometry, and liquid and vapor properties. The flooding point refers to the location in the column where flooding begins and above which liquid accumulates.

Among the flooding indicators:

- Surges of liquid overhead
- Erratic (including high and rising) pressure drop through the column
- Fluctuating level in column bottom
- A high-temperature profile
- Falling base level or reduced bottoms flow (Figure 2b)

In a flooding column, an overhead surge enters the overhead vapor line, condenser and reflux drum, causing the level in the last to rise abruptly. Reflux to the column cannot reduce the

tion and liquid unloading. This unstable condition, in which flooding appears and disappears, produces an irregular pressure-drop profile. Alternating liquid accumulation and dumping may also cause swings in the base level or the bottoms flowrate.

When the column below the flooding

A systematic procedure can guide you to a speedy diagnosis of the problems that beset distillation columns

point is starved for liquid, continued

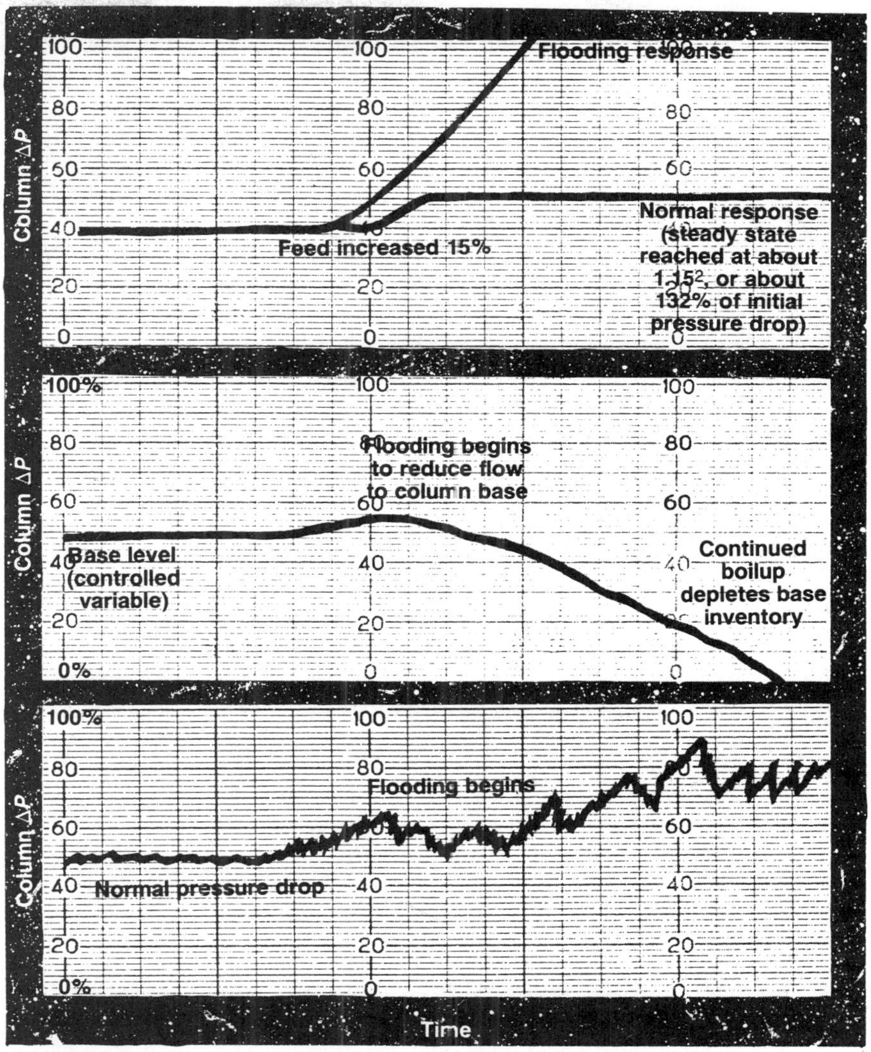

FIGURE 2A. Excessive pressure drop represents one of the surest signs of flooding; **2B.** Liquid holdup on upper trays starves bottom of a column; **2C.** Rising and erratic pressure drop signals the onset of column flooding

surge of heavier-than-heavy-key components in the distillate. However, the first sign of flooding may be the loss of separation caused by massive entrainment — often referred to as an "efficiency flood" or a "mass transfer flood." This is most common in columns in which the vapor rate is high and the liquid rate low, as in vacuum towers. An increase in reflux rate may further diminish the separation quality, because the additional entrainment caused by the higher liquid and vapor traffic may more than offset any improvement in fractionation. As flooding conditions are relieved, liquid will begin moving down the column. This downflow can be detected as a high-pressure-drop bulge moving down the column trays. When the bulge reaches the bottom of the column, it can be observed as a surge in base level or bottoms flowrate.

Pressure-drop clues

Excessive pressure drop (higher than design but less than flooding) commonly limits the capacity of vacuum towers, which are usually designed to operate at low pressure drop. If the overall column pressure must be elevated or the top pressure reduced, because of excessive column pressure drop, in order to maintain a certain base pressure, the base temperature may become unacceptably high or the top temperature too low. This can limit the capacity of a column, if the product could be degraded by heat, or if the heat source and heat sink are limited.

Elevated column temperatures, especially at the base, may indicate a higher base pressure. A high base temperature can reduce column capacity by limiting heat exchange in the column reboiler. This will be indicated by high reboiler steam pressure or a wide-open steam-control valve.

If the base temperature is controlled, excessive pressure drop will lower the top pressure and temperature. Also, any increase in base pressure must be compensated for by a hike in the composition of lower-boiling components in the column underflow. If the top temperature is controlled, any reduction in the column top pressure will be compensated for by a boost in the higher-boiling components in the distillate.

Inefficiency indicators

Poor efficiency usually results from inadequate liquid and vapor contact or insufficient disengagement of liquid from vapor. The first can be attributed to liquid and vapor bypassing each other due to maldistribution, such as by the loss of contacting medium. The key indicators of efficiency problems are off-specification product, excessive reflux and a high rate of boilup.

A drop in column efficiency is often countered by increasing the reflux ratio. If the efficiency deteriorates to the point at which the column approaches the minimum number of stages necessary to achieve the desired separation, a significant increase in reflux ratio and boilup will indicate a problem. This indication may be overlooked if the column continues to make the desired separation. However, the energy penalty and wasted column capacity can be costly.

Now, the detailed evaluation

The operators who are familiar with the column and who watched the problem develop should be consulted. Among the things they can help you determine: When did the trouble start? What were the first signs of it? What was done to correct the problem? Was the process upset prior to when the problem appeared? Did the problem follow any process changes? Have the symptoms appeared before? Are operating procedures being followed?

Always consider that the problem may have occurred before, especially if it is a process, not a mechanical, one. Chronic problems may justify warning indicators, such as feed temperature alarms and composition analysis.

Without information, troubleshooting can only be guesswork. Information sources include operating charts and logbooks, column drawings, piping

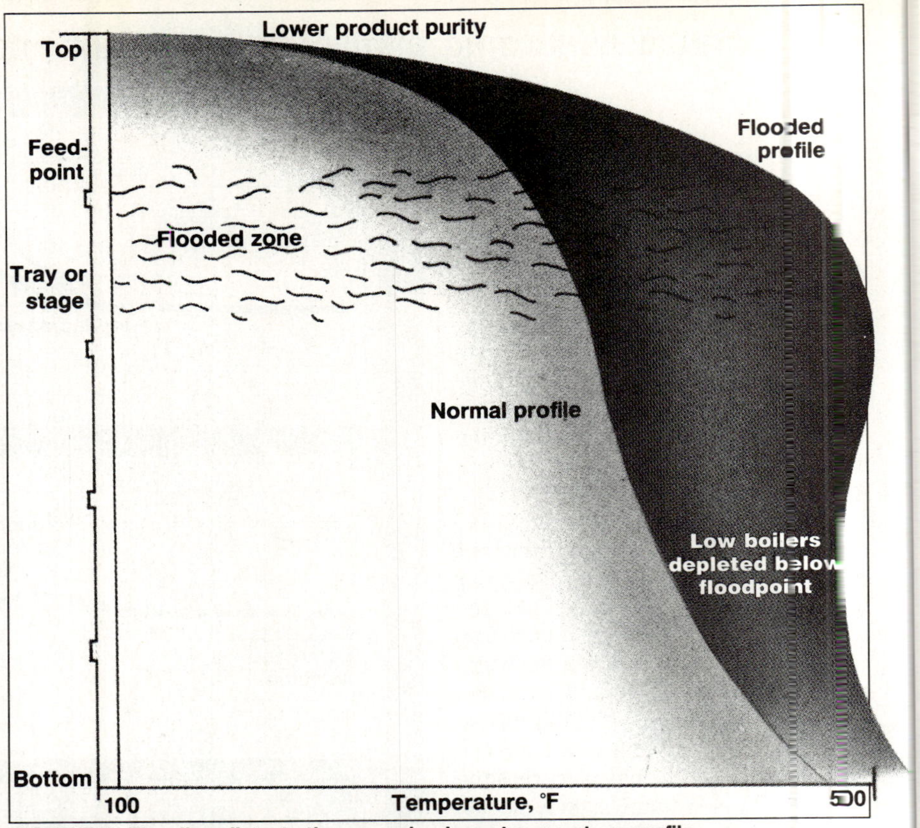

FIGURE 3. Flooding distorts the normal column temperature profile

and instrument diagrams, and equipment manuals. Charts enable the troubleshooter to study the timing of events so as to trace the development of a problem, and to differentiate between automatic controller and operator actions. From the logbook, the troubleshooter can spot changes in operating procedures that may have initiated the problem.

Differential-pressure measurements of sections of the column are more useful than a single overall measurement for detecting and locating hydraulic problems. For best results, such measurements should span every 10 trays or each packed bed, or at least the rectifying and stripping sections. For example, the increase in differential pressure may be several-fold across a flooded 5-ft section of a 20-ft column. The same flooding in a 100-tray column gauged only as an overall pressure drop would be equivalent to an additional 0.3 in. of pressure drop per tray, and might not be perceived as a problem.

Because of the importance of pressure-drop measurements in column troubleshooting, the proper operations of instruments must be verified. When safety considerations permit, manometers or pressure-drop cells can be temporarily rigged between valved nozzles for taking pressure-drop measurements. In some cases, water-filled flexible transparent tubing can serve as a manometer. Such a manometer can also serve as a level sightglass where a permanent one has not been installed.

The troubleshooter should also examine controlled variables and their setpoints, as well as controller outputs and the values of manipulated variables. Also check for improperly ranged instruments. Problems with the condenser, vents and the backpressure controller can be detected via deviations from the normal top pressure. Field pressure gages are notoriously inaccurate, and should be checked with test gages.

With a column temperature profile and temperature measurements of the feed and reflux streams, the troubleshooter can detect energy balance shifts and such problems as overheated and flashing feeds. Process flows, including reflux, are needed for calculating material balances. Readings should represent steady-state or averaged values. Controller changes, or changes in column inventory, can skew streamflow measurements.

A heat-and-material balance furnishes the means for checking the consistency of key data, such as process and utility flows, and product compositions. The data collected should be averaged so as to represent steady-state conditions. Flowmeter readings may need to be corrected for actual fluid densities. Of course, a material balance based on flooded conditions could be in error because of the accumulation in the column.

A component material balance will often pinpoint material-balance problems caused by the ratio of distillate to feed making the desired separation impossible. For example, a column that is supposed to be fed a 50/50 mixture of low/high boilers will not produce a bottoms flow that is free of low-boiling components at a fixed distillate-to-feed ratio of 0.6 when the feed composition shifts to 70/30. If the overall and the component material balances will not close to within 5% to 10%, one, or more, of the flow or composition measurements is probably inaccurate.

Analyses of the composition of distillate, bottoms, feed and sidedraw streams are particularly useful. The reliability of onstream analyzers, as well as of manual sampling procedures and analyses, must be checked. Random sampling may not disclose some composition problems, such as slugs of water in the feed stream, that may be unsettling the column. A column composition profile, especially in the case of trayed columns, will enable the troubleshooter to evaluate the separation taking place in different sections of the column and to detect buildups of impurities. Recordings of the reflux-drum and column-base levels

FIGURE 4. Corrosion damage causes inefficient operation

can be especially valuable to the troubleshooter, as well as of utility flows, and supply and return pressures and temperatures can provide helpful insights.

Besides showing the placement of control-room and field instruments, piping and instrument diagrams also indicate the location of nozzles and valves that may be suitable for sampling and for serving as alternative feedpoints. However, the troubleshooter should not regard such drawings as necessarily representing actual installations. Nevertheless, if mechanical problems are suspected, drawings of the vessel and its internals will provide information on tray or packing assemblies, attachment methods, nozzle locations and special features. Such information can be critical for interpreting data from column instruments and for identifying and visualizing possible mechanical problems.

Operations and maintenance records and files of photographs can be valuable in detecting long-term changes in column performance. A well-documented test run can also be useful, particularly if the file contains — in addition to the usual record of the primary process data — such information as the rectifying and stripping section pressure drops and temperature profiles, reboiler condensing pressure and temperature, condenser and other exchanger temperature differences and pressure drops, control-valve positions and controller settings. Ideally, the test run had been conducted at normal operating rates, as a capacity or acceptance test.

The column specifications and design calculations will enable the troubleshooter to compare current with intended operating conditions. Stage and tray hydraulic calculations can provide guidance on the column's limits. Corrosion data, together with the replacement history of column internals, will provide a basis for considering the probability of corrosion damage (Figure 4).

FIGURE 5. Multiple pressure taps enable troubleshooter to locate floodpoint

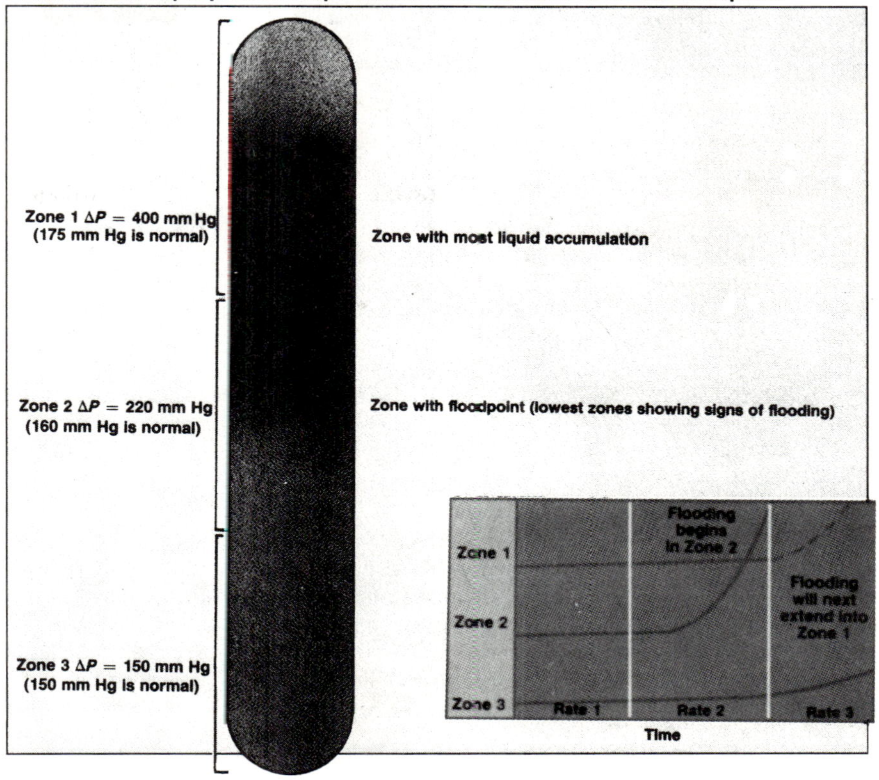

Checking a new column

A new column will sometimes not perform up to operational requirements. The causes may be the result of process-design error — such as the column, or the auxiliary equipment, being under- or oversized, the feedstream being poorly located because the presence of certain components were not anticipated, mistaken calculations, wrong assumptions about the efficiency of certain elements, or poor control logic.

Mechanical-design mistakes can result in such problems as nozzles located improperly with respect to packed-column distributors or tray downcomers, and the early failure of internals because the wrong material was selected (See Box, p. 121).

Improper installations, such as of internal structures and feedpipes, are frequent causes of problems. When different types or sizes of trays, downcomer panels, packing or liquid distributors are being installed in a column, the locations of these internal structures can be interchanged. These structures can also be incorrectly installed, or they and their fasteners even omitted. Debris (such as coats, rags, even boards and boxes) can be left in a column, restricting liquid and vapor traffic. Improperly installed control loops can cause tasks to be performed in the wrong sequence, or even incorrectly (such as opening a valve instead of closing it).

Establishing cause

Sometimes, the cause of flooding or poor efficiency will be obvious from an examination of operating variables and heat- and material-balance calculations. At other times, the troubleshooter will have to evaluate all the available information to determine the cause. Operating tests may have to be run and additional data collected to confirm or rule out explanations of problems.

With guidance from the trouble indicators and data, the troubleshooter can do a better job of isolating problem areas and pinpointing the specific cause of poor performance. For example, if flooding occurs with overhead liquid surges but without an appreciable hike in column pressure drop, the engineer can focus on flooding at, or near, the top tray or packing. A low pressure drop over a column segment provides strong evidence of trays not being in place.

Multiple pressure-drop taps enable

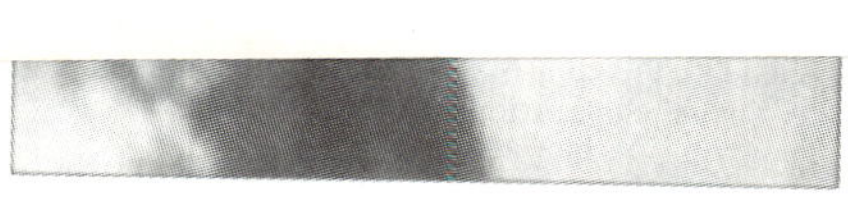

In addition to detecting and evaluating plugging, corrosion and damage, the video camera has been used to oversee column hydraulic tests (such as fol-

FIGURE 9. Cable allows camera to be snaked into tight quarters in column

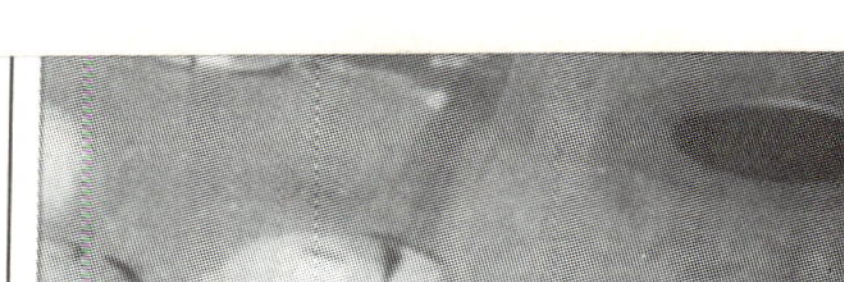

Why the packed column did not scrub: The case of the hatless vapor riser

How not to modify a column: The case of the troubleshooter who missed the target

A further hike in vapor rate may trigger flooding by expanding the liquid holdup to the point at which the phases begin to invert, with the liquid phase becoming the continuous phase. After flooding has begun, the liquid can continue to build up above the floodpoint in the column.

Vapor, being the continuous phase in a packed column, does not bubble through the liquid, as in a trayed column. Because of this and the absence of vapor-flow orifices (as in a tray), packed columns operate at much lower pressure drops than trayed columns. Because liquid and vapor contact in a packed column is less agitated than in a trayed column, the contents of packed columns are less likely to foam.

Pressure drops through random packing typically range from 0.3 to 0.9 in. water/ft of packing. At the higher end of these ranges are smaller packings and packings having less void space (such as ceramic packing); at the lower end, larger and more-open packings (Figure 2).

Low pressure drops, 0.3 to 0.5 in./ft, are normal for absorption, scrubbing and vacuum columns. Structured packing usually operates at lower pressure drops, typically 0.1 to 0.7 in. water/ft of packing.

Between 0.8 and 1.4 in. of pressure drop/ft of packing, liquid entrainment may begin to reduce the apparent efficiency of the packing. Flooding often occurs when the pressure drop exceeds 1.3 to 2.5 in./ft of packing.

Predicting the efficiency of packing, or HETP*, in a particular column with accurately is difficult. HETP is a function of packing type and size, and the physical properties, relative volatilities, distributions and flowrates of the liquid and vapor. Efficiency can best be assessed be means of plant columns in the same service, models scaled from such columns, software models based on data from such columns, or measurements from laboratory models.

*Height equivalent per theoretical plate is a measure of mass-transfer efficiency; in a packed distillation column, it is the height of packing that makes a separation equivalent to that of a theoretical plate.

For other recent articles in Chemical Engineering on packed distillation columns, see "Practical Tips on Tower Packing," Dec. 7, 1987, p. 101; "Boosting Tower Performance by More Than a Trickle," May 27, 1985, p. 22; "Structured is the Byword in Tower-Packing World," Mar. 4, 1985, p. 22; and "Packed Column Internals," Mar. 5, 1984, p. 40.

For troubleshooting purposes, however, a particular packing type and size can be categorized by a typical HETP. As a general rule, one can assume for random packings an HETP of 18 in. for 1-in. packings; 26 in. for 1½-in. packings; and 36 in. for 2-in. packings. For structured packings, typical HETPs are 9 in. for 0.25-in. crimp; 18-in. for 0.5-in. crimp; and 33 in. for 1-in. crimp.

Types of packings

Random packings, of which there are many types, sizes and materials, are so called because each packing in a bed need not be oriented in any specific way — i.e., they can be dumped into a column. Common structured packings are manufactured into mats of thin, corrugated metal sheets, layered together and custom installed. Both generic and proprietary packings are available.

Once chiefly solid-walled (such as Raschig rings and Berl saddles) and constructed of ceramic, metal or plastic, random packings are now predominantly the slotted-ring type fabricated mostly from metal or plastic (Figure 2). Newer proprietary packings having aspect ratios that differ from those of standard ring packings have improved column performance. However, older packings, such as ceramic rings and saddles, are still used in corrosive services.

The capacity of a particular packing depends on its geometry, and liquid and vapor flowrates and physical properties. In general, capacity increases with packing void-space.

As might be expected, corrosion is common with metal packings. For the typical metal-ring wall thicknesses of 15 to 35 mils, a corrosion rate of 10 mils/yr is severe because packing is attacked from two sides. Even if corrosion rates are moderate, it is not unusual for packing to be replaced frequently. Corrosion is an even greater problem with structured packing. The corrugated-sheet and knitted-wire packings are manufactured mostly from 5-to-8-mil-thick material.

Plastic and ceramic packings usually resist corrosion better, but present other problems. For example, heat and solvation can soften plastic packing, and me-

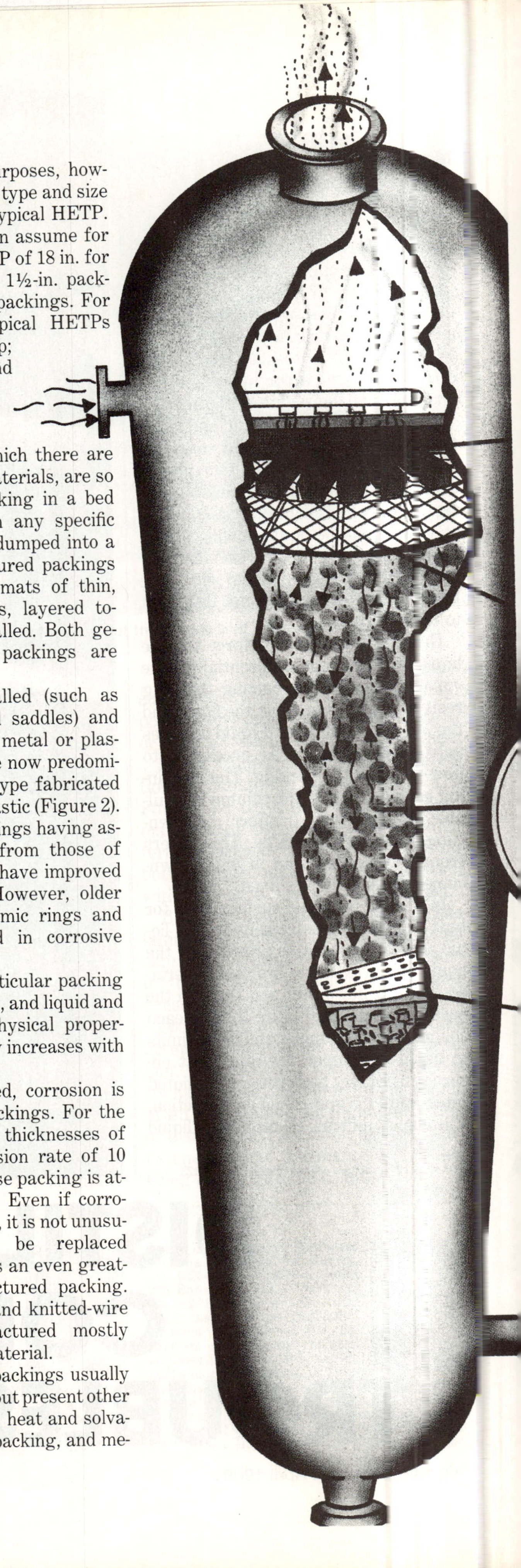

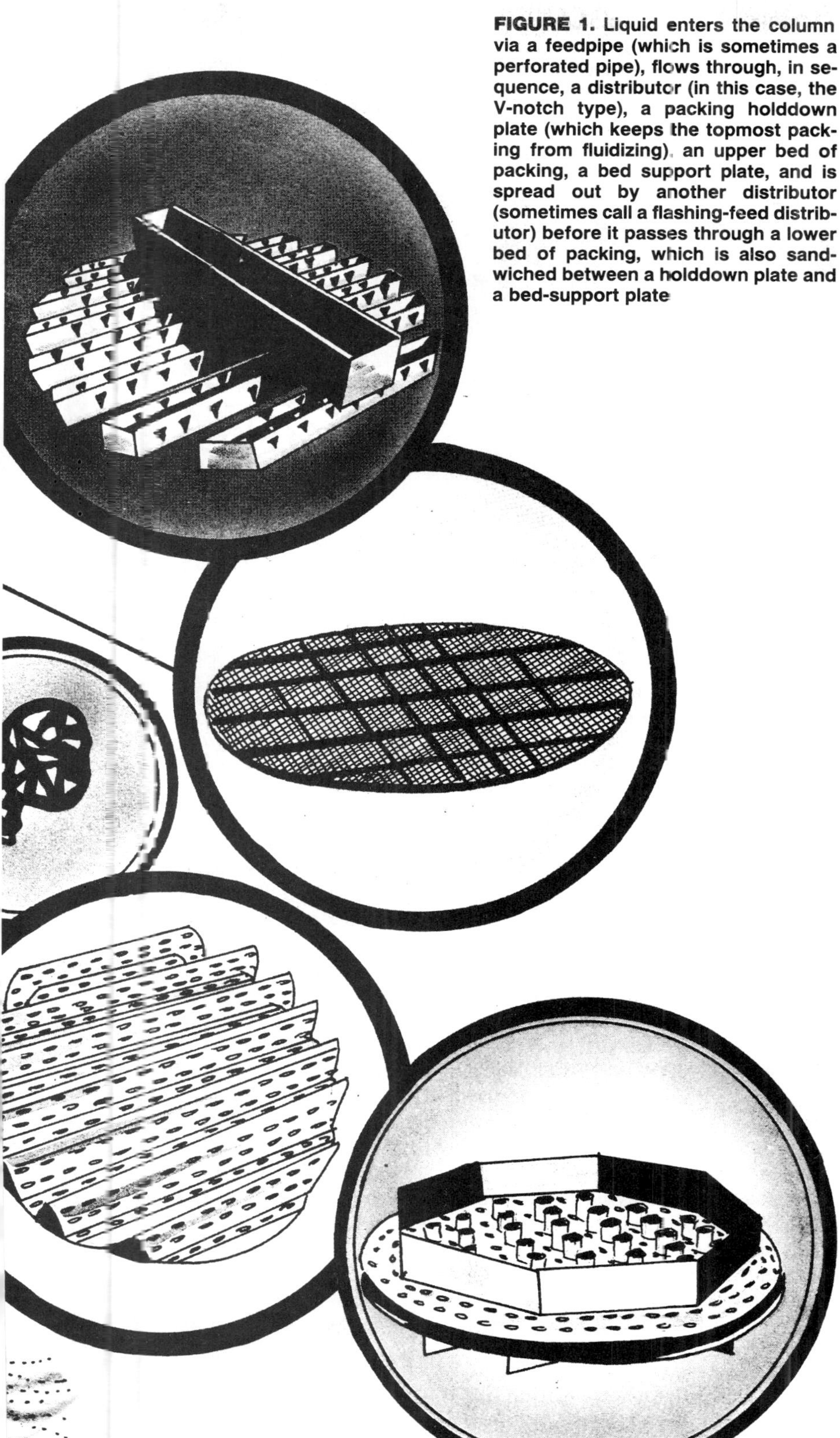

FIGURE 1. Liquid enters the column via a feedpipe (which is sometimes a perforated pipe), flows through, in sequence, a distributor (in this case, the V-notch type), a packing holddown plate (which keeps the topmost packing from fluidizing), an upper bed of packing, a bed support plate, and is spread out by another distributor (sometimes call a flashing-feed distributor) before it passes through a lower bed of packing, which is also sandwiched between a holddown plate and a bed-support plate

chemical shock and thermal cycling can disintegrate ceramic packing. Distorted packing and packing chips can reduce bed voidage, limiting column capacity.

Potential troublemakers

The liquid distributor is the single most important internal structure of a packed column. Although a packed bed could not exist without a support plate, the distributor is critical for initiating uniform contact between liquid and vapor. It, therefore, strongly influences packing efficiency. Packings with fair to poor ability to redistribute liquid will not correct poor initial distribution.

Local fluctuations in the ratio of liquid to vapor flowrates can vary component separation, as if the column were divided into many smaller ones, each operating at a different reflux ratio. In the same way, a malfunctioning distributor can undermine the performance of the entire bed. This cannot happen in a trayed column, because each tray redistributes the liquid from the tray above, minimizing the effect of a single bad tray.

The liquid distributor must spread the liquid uniformly, resist plugging and fouling, provide free space for gas flow, and allow operating flexibility. Distributors fall into two categories: gravitational — such as orifice and weir types, and pressure-drop types — such as spray nozzles and perforated pipes.

Orifice-type distributors, perhaps the most prevalent in smaller columns, consist of flat trays having risers for vapor flow and orifices for liquid flow (p. 124). Common in columns larger than four feet in diameter are troughs having bottom orifices. The feed liquid and liquid downflow from upper beds should build a uniform head over all of the orifices, so as to balance the flow from each orifice (Figure 4). The key to orifice-distributor operation is the liquid head and flow relationship that limits a distributor's liquid flowrate flexibility. The basic orifice-distributor hydraulic relationship is given below:

$$Q = 3.86nd^2H^{1/2}$$

Here, Q = distributor flow, gal/min; n = number of orifices in distributor; d = orifice dia., in.; and H = liquid head loss, in. of the liquid.

The actual liquid level on the distributor is the sum of liquid head loss

flooding starts in the transfer area or in the downcomer will depend on which area is limiting. Tray flooding is often differentiated into jet flooding and downcomer flooding.

Jet flooding occurs when the vapor passing through a tray generates a froth that carries excessive liquid into the tray above. This rapidly increases the tray pressure drop and the liquid recycle across the tray. The higher pressure drop and liquid flowrate backs liquid higher into the downcomer until liquid spills onto the tray above. The liquid continues building, flooding in order the higher trays. This is called jet flooding because the flooding is initiated by the jetting of liquid from the tray transfer area (Figure 3).

Jet flooding results from more energy being transferred from the entering vapor into the liquid on a tray than can be dissipated in the space above the tray. Most common in vacuum and low-pressure columns (in which the vapor flowrate is much higher than the liquid flowrate), jet flooding can also occur on trays having a low percentage sieve-hole area or a small number of valves or bubble caps per tray transfer area. The higher vapor velocity resulting from the reduced contacting area can initiate jet flooding.

The variables that primarily affect jet flooding are the tray vapor area, tray spacing, and vapor flowrate and density. Among the less influential variables are tray geometry, type of contacting medium, liquid flowrate and other liquid and vapor properties.

Downcomer flooding refers to flooding initiated liquid from the downcomer. Several factors contribute to setting the operating liquid level in a downcomer. The dominant one is the liquid backup that balances the pressure drop across the tray. (This same pressure drop exists across the downcomer.) The equivalent clear liquid level on a tray below, which is influenced by its outlet weir height and the weir's crest of flowing liquid, generates a "static" liquid backup into the downcomer. The head loss caused by the liquid flowing through the restricted space between the bottom of the downcomer panel and the seal pan also contributes to setting the level in a downcomer. Thus, the total liquid head from downcomer backup is the sum of the tray pressure drop, the liquid level on the tray below, and the head loss from the liquid flow under the downcomer. The liquid entering the downcomer is aerated, however, and this lowers its density, increasing the liquid height in the downcomer above that equivalent to the clear liquid head.

If any of the foregoing contributors becomes excessive (for instance, too much head loss under the downcomer or aeration due to foaming), liquid will back up in the downcomer and spill onto the tray. This will increase the tray pressure drop,

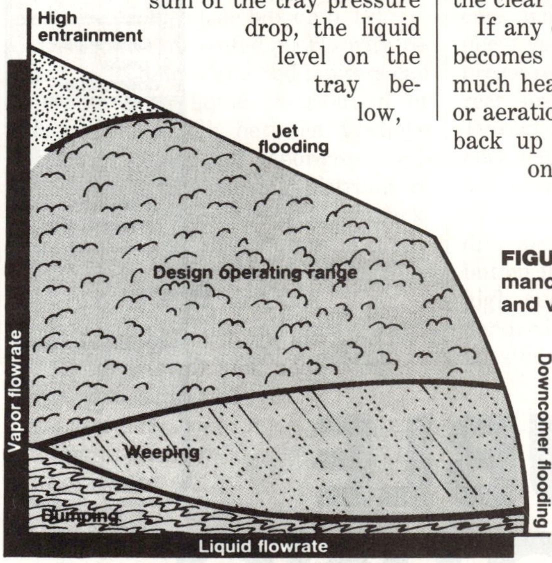

FIGURE 2. Sieve tray performance at various relative liquid and vapor flowrates

FIGURE 3. Sequence of jet and downcomer flooding in a trayed column

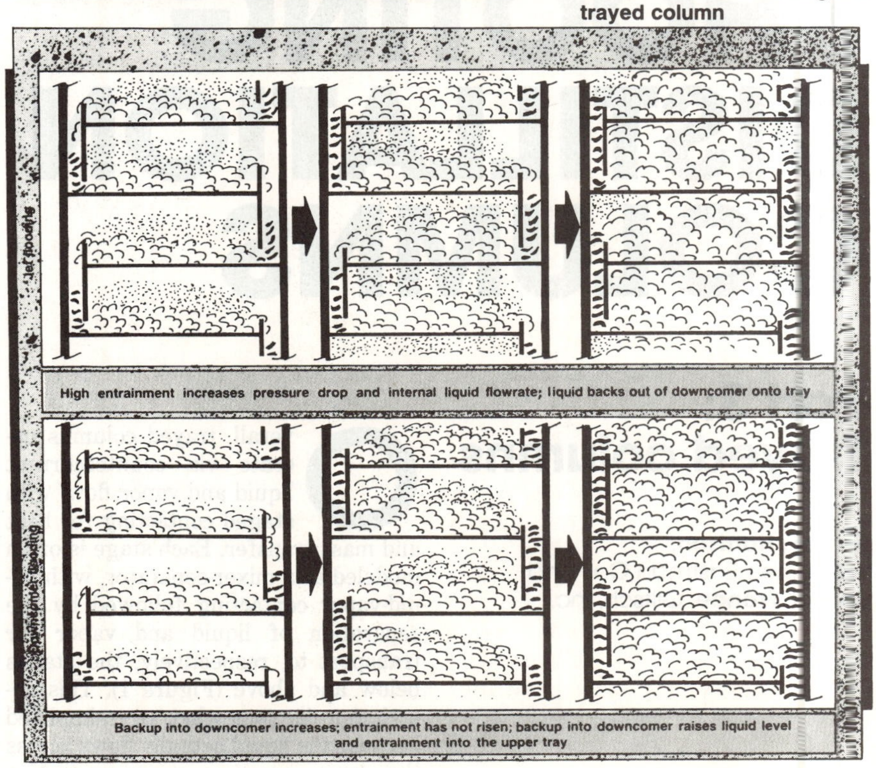

How a good turn eliminated hydraulic hammer: The case of the noisy feedpipe

Problem: A water-hammer-type pounding at the column feedpoint was violently shaking column and the connecting piping. The column was operating at about only 30% of design rate.

Troubleshooting: The location of the noise suggested a problem with the feedpipe. A check of the design drawings indicated that the feedpipe and feed sparger might be somewhat oversized, especially at the 30% feed rate. The sparger discharge-orifice velocity was calculated to be less than 1 ft/s. The feed was cool and far from its bubble point, so flashing in the sparger could be ruled out. One postulation was that feed liquid was running out of the upstream orifices, allowing vapor to enter the feed sparger through open downstream orifices, and that the condensation of this vapor in the feed sparger was causing a hydraulic hammering.

Corrective action: One solution might have been to plug some of the orifices to raise the discharge velocity to several ft/s. However, to keep velocities below 6 ft/s at design feed rates, the following remedy was implemented: the feedpipe was turned so that the discharge orifices were on top of the pipe; this ensured that the sparger remained full of liquid at low feedrates; additionally, a deflector bar was installed above the orifices to keep feed from impinging on the tray above.

Outcome: The hydraulic hammer was eliminated.

Only an inspection disclosed the problem: The case of the top-flooding column

Problem: After several months of operation, an amine stripper in a natural gas treating plant designed to remove CO_2 from a rich amine solution became hydraulically unstable, surging reflux into the overhead accumulator.

Troubleshooting: The surges and instability indicated flooding. The bottom product continued to meet specifications, suggesting that the flooding was starting near the top of the column. The relatively small pressure-drop rises at the time of surging reinforced this conclusion. Foaming became suspected, but the addition of antifoam agents brought little improvement. Raising the tower pressure also produced marginal benefit. The troubleshooter decided to inspect the tower. Opening the column revealed that the top few valve trays were coated with a 1-inch thick layer of black crystalline material, which analysis established to be mostly iron. Because the tower did not show signs of corrosion, it was concluded that the material was being carried into the plant by the natural gas stream.

Corrective action: The material was found to be readily soluble in an acid solution, so the trays were washed with acid.

Outcome: After being cleaned, the column operated as expected. Washing the column with acid once a year was found to be sufficient to maintain the column capacity.

Debris in the downcomers: The case of the clearances made too narrow

Problem: The carbon-steel valve trays of a naphtha splitter were being severely scaled, often limiting capacity. At each turnaround, the trays had to be cleaned, a difficult and time-consuming task. To minimize scaling and facilitate cleaning, the decision was made to replace the valve trays with stainless-steel sieve trays. Upon startup with the sieve trays, the column turned out to be hydraulically unstable. The overall pressure drop was erratic and high, and the column did not properly separate the key components. Only reducing the boilup rate to a fraction of design rate kept the tower stable.

Troubleshooting: The column pressure drop indicated flooding. Because the pressure drops in the stripping and rectification sections could not be gauged separately, it was not possible to locate the floodpoint. The drawings revealed the following: the flow from reboiler jetted directly into the seal pan; an inlet weir had been placed on the feed tray; the weir height and clearance under the downcomer had been lowered to provide greater turndown (calculations indicated the 1-in. downcomer clearance specified was more than adequate); and a reboiler overflow weir had been raised to within seven inches of the reboiler return nozzle, and this was possibly limiting the disengagement space.

The reflux rate was lowered until the column became hydraulically stable. Raising the reflux only a little made the column unstable. This sensitivity to liquid rate suggested a liquid flow-path problem, such as downcomer flooding. Feed to the column could be hiked if the total reflux was left unchanged. This indicated flooding in the rectification section. Suspecting a problem with downcomers in the rectifying section, the troubleshooter checked to see if the design tray pressure drop could be a significant contributor to downcomer backup, and found the pressure drops to be reasonable. Unable to attribute the stability problem to a definite cause, the troubleshooter recommended an inspection. The inspection revealed that the clearance under the downcomers in the rectifying section ranged from ¾ inch to ⅝ inch, vs. the 1 inch specified. Additionally, some of the downcomers were completely filled with loose scale and fiberglass tray gasketing material. The clearance problem was mainly attributed to scale left on the tray support rings during the installation of new panels that raised the height of the tray panels. Thus, tray improper installation restricted downcomer clearance, which limited the liquid throughput and caught the larger pieces of scale and loosened tray gasketing.

Corrective action: The trays were removed, the support ring cleaned, the trays reinstalled, and the gaskets omitted. The inlet weir on the feed tray was also left out. To eliminate potential problems at higher rates, the reboiler return was deflected away from the seal pan, and the reboiler overflow weir was lowered.

Outcome: Restarted, the column operated properly at design capacity. The switch to sieve trays reduced the frequency of tray cleaning to remove scale.

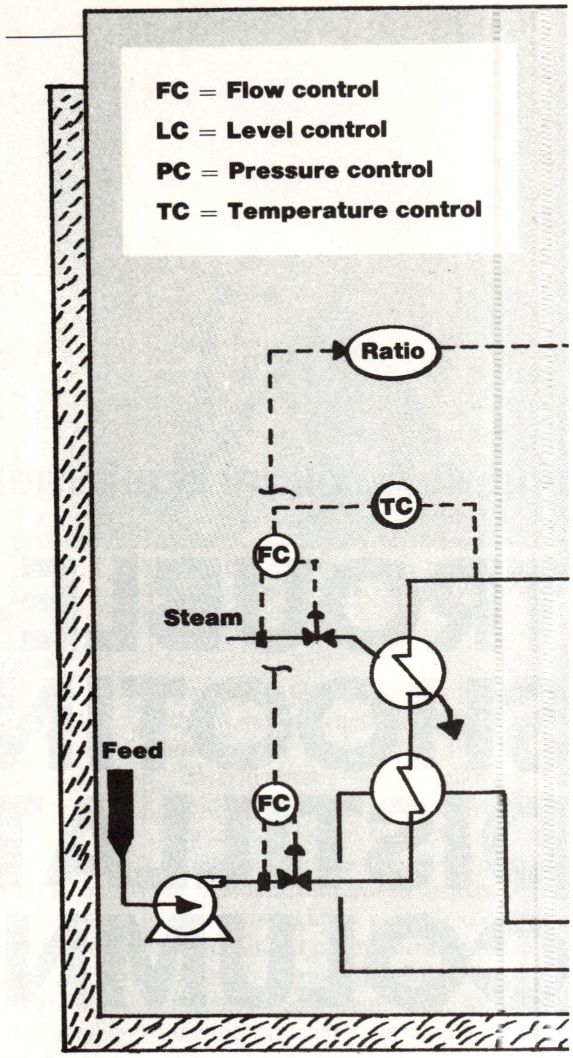

Heat exchangers, vacuum pumps or ejectors, process pumps and instrumentation represent some of the critical components of distillation columns (Figure 1). Often, the problems of the column and its auxiliary equipment are inseparable. A problem with one item of the auxiliary equipment can impose a limit on a column's capacity or separation ability, or both. For example, an undersized reflux valve may constrain the reflux rate, or a fouled reboiler may suppress column boilup.

The troubleshooter can pinpoint many problems with the auxiliary equipment by comparing the actual state of critical variables against values for design, or normal, operating conditions (for example, the ratios of steam rate or reflux rate to feedrate) and noting signs of control limitations (such as a wide open control valve).

How reboilers fail

Reboilers are generally classified as natural or forced circulation. The first includes vertical and horizontal thermosiphon types, kettle reboilers, and internal tube bundles (Figure 2). In these reboilers, the density difference between the entering liquid and the exiting liquid-vapor mixture causes the process fluid to circulate through the tubes or the shell. In the case of forced circulation reboilers, a pump circulates liquid the through an external heater or partial vaporizer.

Some problems common to reboilers: fouling, steam-condensate flooding,

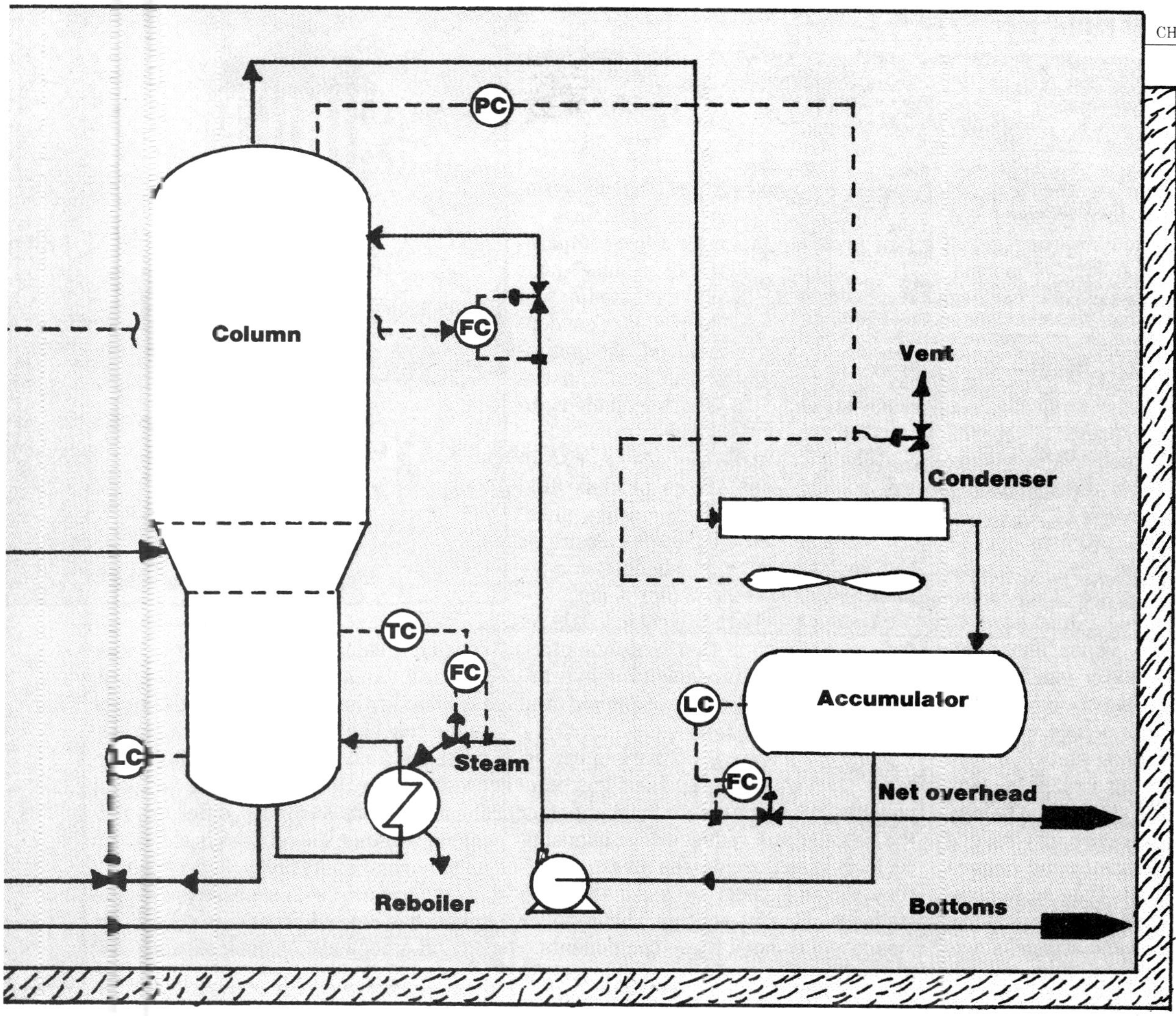

Figure 1. Malfunctioning auxiliary equipment often undermines column performance

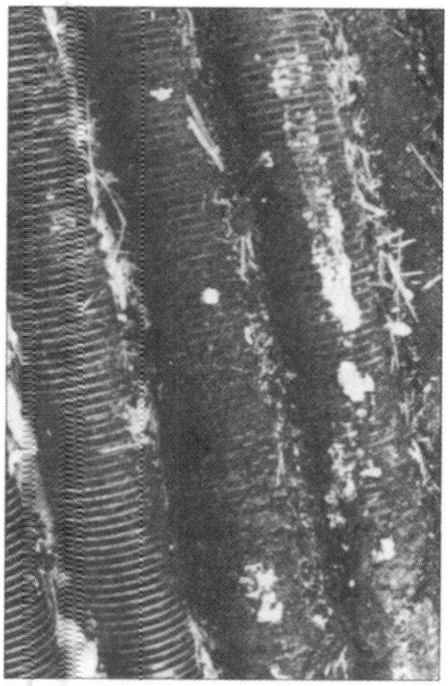

Product that crystallized during the shutdown of a distillation column decorate the fouled finned tubes of a reboiler

steam-condensate-seal blowing, film boiling, heating-medium inadequacy, boiling-point elevating, foaming, inert blanketing, leaking and undersizing.

Fouling — The vaporization of process fluids at tube walls can precipitate dissolved solids onto the tubes. Tar-like materials in the feed may plate out on metal surfaces. Tube-wall temperatures (which are higher than bulk-fluid temperatures) may polymerize or degrade hydrocarbons, which subsequently settle on tube surfaces. Corrosion products also can foul heat-transfer surfaces. Such occurrences will reduce heat transfer and restrict liquid circulation. (See the example of tube fouling in the photograph to the left.)

Vaporization-induced deposition is most severe in kettle reboilers, because they often operate near total vaporization. Fouling is particularly likely if the liquid level drops below the top of the tube bundle. Forced-flow reboilers are the least prone to fouling because they are frequently designed to operate at high liquid circulation rates and a restricting valve or orifice may be installed in the outlet piping to suppress vaporization. The higher fluid velocity also tends to keep deposits from adhering to surfaces.

A drop in the value of a heat-transfer coefficient is a reliable indication of fouling. A gradual rise in the steam condensing pressure (and temperature) needed in steam-heated reboilers to maintain heat input usually suggests the onset of fouling.

Steam-condensate flooding — Condensate backing into an exchanger submerges surface area, reducing the area available for heat transfer. A malfunctioning steam trap (e.g., plugged or incorrectly installed), or high pressure in the condensate-collection header can cause condensate to back up (Figure 3).

One quick check for condensate flooding is to open a condensate blow-

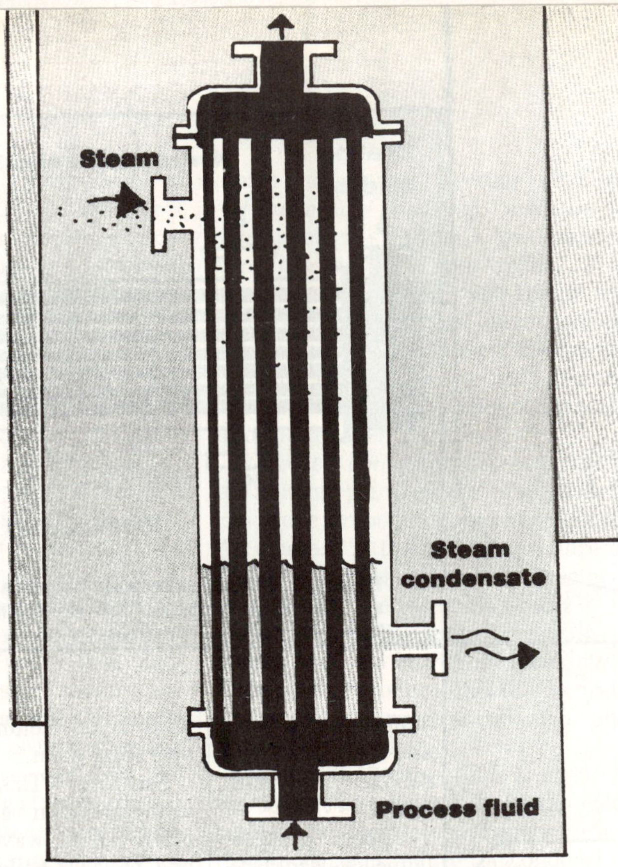

Figure 3. Backed-up condensate reduces heat transfer

reboiler may "stall" — i.e., stop boiling. The dumping of low-boiling components into the column base may lower the base temperature. However, because these components will not enter the reboiler, circulation will not restart. To combat this reboiler "stalling," first lower the base level, then increase it, so that the reboiler will be replenished with low-boiling liquid and can resume normal operation.

Vaporization-induced fouling can also lead to surging reboiler operation. If the column feed contains a dissolved solid that is being concentrated close to its precipitation point, precipitate may foul tube surfaces or plug tubes, restricting heat transfer to the point that the steam flow falls. With reduced vaporization, solvent-rich liquid dumped to the base of the column may dissolve the precipitates, restoring bottoms-flow circulation, until the cycle repeats.

Liquid temperature and distribution — Poor liquid inlet distribution across a horizontal tube bundle can cause liquid to stagnate and low-boiling components to become depleted in parts of the bundle. The blanketing of the bundle, with the high-boiling liquid and the temperature pinch that results, reduces heat transfer. If the problem is severe, consider retrofitting the reboiler with multiple liquid inlets. Briefly lowering, and then restoring, the liquid level may displace the stagnant liquid.

Kettle reboilers

This reboiler presents problems in vapor disengagement, residue accumulation and level control and distribution.

Vapor disengagement — If head room is not provided in the kettle reboiler for disengaging vapor from the liquid, excessive entrainment can choke the vapor return line, causing hydraulic instability and the pressure (and liquid boiling point) in the reboiler to rise. Because the kettle reboiler's disengagement space is so much smaller than that in the column, this reboiler is not recommended for boiling foamy liquids.

Residue accumulation — The usual overflow baffle in a kettle reboiler can trap solids in the shell, fouling tube surfaces. A nozzle for a slip stream can prevent such accumulation.

Level control and distribution — A low liquid level accelerates vaporization-induced fouling of the uncovered tubes. An overflow baffle that sets the liquid level over the tube bundle will prevent this. A high liquid level floods the vapor-disengagement space, causing unstable hydraulics and excessive backpressure. As in horizontal reboilers, poor inlet liquid distribution can result in liquid stagnation and high-boiling-liquid blanketing of heat-transfer surface area (Figure 4).

Forced-flow reboilers

Forced-flow pump circulation suppresses vaporization in a reboiler, reducing fouling. The higher circulation velocity also curtails fouling. Forced-flow reboilers are less sensitive to column liquid level because circulation does not depend on density difference.

Although vaporization is often suppressed by an outlet valve or orifice, a large drop in flowrate (hence, pressure drop) may promote vaporization, which can cause hydraulic problems or vapor bind the heat-transfer area, especially in multiple tube-pass reboilers. If va-

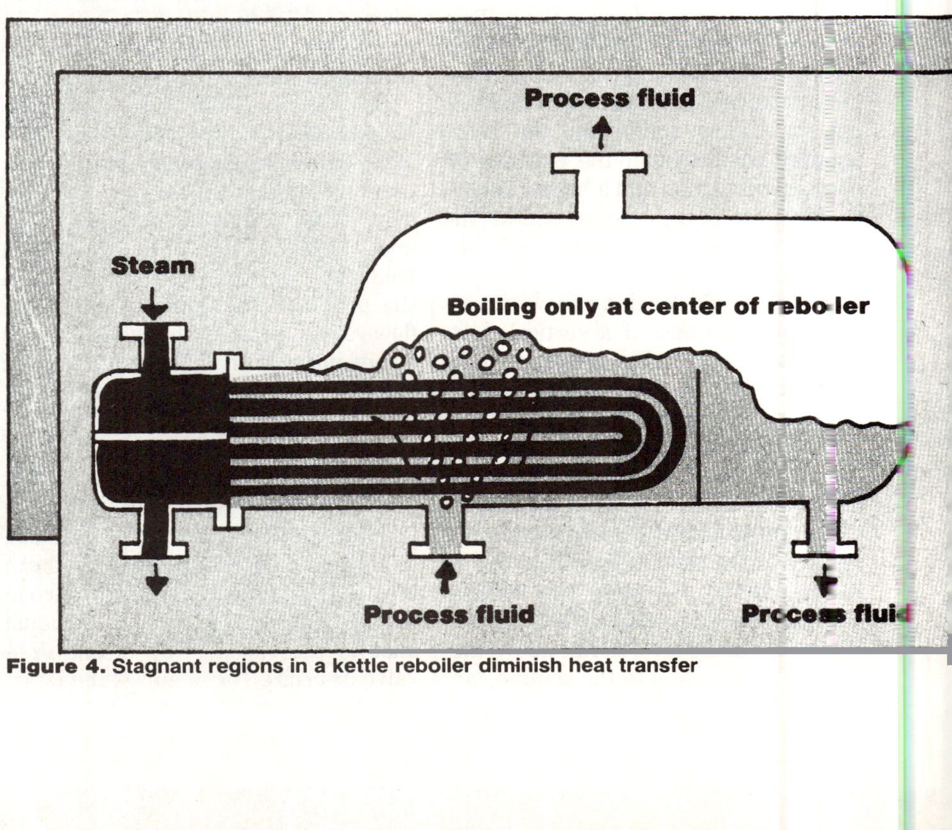

Figure 4. Stagnant regions in a kettle reboiler diminish heat transfer

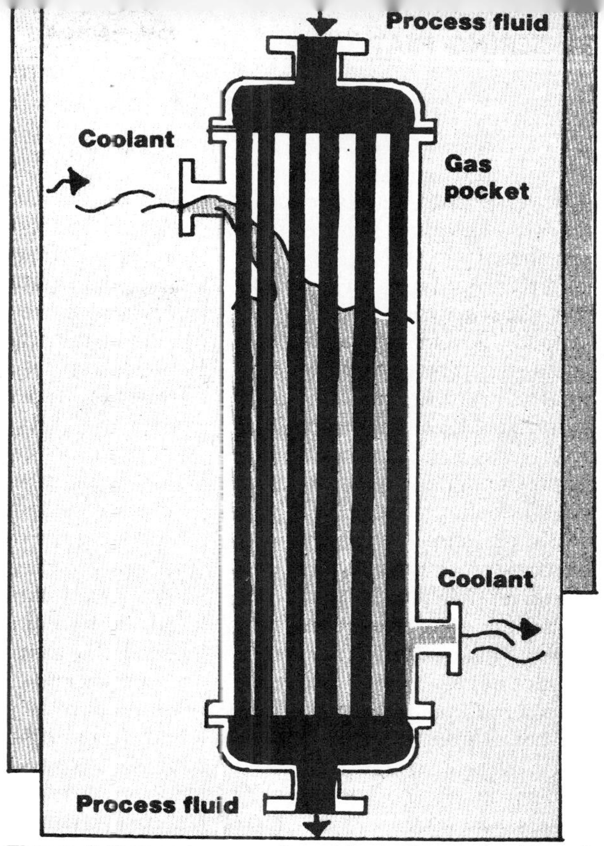

Figure 5. Trapped noncondensable gas lowers heat transfer

porization is suppressed, a reduction in flowrate will also boost the process temperature rise, sometimes pinching the temperature driving force for heat transfer.

Overhead condensers

Column overhead vapor is most often condensed by a water- or air-cooled exchanger. Alternatives include the economizer (in which a process stream serves as the coolant), the steam-generating condenser (in which thermosiphon steam condensate serves as the coolant), and the heat pump (which provides heat for the reboiler).

Typical condenser problems include: fouling, low coolant flow, vapor binding, inert blanketing, vent system overloading, condensate flooding, leakage and undersizing.

Fouling — Condensers are less likely to be fouled than reboilers because the mass transfer in the column keeps solids out of the overhead stream. The washing action of the process condensate usually keeps the condenser tubes clean. Corrosion products or material sublimation can cause fouling. Process condensate can freeze if the coolant temperature is too low, such as when a process upset allows high-boiling components to be driven overhead. Such a freeze-up can plug the condensate drain and flood the condenser, coat the tubes and reduce heat transfer, or restrict vapor flow.

Coolant-side fouling is a common problem. Low water velocity, especially on the shellside, can allow silt to accumulate, and a high water-outlet temperature abets mineral deposition on heat-transfer surfaces. Additionally, algae or fungus can foul or plug the exchanger. Regular backwashing can minimize some of these problems. Of course, the extent of fouling can be estimated from the change in the heat-transfer coefficient.

Low coolant flow — As mentioned, a rise in the temperature of the outlet water of a water cooled exchanger may indicate reduced coolant flow. A high coolant inlet temperature will, of course, limit condenser capacity. The temperature can vary seasonally, and even with the time of day, particularly with air-cooled condensers. A poorly located air cooler can recirculate its air, curtailing its capacity, or exhaust its air into another air cooler, limiting the latter's capacity.

Vapor binding — Vapor or air pockets can block off heat-transfer area from the coolant, especially at startup. This is more likely to occur when the coolant leaves from the bottom of the shell (Figure 5).

Inert blanketing — Inert gases — such as may flow in through instrument purge lines, and through pipe fittings if the column operates under vacuum, or be present in the column feed — can reduce the condenser's heat-transfer coefficient. The problem is more severe near the exchanger outlet, where a large concentration of inert gas can create a mass-transfer resistance to vapor condensation and may even block off the flow of vapors, if the inert gas is not vented. Condensers that are operated flooded for pressure control are prone to accumulating inert gas, if the gas is not shunted via a bypass.

Vent-system overloading — If a condenser becomes overloaded, a backpressure may build up from the vent line or vent condenser, which are usually sized for small loads. A hot vapor flow from an overloaded or poorly performing condenser can be detected by measuring the vented gas temperature.

Condensate flooding — As can steam condensate in reboilers, process condensate can back up into a condenser and cover the heat-transfer surface. Poor baffle orientation in a horizontal condenser can similarly flood the compartments. Excessive entrainment from the column can overwhelm the condenser's process drain, which should be sized for gravity flow to avoid liquid backup.

Leakage — In multiple-tube-pass exchangers, excessive clearance between the tubesheet and the pass-partition baffles can let coolant bypass of portions of the tube bundle, reducing the coolant flow through the tubes and the effective temperature difference for condensing. Common causes of excessive clearance include corrosion, erosion, and bowed tubesheets. The last can be caused by excessive pressure or a thermally-induced tube contraction, which can occur in a condenser having an inadequate shell expansion joint or one that is subjected to temperature differences between the shell and tubes that exceed the design limits.

Undersizing — A condenser that cannot handle the overhead vapor will operate at a higher-than-desired pressure drop or with excessive venting. Sometimes, an inadequate heat-transfer area in a shellside liquid-cooled condenser can be compensated for by changing the condenser's heads and pass-partitions to reduce the number of tube passes, boosting the coolant flow and the temperature driving force. Shellside modifications to reduce the pressure drop are more difficult, because baffle cuts and baffle and tube spacing are difficult to alter.

Vacuum equipment

Air leaking in through fittings and vessel walls (due to the porosity of the metal) and noncondensable gases entering in the feed stream can overload a vacuum system, diminishing condenser performance. Of course, a drop-off in condenser performance will also overload the vacuum system with condensable vapors. Insufficient sealant flow to a liquid-ring vacuum pump can cause the pump to overheat and flash the seal fluid, curtailing the pump's capacity, as can worn seals and bearings. Diffuser throat erosion, intercondenser failure, inadequate cooling water, and low motive-steam pressure can undermine the performance of a vacuum steam jet.

An interruption in the operation of a jet or vacuum pump can result in the complete or partial loss of vacuum. A check valve or seal leg will prevent the loss of column vacuum due to vapor flowing back through the vent system. However, the column boilup will continue to push the normal inert load into the condenser, where it can quickly blanket the tubes and fill the exchanger. The column pressure will begin to rise to the limit capped by liquid boiling temperatures and the temperature of the reboiler heating fluid.

When returned to service, a vacuum pump that may have taken hours to evacuate the inert gases from the column at startup will almost instantly evacuate the few cubic feet of inert gas blanketing the condenser surface area, restoring process condensing, almost immediately regaining the vacuum. This sudden drop in column pressure, compounded by the column having served as a heat sink at the higher, upset pressure, will produce an excessive boilup surge that can damage column internal structures. This surge can be minimized by manually opening the vacuum-control recycle stream before returning the vacuum pump or jet to service and slowly ramping to the desired pressure.

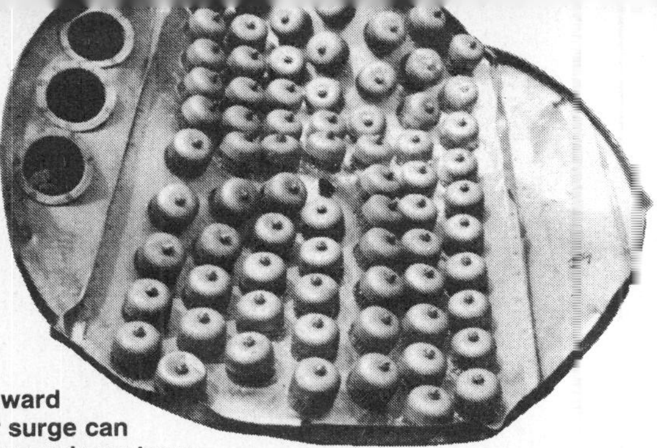

Downward vapor surge can collapse column trays

A sudden loss of vacuum from the top of a column may let vapor flow back through the column. Because of the presence of liquid on column trays, the vapor downflow will force liquid back through the trays at a high differential pressure. The force of this downflow can collapse trays (such as the bubble-cap tray in the photograph above) — and especially valve trays, which by design close against downward flow.

Other equipment

Undersized or cavitating pumps will limit column capacity. Leaking seal fluid can introduce contaminants into the column. Pump discharge pressures should be checked periodically to ensure that impellers have not eroded or suction screens have not become plugged.

Pipes can be plugged by debris from a column (such as packing or loose valves) and process deposits. Vent lines may have low segments that can become filled with condensed vapors and cause backpressure problems. A field check with the piping and instrument drawings can help identify problems, such as inadequate liquid heads and seals, improper routings and open bypass valves.

Pressure changes in upstream or downstream columns, or other equipment, may limit pump capacity. The composition of the feed from a reactor to a column may vary, upsetting column performance. Unexpected or inconsistent byproduct formation can also undercut the column separation.

Variations in the temperatures of fluids from preheaters and interchangers may upset a column's heat balance. Overheating a liquid feed may flash it, and render it incompatible with the column's piping or internal arrangements. Reflux and feed decanters can become upset, and let impurities enter the column. Surges of low-boiling components (common if a column upstream becomes upset) can overwhelm the condenser and vacuum system.

The case of the inconsistent column heat balance and control strategy

Problem: A packed column was designed to remove low-boiling components from much higher-boiling products for recycling to an upstream operation. Some of the products contained a varying mixture of high-boiling components (which caused the base temperature to fluctuate), the reboiler steam was flow controlled instead of base temperature controlled. Some high- or intermediate-temperature boiling components were allowed to pass overhead. Occasionally, the operator trimmed the steam flow to control the amount of high-boiling components taken overhead. At design feedrates, the column operated as expected, but it flooded when the feed was shut off.

Troubleshooting: Because the plant was new, the troubleshooter checked the design basis for the column. The design heat and material balances indicated that the feed was cold, requiring about 40% of the reboiler heat duty just to heat the high-boiling components in the feed to the underflow temperature. The column was sized for a vapor boilup corresponding to the remaining 60% of the reboiler duty.

The troubleshooter realized that flooding occurred when the feed was reduced or shut off only after operation near to the reboiler design steam flowrate. Because the reboiler steam was flow controlled, shutting off the feed removed the heat sink for 40% of the reboiler duty. The entire reboiler duty would then translate into boilup equivalent to 166% of the column design capacity. Thus, shutting off the feed eliminated a significant heat sink, causing excessive boilup into the column.

Corrective action: An upper limit to the column pressure drop was established, based on operation short of flooding. The distributed control system was modified to manipulate reboiler steam flow to maintain a column pressure drop below the limit. This would provide indirect steam flow control during normal operation (i.e., with feed), but would reduce the reboiler steam flow to curtail the higher boilup when the feed is reduced or shut off.

Outcome: Reducing the feed no longer started column flooding.

Instrumentation

Many sensing elements must be installed through nozzles located at specific elevations and orientations with respect to a distillation column's internal structures. For example, pressure transmitters are normally located in a vapor space. Similarly, temperature sensors for composition control should be placed by a tray that will provide sensitivity to composition changes. Sensors also should be designed and constructed to avoid corrosion, fouling and plugging.

Transmitters improperly calibrated will produce erroneous readings, which will create problems or confuse troubleshooting efforts. The troubleshooter should always consider the possibility that readings may be false and verify instrument measurements, especially those that deviate from normal.

An incorrectly calibrated level instrument is often the initiator of column flooding. For example, if an instrument for gauging the base level of a column has been calibrated for a liquid having a density of 64 lb/ft^3 but is sensing a liquid having a density of 53 lb/ft^3, an indicated level of 4 ft would actually be over 7 ft. Thus, the instrument could raise the liquid level higher than the reboiler return nozzle and even into the column, causing entrainment and subsequently flooding.

A similar problem will occur if the liquid level is allowed to rise above the vapor tap. Because there will be a constant head of liquid between the two taps, any addition to the base level will not be detected. If the density of the calibration liquid is greater than that of the actual liquid (as in the foregoing example), the level transmitter will range out at less than 100%, conveying a misleading indication. This problem can be detected by noting a suspiciously constant level indication.

Transmitter legs that are open to process fluids often have to be purged or traced to keep liquid or condensate from accumulating in them and giving false readings. Some reasons why a control valve may not respond satisfactorily, besides being too large or too small, include having a sticking valve stem, an inadequate air supply, a malfunctioning valve positioner and a leaking diaphragm. ■

Additional reading for this four-part troubleshooting report

Part 1: General troubleshooting
(March 1989, pp. 116-123)

1. Drew, J. W., Distillation Column Startup, *Chem. Eng.*, Nov. 14, 1983, pp. 221-226.
2. Frank, O., Shortcuts for Distillation Design, *Chem. Eng.*, Mar. 14, 1977, pp. 111-128.
3. Hausch, D. C., How Flooding Can Affect Tower Operation, *Chem. Eng. Prog.*, October 1964, pp. 55-59.
4. Kister, H. Z., How to Prepare and Test Columns Before Startup, *Chem. Eng.*, Apr. 6, 1981, pp. 97-100.
5. Kister, H. Z., Inspection Assures Troublefree Operation, *Chem. Eng.*, Feb. 9, 1981, pp. 107-109.
6. Lieberman, N. P., Basic Field Observations Reveal Tower Flooding, *Oil & Gas J.*, May 16, 1988, pp. 39-41.
7. McLaren, D. B., and Upchurch, J. C., Guide to Trouble-Free Distillation, *Chem. Eng.*, June 1, 1970, pp. 139-152.
8. Vital, T. J., Grossel, S. S., and Olsen, P. I., Estimating Separation Efficiency; Part 1: Introduction, *Hydrocarbon Proc.*, October 1984, pp. 55-56.

Part 2: Packed columns
(April 1989, pp. 121-123)

1. AIChE, "Packed Absorption and Distillation Columns, AIChE Equipment Testing Procedure," American Inst of Chem. Engrs , New York, 1965.
2. Bolles, W. L., and Fair, J. R., Improved Mass-Transfer Model Enhances Packed-Column Design, *Chem. Eng.*, July 12, 1982, pp. 109-116.
2. Chen, G. K., Packed Column Internals, *Chem. Eng.*, Mar. 5, 1984, pp. 40-51.
3. Eckert, J. S., How Tower Packings Behave, *Chem. Eng.*, Apr. 14, 1975, pp. 70-76.
4. Fadel, T. M., Selecting Packed-Column Auxiliaries, *Chem. Eng.*, Jan. 23, 1984, pp. 71-76.
5. Kunesh, J. G., Practical Tips on Tower Packing, *Chem. Eng.*, Dec. 7, 1987, pp. 101-105.
6. Vital, T. J., Grossel, S. S., and Olsen, P. I., Estimating Separation Efficiency; Part 3: Packed Columns, *Hydrocarbon Proc.*, December 1984, pp. 75-78.

Part 3: Trayed columns
(May, 1989, pp. 126-133)

1. AIChE, "Tray Distillation Columns, AIChE Equipment Testing Procedure," American Inst. of Chem. Engrs., New York, 1987, .
2. Anderson, A. E., and Jubin, J. C., Case Histories of the Distillation Practioner, *Chem. Eng. Prog.*, October 1964, pp. 60-63.
3. Bolles, W. L., Estimating Valve Tray Performance, *Chem. Eng. Prog.*, September 1976, pp. 43-49.
4. Jamison, R. H., Internal Design Techniques, *Chem. Eng. Prog.*, March 1969, pp. 46-51.
5. Kister, H. Z., Design and Layout for Sieve and Valve Trays, *Chem. Eng.*, Sept. 8, 1980, pp/ 119-123.
6. Kister, H. Z., Downcomer Design for Distillation Tray Columns, *Chem. Eng.*, Dec. 29, 1980, pp. 55-60.
7. Kister, H. Z., Guidelines for Designing Distillation Column Internals, *Chem. Eng.*, May 19, 1980, pp. 138-142.
8. Kister, H. Z., Mechanical Requirements for Sieve and Valve Trays, *Chem. Eng.*, Nov. 17, 1980, pp. 283-285.
9. Lieberman, N. P., Checklist is Key to Dealing with Distillation Troubles, *Oil & Gas J.*, Sept. 10, 1979, pp. 121-125.
10. Thorngren, J. T., Valve Tray Flooding Generalized, *Hydrocarbon Proc.*, August 1978, pp. 111-113.
11. Vital, T. J., Grossel, S. S., and Olsen, P. I., Estimating Separation Efficiency; Part 2: Plate Columns, *Hydrocarbon Proc.*, November 1984, pp. 147-153.

Part 4: Column auxiliaries

1. Shah, G. C., Troubleshooting Distillation Columns, *Chem. Eng.*, July 31, 1978, pp. 70-78.
2. Yokell, S., Troubleshooting Shell-and-Tube Heat Exchangers, *Chem. Eng.*, July 25, 1983, pp. 57-74.

Acknowledgments

The authors wish to express their appreciation to John Aycock and Peter Lodal for their editorial contributions, and to J. R. Winter, Jim Dorton and again to John Aycock for supplying useful photographs. They are all with the Tennessee Eastman Co.

The authors

Mark E. Harrison is a senior chemical engineer with Tennessee Eastman Co. (P.O. Box 511, Kingsport, TN 37662. He works in the Process Systems Engineering Dept. of the Engineering Div., where he specializes in distillation column design, troubleshooting and debottlenecking. He has been involved in the commissioning of more than 15 distillation systems. Currently Eastman's representative in the University of Texas' Separations Research Program, he has also served as the company's alternate representative to Fractionation Research, Inc. (FRI). A recipient of B.S. and M.S. degrees in chemical engineering from Tennessee Technological University, he is a registered engineer in the State of Tennessee and a member of AIChE.

John J. France is Southeast Regional Manager for Glitsch, Inc. (8254 One Calis Ave., Suite 250, Baton Rouge, LA 70809; tel.: 504-769-7612). He provides technical assistance on distillation systems. A holder of B.S. degrees in education and chemical engineering and an M.S. degree in chemical engineering, all from New Mexico State University, he has worked for Conoco, Inc., El Paso Natural Gas Co., and Southern Union Refining Co. A registered engineer in the State of New Mexico, he is a member of the New Mexico Hazardous Waste Management Soc.

Section VII

Insuring Against Risks and Liabilities

DO NOT LET A CRISIS CATCH YOU OFF GUARD

STEPHEN WALSH

An up-to-date response plan must involve employees and the community

JoAnna J. Wagschal
Environmental Science & Engineering, Inc.

Unfortunately, there is no foolproof way to prevent a crisis. However, armed with a crisis-communications plan, your firm will be better able to recover from a disaster, minimize the loss of life and property, and resume normal business operations in less time. A well-prepared crisis-communications plan will also help you remain organized in the face of potential disaster.

To be successful, such a plan must be updated periodically. As a manager, you need to recognize the most common weaknesses that undermine most crisis-communications plans, the regulatory requirements that provide the public with the right to involvement and information, and the steps involved in reviewing your firm's existing plan.

Common weaknesses

To minimize vulnerability, an existing crisis-communications plan must be reviewed at least every four years, to update the list of corporate spokespersons and media contacts, and to accommodate any changes in technology, equipment or plant layout that may impact emergency preparedness. Plans should be updated within two years if any significant regulatory or operational changes have impacted the facility or the industry.

You will need to keep the plan practical, useable and only as detailed as necessary. It will help to review similar crisis-communications plans from other organizations to get a better idea of the possible structure and level of detail included.

There are common weaknesses affecting many crisis-communications plans. The following tasks are recommended to avoid them:

- Appoint and train local and backup spokespersons
- Install backup communication power and equipment
- Conduct practice drills
- Maintain community relations
- Establish formal communication channels among your employees, and with the community
- Conduct site-specific training and education for local emergency-response crews
- Clarify the role of various company officials during crisis response
- Establish a database of operations and environmental information for the facility

Regulatory requirements

It is important to review federal, state and local regulations that specify requirements for emergency-response communications or operations. In the U.S., for example, the Comprehensive Environmental Response, Compensation, and Liability Act (CERCLA) requires public involvement during removal actions and remedial responses. Similarly, the Resource Conservation and Recovery Act (RCRA) requires public involvement during permitting and corrective-action procedures.

On October 20, 1993, the U.S. Environmental Protection Agency (EPA; Washington, D.C.) released a proposed rule under Section 112R of the 1990 Clean Air Act Amendments. It requires the devel-

Originally published September 1994

If EPA'S proposed Rule112R is adopted, affected facilities must:

- Submit a risk-management plan to the regulatory agency, provide copies to the Chemical Safety and Hazard Investigation Board and local planning commission, and make copies available to the public
- Conduct hazard assessments
- Perform routine process-hazard analyses
- Generate process safety information

and a formal risk-management plan must include:

- A summary of consequence analyses (including worst-case and other scenarios)
- A five-year history of significant accidental releases for each regulated substance (including the concentration, size, and duration of releases)
- A list of major hazards defined in the process hazard analyses
- A prevention program documenting safety precautions, maintenance and monitoring procedures, and training for employees
- An emergency-response program that specifies certain actions in response to a release, to protect human health and the environment (including informing the public and local agencies, and providing emergency health care, and employee training)

opment and implementation of risk management programs at facilities that manufacture, process, store, or handle regulated substances in quantities that exceed specified thresholds (box).

Review and update the plan

Six steps must be incorporated into the revision process:

- Review past incidents
- Review contact lists
- Identify key issues
- Draft crisis scenarios
- Develop practice drills
- Integrate the crisis-communication plan with the emergency-response plan

The first step is to *review past incidents* that have occurred at your plant or similar facilities. This may reveal sections of the plan that have been overlooked, or areas that need to be better developed. Be careful not to sidestep critical issues with the rationalization "we won't do that again because it was a stupid mistake."

Knowledge gained from previous incidents should be incorporated into a crisis-communication plan. Ignoring past mistakes will make it more difficult to learn from them, particularly if your organization has a high turnover rate.

When reviewing past incidents, analyze the chronological sequence of each event. This may reveal time-critical elements of the crisis. Many crises occur after normal work hours, when relatively few staff or managers are around.

Every crisis-communications plan should include a plan to coordinate action strategies with communication strategies. Action strategies are preventive measures or planned operational responses to a crisis. Communication strategies detail how, and with whom, the organization will communicate during a crisis.

During this review of past events, internal and external factors that may have affected the organization's ability to respond to a crisis in the past must be reviewed. Face-to-face interviews should be conducted with corporate executives, site managers, and supervisors to obtain information on additional vulnerabilities that should be addressed in the plan. An assessment of community involvement may uncover perception problems in the surrounding community. This will allow you to address these problems before a crisis occurs.

Involve community members, particularly emergency-response personnel from hospitals, fire departments, and ambulance crews. Allow them to review corporate actions and planned responses to an emergency. They will provide managers with an invaluable perspective during the finetuning of a facility's emergency-response plan.

Next, *review the contact lists* in the crisis-communications plan. Each person on the media list should be contacted to verify that the title, address, and telephone number have not changed. This task will offer an opportunity to reestablish or develop a relationship with the press.

All the other contact lists should be updated as well. This includes a list of elected and regulatory officials, emergency telephone numbers (fire department and hospitals), and the notification list for company personnel.

Identify key issues that may create considerable exposure or liability for your organization. These include issues that are likely to:

- Affect large numbers of people or many audiences
- Harm people or the environment
- Interrupt or halt operations
- Affect property values or businesses
- Violate community beliefs and opinions

With each key issue you need to document who are the potential audiences or communities. This listing of key issues, and the potential audience for each, should be reviewed by corporate and facility personnel to make sure that an audience wasn't overlooked.

Draft crisis scenarios for each of the key issues identified in the previous step. Imagine yourself walking through the scene. What's happening and when? Write out the the events that are likely to take place in chronological order. Again, a review of previous incidents, at yours or other sites, may be helpful.

Identify the key issue and describe the situation and your analysis. Then generate at least three alternatives that may have been instrumental in managing the situation more quickly and effectively. Evaluate each alternative to determine its consequences.

Make sure the scenarios are specific. If the scenarios aren't specific, they will not be effective in preparing your organization for a crisis.

Develop practice drills for the crisis scenarios. Pay close attention to the top two key issues — the one most likely to happen, and the one that will affect the most people. Practice drills, while time consuming to prepare for and carry out, may uncover unforeseen consequences that the hypothetical draft scenarios may have been overlooked.

Once the draft scenarios have been established, priority must be given to employee training and practice drills. To be as close to reality as possible, and to emphasize the importance of responding to the element of surprise, the practice drills should be a surprise.

During the post-drill evaluation, answer the following questions:

- Did we respond in a timely manner?
- Were all the contacts needed listed in the plan?
- Were the logistics organized? For instance, were there enough people to answer the phone, was a briefing room established, was there site security?
- Did employees know the specifics of the plan?
- What audiences were overlooked?
- Were emergency information and background materials readily available?

This is the time to judge employee reaction. Do the employees need more equipment or support to manage their emergency-response roles effectively?

A background information kit, prepared ahead of time, should be made available to the media in the event of a crisis. The kit should contain information on the organization's history, the type of work done or product produced at the facility, the size of the facility and number of employees. It should also include plant photos, and a description of any training, environmental and safety programs currently in place.

Most chemical process operations have an emergency-response plan in place. To be most effective, the crisis-communication plan must be integrated with the emergency-response plan. Without organized communication procedures, any crisis can grow larger, gain more media attention, and become less manageable.

To integrate both plans, a concise index should be developed to cross-reference the relevant sections. Each section of each plan should be succinct, logical and informative.

By scrutinizing both plans together, it may become apparent that certain sections will be more effective and less redundant when combined. For instance, maintaining one telephone list for key corporate, emergency-response and media contacts is more effective, and more easily updated, than maintaining two separate lists.

To fully integrate both plans, employees must receive training in both emergency response and crisis communications. Employees who receive training in only one specialty will not understand the full process or how their actions could influence the outcome.

Facing the public

To expedite communications with the media, prepare generic news releases in advance, leaving blanks to fill in the appropriate information. Train more than one crisis-response team member as an authorized spokesperson. Appointed spokespersons should be from a variety of job functions, and should possess different types of backgrounds and knowledge of the site (e.g., site manager, regulatory director, and corporate press officer).

You should also identify which team members shouldn't be spokespersons. Such team members can be assigned to other essential positions, such as:

- Community coordinator, especially in the event of an evacuation
- Investor- and customer-relations coordinator
- Logistics coordinator
- Emergency site-remediation manager
- Manager for continuing site operations
- Personnel to field telephone calls, document questions and to give general information

In the event of a crisis, swift response and effective communication will minimize the potential loss of life and property and protect the surrounding community. Such decisive leadership will maintain your facility's image among your employees and the surrounding community. ■

INSURING AGAINST ENVIRONMENTAL RISKS

Environmental managers now have many choices for insurance

Kenneth Anspach

As the chemical process industries now know all too well, environmental damages represent a significant risk to the firms and individuals in it. Whether the cause of the damage is a sudden spill or the gradual contamination of a site through underground leakage, major financial losses are a constant threat.

U.S. insurance companies are also aware of these risks. During the late 1970s and early 1980s, many limited environmental coverage in the policies that they sold; later, such coverage was dropped entirely. Recently, many new types of coverage have become available, but often at high expense.

To get adequate insurance coverage at a reasonable price, CPI firms can pursue several options: general liability insurance, self-insurance, specialized environmental insurance — or no insurance at all. Each of these options raise certain risks and costs.

At the same time, individual engineers or consulting engineering groups that service the CPI have their own set of insurance options. Most independent engineering consultants carry some type of liability insurance; now, as the potential consequences of their work on the environment become clearer, some have invested in various types of professional insurance (box).

Pollution exclusion

In 1973, a "pollution exclusion" was added to the standard general liability policies for businesses. This exclusion relieved insurance companies from paying for personal injury or property damage due to any release of pollutants or contaminants unless such release was "sudden and accidental," a phrase meant to cover rapidly evolving spills or explosions.

However, the courts have found that the exclusion contained an ambiguity which would allow coverage if the pollution damage was caused by a party other than the insured, or if the pollution merely was not expected or intended by the insured. Many insurance companies have now incorporated new exclusions into their general liability policies, most of which purport to deny coverage for any pollution damage.

To get the coverage they want, companies can self-insure. This is prevalent among larger firms with substantial financial reserves. Many of the members of the Chemical Manufacturers Assn. (Washington, D.C.), in which larger companies tend to congregate, have taken up this option. When self-insuring, companies generally set aside — or book certain reserves — each year in a special account. That money can then be used to pay for environmental damages.

STEPHEN WALSH

Cooperative ventures

Just as individual companies can self-insure, so can several like-minded manufacturers form their own insurance "pool." Insuring environmental risks in this way has helped small- to medium-sized companies. One such group of chemical companies got together during 1977–1978 to found their own insurance organization, Affiliated Chemical Group Ltd. (ACG).

The group was formed to provide "first dollar" coverage of risks, with the insurance industry providing excess coverage, according to Robert Hixon of John L. Wortham & Son, L.L.P. (Houston), the broker for ACG. "First dollar" refers to the practice whereby multiple insurers provide coverage only when a loss exceeds a predetermined limit. ACG provides the initial coverage and then, depending on other policies a company might have, those policies take over.

ACG started with just 12 member companies, and has now

grown to 34, while never losing a member. Due to this popularity and the current demand for environmental insurance, ACG began offering coverage to non-equity members, who are offered the same rates and privileges as members except they are nonvoting and do not receive a return on their investment. The first time the company did this, in 1985, it had 220 applicants for coverage but could only accept 16.

A hard-to-overlook reason for ACG's popularity is that their rates and premiums are said to be 50% less than market rates. (The intricacies of site-specific conditions and differing policy options make direct comparisons problematic.) Also, ACG offers policies with a wide range of prices and coverage options. Policies offered by ACG have deductibles ranging from $20,000 to $250,000.

The commercial general liability policies offered by ACG to CPI firms provide coverage for pollution emanating from products or sites not owned by the insured (but which might have pollution originating in the insured's products). These policies provide coverage in the range of $1–$5 million.

In addition to this basic coverage, ACG offers special pollution coverage to provide for higher upper limits (that is, the maximum amount the insurer will pay out), and to extend pollution insurance for sudden events. This coverage can be provided to those who self-insure to a limit as low as $50,000.

Specialists target niches

Meanwhile, a number of insurance firms have introduced policies that provide coverage for specialized situations in environmental liability. Companies currently providing specialized environmental insurance include Commerce & Industry Insurance Co. (C&I), a member of American International Group (New York), Reliance National Insurance Co., Zurich American Specialties (New York) and United Coastal Insurance Co. (New Britain, Conn.).

Competition among these and other firms for CPI business is intensifying.

C&I has several types of policies, each with a variety of options to customize the coverage available to client companies. C&I's environmental impairment policy provides coverage for third-party, offsite bodily injury and property damage losses. This policy can also cover cleanup and legal-defense costs. Additional coverage is available for sudden or gradual occurrences.

The policy provides an aggregate liability limit of up to $20 million, and has a minimum deductible of $50,000. "The premium for such a policy varies according to the exposure," says Marie Raninger, a vice president at Mack and Parker, Inc., a broker for C&I and other insurers.

C&I also offers a legal liability policy that has limits of up to $40 million available and covers waste treatment and disposal sites, landfills, utility companies, chemical, photochemical and other manufacturing facilities. The policy covers liability and cleanup costs.

An optional provision is available for "non-owned disposal site coverage," which insures against damage due to offsite disposal of wastes. The premium can vary depending on the coverage needed and the exposure at the site.

Another insurance company specializing in environmental insurance is Reliance National, in conjunction with ECS Underwriting, Inc. (ECS; Exton, Pa.), its managing general agent. Examples of ECS's products include a Chemical Distributors Program under its general liability policy.

Unique for this type of policy, there is no absolute pollution exclusion that would exclude coverage due to a chemical release. The policy has been tailored to cover chemical distributors, including those involved in blending, repackaging and wholesaling. It can be extended to cover some manufacturing operations. It offers limits of liability from $1 to $5 million with a minimum annual premium of $20,000 and a $10,000 minimum deductible; even higher limits are available.

In addition to its general liability policy, Reliance National has a legal liability policy, which provides coverage for pollution emanating from a covered site. These sites must be listed within the declarations section of the policy, and a risk-assessment survey must be done on each site before it can be covered. This survey must be performed by an engineering firm approved by ECS, and could be paid for by ECS. The policy has liability limits of $10 million per loss and $20 million aggregate, with a $20,000 minimum premium and a $25,000 minimum deductible.

However, the legal liability policy does not cover onsite damage; it only covers that which occurs offsite. To fill this gap, Reliance National has now added policies for pollutant and contaminant cleanup to its property insurance business. These offer coverage to $1 million, with higher limits available.

Minimum premiums for this coverage extension range from $1,000 to $7,500, depending on the risk, and have a minimum deductible of $2,500. "This extension allows CPI firms to have an alternative method of protection from cleanup costs," notes Jeffrey Lejfer, a vice president of ECS's Property and Inland Marine Dept.

Another perspective on the range of specialized insurance policies available today comes from United Coastal Insurance Company. United offers a legal liability policy which covers either onsite (first-party) or offsite (third-party) damages and can be written to cover both. These sites must be specifically listed in the policy, and an environmental risk assessment must be completed before the policy is issued.

STEPHEN WALSH

INSURING THE ENVIRONMENTAL RISKS OF PROFESSIONAL WORK

Just as CPI firms must be concerned with liability for environmental damage, so must the many individuals and consulting groups serving the CPI. To address the need to insure these individuals, a growing number of policy types and vendors has arisen.

One common type of environmental insurance is a traditional policy broadened with new environmental features. Evanston Insurance Co. (Evanston, Ill.), for example, offers a policy for architects and engineers that narrows and, in some cases, eliminates, the pollution exclusion. Liability limits of up to $5 million are available, with minimum deductibles as low as $5,000. Evanston's underwriting manager is Shand, Morahan (Evanston, Ill.). Another firm offering such a policy is Lloyd's of London, through a U.S. firm called Avreco (Chicago).

Reliance National Insurance Co. offers a policy for environmental consultants' professional liability, which provides coverage for acts, errors or omissions resulting from specified professional services. The policy offers liability limits of up to $15 million per claim (or $30 million in aggregate) and a minimum deductible of $25,000. The minimum premium available is $10,000 per year. Reliance's managing agent is ECS Underwriting, Inc. (Eaton, Pa.). ECS' policies provide coverage for both sudden and gradual pollution releases.

The errors-and-omissions policy is a common instrument in professional insurance. Another company offering such a policy is Commerce & Industry Insurance Co. (C&I, of New York). It has a liability limit of about $2 million, and a minimum deductible of $5,000. □

Additionally, the policy can be extended to cover non-sudden occurrences, or sudden and accidental occurrences, or both. The policy can also be combined with a general liability policy for even more-complete coverage. This policy has a $30,000 minimum premium and a $5,000 minimum deductible.

No safety net

Some CPI firms have elected not to pay the price for environmental insurance. For example, W.R. Meadows, Inc., (Hampshire, Ill.), has staked its business on careful compliance. "The company has taken a very proactive approach to minimizing environmental problems," says David Carey, environmental specialist for the firm.

Despite the cost and related financial issues, the insurance industry's stance toward the CPI seems more reasonable than in the recent past. "As both the insurance industry and the CPI have become more familiar with environmental risks and liabilities, they have become more comfortable with their exposure," notes Myra Tobin, a managing director at Marsh and McLennan (New York), an insurance broker. Then, too, the CPI's level of expertise in analyzing risks, and reducing them, has risen as well. ■

Edited by Nicholas Basta

The author

Kenneth G. Anspach is an attorney (Suite 1435, 111 West Washington St., Chicago, IL 60602; tel: 312 407 7888). He concentrates in environmental and commercial law, representing industrial property owners, manufacturers, banks and insurance companies. Mr. Anspach is also a hearing officer for the Illinois Pollution Control Board and a former Illinois Asst. Attorney General in the Environmental Control Div. Heather Lockhart, a paralegal concentrating on environmental insurance issues, assisted in the preparation of this article. This article is not to be interpreted as a substitute for an attorney's competent legal analysis and advice.

U.S. laws raise the possibilities of severe legal actions against CPI managers. Rigorous audits are a preventive

AVOIDING CRIMINAL LIABILITIES

STEPHEN WALSH

We all chuckle over the oxymoron "I'm from the Government and I'm here to help you." Not so funny is the ever more common refrain, "I'm from the EPA and I'm here to put you in jail." Armed with more than 120 investigative agents, the U.S. Environmental Protection Agency, through its attorneys at the Dept. of Justice, charges 5 to 10 engineers and business people with criminal violations of the nation's environmental regulations in any given week.

For instance, recently several former officials of a public utility company were indicted on criminal fraud charges for allegedly failing to report the release of 200 pounds of asbestos during a steam manhole explosion more than 4 years ago. Apparently they failed to live up to the requirements of the Comprehensive Environmental Response, Compensation, and Liability Act (CERCLA) relative to the release of more than 1 pound of friable asbestos. These men risk spending part of their retirement in federal prison.

Equally worrisome is the case of the environmental engineer working for one of the nation's largest corporations. He received a 16-count indictment alleging violation of his facility's water-pollution permit and the making of false reports to EPA relative to the alleged discharges. This engineer faces a series of trials and the risk of imprisonment in addition to a multimillion-dollar fine.

Often, the FBI assists EPA enforcement agents in environmental prosecutions. In Florida, a woman was arrested and indicted, charged with aiding and abetting the illegal dumping of acid and lead from thousands of used batteries onto a multi-acre site. If this woman goes to jail, it will be for ignoring or misunderstanding the provisions of the Resource Conservation and Recovery Act (RCRA).

Think this can't happen to you or your company? Think this is something that only happens to the "other guys?" Think again. Gone are the days when only evil "midnight dumpers" heard the cold clank of jail doors closing behind them and felt the strong bite of criminal fines. All managers in the chemical process industries (CPI) face the threat of criminal environmental enforcement.

The stark reality

Criminal environmental prosecutions are on the rise. Over the past five years, pleas and convictions are up 70%, fines 80%, and jail time 35%. From 1992 to 1993, the number of federal criminal cases has doubled.

As environmental laws are updated, Congress typically creates a more fertile setting for criminal prosecutions. The Clean Air Act Amendments of 1990, for example, mandate state operating permit programs. One of the required aspects of the program is periodic "compliance certifications" by a responsible corporate official. If such a certification turns out to be erroneous, despite the best efforts of the company to comply with the permit terms, the government could well find that the certification constitutes false reporting or concealment of wrongdoing. Those are prime elements of a criminal enforcement action (box).

Corporations are not the only targets of enforcement actions these days. The Department of Justice's stated policy is to conduct environmental criminal investigations with an eye towards "identifying, prosecuting, and convicting the highest ranking responsible corporate officials."

In 1992, 61% of the defendants in federal criminal cases were corporations. In 1993, the tables have turned: cases against individuals have leaped to 75%. Individuals are also routinely pursued by state authorities today. Violations can occur with any of a host of

By **J. Wray Blattner, Esq.**, Thompson, Hine and Flory, and
Gary M. Bramble, P.E., Environmental Science and Engineering, Inc.

Originally published June 1994

regulatory obligations that states have imposed (box, p. 130).

As a general rule of law, a person must "knowingly" or "willfully" violate a law to be found criminally liable. In the environmental context, however that isn't exactly so. First, both the federal Clean Air Act and Clean Water Act impose criminal liability for certain negligent acts. It is also important to remember that state laws may require less of a showing than "knowing" or "willful" violations to trigger criminal charges. For instance, "reckless" violations of the state of Ohio's law for hazardous waste result in criminal liability exposure.

Second, courts have demonstrated a willingness to aggressively apply criminal sanctions to environmental offenders, and to extend the long arm of the law to reach individuals. The good news for corporate managers and officers is that a position of authority does not, in and of itself, create criminal liability based on the misconduct of subordinates.

In some cases, however, courts have inferred knowledge of illegality (and hence attached criminal liability) to responsible corporate officials based on circumstantial evidence. For instance, the boss informs the new plant manager that the company's policy is to "keep the drum count down." When the plant manager seeks further elaboration, the boss cuts him off with, "I never had problems with my former plant manager, do I have a problem with you? Handle it!" In the subsequent criminal enforcement action that arises out of on-site dumping of wastes, the court determined that a jury may infer that the boss effectively ordered the manager to dispose of wastes illegally.

If convicted in federal court, defendants find themselves facing the harsh edicts of the "Federal Sentencing Guidelines" for environmental crimes. The Guidelines establish base levels of punishment, with the prospect of upward adjustment if the offense involves such factors as repetitive spills or releases, disposal without or in violation of a permit, significant cleanup requirements, or serious injuries. There are few circumstances where the court may depart from the Guidelines. The result has been a significant increase in penalties over sanctions imposed previously for similar offenses.

TOP TEN INGREDIENTS FOR AN ENVIRONMENTAL CRIMINAL PROSECUTION

1. Reckless, knowing, or willful disregard of environmental laws or regulations
2. Evidence of deceit or concealment of wrongdoing (for example, false reporting, fraudulent certifications or failure to report)
3. Failure to comply with agency orders
4. Significant harm to the environment
5. Economic benefit gained by the offender as a result of the wrongdoing
6. Complicity of a corporate manager or other responsible corporate official
7. Strength of the evidence
8. Offender's history of noncompliance
9. Perceived deterrent value to the offender and others in the regulated community
10. High level of citizen involvement or media attention

There are some 10,000 pages of federal (let alone state) environmental regulations. The rules apply to large and small companies alike. As a practical matter, the sheer scope and complexity of environmental regulatory programs make 100% compliance virtually unattainable for most industrial enterprises. Where it is no longer a defense to claim lack of knowledge of one's regulatory obligations, and where courts allow the inference of criminal knowledge based on what the defendant should have known, what is a company to do?

A way out

The environmental audit provides a solution to this problem. Even where audits have long been in place, there is increased emphasis being placed on these programs through commitment of additional resources and increased emphasis on managing the results of these audits.

Progressive audit programs are established with three goals in mind:

- To ensure that programs and practices at facilities are in compliance with applicable rules and regulations
- To affirm that management systems are in place at the facilities to support ongoing compliance
- To identify needs or opportunities where it may be desirable to go beyond compliance to protect human health and the environment

In addition to having well defined goals, an audit program must have the support of the senior management of the company. Establishment of an environmental audit program obligates the company to take action on the findings and recommendations which result from conducting the audit. Senior management's active support is essential to ensure that adequate resources are in place for forthright reporting and to correct program deficiencies.

The individual engineer's situation where management approval is lacking is a difficult one. Since so much of an organization's resources are involved in an audit, undertaking one may not be done by one staff person, operating independently.

Independence is key

Once the goals of the program are established and top management support is in place, the next question that must be answered is who will lead the audit effort. There are several considerations that must be made in making this choice, but the most important is independence from the facility that is being audited. This can be accomplished through use of outside consultants or legal counsel, or can be accomplished using in-house resources if the reporting relationship of the individuals leading your audits is not directly aligned with operations (e.g., corporate staff function).

Ideally, the audit leaders for your company should also have seasoned technical, regulatory and legal knowledge. It is rare to find a single individual with sufficient knowledge and experience in all three of these areas. It is also unusual, and in fact not recommended, that an individual be responsible for a complete audit. The team approach also ensures that the appropriate mix of technical, regulatory and legal skills are available.

The next step in establishing an audit program is to identify the desired scope of the audits by means of "protocols" that specify what compliance issues are relevant. The type and number of protocols needed are obviously a function of what regulatory programs apply to your facilities. One word of caution however. It is dangerous to assume that a particular regulatory program does not apply based on previous assumptions made about the kind of operations at a particular facility.

For example, based on prior knowledge it may be understood that the facility does not generate hazardous wastes. It would be shortsighted and could lead to serious consequences if the audit team does not verify that this is in fact the case. Consultants and legal firms have developed protocols for use in conducting audits; it is also possible for a plant to develop its own protocols in-house which are tailored to specific operations.

Performing the audit

Once the goals of the environmental audit program have been established, top management has bought in, and the appropriate protocols are in place, the real work begins. Most environmental audit programs can be divided into three phases. First is pre-audit activities, then the actual on-site audit, and finally the post-audit activities.

Pre-audit activities consist of identifying the team members who will conduct the on-site audit, keeping in mind the need for the appropriate mix of skills. The audit team should generate a previsit questionnaire, which identifies the regulatory programs that will be audited. This also allows the facility time to organize any files and documents needed for review while the audit team is onsite and setting up interview appointments. Having this done beforehand is an absolute must in order to complete the audit in a reasonable period of time.

Upon arrival at the facility the team should sit down with the key facility personnel, explain again the goals and expectations for the period of time that the team will be onsite, and answer any questions or clear up any last-minute details. The team should then break into members' respective regulatory areas and examine the compliance programs.

A word of caution: It is often easy to examine a written program, and if it looks good, to stop there. The best written program in the world is of little value if it is not being practiced.

During the onsite review, communication is key. This includes communication both among audit-team members and between the audit team and facility personnel. If an area of concern is identified, make sure facility personnel know about it immediately. Often the issue can be quickly addressed; sometimes, once the concern is clearly understood, the problem is simply a case of the right information not being made available to the audit team.

Finally, near the end of the onsite visit, the team members must gather and reach an understanding of what compliance issues exist. The team should then have a closeout meeting with facility personnel to communicate these issues; a preliminary writeup might also be presented. Subsequently, in the post-audit phase, the team issues a formal audit report to management, with recommendations for attaining compliance.

Now, management has the information in hand to make judicious decisions about improving environmental performance. Too often, companies will complete audits but will not do an adequate job of ensuring that all the issues identified by the audit team are appropriately addressed. Simply put, if there is any hesitancy on the part of your company to quickly and completely address the compliance issues identified during the audits, do not start an audit program. While ignorance is no excuse, your legal problems will be further compounded by identifying non-compliance situations and then not adequately dealing with them. An audit followed by inaction may well create the basis for a criminal enforcement action.

Let's face the facts. Compliance with regulatory standards in the 1990s is difficult but the signals are clear. Companies or individuals who "choose" not to have effective and complete compliance programs in place will ultimately pay a price — substantial fines and possible jail time. Don't bury your head in the sand. ■

Edited by Nicholas Basta

REGULATORY HOT SPOTS TO WATCH FOR

Illegal transport, storage, or disposal of hazardous waste
Hazardous-waste inventory reporting (SARA 313)
Failure to report releases of hazardous substances
Demolition or renovation involving asbestos
Mishandling of solvents or degreasers
Polychlorinated biphenyl releases
Dredging or filling of wetlands without a permit
Contamination of groundwater
Falsification of reporting required by permit
Air-emission monitoring requirements
Underground storage-tank mismanagement

The authors

J. Wray Blattner, is a partner at Thompson, Hine and Flory (POB 8801, Dayton, OH 45401; tel: 513 443 6539). His practice focuses on such environmental issues as permitting, regulatory compliance and audits, as well as defense against private and governmental enforcement actions. He is a Phi Beta Kappa graduate of Denison University, and obtained his J.D. from Georgetown University in 1980. A member of the bar associations of Dayton, the state of Ohio and the American Bar Assn., he has published a variety of environmental manuals and handbooks.

Gary M. Bramble, P.E., is chief engineer at Environmental Science & Engineering, Inc. (Miamisburg, Ohio; tel: 513 859 3600). He has more than 20 years' experience in a broad range of environmental and energy management services, including environmental audits. He holds a B.S.E. in mechanical engineering and an M.S. in civil and environmental engineering from the University of Cincinnati, as well as an M.B.A. from the University of Dayton. He frequently addresses environmental compliance issues at seminars throughout the country.

Section VIII

Instrumentation Support for Testing and Diagnostics

IMPROVING PLANT MAINTENANCE VIA NONDESTRUCTIVE TESTING

Frank J. Sattler*
Giitsch Field Services/NDE, Inc.

A CAPSULE SUMMARY

This fifth installment of CE's Refresher on nondestructive testing presumes that the reader has a working knowledge of the NDT methods covered in previous articles. These were:

Part 1: "Improving Quality Through Nondestructive Testing," April 1989, pp. 116–119, covers NDT in general, describes each technique, and explains how NDT can be used for evaluating materials and monitoring process operations.

Part 2: "Finding Surface Defects with Nondestructive Testing," July 1989, pp. 125–133, details methods that detect surface or near-surface discontinuities, specifically visual, liquid-penetrant and magnetic-particle methods, as well as proof and leak tests.

Part 3: "Finding Below-Surface Defects with Nondestructive Testing," September 1989, pp. 161–167, describes methods that can detect discontinuities within the volume of a material, including penetrating radiation, ultrasonic testing, and eddy-current testing.

Part 4: "Nondestructive Testing via Optical and Acoustical Methods," November 1989, pp. 191–198, discusses several newer and lesser-known NDT methods, namely optical holography, acoustical holography, and acoustic emission testing.

Applications of nondestructive tests to monitor different aspects of chemical-process-industries (CPI) plants are extensive. Most involve finding discontinuities in plant components or measuring the thickness of containment systems. Discontinuities in materials may result from defects in the original ingot, from fabrication or repair operations (such as welding), from the improper design of equipment or selection of materials, or from fatigue, creep (i.e., material elongation) or other failure mechanisms.

Often, in preventing one problem, another is created. For example, if corrosion is expected to be a problem, rather than specifying carbon steel piping, one might consider using an austenitic stainless steel. But care is not always taken in selecting the alloy or the welding procedure for joining it. Merely specifying stainless steel may not result in an alloy low in carbon or a welding procedure with low heat input.

Depending upon the stresses on the pipe and the process material in it, intergranular stress-corrosion cracking may occur; this happens when the heat-affected zones of the welds are depleted of chromium, which normally provides the corrosion protection. Thus, one form of corrosion is substituted for another.

Failures in CPI plants have a variety of origins and causes, and depend upon the component, materials, stresses, service, process materials, presence of discontinuities, and other factors. Usually a primary cause can be established, but other factors will contribute to the failure. The table summarizes a few common failure modes, their characteristics, and possible causes. References [1–5] provide in-depth descriptions of failures.

It is not economically feasible to set up programs that look for every type of possible failure in a plant. However, experience at similar plants or with similar components should be evaluated to determine what types of failure mechanisms might be present. Any NDT program will, to some extent, involve a "wait-and-see" approach. But it must also be able to recognize and respond to problems that do occur,

Defects in materials of construction, faulty welds, and fouling of heat exchanger surfaces are a few of the problems NDT can detect

* For information about the author, see *Chem. Eng.*, April 1989, p. 119.

Originally published December 1989

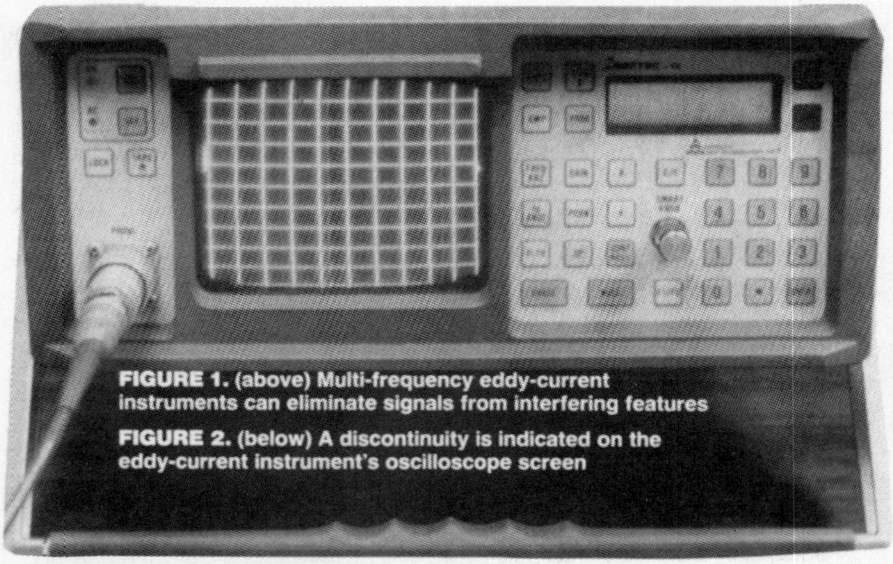

FIGURE 1. (above) Multi-frequency eddy-current instruments can eliminate signals from interfering features

FIGURE 2. (below) A discontinuity is indicated on the eddy-current instrument's oscilloscope screen

since not every problem can be anticipated.

One should not wait until after construction is finished to start a quality-assurance program. Rather, such activities should begin in the design stage and continue through construction and into operation. If there are problems caused by plant or process design or by selection of materials, maintenance personnel will always be applying stop-gap measures to a problem that only gets worse. The sooner severe-service environments are accounted for, the smaller will be problems once the plant is operating.

Discontinuities in materials are ever-present — perfect materials simply do not exist. The consequences of detrimental discontinuities remaining in plant components can be prevented by using proper inspection methods and fracture mechanics analysis (if needed) before a component arrives onsite. This may mean that additional inspection methods are needed to supplement any required by construction codes.

Fabrication methods used onsite during construction and erection should not be permitted to introduce any additional discontinuities. Even discontinuities of acceptable size or shape but at the wrong location, if left unrepaired, can grow (e.g., through corrosion, erosion or fatigue) to a size where catastrophic failure may occur.

The use of NDT for specification and procurement will be covered in a future installment. The remainder of this article discusses NDT for plant maintenance, covering some component problems and NDT methods for detecting discontinuities or measuring other features of equipment or materials.

Heat exchangers

Any time that heat is transferred between fluids, problems can result. If a vapor is formed, soluble or suspended materials that were in the liquid can be left behind on the heat transfer surfaces. These materials may not have been a problem when they were in the liquid if their concentration was low. But once they are deposited, they can, at the least, impede heat transfer; even worse, these deposits may lead to severe corrosion and degradation of the exchanger materials. Fouling is another problem that has a detrimental effect on operation because it alters heat transfer as material on the exchanger surface becomes thicker.

Heat exchangers are also subject to crevice corrosion between the tube and tubesheet because of the way tubes are pushed through the tubesheet, expanded and welded. Once the exchanger is in service, contaminants can concentrate in these crevice areas, causing corrosion to proceed more rapidly.

Heat exchanger leaks stem from a variety of causes (Reference [*6*]). Some problems cause damage to the tubes that can penetrate the wall; others involve the tube-to-tubesheet weld (i.e., root cracking, tube pullout or pushout, relaxation of the residual stresses from the tube expansion in the tubesheet, tube-hole finish problems or damage, and distortion). Leaks through the tubesheet around the tubes can usually be detected by either a hydrostatic pressure test or a leak test using a gas at low pressure.

Normally, the tube-to-tubesheet weld undergoes a liquid-penetrant test after it is made. If leaking occurs during operation, the test can be performed during a shutdown (however a clean dry tubesheet is needed in order to perform the test).

Tube leaks, or areas of nonferromagnetic tubes that might become future leak sites, can be located best by an eddy-current test using a differential probe (i.e., a probe containing two coils) inside the tubes. Special multifrequency equipment allows for mixing and subtracting signals resulting from the baffle, tubesheet and support plate, thus revealing any signals due to discontinuities.

Figure 1 shows a multifrequency eddy-current instrument. When the output signal from the test indicates that the defect does not fully penetrate the wall, then the phase angle of the signals determines the wall penetration of the defect and from which wall (i.e., the inside or outside wall) the defect started. Figure 2 illustrates a discontinuity indication on the eddy-current instrument's oscilloscope screen.

If general corrosion of the tubes is a problem, an absolute (i.e., single-probe) eddy-current test should be conducted. This can be done with a multifrequency instrument by using one of the differential coils as an absolute coil to measure thickness at an appropriate frequency at the same time the defect scan is made.

If the tubes are ferromagnetic, special techniques and equipment, such as a Ferroscope* (Figure 3), will be needed because of the limitations imposed by the magnetic permeability of the material. (Recall that increases in magnetic permeability cause the eddy currents to penetrate only to very shallow depths, requiring the use of very low frequencies.) The newest instruments for testing ferromagnetic tubes use remote field eddy-current techniques.

Other causes of leaks include erosion and ligament cracking in the tubesheets, erosion damage in the shell inlet, cavitation damage on the shell outlet, and inadequate support of tube bundles at the floating end. Liquid-penetrant tests will detect ligament cracking. Erosion and cavitation damage can be found with visual tests, although a penetrant test may be necessary to determine whether any cracking is also present.

Ultrasonic thickness measurements from external surfaces can be used to locate erosion and corrosion problems if access to the inside surface is restricted. Or, a remote visual examination, using fiber optic probes or a small-diameter closed-circuit television (CCTV) system, may be necessary if access is limited. Eddy-current tests may be used to assess any tube damage in the last baffle or floating tubesheet.

Determining the extent of fouling on heat transfer surfaces is usually done visually or by analyzing samples scraped from the tubes. Flexible fiber-optic scopes and CCTV systems can aid in getting beyond the tubes accessible from the outside of the exchanger. If the fouling material has a consistent electrical conductivity or magnetic permeability, an eddy-current test might be used to measure the extent of fouling.

Alternatively, one could assess the total extent of fouling by using a radiation source and detector in adjacent tubes as shown in Figure 4. The source and detector could be placed anywhere on the tubesheet in adjacent tubes and the entire length of the tubes could be scanned by pulling both devices through the tubes in parallel. The transmitted radiation would be a function of the tubewall thickness and the amount of fouling material on each tube in the radiation path. Depending upon the density and thickness of the fouling material and the tubewall thickness a low-energy gamma-ray source (such as thulium 170) may be necessary to provide better sensitivity.

* Trademark of Russell Technologies, Inc., Edmonton, Alta.

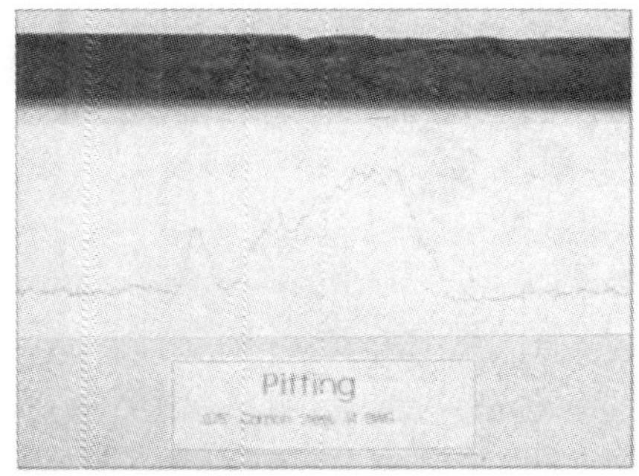

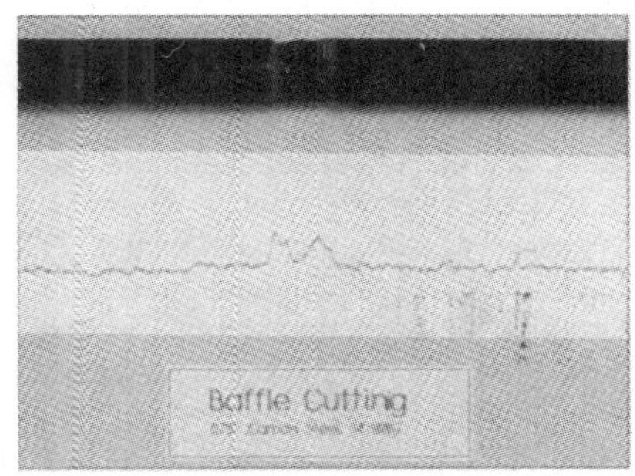

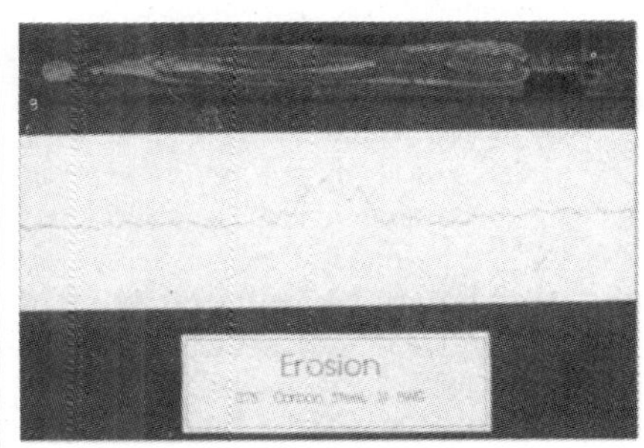

FIGURE 3. (a.) A Ferroscope permits eddy-current testing of ferromagnetic materials; (b.) Ferroscope indication of baffle cutting; (c.) Ferroscope indication of pitting corrosion; (d.) Ferroscope indication of erosion

Metal vessels

In any vessel, weld integrity is important. A weld can be a weak link because it may have undetectable discontinuities introduced during fabrication. Welding discontinuities, such as narrow cracks and incomplete fusion, may be very difficult to detect with radiography in thick sections of equipment. Ultrasonic angle-beam tests should be used to supplement any radiography required by codes. Experience in the nuclear industry has shown that unacceptable discontinuities can be present in vessels even though they are deemed acceptable based on radiographic and surface-examination standards.

Other areas of concern in a vessel include nozzle and entry-port welds, and the inner radii of these structures. The nozzle or entry port is already a hole in the vessel and even small cracks aligned across the principal stress axis of the vessel can be cause for concern. The crack should be regarded as having not just its own dimensions, but an effective crack length equal to the crack's actual length added to the nozzle or entry-port bore diameter.

Liquid-penetrant and magnetic-particle tests will detect surface cracks in a vessel very well. Access will be needed to the area to be examined, and the surface must be thoroughly clean and free of any contaminants. Vessels in severe service should have examinations scheduled more frequently (the specific frequency determined by such considerations as design conditions, materials and safety factors), and fluorescent magnetic-particle and liquid-penetrant tests should be specified.

The inside and outside surfaces of welds should be examined by either liquid-penetrant or magnetic-particle tests, as should the areas adjacent to the welds. Welds subjected to higher stresses, such as those on nozzles or transition sections, should be examined more frequently. Consideration should be given to ultrasonic angle-beam examinations of the welds or monitoring them with acoustic emission testing.

Vessel internals and supports are usually monitored visually (or by remote visual techniques, depending upon the access restrictions and internal conditions). A thorough cleaning may be needed before a visual exam can be conducted. The inspector should know what the vessel internals should look like and where potential problems may occur. For example, inlet and outlet nozzles are areas where erosion and cavitation might be possible, respectively.

Instrumentation penetrations that do not have full-penetration welds may be a source of leaks. Such welds should be observed visually during operation if possible; otherwise, a hydrostatic or leak test may be necessary.

Nonmetallic vessels and piping

Before the development of acoustic emission testing, there was no nondestructive method of assessing the quality of fiber-reinforced plastic (FRP) components. Prior to initiating an extensive acoustic-emission testing program in 1980, Monsanto Co. (St. Louis) experienced an average of over two failures per year. From 1980 to 1987, a total of only two failures occurred, and these vessel failures were predicted by the acoustic emission tests (Reference [7]).

Constant improvements in acoustic emission equipment have reduced the cost of such testing, as well as the skill level required by the technician to operate the equipment. Figure 5 shows an acoustic emission technician analyzing test data.

Because of the nature of sound propagation in composite materials, locating the exact source of the acoustic emissions can be inaccurate. General areas identified by acoustic emission testing should therefore be examined by other nondestructive methods, such as visual, penetrant and ultrasonic tests.

Standard acoustic-emission test procedures have been adopted by the Soc. of the Plastics Industry for fiber-reinforced materials. These include: ASTM E1067, "Recommended Practice for Acoustic Emission Testing of Fiber Glass Tanks/Vessels," and ASTM E1118, "Recommended Practice for Acoustic Emission Testing of Reinforced Thermosetting Resin Pipe."

Metallic tanks

Liquid storage tanks are not usually thought of as pressure vessels, since they normally contain fluids under no external pressure. However, in a very large tank, significant loads can develop in the tank wall next to the floor, due to the static head of liquid and the tank diameter.

Many environmental problems have occurred in the past few years from tanks that have failed to hold their contents. Storage tanks are subject to the same corrosion problems experienced by other metallic containment devices. However, since they often are not subject to the same quality standards as pressurized vessels, problems with materials and bad welds can be common.

Leakage paths in the plates that form the floor and walls of the tanks are not always easy to identify. Detecting leaks from a tank where the bottom has been replaced by adding a second floor above the original one is particularly complicated, since the visual evidence of leakage outside the tank may have no direct relation to the inside leakage site. Finding leakage cavities is also often complicated by paint on the external surface and the contained liquid on the inside surface.

Two types of liquid-penetrant tests can be used to locate leaks. In the first a visible dye is applied to one surface of the tank and is allowed to penetrate for up to several hours. The side chosen for the application of the penetrant, as well as the side from which the penetrant will exit, must be very clean and free of contaminants and surface coatings. The penetrant must not be allowed to dry during the examination (otherwise, it will not be able to flow out to be detected).

In the second method, the penetrant is applied to one of the surfaces (which should have been cleaned prior to application). The test is then conducted in the conventional manner, with a developer

FIGURE 4. Gamma-ray source and detector in adjacent tubes can measure the extent of fouling

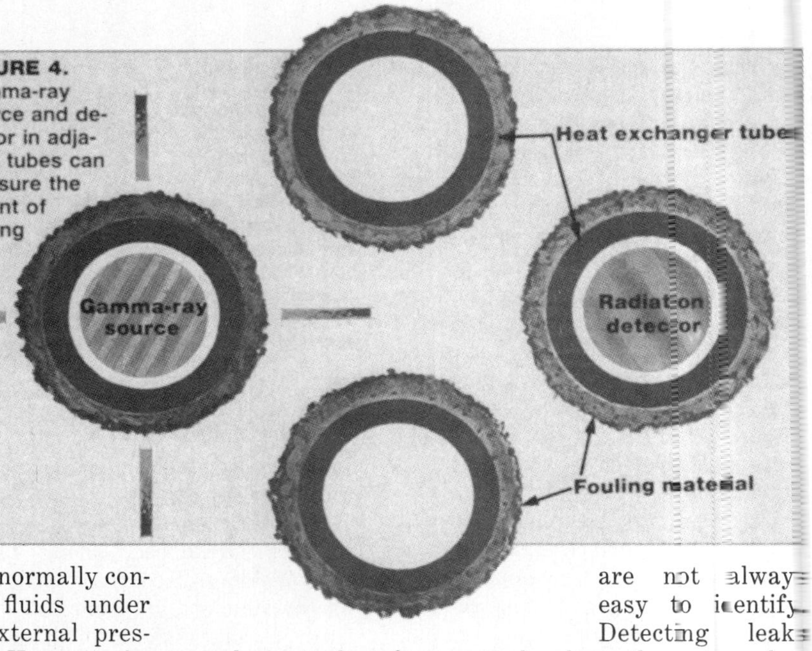

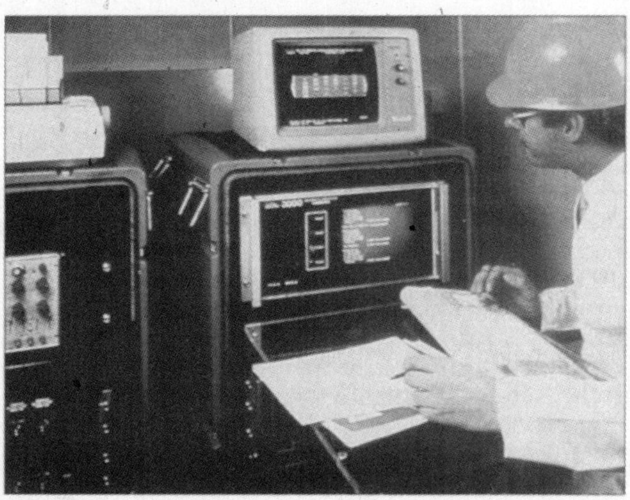

FIGURE 5. Physical Acoustics Corp.
Technician analyzes data from an acoustic emission test

used to aid the detection of any surface openings.

A problem with using penetrants is that the leakage path may be blocked by some of the contained material (in liquid or solid form). This prevents the penetrant from indicating leaks.

Another problem is that numerous indications may appear on the developer coating. Each could indicate a leak site, or any other surface discontinuity.

If the tank is ferromagnetic, surface and near-surface discontinuities that might be part of the leak path can be identified with magnetic-particle tests. However, again numerous indications may result. One is never sure that the leaks have all been identified and corrected until the tank is refilled and no longer leaks.

One of the most effective ways of locating leaks is to use a vacuum box on suspected areas. Usually welds are a good place to start. The surface must be smooth so that the vacuum box can seal, and it must be free of contaminants that might cause false indications. Leaks can be repaired and reexamined with minimum effort.

One of the problems with vacuum-box leak testing is that a separate vacuum box is needed for each curvature or geometric profile. A large number of different-sized and -shaped tanks could require a significant investment in vacuum boxes. Another problem is that, as with other leak tests, liquid or solid material in the leakage path can make the test ineffective.

Depending upon the liquid contained, ultrasonic thickness measurements may be necessary to verify that corrosion is under control. Any corrosion products present may need to be removed from both sides of the tank in order to not influence the thickness measurement. Corrosion products must be removed from the surface where the transducer will be placed.

Moreover, when the corrosion products on the wall away from the transducer are acoustically similar to those of the metal being measured, they will need to be removed. Corrosion does not always occur on the contained material side of the tank or other component, but can result on the tank bottom if it is in contact with groundwater.

SOME COMMON FAILURE MODES... AND THEIR CHARACTERISTICS	
Permanent deformation	Elongation
Distortion	Misalignment
Surface damage	Low-thickness areas
Fracture	Separation or opening

Metallic piping and supports

Many of the failure causes outlined in the table apply to piping. However, additional problems can occur because of piping supports and other components (such as pumps, valves and vessels) to which the piping is attached.

If no flexibility has been built into the piping system to allow it to expand and contract as a function of temperature, very high stresses can develop. These are added to the stresses caused by fluid pressure within the pipe and any other external loads or residual stresses. Although degrees of freedom can be built into component supports, the attached piping will usually have to move to accommodate the larger component.

Pipe supports, restraints and snubbers should be checked visually to make sure settings, configurations and attachments have not changed. Bolts and nuts attaching support flanges or other members to building structures should be fully engaged. Attachments should not be loose from the support member.

Pipe-support welds should undergo a careful visual examination, or liquid-penetrant or magnetic-particle testing (depending upon the material), should be performed. Where the materials are ferromagnetic, magnetic-particle tests provide a fast means of assessing quality.

At room temperature, plastic deformation (or creep, which is the elongation of a part) of most materials does not begin until the material is stressed beyond its yield strength. At elevated temperatures, some degree of creep exists at almost any level of stress. The rate of creep is dependent upon the stress, the temperature, and the amount of time the material is exposed to the temperature and stress. The higher the temperature or the higher

SOME COMMON CAUSES OF FAILURES

Design		Materials	Discontinuities and Defects	Service		
LOADS	CONFIGURATION	SELECTION		ABUSE	ENVIRONMENTAL	EROSION
Static	Weld joint	Properties	Inherent	Excess Loads	Corrosion	Cavitation
Dynamic	•Location	Handling	Primary-processing	Improper maintenance	General	Hydrogen embrittlement
Vibrational	•Shape	Formability	Secondary-processing	Finishing	Pitting	Temperature
Residual stresses	•Preparation	Quality Assurance	Service	Cleaning	Intergranular stress-corrosion cracking	•Low
Bending moments	Notches			Lubrication	Corrosion fatigue	•High
						••Creep
						••Stress rupture

Krautkramer Branson

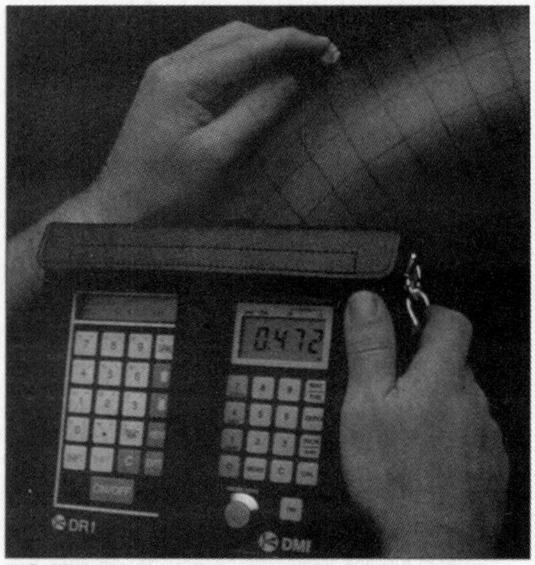

FIGURE 6. A digital ultrasonic thickness instrument and data logger can record data in a format suitable for computer processing

the stress, the more extensive the creep will be.

Creep normally occurs in three different stages at high temperatures and stresses. The first stage is characterized by a rapid increase in the creep rate. The second stage involves a constant or decreasing rate of creep. Depending upon the stress and temperature, the creep rate during the second stage can be a constant positive value, negative or zero. As long as the material remains in second-stage creep, fracture will not occur.

In the third stage, the creep rate increases rapidly and fracture occurs. Necking (i.e., thinning, as if to form a neck) occurs in the third stage, further increasing stress as failure approaches. The fractures that occur are small initially. The higher the temperature or the higher the stress, the faster the third stage of creep occurs. In addition, the older the plant, the greater the chance that some of its piping and other components are already in the third stage.

Material that is suspect (because it is subject to high temperatures or stresses or both for fairly long periods of time) should be examined by either liquid-penetrant or magnetic-particle methods to locate any fractures. One can then employ a surface-replication technique using a plastic material to help determine the extent of creep. This is done by evaluating the replica under a microscope and determining the expected remaining life based on the amount of cracking present. Ultrasonic angle-beam examinations can also be used to detect discontinuities and measure their size.

Erosion and corrosion problems

In some piping components, the impingement of the liquid, or of a liquid containing solid particles, onto the inside surface can cause erosion. Areas where erosion is a possibility should be examined for changes in wall thickness. Usually straight-beam ultrasonic tests can be used if access can be obtained.

It is also possible to determine whether there is any surface erosion by radiation gauging techniques. This can be done through any insulation, as long as the radiation absorption by the insulation is not a large factor — no more than about 5%–10% of the metal's absorption. Water or other liquid in the pipe can also be present, as long as the pipe is full. Such a scan will identify any lower-thickness areas. Since this type of test measures total thickness (i.e., both walls are being measured), the distribution must be determined by using multiple scans at fixed increments, or ultrasonic measurements.

Cavitation damage often occurs in an elbow or pipe bend if the component is immediately downstream of a valve or other constriction in the pipe. The same testing methods are applicable to detecting this type of damage.

In general, corrosion damage can be evaluated by ultrasonic measurements. General corrosion can be determined using straight-beam ultrasonic search units. Thin pipe or tubing might require the use of dual search units whose beams are focused (or directed to account for refraction) to cross at a given depth. This two-transducer technique overcomes the limitation of poor near-surface resolution that prevents the use of single-transducer units.

Pitting corrosion can also be seen with ultrasonic straight-beams, but determining its extent may require some special scanning techniques or film radiography to provide a map of the pits. If surface build-ups are present, such as the byproducts of corrosion, these will need to be removed to expose the bare metal in order to accurately measure the amount of remaining metal. Calibration specimens made of the same materials and geometries as the materials being tested are very important.

Care must be taken in making ultrasonic measurements where pitting is a problem. Digital ultrasonic units should only be used in conjunction with the conventional A-scan presentation (amplitude vs. time).

Digital ultrasonic gages

When evaluating damage due to corrosion, erosion or cavitation, it is important to find the signal shortest in time (corresponding to the lowest material thickness) and maximize it before making any other measurements. Digital ultrasonic gages measure the shortest time signal above some preset amplitude. If the signal of interest is not greater than that threshold, the device will not trigger, but will only do so on the next signal that does go above the trigger amplitude. Thus, the measured value will be greater than the actual minimum thickness, and one might get a false sense of security from the measurement.

On the other hand, finding the minimum thickness with a digital gage can be very easy in some applications. Instrumentation is available today to take readings in a digital format suitable for computer processing. Once the reading is fixed, it can be sent to a recorder where it is logged in a data array that can then be read by a computer (Figure 6). Two to three hundred readings per hour (or more) can be taken by a single technician. Once transferred to a computer disk, a variety of programs can be used to analyze the readings and plot thickness values.

Ultrasonic scanners and instrumentation with color displays or printouts can represent different thicknesses using several colors to represent different locations on a vessel, rows of tubes, or a single tube. These displays can provide a vivid visual presentation of thickness and make the interpretation of data quicker, as the colors call attention to specific problem areas. Figure 7 represents tubewall thickness readings for

Thomas C. Ridgeway, Inc.

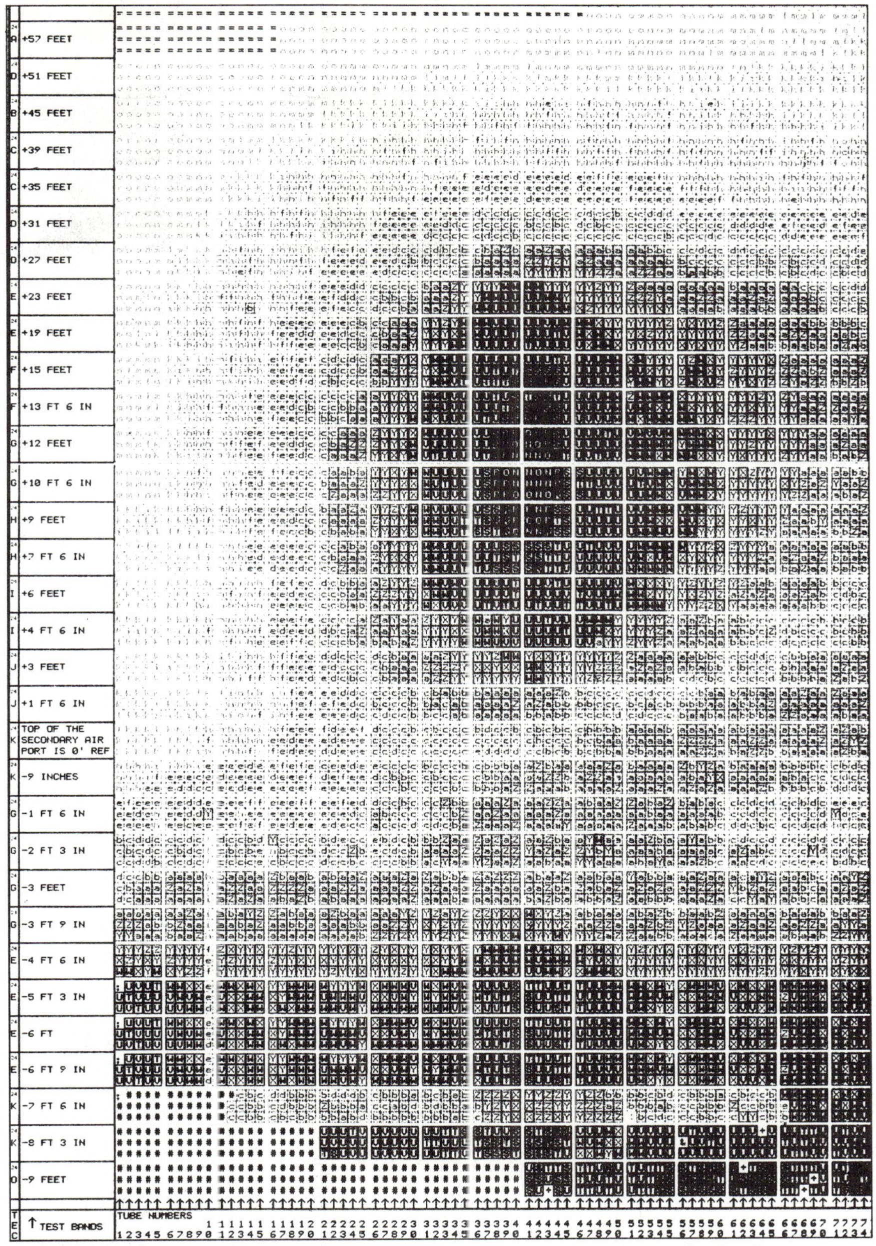

FIGURE 7. Changes in tubewall thickness are indicated by color or shade changes on a printout of an ultrasonic thickness test

one wall of a heat-recovery boiler at various elevations. Note how easy it is to identify areas where wall thickness has changed, as denoted by different colors and shades.

Intergranular stress corrosion and corrosion fatigue can be detected by ultrasonic angle-beam testing. Special techniques — using diffracted signals from cracks that originate at the crack tip — can be used for both detection and measurement, and have shown good accuracy. Further work in using crack-tip diffraction techniques that are not amplitude-dependent should make measurements more accurate and yet simpler to use.

In piping or any other type of cylindrical vessel, the stresses due to the internal pressure are about twice as large in the circumferential direction as they are in the axial direction. This means that any axially oriented discontinuity will be seeing a much higher stress than a circumferentially oriented discontinuity. The critical crack size (i.e., that can cause failure) will be smaller in the axial direction and larger in the circumferential direction, at least from internal pressure. (External load stresses and any residual stresses within the material will have to be added to these crack stresses to determine the actual critical crack-size at a given location.)

Any piping that has longitudinal weld seams should have those seams, as well as the circumferential seams, examined using ultrasonics and surface methods. A discontinuity across a circumferential weld can have a smaller critical crack-size because of the higher circumferential stress.

Pumps and valves

Most pump and valve casings are generally made from castings. However, in large pumps and valves, sections of the casing may be assembled by welding. Ordinarily, radiography is used to evaluate the initial quality of both the casting material and welds. If discontinuities are detected, they are ground out and the cavities repaired by welding. Often the casting has more weld metal than cast metal by the time the casing is satisfactory. (Recall that radiography is not a good method to detect tight discontinuities such as cracks, and therefore tends to find only large discontinuities with greater thickness changes.)

Radiographing a component at high energy causes the contrast in any image to be lower. The net effect of a low-contrast image is that only the larger discontinuities (thicker cracks) will be found.

Using ultrasonic testing on a casting to find crack-like discontinuities often is not successful because the beam cannot penetrate a large-grain casting. Such discontinuities can be detrimental later in the life of the component. Therefore, radiography should generally be supplemented with magnetic-particle and liquid-penetrant tests on the surfaces of the component, and with ultrasonic testing where possible.

Bolted joints have been referred to as "where machinery leaks, wears, slips, ruptures, loosens up, or simply fails" [9]. These statements also apply to bolt or stud and nut connections on valves and pumps (and on almost any other item one could name). Many of the problems occur because of damage to sealing surfaces, due to improper handling of the bolt or of the pump or valve.

Another common problem with bolted joints is that nondestructive testing was not specified when the studs, bolts and nuts were purchased. Seams and cracking in the bolting materials are common problems, and an engineer who does not specify nondestructive examinations may be getting someone else's problems.

Fatigue is a common cause of service failures in studs and bolts. The greatest stress in a bolted connection is at the thread at the flange face; there, the thread forms a notch where a fatigue crack can start if the cyclical stress is high enough.

If a corrosive environment is present, stress corrosion cracking could occur. Ultrasonic tests can be performed on bolts and studs if their diameter is large enough (i.e., the size of the transducer or larger), and this can often be done in the assembled condition. Magnetic-particle tests work well on ferromagnetic studs and bolts if fluorescent particles are used; this must be done when the components are disassembled. Some nonrelevant indications from the threads may be present, but the glow from a thread crack will be very brilliant. Visual examination with the valve or pump under internal fluid pressure should be used for leak tightness.

Pump and valve configurations that restrict the flow of a fluid (or for that matter even a reducer) are sources of pressure drop that can result in cavitation, which leads to erosion of the surface metal. Pitting is an indicator of cavitation; advanced cavitation damage may leave the surface in a spongy condition. Any surface in the flow path downstream from where the cavitation initiates is subject to damage if the pressure there is greater than the fluid's vapor pressure. Impingement plates are sometimes necessary in heat exchangers to avoid damage to the tubesheet or tubes.

Pump impellers can create the very conditions that lead to cavitation, and then they suffer the consequences. Radiographs of impellers and other areas subject to cavitation should show the metal loss. Elbows or pipe bends downstream can also suffer the consequences of cavitation, and can often benefit from radiography (although ultrasonic thickness measurements will usually be cheaper).

Rotating-equipment shafts

A critical part of rotating equipment is the shaft about which the rotation takes place. In order to transfer power from a shaft, splines and keyways are necessary. If the grooving has been made without a generous radius, torsional fatigue cracking with axially oriented cracks can result.

This is one of the most difficult types of discontinuity to find and measure when the equipment is assembled. Even if one could get access to much of the external surface of the shaft, any crack that might be seen with an ultrasonic angle-beam could be difficult to identify. This is because the reflection from the groove surfaces would be in the same time frame and, if the crack is

FIGURE 8. A crack at a thickness-transition point on a shaft is best detected from the small (top) rather than the large end (bottom)

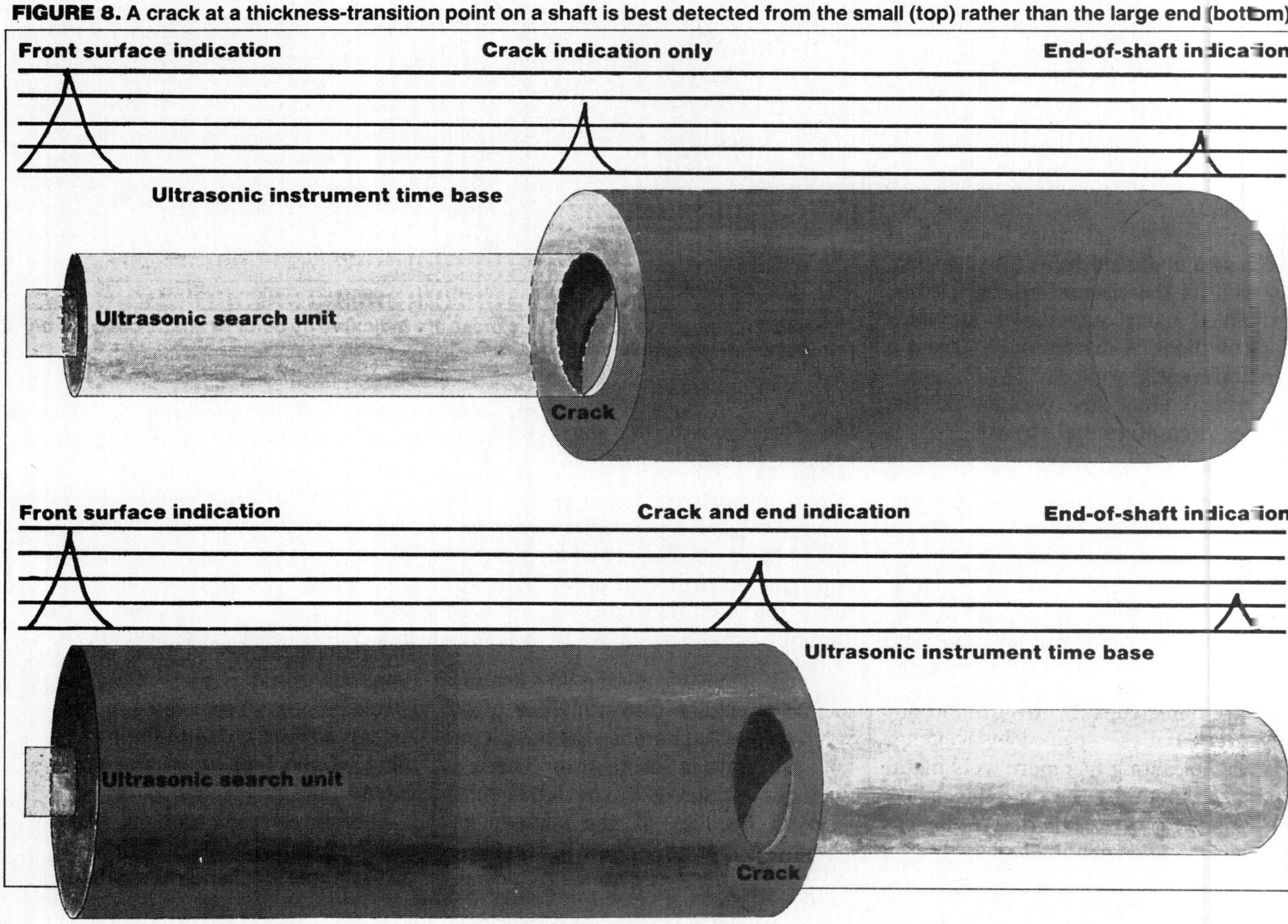

small, would be much larger than the crack signal.

The best examination for this type of fatigue cracking is a fluorescent magnetic-particle or liquid-penetrant test, with the equipment disassembled. Even then it may be difficult to distinguish a relevant indication due to a crack from a nonrelevant indication due to a sharp corner.

If one is performing a magnetic-particle examination, the part may need to be demagnetized. In addition, low currents, one-fourth to one-half the recommended values (usually 250 to 500 amps per inch of diameter), may be required for the circular magnetization in order to distinguish any cracking from a sharp corner. High-sensitivity fluorescent penetrants that are water-washable might also be used effectively to see this type of fatigue crack in either magnetic or nonmagnetic material.

As with other components, the type and orientation of any flaws generated during service is dependent upon the resultant stresses and their orientation. Torsion and bending moments on shafts place the maximum stress on the external surface, while tension and compression loads result in a uniform distribution across the shaft.

The orientation of a discontinuity will also depend upon whether the material is brittle or ductile. If the shaft is subject to only one form of load, the discontinuity that results will be axial, circumferential or at a 45-deg angle to the axis or circumference. Combinations of loads generally result in maximum stresses, in other orientations.

Using fluorescent magnetic particles with both circular and longitudinal magnetization should reveal the axial and circumferential discontinuities clearly, and pick up the off-axis discontinuities as well. Liquid-penetrant tests do not have any sensitivity-direction problems (i.e., unlike magnetic-particle tests, they are not affected by the orientation of the magnetic field with respect to a discontinuity), and can, therefore, detect any surface discontinuity.

When the equipment is assembled, ultrasonic tests can be performed from the shaft ends (if they are accessible). One would see best any circumferentially oriented discontinuities. Other discontinuities might also be picked up — by monitoring the reflection from the opposite end of the shaft, or by seeing a reflection from the surface opposite the one containing an angular crack.

This type of ultrasonic exam does not work well if the shaft diameter decreases as the beam travels from the inspection surface. It will work well, however, if the beam is traveling from a smaller to larger diameter, except that the larger shaft may not be adequately covered by the beam. In that case, an inspection from both ends of the shaft is warranted, with testing from the large end only to the diameter-transition point. Any abrupt change in shaft dimensions is a place where discontinuities can be expected.

Figure 8 illustrates the ultrasonic detectability of a crack at the shaft's thickness-transition point. The crack may be detected from the smaller-diameter end of the shaft more reliably than from the larger end. ■

Looking ahead

The next installment of this Refresher series, Part 6, will cover applications of nondestructive testing for improving process operations.

References

1. Metal Handbooks, 9th ed., Vol. 11, Failure Analysis and Prevention, American Soc. for Metals, Metals Park, Ohio, 1986.
2. Lee, R. P., How Poor Design Causes Equipment Failures, *Chem. Eng.*, Jan. 31, 1977.
3. McGuire, R. E., Knowledge is Key to Stress Corrosion Battle, *Oil and Gas J.*, June 9, 1986.
4. Lee, R. P., Tracing the Causes of Metal Failures in Process Equipment, *Chem. Eng.*, Sept. 13, 1976.
5. Thielsch, H., Defects and Failures in Pressure Vessels and Piping, Reinhold Publishing Corp., New York, 1965.
6. Yokell, S., Trouble Shooting Shell-and-Tube Heat Exchangers, *Chem. Eng.*, July 25, 1983.
7. Fowler, T. J., Acoustic Emission Testing of Vessels, *Chem. Proc.*, September 1988.
8. Thielsch, H., Why High-Temperature Piping Fails, Pressure Vessels and Piping: Design and Analysis-A Decade of Progress, Vol. 4, Quality Assurance, H. Waite, ed., American Soc. of Mechanical Engineers, 1976, p. 350.
9. Bloch, H. P., and Geiter, F. K., Practical Machinery Management for Process Plants, Vol. 2, Machinery Failure Analysis and Trouble Shooting, Gulf Publishing Co., 1983.
10. Turner, M., and King, R., Should You Worry About Corrosion Monitoring?, The Chemical Engineer, February 1986.

NONDESTRUCTIVE TESTING VIA OPTICAL AND ACOUSTICAL METHODS

Frank J. Sattler*
Glitsch Field Services/NDE, Inc.

Novel techniques can measure surface displacements, evaluate discontinuities, and locate flaws as they form

Some relatively new and lesser-known nondestructive testing (NDT) techniques that rely on optics or acoustics to detect discontinuities in materials are covered in this fourth installment of CE's Refresher†. These methods include: optical holography, which is used to measure and compare surface displacements; acoustical holography, which usually serves as an evaluation tool once a discontinuity has been located by some other method; and acoustic emission testing, which can locate discontinuities as they begin to form, before they can be detected by other methods.

Frequently more than one NDT method must be used for a particular material in order to find and evaluate discontinuities. Each technique has its own limitations and advantages, and often an advantage in one application is a limitation in another. Because of the large number of tests available and the tests' sensitivities, care must be taken in selecting a method for a given application.

Holography is a technique for form-

* For information about the author, see *Chem. Eng.*, April 1989, p. 119.

† Previous installments of this series on nondestructive testing covered: NDT in general, April, pp. 116–119; surface examination methods, July, pp. 125–133; and techniques for finding below-surface defects, September, pp. 161–167.

ART: STEVE HAIMOWITZ

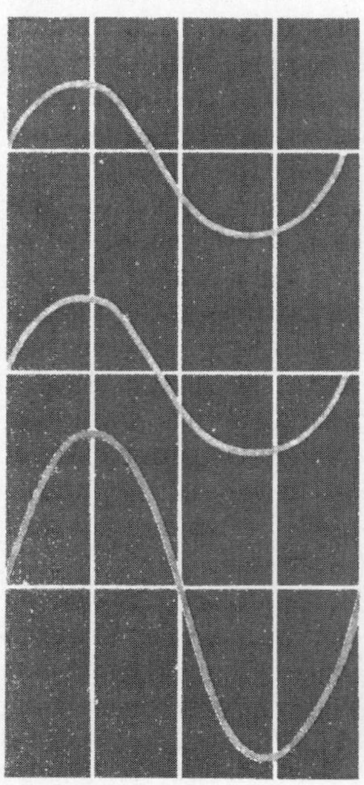

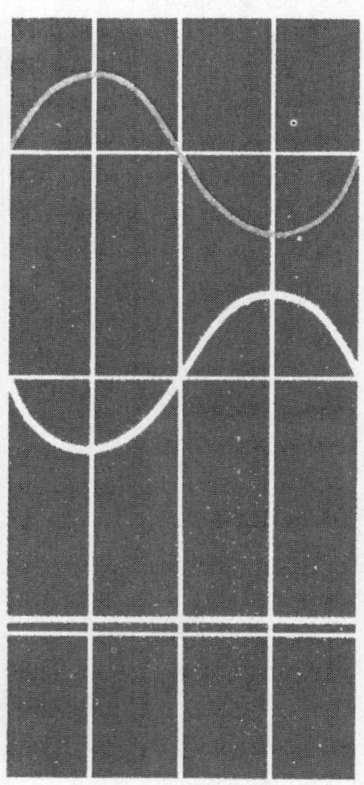

FIGURE 1A. Constructive interference results when two waves are identical and in phase; the amplitude of the combined wave (shown at the bottom of each panel in Figures 1a, 1b, and 1c) is twice that of the other two

FIGURE 1B. When two waves are identical but 180 deg out of phase, destructive interference results and the two waves cancel each other out

FIGURE 1C. When two waves are out of phase by an amount other than 180 deg, the waves are still additive, but the amplitude of the combined wave depends on the phase difference

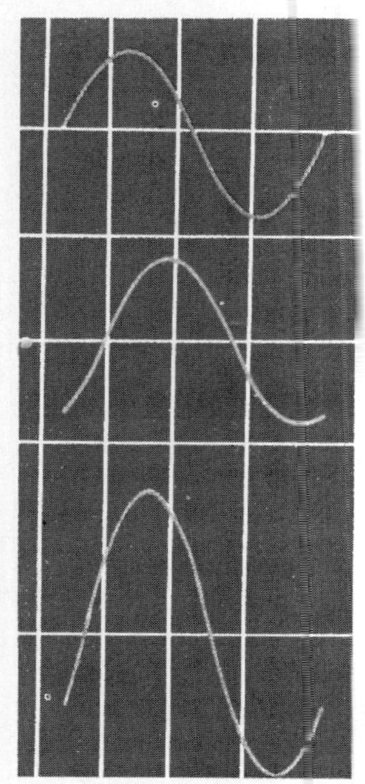

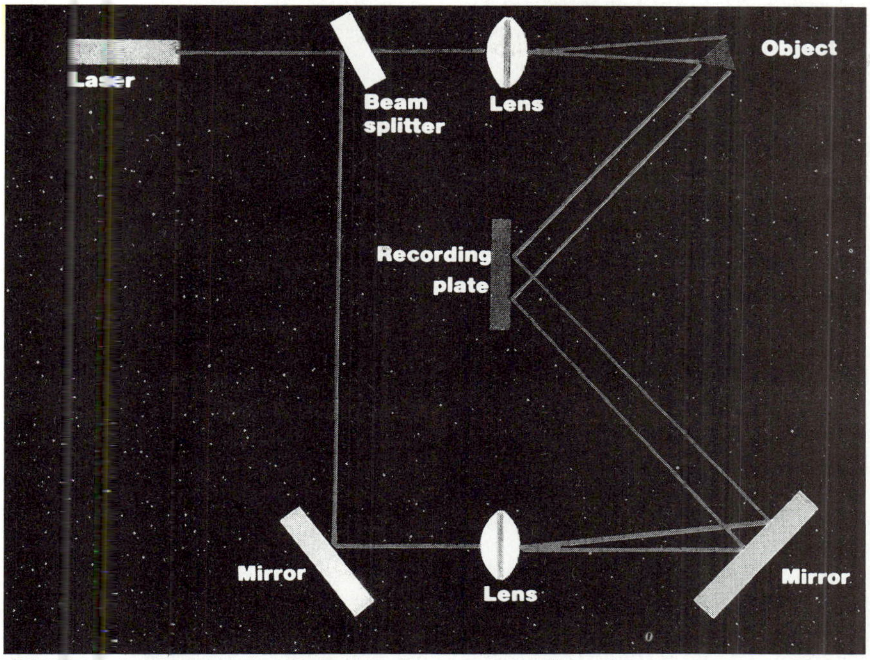

FIGURE 2. Interference between the object wave and a reference wave produces a holographic image on a recording plate

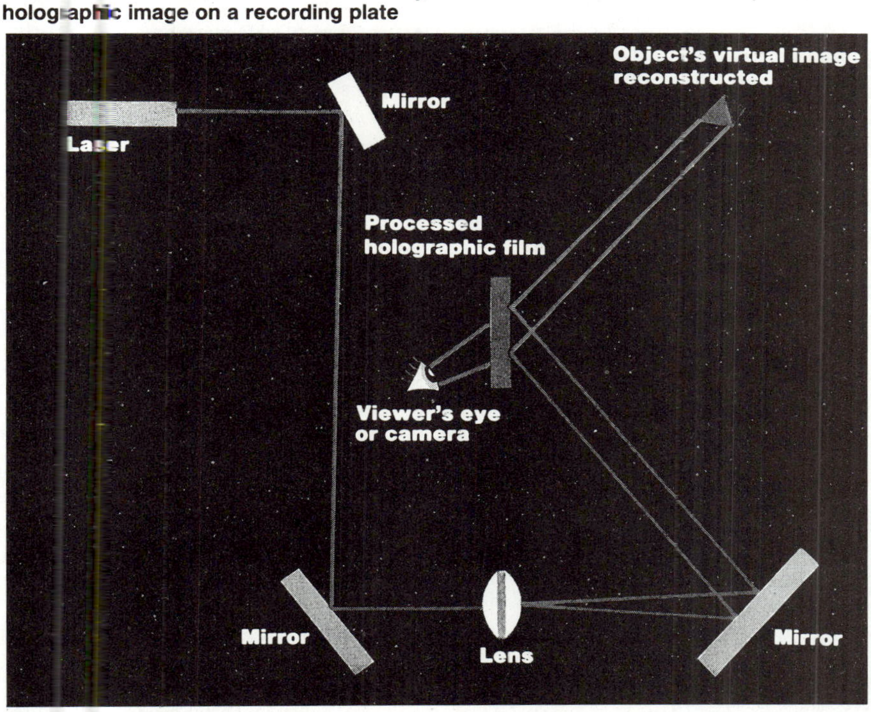

FIGURE 3. To view the hologram, the recording medium is illuminated by coherent light, which reconstructs a virtual image of the test object

ing a three-dimensional image by recording and later reconstructing the amplitude and phase distributions of a coherent-wave disturbance. The wave may be optical, acoustical, ultraviolet, infrared, microwave or X-ray, so long as it is coherent (i.e., monochromatic, or single-wavelength). Nondestructive testing involves either optical or acoustical holography. The image produced by holography, generally on a photosensitive medium by light, is called a hologram. Many people are familiar with holograms, as they are often incorporated into their bank credit cards.

Optical holography

Optical holography is based on the interference between a reference wave and an object wave (or between two parts of an object wave). Interferometers are used to measure wavelengths, wave velocities, distances and directions. In optical holography, distance differences show up as fringe patterns — i.e., series of light and dark bands produced by interference of light waves — in the hologram.

Interference occurs when two waves pass a point at the same time. If the two waves are in phase and have the same frequency and amplitude (and therefore the same wavelength), they combine to form a third wave having the same wavelength but twice the amplitude (Figure 1a). If the two waves are out of phase by 180 deg, destructive interference results and the two waves cancel each other (Figure 1b). When the phase difference is other than 180 deg, the waves will form a combined wave with the same wavelength, but a different maximum amplitude (Figure 1c).

A typical system for producing optical holograms is shown in Figure 2. A beam of coherent light from a laser is split into two separate beams. One of the beams is directed towards the object being examined and reflected toward the recording medium (usually a high-resolution photographic plate). The second part of the beam takes a different path and is reflected toward the recording medium without striking the object. The interference of the two light beams is a function of the path length of the light beam reflected by the different features of the object and the path length of the reference beam.

The holographic image (once the photographic plate has been developed) is an irregular pattern that consists of a series of interference fringes, most of which are not visible to the unaided eye. To visualize the original scene recorded in the hologram, the image must be reconstructed. This is done (Figure 3) by illuminating the hologram with coherent light, generally from the same source as was used to make the hologram. The laser light reconstructs the wave relationships and produces a virtual three-dimensional image (an image from which rays of reflected or refracted light appear to diverge) of the original object.

When looking at the reconstructed image, it is as if one has the original object to view. Changing the angle at which the hologram is viewed has the same effect as if one were looking at the object itself from different angles.

In a nondestructive test, the reconstructed image and the object surface are superimposed. Loads are then applied to the object. Any discontinuities in the object will cause uneven displacements, and these can be visualized as differences between the object and the holographic pattern superimposed upon it (Figure 4). Even though such displacements may be very small, movements on the order of a fraction of the light wavelength can be detected. For example, the light from a helium-neon gas laser (which is deep red in color) has a wavelength of about 6.3×10^{-7} m, or 0.25×10^{-6} in. The sensitivity of the technique to changes in contour are a fraction of this wavelength.

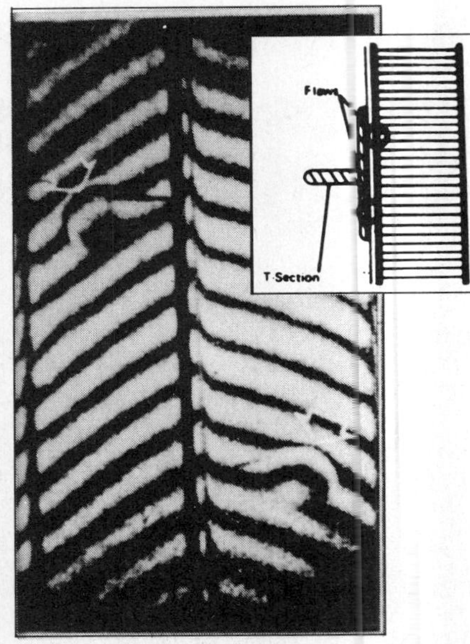

FIGURE 4. In this hologram of a honeycomb laminate, flaws show up as changes in the interference fringes

Some limits to testing

One of the limitations in performing this type of test is that external sources of displacement (other than those loads applied intentionally to conduct the test) must be prevented from moving the object. In the laboratory, this means that a special table must be used to isolate the holographic equipment and the object from external vibrations. This limitation often precludes the use of optical holography in a plant environment. However, a relatively new technique called shearography, developed by Laser Technology, Inc. (Norristown, Pa.), can be used on large structures or parts of large structures.

Electronic shearography was originally developed for surface strain measurements. It can produce an image of an object as large as 15 × 15 ft, or as small as microscopic.

The surface to be examined is illuminated with a divergent laser beam (Figure 5). The light reflected by the discontinuity on the surface passes through special image-shearing optics and is picked up by a video camera. Light reflected from adjacent points on the surface also passes through these image-shearing optics, producing interference that is recorded as the hologram (or shearogram).

The brightness of the video image is a function of the interference of light reflected from matched pairs of points on the surface of the test object (i.e., a point on the discontinuity and a corresponding point elsewhere on the surface). Where the interference is constructive, the image will be light; where the interference is destructive, the image will be dark. Since comparisons are being made between points on the same surface that are subject to the same external vibrations, isolation of the components is not needed.

The application of stress to the com-

FIGURE 5. Electronic shearography can be used on large objects

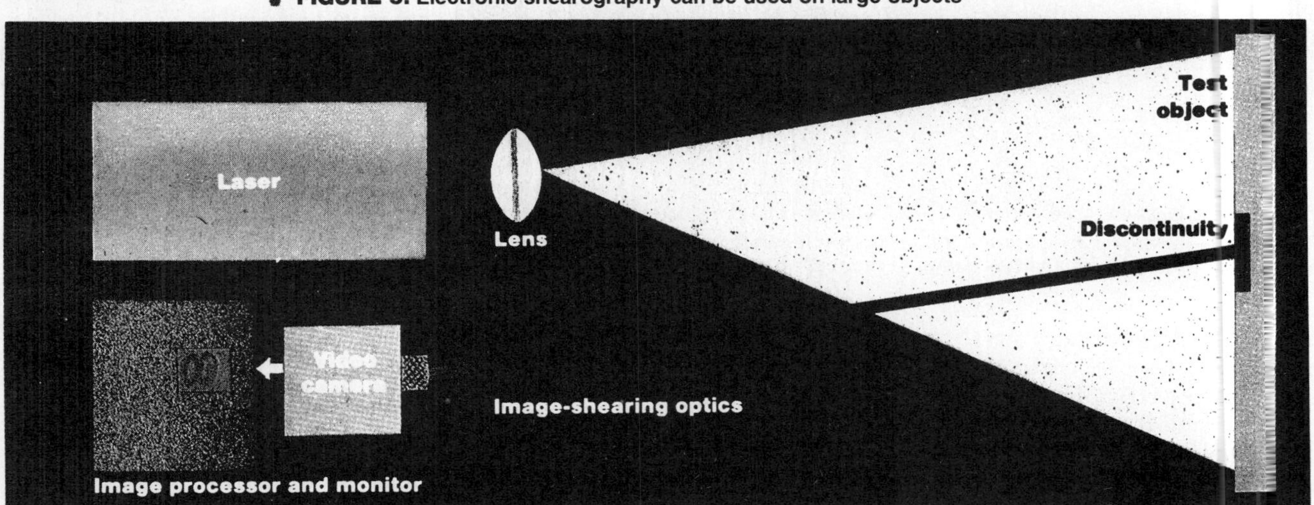

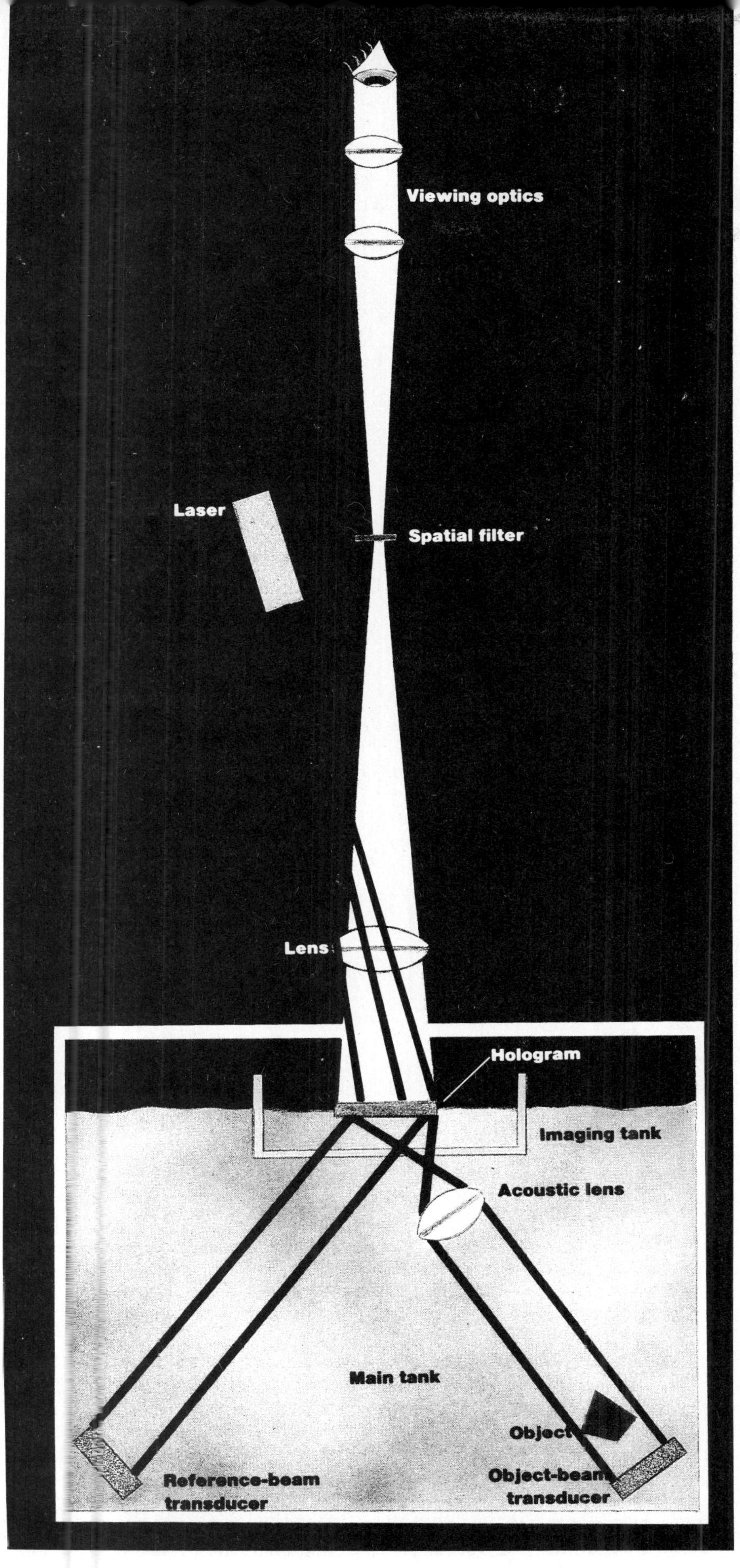

FIGURE 6. **Acoustical holography by the liquid-surface technique involves transmitting an ultrasonic beam through the test object**

ponent results in different surface displacements wherever there is a defect. The image processor compares the original hologram of the surface with the present stress conditions, and any nonuniform displacement will appear as light and dark fringe patterns on the television monitor.

Acoustical holography

Acoustical holography is similar to optical holography. However, in the former, images are created using a coherent acoustic or ultrasonic beam (rather than a coherent light wave) that either passes through or has been reflected from within an object. Although confined to the laboratory for many years, the technique has been used in manufacturing, nuclear and hydroelectric plants, and on vessels and piping similar to those found in chemical-process-industries plants. It is often used as a supplement to conventional ultrasonic testing (*Chem. Eng.*, Sept., pp. 164–166), or to other NDT methods.

In pulse-echo ultrasonic testing, a discontinuity (or other feature of the object being tested) is characterized by the amplitude and travel time of the wave reaching the receiving transducer. The returning beam (reflections from different features of the test object) forms signals called indications; an indication can actually consist of reflections from several discontinuities, or different parts of the same discontinuity. (The discontinuity, or whatever else within the object reflects the ultrasonic beam, is called a reflector.) Characterizing the reflectors causing the indication can be very difficult.

The principal advantage of acoustic holography is its ability to give a three-dimensional view of the reflectors within a solid body so that in reconstructing the image, the various size and depth relationships can be seen. Acoustical holography overcomes one of the main limitations that hamper the interpretation of ultrasonic results.

Two types of recording have been

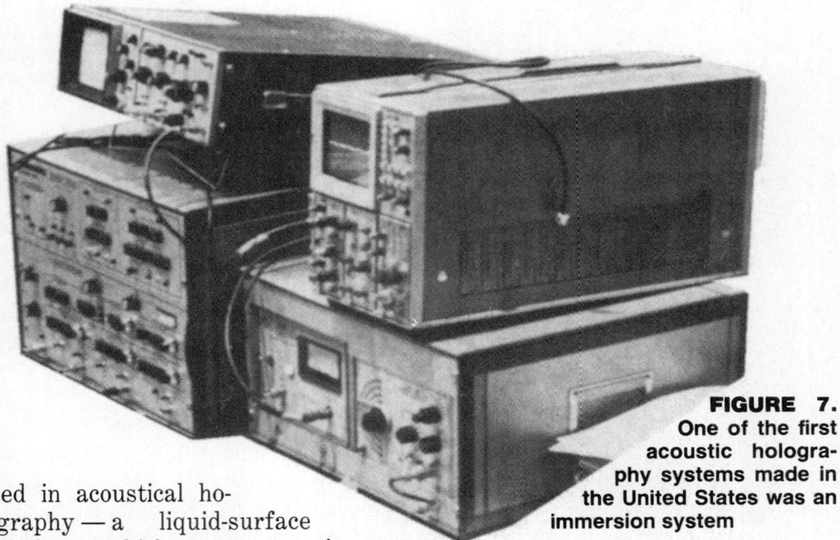

FIGURE 7. One of the first acoustic holography systems made in the United States was an immersion system

used in acoustical holography — a liquid-surface technique, which creates an image based on the sound intensity of a beam transmitted through the test object, and a scanning system, which creates an image using a reflected beam. With the liquid-surface technique, one must have access to both sides of the test object, whereas the scanning system requires access to only one surface.

The wave theory describing acoustical and optical holography is the same. The interference of an object beam and a reference beam is the same, except that the wavelength of the ultrasonic beams is somewhat larger than the wavelength of the light used in optical holography. The use of an ultrasonic beam, however, allows the object to be characterized by more than just a surface reflection. Since ultrasonic waves can penetrate an object, internal conditions that reflect, scatter and absorb ultrasonic waves can be visualized based on changes in the intensity of the beams. Because the waves are coherent, the positions of the internal features causing the changes can be determined from the hologram.

Two basic techniques are used to obtain an acoustical hologram. In the first — the liquid surface technique (Figure 6) — the ultrasonic beam is transmitted through the test object. The hologram is formed by combining the transmitted beam with a reference beam emitted from a separate transducer.

To do this, the transducers are placed in a tank containing water that also holds the object being tested. A second, smaller tank, made of a thin, rubber-like material, is suspended on the surface of the larger tank where the beams interfere. To compensate for distortion that occurs because of the difference in the ultrasonic and laser-light wavelengths, an acoustic lens is used to focus the object beam on the liquid surface in the second tank. This produces a focused hologram in which one plane of the object (about 1/8th in. in distance along the ultrasonic wavepath) can be distinguished. Various planes can be focused on by moving the lens or the object, thus allowing the entire thickness of the object to be imaged.

The interaction of the reference and object beams produces a ripple pattern on the liquid surface in the smaller tank. This ripple pattern is the acoustic hologram, and is a function of the sound beam intensities and phase, and the impedance mismatch at the surface between the liquid and air. (This refers to the difference in the acoustic wave's impedance in liquid and air that occurs at the liquid-air surface, resulting in almost total reflection of the wave.)

The information in the hologram is revealed by exposing the surface to a divergent beam of laser light. When viewed with the proper optics, the reconstructed image is a real-time image, which can also be photographed or reconstructed at a later time. One of the first acoustic holography systems made in the United States is shown in Figure 7.

The liquid-surface technique has severe limitations for use in a plant environment: The tank must be large enough to contain the object to be tested, and both surfaces need to be accessible. For small objects, such as studs, bolts and nuts, these constraints are not a problem, but for vessels and piping, such conditions usually cannot be met. Instead, for CPI plant applications, a scanning system (Figure 8) can be used.

In the early scanning systems a separate, fixed-position transducer sent ultrasonic waves into the object while a small-diameter receiving transducer was scanned over the surface of the object. The signals arriving at the receiving transducer were a combination of the interactions of the wavefronts from various areas of the object. This received signal was converted to an electrical signal by the transducer.

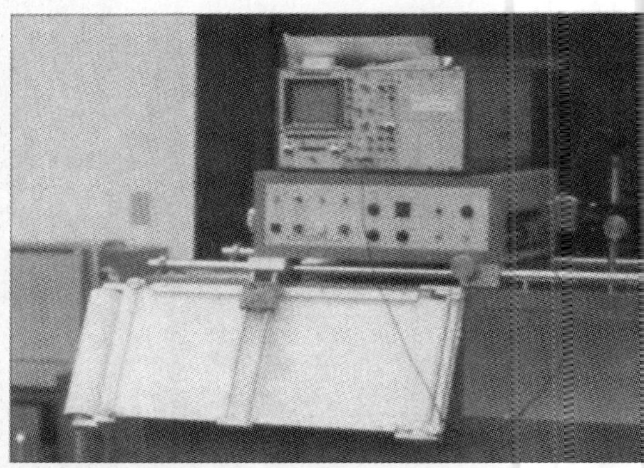

FIGURE 8. Scanning acoustic holography electronics.

The reference wave was not obtained from a reference transducer, but was simulated by obtaining an electrical signal from the oscillator driving the sending transducer.

The received signal and the reference signal were used to modulate the intensity of a light spot that was linked (either mechanically or electrically) to a second scanner that served as a light source. As the scanner and the light source scanned over the object, a hologram was recorded containing phase and amplitude information about the ultrasonic beam. Size and location information about any discontinuities in the image was then obtained by using a laser light source and optical system.

Because the ultrasonic waves and the reconstructing laser light had different wavelengths, distortion in the image was present. This problem has been overcome in the latest scanning acoustic holographic systems, where the holographic information is obtained through computer analysis using Fourier transforms. The computer printouts of discontinuity information (Figure 9) yield better images with less distortion than can be obtained using the techniques discussed previously, such as the liquid-surface or the mechanical scanning techniques.

Since many scan lines are required to obtain the holograms, a 6-in. × 6-in. area might take as long as 1 h to scan. This limits the use of the scanning system to areas where the hologram can provide additional information to supplement acoustic emission, conventional ultrasonic or other nondestructive test data. This time, however, has been very worthwhile in several instances where the true size of the indication was in doubt.

Scanners can operate in a variety of positions, and can use immersion-coupling ultrasonic techniques with bubblers (i.e., squirters) that provide the water couplant for the ultrasonic beam. A variety of pipe and vessel geometries have been scanned in field applications with such a system in many different orientations.

Acoustic emission testing

All the NDT methods covered so far in this series of articles use either a probing beam of energy or the flow of matter to form indications of discontinuities. Acoustic emission testing involves neither of these active methods of locating flaws, but rather depends upon the material itself for the signals that indicate the presence of a flaw. Acoustic emissions are the sounds (and ultrasonic waves) that a material emits when it is under stress, and these sounds can reveal many things about the material.

This method of nondestructive testing has grown rapidly over the last 20 years, with many of its applications involving containment systems in CPI plants. It can be used to monitor vessels, piping or other structures during operation, and can identify the areas where problems exist in a structure in a short period of time. This method of examination has also been used to monitor hydrostatic tests of vessels with dependable results.

The application of loads to a material results in either the release of strain energy as a crack grows or the release of chemical free energy if a phase transformation occurs. Both forms of energy are released as acoustic waves. These emissions provide a sensitivity that cannot be achieved by other methods of nondestructive testing. For all other NDT methods, the flaw must be large enough so that the discontinuity interacts with the probing energy beam or is capable of accepting some flow of matter. For acoustic emission testing, even the onset of cracking is detectable, and the emission rate increases as the flaw grows. Both static and dynamic load conditions can generate emissions, as can residual stress conditions in a member such as a weld. The acoustic emission test is a real-time test, and it can only detect discontinuities that are affected by some applied or existing stress.

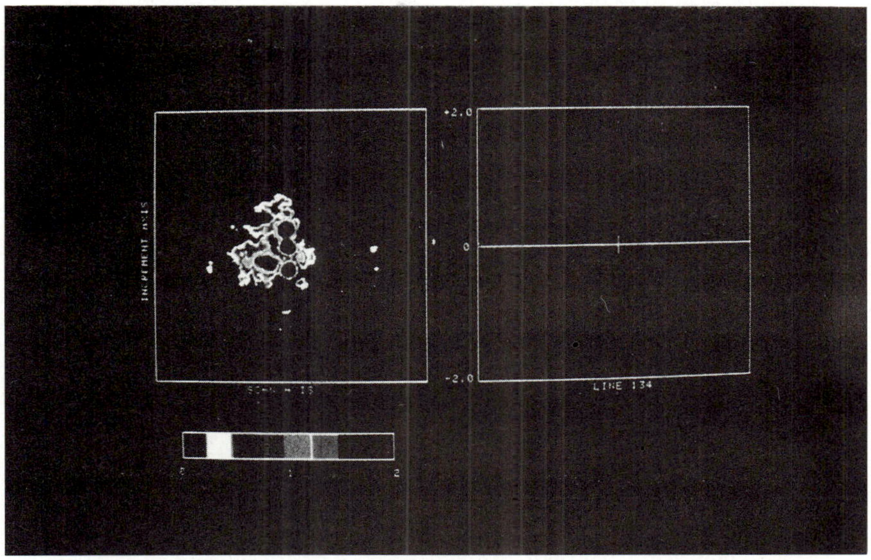

FIGURE 9. Reconstructed acoustic holographic color image of a flaw, showing more intense reflections as the black part of the image.

Active flaws detectable

One of the main advantages of acoustic emission testing is that it locates active (i.e., growing) flaws. Discontinuities in a structure that are unaffected by the stress conditions in local areas do not release any energy and are not significant. Conversely, if a discontinuity is emitting stress waves as a result of the loading conditions, its location and significance can be determined by the characteristics of the emission.

FIGURE 10. Acoustic emissions from a discontinuity propagate as a spherical wave front

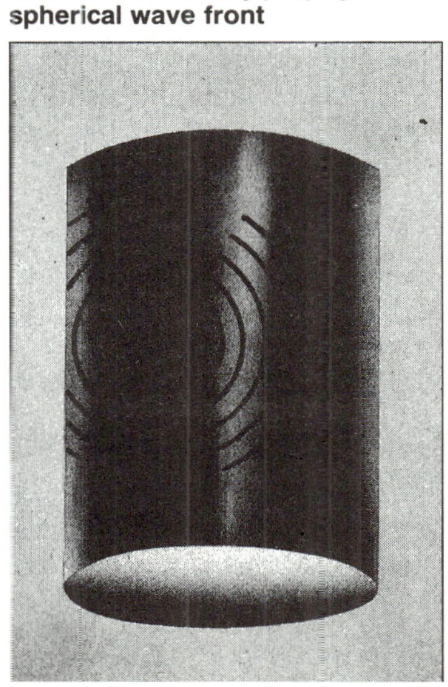

This ability allows one to concentrate repair efforts on areas where signifi-

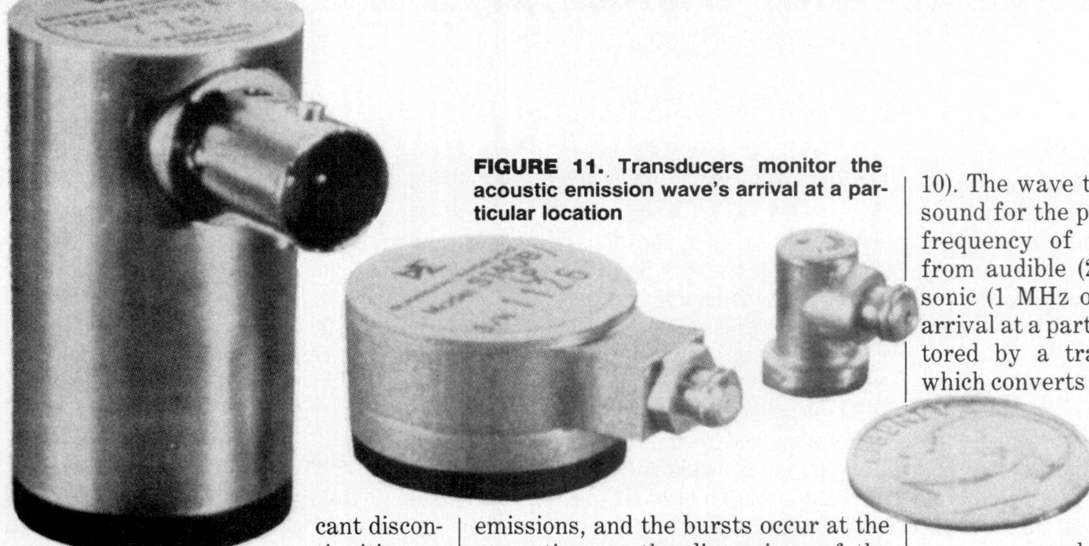

FIGURE 11. Transducers monitor the acoustic emission wave's arrival at a particular location

cant discontinuities are present. In all the other methods of nondestructive testing, only the flaw's presence and size can be determined, and the significance of the discontinuity must be determined from stress-analysis and fracture-mechanics calculations.

There are various causes for acoustic emissions. In a material, certain areas will contain stress concentrations before the overall structure is stressed to that same high level. These high stress concentrations result in local movements that cause plastic (i.e., permanent) deformation, slip (i.e., the movement of small defects in a metal's crystalline structure), and crack formation and growth. The flaw is the source of the acoustic activity.

Two types of acoustic emissions have been identified — continuous and burst. When an unflawed metal specimen is monitored during loading, the acoustic emission rate continuously increases to a maximum near the material's elastic limit. As additional strain is added (taking the material into the plastic region), there is an exponential decay and the emission rate approaches zero. This is termed continuous emission, since the activity is characterized by a very high number of pulses rather than discrete events. Once a material has been stressed to a certain level and the load is then reduced or removed, the material will not exhibit continuous emissions (when loading is reapplied) until the prior maximum stress level is again reached. This phenomena is called the Kaiser effect.

When a metal specimen containing a discontinuity is stressed, the acoustic emissions are produced as high-amplitude bursts. These are called burst emissions, and the bursts occur at the same time as the dimensions of the change. These burst emissions contain much more energy than do the continuous emissions. As the flaw grows, the rate of acoustic emissions continuously increases, up to where fracture occurs.

The acoustic emission pulse that is generated (by either continuous or burst emissions) radiates from the site as a spherical wave front (Figure 10). The wave travels at the speed of sound for the particular medium. The frequency of the pulse can range from audible (20-Hz range) to ultrasonic (1 MHz or higher). The wave's arrival at a particular location is monitored by a transducer (Figure 11), which converts the mechanical energy of the pulse to an electrical signal.

These signals are characterized as damped sinusoidal pulses with very fast rise times (Figure 12). The wave itself resembles the wave shown in the top part of Figure 12, but the wave that is seen by the instrumentation is a function of the characteristics of the pickup transducer. After the signal is amplified, it is further processed into a digital signal, as shown in the lower part of Figure 12. These digital signals can then be counted, and the signi-

FIGURE 12. Stress-wave emission signals are seen as a damped sinusoidal wave by the transducer (top), and as a digitized signal after processing (bottom)

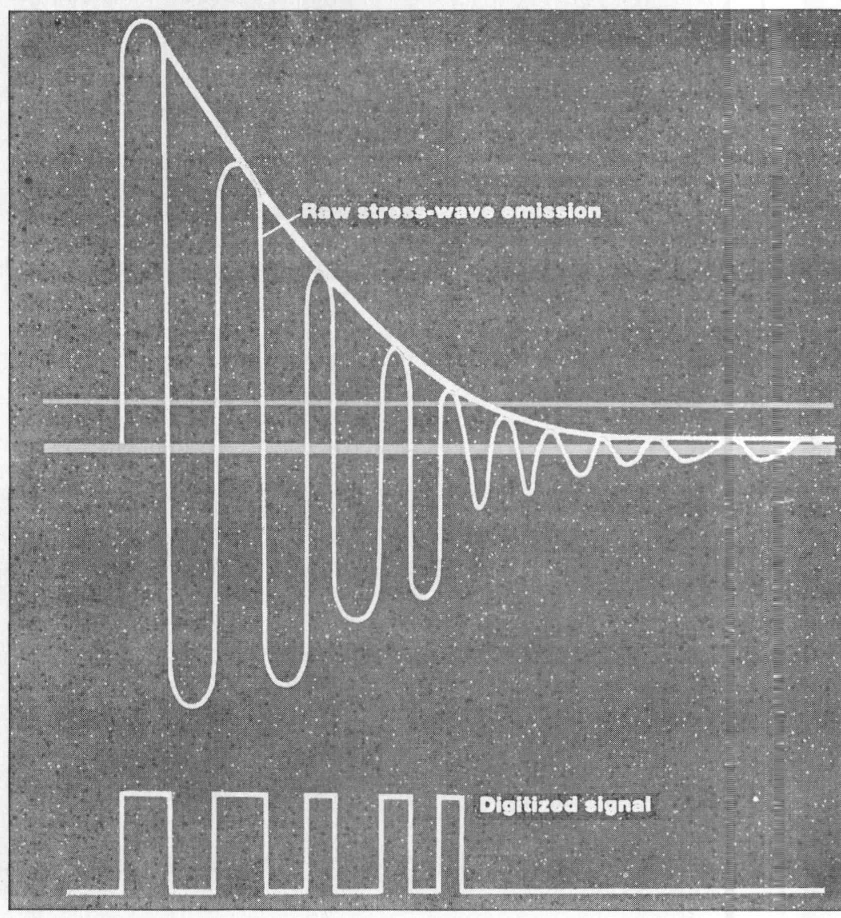

Physical Acoustics Corp.

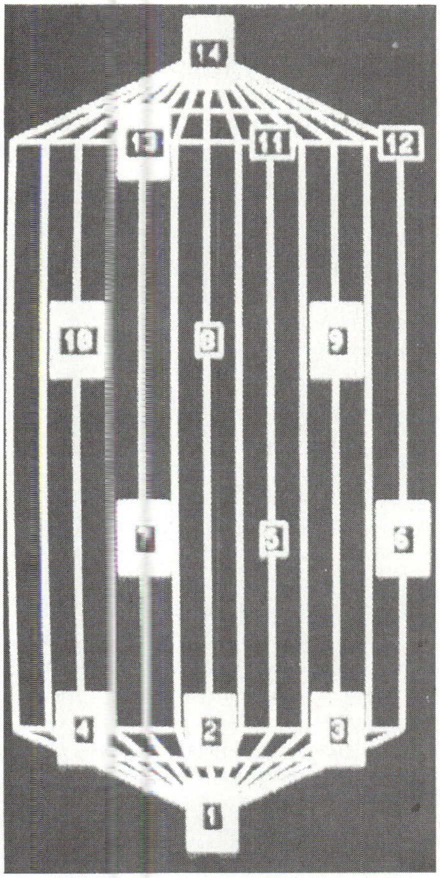

FIGURE 13. A display of acoustic-emission test results is superimposed on a vessel drawing to show locations of emission sites

cance of the discontinuity determined.

Several other types of readouts and signal presentations may also be used. In addition to emission rate, common displays show the sum of emissions, amplitude and power; the sum and rate of emissions are the most common. The data can appear in analog or digital form, and can be read via meters or permanent (for example, stripchart) recordings.

When performing an acoustic emission or other test, one wants to know where any discontinuities are located. Since the acoustic emission signals travel in spherical wave fronts at a constant velocity, the time of travel between the emission of the acoustic wave to the arrival of the pulse at a transducer can be measured. With suitable multichannel instruments, several transducers can be placed around the system being monitored. The arrival time of a particular pulse at the various transducers can then be used to determine the location of the emission source.

Because the stress waves travel at about 13,000 miles per hour in steels and aluminum, time resolutions must be in the microsecond range in order to locate discontinuities within 1 in. of their actual location. In addition, quick turn-around times for the testing require that these calculations be made with a computer-based system with an appropriate data display, such as that in Figure 13.

Many applications for acoustic emission testing in CPI plants exist. Acoustic emission test are regularly performed on critical vessels, such as those holding hazardous fluids, and on other large storage tanks. Monsanto Co. (St. Louis) has used the technique on various types of vessels [*1*]. The company's early efforts were more successful on fiber-reinforced plastic

FIGURE 14. A hydrostatic test on a zirconium column is monitored by acoustic emission testing

tanks than on metallic vessels. With titanium and zirconium vessels, acoustic emission testing had the advantage of detecting a specific type of weld contamination. ■

Looking ahead

We have now covered the primary methods of nondestructive testing, in Parts 2 through 4 of this Refresher series. The next two installments will cover applications of NDT. Part 5 will deal with NDT for plant maintenance, and Part 6 will cover NDT for improving process operations.

References

1. Fowler, T. J., Acoustic Emission Testing of Vessels, *Chem. Eng. Prog.*, September 1988, pp. 59–70.
2. Boyer, H. E., Nondestructive Inspection and Quality Control, *Metals Handbook*, 8th ed., vol. 11, 1976.

FIGURE 15. A newly constructed butane sphere undergoes testing prior to plant startup, and fabrication discontinuities were detected in the welds

FINDING BELOW-SURFACE DEFECTS WITH NONDESTRUCTIVE TESTING

Three methods that use 'probing energy,' such as penetrating radiation, ultrasound or electric currents, can characterize defects in a material

Frank J. Sattler*
Glitsch Field Services/NDE, Inc.

This third installment of CE's Refresher† on nondestructive testing (NDT) covers the principles of "volumetric" testing methods — those techniques that can detect discontinuities below the surface of a material (i.e., within its volume). These methods include penetrating radiation, ultrasonic testing, and eddy-current testing.

Some surface or near-surface discontinuities detectable by the surface methods (covered in the second installment) can also be found by the volumetric methods covered here. Different NDT methods can often find the same discontinuities. However, it is often best to apply more than one method of testing, since not all have the same sensitivity for a particular type of discontinuity. The variations that can occur in a discontinuity's appearance, location, angle or other features, or in how easily the inspector and equipment can access the surface for testing, can make one method better in a given application.

Penetrating radiation methods are based on the ability of short-wavelength radiation to penetrate matter. Most of us are familiar with penetrating radiation in medical and dental applications. There, film is used as the detecting medium, upon which is an image that represents the intensity of the radiation beam as it exits the part of the body being X-rayed. As the film is viewed with the aid of an illuminator, the intensity of the X-ray beam after passing through the part is seen as light or dark areas on the film. Light areas correspond to areas of greater density or thickness in the test object, dark areas to lower-density or lower-thickness. A crack in a bone or tooth decay will show up on a radiograph as dark lines or dark areas against the more dense bone or tooth, since more radiation passes through the damaged area of the test object.

Similarly, materials of many different types can be tested by radiography, utilizing a penetrating radiation beam — either X- or gamma-radiation. Film is just one means of measuring the radiation intensity; other applications will be discussed where nonfilm methods of detection are used.

Ultrasonic testing employs a different form of probing energy. Here, a beam of mechanical vibrations is sent into a test object to perform different functions, such as discontinuity detection or thickness measurement. (The ultrasonic beam is caused by the vibration of a special crystal in a transmitter. These vibrations are coupled through some medium (water or another liquid) to transmit them from the generating crystal into the test object. Once the ultrasonic beam is inside the test object, it will be reflected and scattered by discontinuities and geometric features within the test object. Either these reflections or the transmitted beam are monitored by reconverting the mechanical wave to an electrical signal whose amplitude, time of arrival, frequency or another attribute can reveal some property of the material being tested.

Eddy-current tests — which can be used to identify different alloys, perform thickness measurements or locate discontinuities — employ yet a third type of probing energy. Eddys are currents that move contrary to the direction of motion of the main current, typically in a circular fashion. The eddy currents themselves are electrical currents generated in an electrical conductor because of a changing magnetic field. Circular (helical) coils of wire are used to generate eddy currents. A high-frequency elect-

* For information about the author, see *Chem. Eng.*, April 1989, p. 119.
† Previous installments of this series on NDT appeared in April, pp. 116–119, and July, pp. 125–133.

Originally published September 1989

cal current is applied to the coil, which in turn generates a magnetic field external to the coil. When this magnetic field is brought into proximity of an electrical conductor, eddy currents can be generated within the conductor. This is because certain properties of the conductor, such as its electrical conductivity, thickness and magnetic permeability, cause an interaction with the magnetic field, and this in turn causes the eddy-current flow. The interaction of the currents with discontinuities, thickness changes and different materials can then be determined on a readout device that is part of the eddy-current testing instrument.

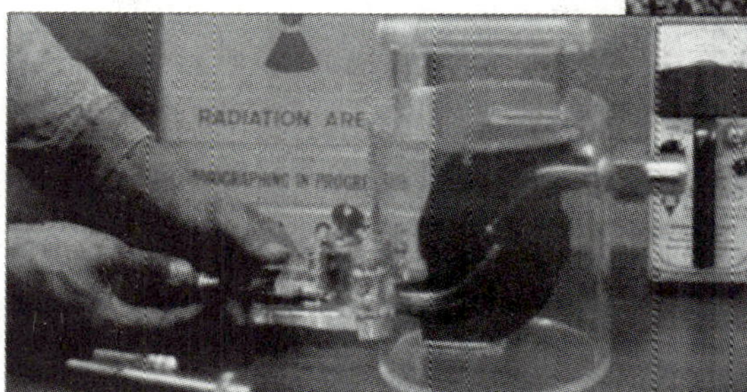

FIGURE 1. Isotope camera (demonstration model, used for training) showing how the drive cable is hooked up to the source cable. Note the safe location of the radioactive source inside the center of the S-tube

FIGURE 2. An isotope camera is set up to examine a weld on the opposite side of the pipe (at right)

The location of a discontinuity below the surface or on a hidden surface makes the discontinuity invisible to some methods of detection, depending upon how far below the surface it might be. These types of discontinuities cannot be seen by visual or liquid penetrant methods. In ferromagnetic materials only, magnetic particle methods can find some deeper discontinuities, but then the depth of penetration is limited to just below the examination surface. If a fabrication error or service degradation has caused a discontinuity within the material, on a hidden surface, or at a location too small for access, only a volumetric method will be able to locate it.

Penetrating radiation

X-radiation and gamma-radiation have the ability to penetrate solid matter. This ability is related to three factors: the thickness and density of a material, the type of material (which will determine how many electron shells, electrons and nuclei are present), and the energy of radiation (the absorption effect of which is influenced by the type of material). The radiation energy is an important factor in the sensitivity and penetrating ability of the radiation beam. For very dense and thick materials, special high-energy accelerators are necessary to generate radiation of the proper wavelength.

Two of the primary absorption mechanisms involve interactions of electrons with X- or gamma-rays. These are called the photoelectric effect and Compton scatter. In photoelectric-effect absorption, all of the energy of the photon (the X- or gamma-ray) is given up to an orbital electron in the material being examined. In Compton-scatter absorption, only part of the photon energy is given to the electron, with the remainder given off as a secondary X-ray scattered at an angle to the original radiation. A third absorption mechanism, pair production, involves higher energies, generally over 1.02 million electron volts (MeV) of radiation energy. Pair production results in the generation of a positive and a negative particle as the photon approaches a nucleus of the absorbing material.

Each of these mechanisms affects absorption differently as a function of energy and the absorbing material. In general, at low energies, photoelectric effect is the dominant mechanism. From 400–1,200 kV, Compton scatter is more dominant. At energies above 1,000 kV, pair production is the dominant absorption mechanism. When scatter dominates, poorer radiographs result.

The wavelength of radiation is determined by the energy forming the X- or gamma-ray. In an X-ray tube, the minimum wavelength is determined by the maximum energy the electrons receive when accelerated through the high potential from the cathode to anode in the tube. Special high-energy generators, such as betatrons and linear accelerators, are used to produce very-small-wavelength radiation for evaluating thicker or denser materials.

Gamma-ray energy is determined by the particular radioactive isotope involved and its decay scheme. A radioactive material is one that is unstable be-

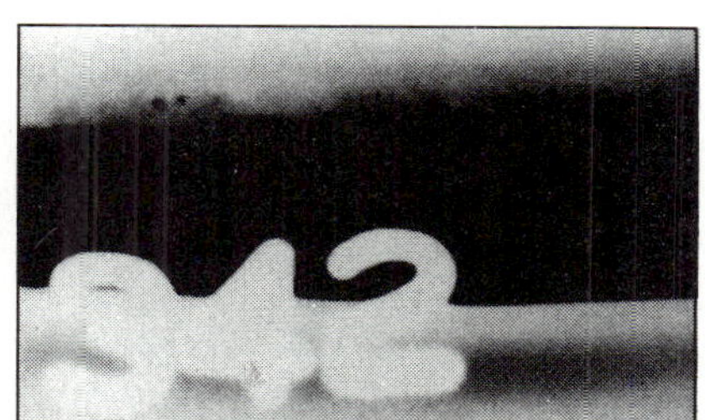

FIGURE 3. Radiograph of a heat exchanger's tube-to-tubesheet weld. Notice the porosity above the numbers 3 and 4 of the identification number 342

cause it has too much or too little energy. The radioactive isotope decays to a stable energy state through one or more releases of energy — either particulate radiation (alpha and beta particles) or gamma rays or both — or it captures a k-shell electron and decays to a lower-energy state. Only the higher-energy gamma-rays with energies above 50 keV are generally useful for radiography. A particular radiographic isotope may have a single gamma wavelength or several wavelengths.

Although radiographic isotopes exist in nature (e.g., radium), commercial isotopes today are obtained through activation in a nuclear reactor. Neutrons freed through fission are captured by the nucleus of certain atoms to create new isotopes. The new isotopes are unstable and will periodically give up their excess energy spontaneously. The rate of decay and the number of unstable atoms determines the strength of the isotope. The number of gamma-rays given off decreases as time goes by, but their energy (and their wavelengths) will always be the same. Cobalt-60 (^{60}Co) and iridium-192 (^{192}Ir) are common radioactive isotopes used in industrial radiography and radiation gauging.

In a radiographic NDT examination, a source of radiation is placed on one side of the item being tested. After the radiation passes out of the material, it can then be detected by either a special detector or X-ray film. The intensity of the radiation passing through to the detector can then be related to the density or thickness of the absorbing material.

About the only way to sense these penetrating radiations is to detect the ionizations they cause. No human sense can detect the presence of radiation. However, various means of measuring ionizations can reveal the radiations, either in real time or with an integrated image formed over some time period.

X-ray film reveals the presence of the radiation in the same way that photographic film reveals the presence of light: The higher the radiation intensity, the darker the film. Thus, dark areas in the radiograph correspond to either low-thickness or low-density areas in the test object. A change in density can occur, for example, as a result of a segregation, which is a pooling of certain elements of an alloy in a casting, or high-density inclusions in welds due to contamination from welding electrodes.

Figure 3 is a radiographic image of a crevice-free tube-to-tubesheet weld in a heat exchanger, showing three pores, the largest being 0.030 inch, immediately above the numbers 3 and 4 in the identification number 342. The largest pore exceeded the acceptance standard for this heat exchanger.

Radiation can also be monitored by gas detectors, where ionization takes place in a captured gas and the number of ionizations can be measured and calibrated against the strength of the radiation field. The number of ionizations per unit time is proportional to the radiation intensity. To measure low-intensity radiation beams (and to select the energy range to be measured), special light-emitting crystals are used with high-gain photomultiplier tubes. Low intensity sources can be used to measure the transmission through large objects such as a distillation tower. This helps limit the exposure to personnel in the immediate vicinity to very safe levels.

Using X- or gamma-radiation, both thickness and density affect radiation transmission, as follows:

$$I_x = I_o e^{-\alpha\omega t} \quad (1)$$

where: I_x = intensity of the transmitted radiation, roentgen/time; I_o = intensity of the initial radiation, roentgen/time; α = mass attenuation coefficient cm^2/g; ω = material density, g/cm^3; and t = material thickness, cm.

The radiation transmitted is a function of the initial radiation intensity with no absorber, I_0. If any of the variables in the exponential equation (α, ω or t) increase, the radiation absorption increases.

For a given energy of radiation and a specific material, α will be a fixed value that can be calculated or obtained from a handbook (such as the NDT handbooks published by the American Society for Nondestructive Testing). Its value can only change when the radiation energy or material is changed.

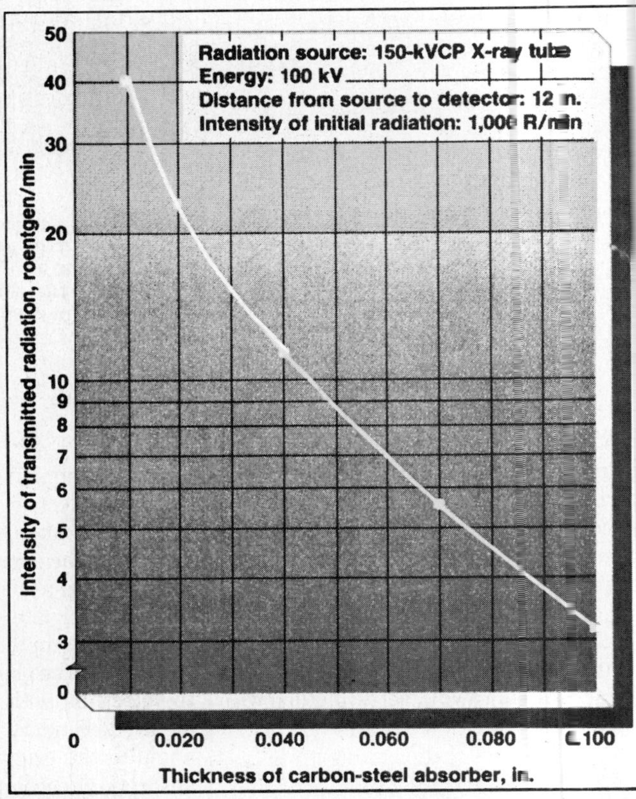

FIGURE 4. The intensity of transmitted radiation is a function of material thickness

Either of the variables t or ω can be measured if the other is held constant. If thickness is known or can be measured (e.g., with a micrometer), density can be determined as follows:

$$\omega = \ln(I_x/I_o)/(-\alpha t) \quad (2)$$

If the density is constant or is known, thickness can be determined from a similar equation:

$$t = \ln(I_x/I_o)/(-\alpha\omega) \quad (3)$$

Another way to obtain a thickness correlation is to plot, as shown in Figure 4, the intensity of the transmitted radiation, I_x, as a function of thickness or density

Radiation thickness gages are used in rolling mills to control the thickness of metal sheet. Automatic feedback control brings the rollers closer together or farther apart depending upon the reading.

In analyzing a process tower, the readings taken represent the radiation transmitted through the total mass between the source and the detector. This transmission is inversely proportional to the mass. High radiation readings mean low mass, and vice versa. Thus, the process engineer with a knowledge of the particular tower and the radiation profile at each elevation can determine what is or is not functioning properly in the tower.

In this application, density and thickness are measured using a source of radiation and a radiation gage. Typically, one uses an isotope source, such as ^{60}Co, that has the proper energy to penetrate the walls of a thick vessel and the material within the vessel or tower at different locations. Such an examination in progress is shown in Figure 5.

FIGURE 5. Radiation-transmission equipment can be used to troubleshoot a tower

Ultrasonic examinations

Ultrasonic testing uses mechanical vibrations to probe material. As with radiation examinations, there are many different types of material that can be analyzed using this method.

Ultrasonic tests are performed with transducers, which convert a pulse of electrical energy to a mechanical vibration. The piezoelectric crystal converts electrical signals to mechanical vibrations and mechanical vibrations back to electrical signals. In this way, the ultrasonic wave is generated, sent out and bounces back from reflectors. On its return, the signal is converted back to an electrical signal that is amplified and displayed.

A discontinuity causes the ultrasonic beam to be reflected within the material. The reflected beam is a function of the incident beam's wavelength and the discontinuity's size and orientation. If the discontinuity is small (with respect to the wavelength), the reflected energy spreads out and is weak; if the discontinuity is large, a direct, high-intensity wave is returned to the detector.

The examiner measures the reflected beam's travel time and amplitude. Travel time reveals the discontinuity's location — i.e., how long it took for the incident beam to travel to the discontinuity and the reflected beam to return. The amplitude of the returned wave is proportional to the size of the discontinuity that caused the reflection.

There are several modes of sound and ultrasonic waves. The first mode is the longitudinal (or compressional) wave. Longitudinal waves are so called because the particles transmitting the waves vibrate in the same direction as the wave propagates. Longitudinal waves travel at the highest velocities of any of the common wave modes.

FIGURE 6. Ultrasonic search unit contains a transducer for converting electrical and mechanical signals

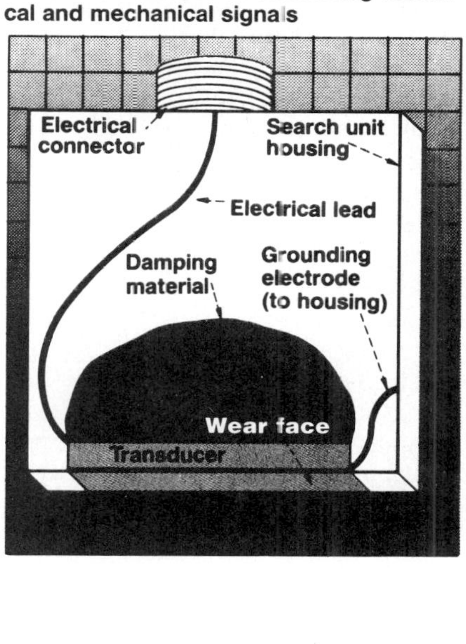

In ultrasonic testing, ultrasonic waves are generated by longitudinal-wave crystals (transducers), and a liquid couplant is used to aid the transmission of the waves into a material. This is needed because sound and ultrasonic waves need some medium through which to transmit vibrations. Air is the couplant for voice transmission. But for ultrasonic waves, air is a poor couplant, and a denser fluid is needed, such as water, oil or glycerin.

Thickness measurements can be made and laminations and planar discontinuities can be detected using longitudinal waves. Figure 7 shows a digital ultrasonic thickness-gage being calibrated on a step block (i.e., a block having steps that represent different thicknesses).

Another mode of ultrasonic wave that is used for discontinuity detection and measurement is the shear wave. With shear waves, the vibrations of the molecules transmitting the wave are transverse, or at right angles, to the direction of wave propagation. Although shear waves can be generated by special shear search-units, they are usually produced by the angular incidence of longitudinal waves onto a surface. Where waves traveling in two different materials have different velocities, refraction — or a change in the direction of the wave's propagation — occurs; in many instances, this is accompanied by

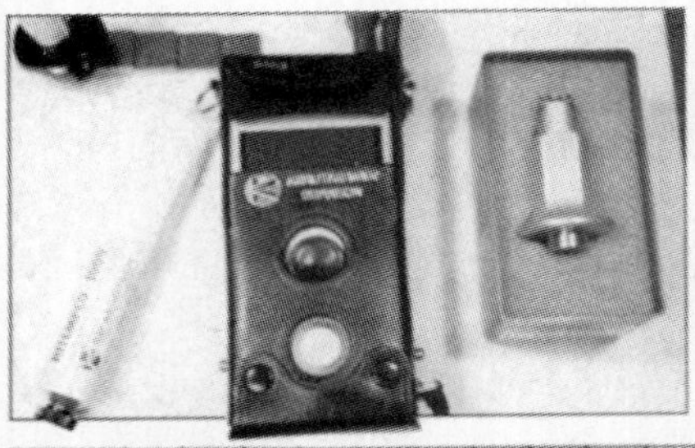

FIGURE 7. Digital ultrasonic thickness-gage is being calibrated on a step block

FIGURE 8. The angular relationships of ultrasonic waves in two media are governed by the velocities and angle of the incident wave (below)

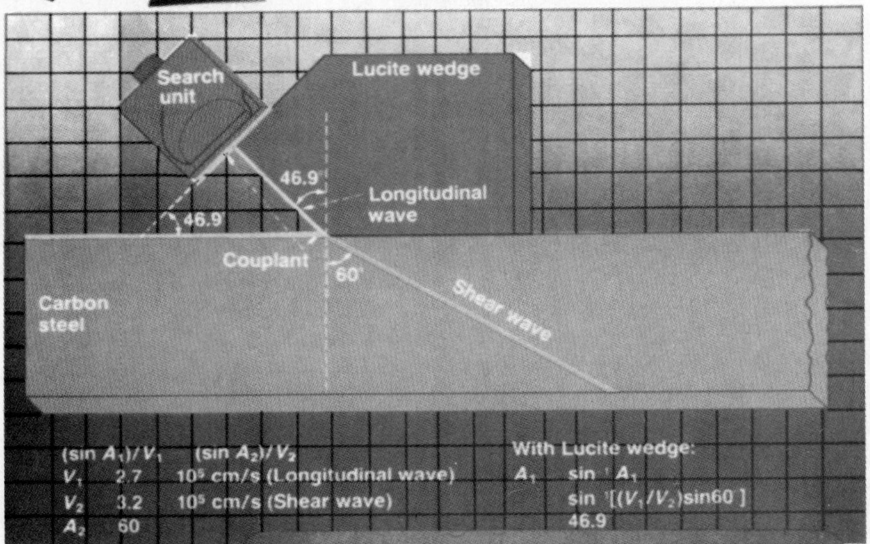

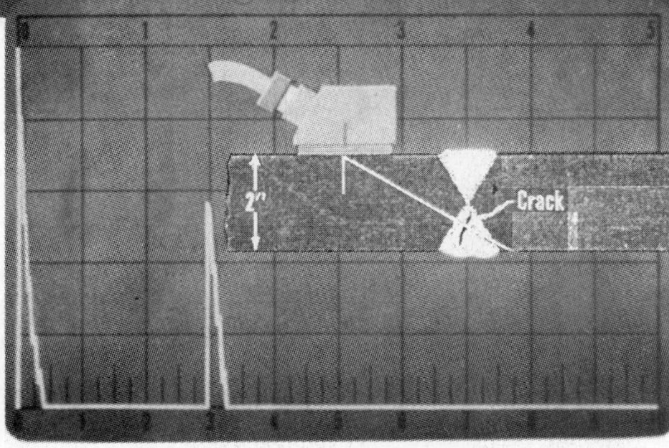

FIGURE 9. Results of a shear-wave inspection of a butt joint are displayed on an oscilloscope screen (at right)

conversion to a different wave mode. This refraction and mode conversion occur when a search unit is used in water, or, for contact methods of testing, by using a special wedge cut at a particular angle (as shown in Figure 8).

The angular incidence of waves in materials is governed by Snell's Law:

$$(\sin A_1)/(\sin A_2) = V_1/V_2 \quad (4)$$

or:

$$(\sin A_1)/V_1 = (\sin A_2)/V_2 \quad (5)$$

where: A_1 = angle of wave in Medium 1, A_2 = angle of wave in Medium 2, V_1 = velocity of wave in Medium 1, and V_2 = velocity of wave in Medium 2.

The angular relationships of the waves in media with two different velocities are shown in Figure 8. Because of refraction and mode conversion, two waves may be propagating in the second material at one time. The degree of angular change (between the incident and the refracted waves) is related to the ratio of the wave's velocities in the two materials. The mode-converted wave's angle is determined by the same relationship, but substituting the velocity of the mode-converted wave for V_2. By controlling the wave's angle of incidence, the second material can be examined at various angles to look for discontinuities that may be nearly perpendicular to the particular angle beam. (An angle beam is a beam that travels at an angle other than normal, or perpendicular, to the surface being examined. It is applied at an angle to the surface, enters the material at an angle to the normal, and refracts at a different angle.) Figure 9 illustrates an angle-beam shear-wave inspection of a butt joint.

The sensitivity of the ultrasonic method is dependent upon the wavelength of the ultrasonic beam. The wavelength is determined by:

$$\lambda = V/f \quad (6)$$

where: λ = wavelength; V = ultrasonic velocity; and f = frequency of the wave.

The smaller the wavelength, the greater the sensitivity or the ability of the examination method to find small discontinuities. A rule of thumb states that the wavelength must be at least approximately one half the size of the discontinuity sought. (The wavelengths in carbon steel of longitudinal and shear waves with a frequency of 1.0 MHz are about 0.23 and 0.13 inch, respectively.)

However, the smaller the wavelength, the less the beam penetrates the material. In some cases, the beam may not be able to penetrate to the depth required in the test object to do a valid examination. Thus, the frequency selected is a compromise between penetration and sensitivity.

The most common form of ultrasonic testing is the pulse-echo technique, where the same search unit produces the initial wave and then detects the wave upon its return to the unit. The time between the initial signal and returning wave is directly related to the travel time of the wave, which in turn is a direct function of the thickness of the material, couplant and wedges through which the wave has traveled. The portion of the wave that is returned to the search unit is a function of the size of the reflector that the wave struck and the reflector's orientation.

The ultrasonic test is calibrated by using materials of known thicknesses to calibrate for travel time (distance or thickness) and reflectors of known size to relate amplitude to the size of a discontinuity. Such reflectors could include flat-bottom holes (where the bottom of the hole is flattened rather than left as a point made by a drill bit), side-drilled holes parallel to the surface being scanned, so that one views the side of the hole), or notches.

Pulse-echo techniques are commonly used to make thickness determinations. Longitudinal waves are coupled into a material at normal incidence (i.e., perpendicular to the entry surface). In a homogeneous material, the wave travels at a constant speed. Although the wave travels at a relatively high velocity (over 13,000 mph in iron or steel), its travel between surfaces can be measured with high precision. Thicknesses as small as 0.020 in. up to several feet can be measured. The pulses of ultrasonic vibrations that are transmitted into a material are returned to the same or a second search unit (containing the transducer(s)), where they are converted to electrical pulses. The times involved in the travel of the waves are displayed on an oscilloscope screen or measured with special circuits and displayed on a meter. These times can be directly related to the thickness, so that the time measurements are actually displayed in thickness units. The display may be a meter or digital gage, or it may be shown as divisions, calibrated in thickness dimensions, on the oscilloscope screen.

Often for corrosion measurements, it is better to have an instrument that has both an oscilloscope screen and a digital meter. Many digital devices will not work until an ultrasonic signal reaches some preset amplitude. By observing the signals on the oscilloscope screen before making the digital measurement, one is assured that lower-intensity signals are not missed during the test. This, in turn, means that small minimum wall-thicknesses will not be missed.

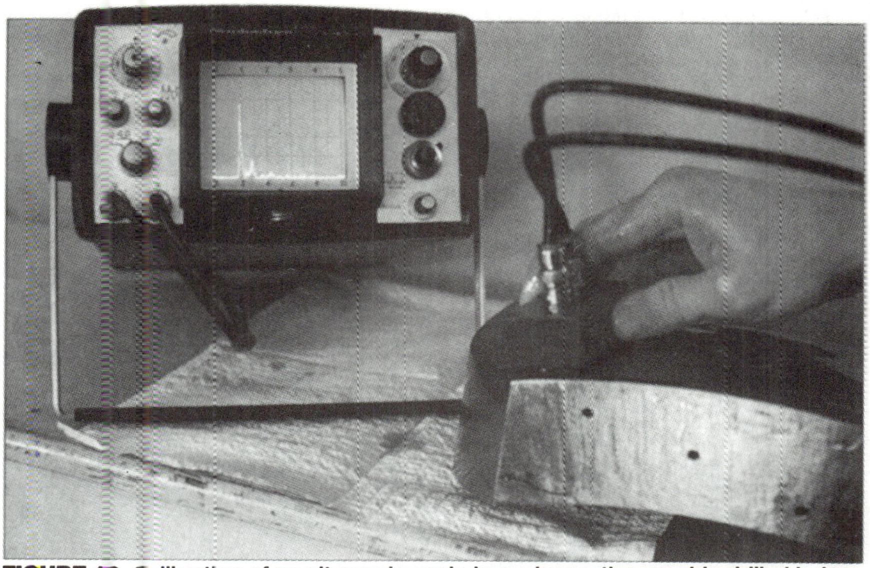

FIGURE 10. Calibration of an ultrasonic angle-beam inspection on side-drilled holes

Eddy-current testing

Eddy-current testing has a variety of applications in the chemical process industries. It is often specified for evaluating pipe and tubing during its manufacture, and it has many uses during the operational phase of a plant's life. A limitation of this test method is that it can only be performed on electrically conductive materials.

Eddy-current testing is applied to conductive materials using helical coils. High-frequency electrical current flowing within a coil induces a magnetic field, which in turn induces eddy currents within the wall(s) of the material being tested. Coils can be made so that they are placed flat on a material, encircle that material, or are put inside a part (e.g., a pipe). A single internal coil (Figure 11) can measure tubing thickness. A differential-coil probe (Figure 12) — consisting of two coils, for which the tester compares the results — can find discontinuities in tubing.

The test circuit can be thought of as a transformer circuit. The high-frequency current in the coil produces a magnetic field in the part being tested. The eddy currents induced by this magnetic field within the part flow in opposition to the path and direction of the coil currents. The eddy currents, in turn, induce a magnetic field that opposes the magnetic field of the coil. The coil can be thought of as the primary circuit of a transformer, while the eddy currents flow in the secondary circuit of the transformer.

The flow of eddy currents will change due to discontinuities or thickness changes. The penetration of eddy

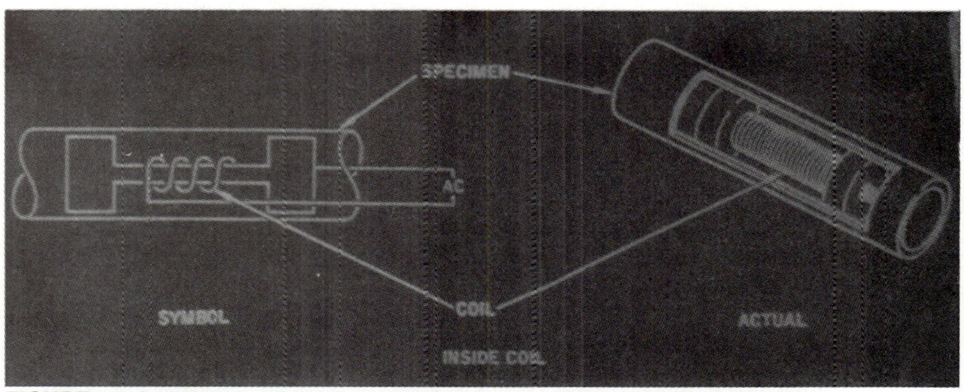

FIGURE 11. An internal coil can measure the thickness of heat exchanger tubing

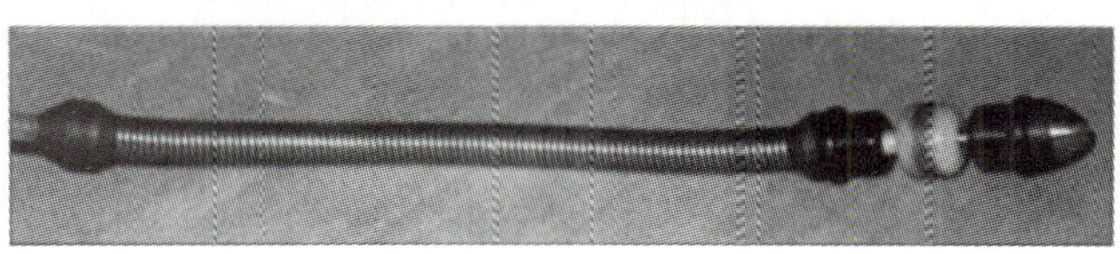

FIGURE 12. Differential coils are used for detecting discontinuities in heat exchanger tubing

currents is inversely proportional to the square root of the product of magnetic permeability, electrical conductivity and frequency of the inducing currents.

The electrical conductivity and magnetic permeability are fixed for a particular material; therefore, the frequency must be selected based upon their values and the thickness of the test object. If one wants to measure thickness, the magnetic permeability and electrical conductivity of the material must remain constant. If the magnetic permeability, electrical conductivity or the frequency increase, the depth of penetration will decrease. The higher the electrical conductivity, the lower the frequency that must be used and vice versa, as their product must remain constant in order to maintain the same penetration of the eddy currents.

Eddy currents are densest adjacent to the coil that generated them, since the magnetic field is most dense at this point. The eddy currents that are deeper in the test object are both smaller in strength and lagging in phase (compared to the currents near the surface). This so-called "skin effect" is even more pronounced when frequency, magnetic permeability or electrical conductivity is increased.

The test is more sensitive to discontinuities on the surface nearest the coil than to ones starting at the opposite wall. In tubing examinations, the discontinuity penetration of the wall (i.e., the percentage of the wall thickness through which the discontinuity extends) is determined by measuring the phase angle. If the phase angle is less than the angle representing a 100%-through-wall defect, the analyst can tell from which surface the discontinuity originated.

In addition to being a good inspection method for heat exchanger tubing, the eddy current method can often help in determining the cause of problems. Small cracks originating from the outside or inside surfaces can be separated from each other. Also, the analyst can determine the depth of wall penetration. If the notch shown in Figure 13 had originated from the inside surface, it would be located counterclockwise from the through-wall hole. The exact location of an indication along a tube can be determined from a stripchart recording, as shown in Figure 14.

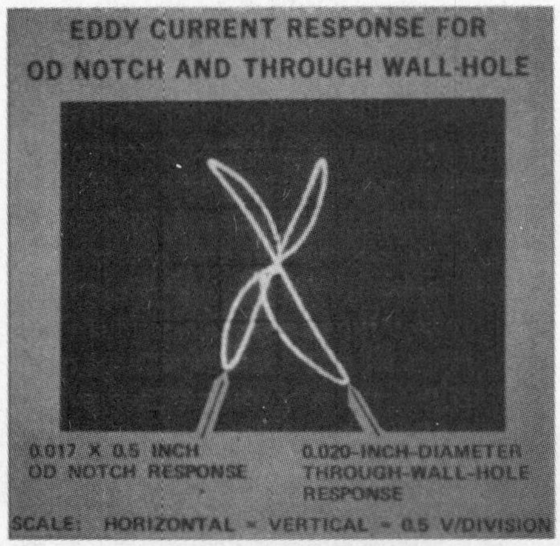

FIGURE 13. Eddy-current signals of a through-wall hole and a 50%-of-wall-thickness notch on the outside surface of a tube (left)

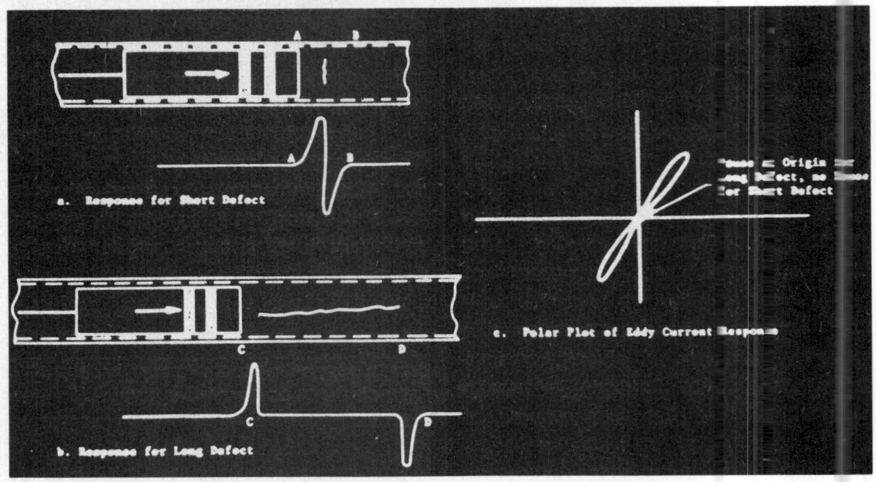

FIGURE 14. Stripchart recordings and displays on an oscilloscope screen reveal circumferential and longitudinal cracks (below)

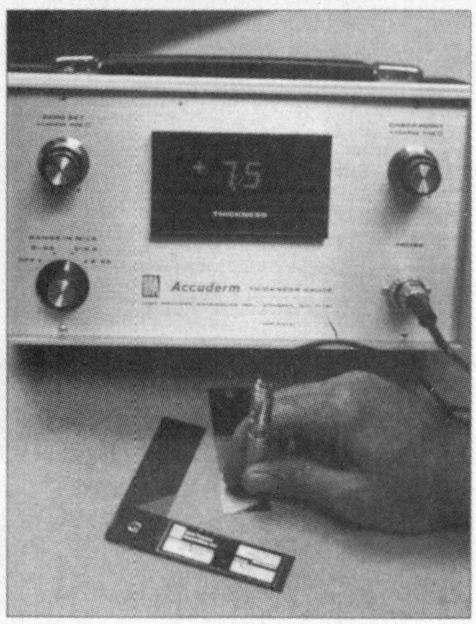

FIGURE 15. Calibration of a lift-off eddy-current test using plastic sheets of varying thickness (left)

Normally, cracks, voids and other lack-of-material discontinuities will decrease the material's conductivity by causing the eddy currents to flow around the void, thus increasing the resistance path for the eddy currents. Some eddy-current tests can measure a material's electrical conductivity. In this way, materials with sufficiently different electrical conductivities can be sorted. However, when sorting materials, the penetration of the eddy currents must be selected (by choosing the appropriate frequency) so that thickness does not become a variable.

The thickness of a nonconductive coating can be measured by using nonmagnetic spacers between the probe and the surface to keep the probe a fixed distance from the material. This distance is called lift-off. The use of spacers limits the amount of field that can be induced in the material. As the probe is kept farther from the material, the induced magnetic field becomes less and the eddy currents flowing in the material are lower in strength. ■

Looking ahead

The next article in this series is scheduled to appear in the November 1989 issue. It will cover newer nondestructive techniques, such as optical holography, acoustic holography, and acoustic emission testing.

FINDING SURFACE DEFECTS WITH NONDESTRUCTIVE TESTING

Visual, liquid-penetrant and magnetic-particle methods can detect surface or near-surface discontinuities

Frank J. Sattler*
Glistch Field Services/NDE, Inc.

This second installment of CE's Refresher† on nondestructive testing (NDT) covers methods that detect surface or near-surface discontinuities, as well as proof and leak tests. The surface-defects category includes visual, liquid-penetrant and magnetic-particle methods. Proof tests are generally hydrostatic and pneumatic methods that are used to "prove" that piping or a vessel can withstand certain conditions. Leak tests include several techniques to detect and locate leaks, as well as determine leakage rates.

Visual inspections can be done in a variety of ways, depending upon which code or standard governs the test. Generally, a visual examination involves observing conditions on the surface of a component, the alignment and separation of surfaces being joined, the component's shape, and evidence of leaking. Evidence of leaking is usually detected by pressurizing the component with water; this type of visual examination is covered under proof and leak tests.

Liquid-penetrant examinations use a carrier oil containing a dye to detect discontinuities. This liquid, called the penetrant, flows into discontinuities and is then revealed in subsequent processing steps. The various types of penetrants and dyes available, as well as different types of developers, provide a large range of sensitivities. Very small discontinuities can be detected when needed.

Magnetic-particle examination is a fast and sensitive method for examining ferromagnetic materials, such as iron, nickel and cobalt and some of their alloys. To perform the test, the material is first magnetized. Where discontinuities exist on and near the surface, a magnetic field (called the "leakage" field) is formed and acts like a tiny bar magnet. This leakage field is similar to the earth's magnetic field. Instead of a compass needle to detect the leakage field on the surface of the ferromagnetic material, small magnetic particles collect at the discontinuity, forming an indication that can be detected visually.

Proof and leak tests are other types of nondestructive tests used to determine structural integrity or leaktightness. In a proof test, usually a hydrostatic or pneumatic pressure is used to stress the component to some percentage of its design working pressure. During this test, a visual inspection is made of all joints and connections to make sure there are no leaks, and system pressure is monitored for any pressure drop.

Several different types of leaks tests are available, such as bubble emission tests, pressure change measurements, and inpsection with such instruments as halogen leak detectors and mass spectrometers. These techniques have different sensitivities (with sensitivity defined in terms of the smallest leak rate that can be detected), and some can detect very low leak rates — down to 10^{-11} std cm^3/s.

Visual NDT methods

Visual methods are commonly employed to inspect welds. They require several different tasks: These include the measurement of weld dimensions,

*For information about the author, see *Chem. Eng.*, April 1989, p. 119.

† The introductory article of this series on NDT — which covered what NDT is, what the various NDT methods are, and some of the kinds of problems NDT can discover — appeared in the April issue, pp. 116–119.

Originally published July 1989

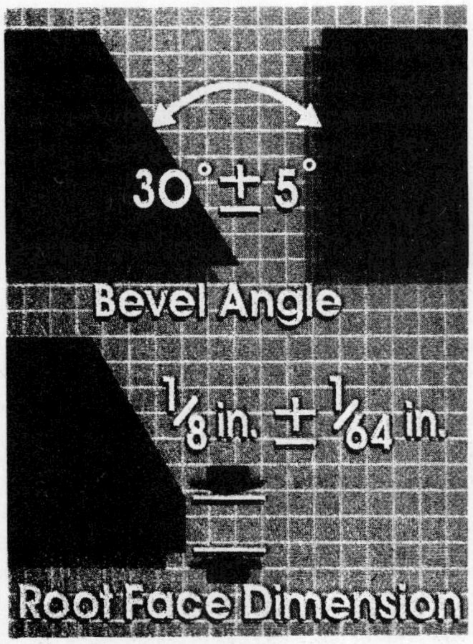

FIGURE 1. Bevel angle and root face dimension requirements for a weld butt joint are specified in the welding procedure

clearances, and other physical features prior to welding, as well as the verification that certain actions have been taken, which is done by observing the welding and by inspecting the weld afterward. Different requirements exist depending upon which code or standard is being referenced.

For example, the verification of joint preparation and cleanliness are two of the requirements of ASME/ANSI B31.3, Chemical Plant and Petroleum Refinery Piping [*1*], which covers the in-process visual examination of welds. In the case of joint preparation, the inspector is interested in such things as the groove bevel angle and the root face (also known as land, nose or flat) dimension defined in the written welding procedure (Figure 1). If the bevel angle is not sufficient, any resultant weld might contain incomplete fusion areas (Figure 2a). A low weld-preparation angle can make for a difficult tie-in of the weld metal to base metal. If the root face is too small, a burn-through of weld metal on the root pass (i.e., the first layer of the weld) can occur (Figure 2b); if the root face is too large, incomplete penetration of the weld (Figure 2c) could result.

Cleanliness of the weld joint and the absence of any contaminants are also essential to preventing porosity (or gas pockets) in the finished weld. If oxides, dirt, grease or oils are present on the surfaces to be welded, the high heat of the welding arc will cause vaporization and the formation of porosity (Figure 3). If the porosity does not have time to escape, or an overhead escape route from the molten weld pool is not available, the gas pockets become trapped, resulting in areas where there is less metal in the weld. Depending upon its size, shape, distribution, orientation and location, porosity can be detrimental to the weld.

These types of discontinuities might go undetected if only visual tests are performed on the finished pipe weld, so visual exams are often combined with other types of tests. It is essential to minimize the possibility of these and many other potential defects from occurring. The early detection of factors that can cause or contribute to them is just one part of a good inspection program. These inspections should apply to new construction welding, as well as to repairs of existing systems.

These previous examples are only a few of the types of things looked for in visual examinations. In addition to the above-mentioned tests, other welding visual exams (some of which are performed before welding, some during, and some after) might include the checking of:

1. Fit-up. This refers to the geometry of the joint (i.e., how far the faces to be welded are separated [joint clearance], and the parallelism between them [alignment])
2. Effectiveness (i.e., uniformity) of preheating, if required. Temperatures may be measured with a direct-reading pyrometer, or with special crayon-type detectors, paints or pellets that are placed on the preheated material and change color when the correct temperature has been reached
3. Welding position. The inspector watches the welder, specifically observing the direction the torch or rod takes during the welding, compared to what has been specified in the procedure for the weld
4. Electrode type, size and baking oven requirements
5. Condition of the root pass (i.e., the first bead of the weld, which is the most critical to the integrity of the weld), observed after cleaning
6. Slag removal and weld condition after each pass
7. Allowable weld reinforcement
8. Size, shape and appearance of the finished weld joint
9. Stress relieving (if applicable). To evaluate this, the inspector observes the stress-relieving operation to make sure that the correct temperature is reached and is held for the proper amount of time

Visual examinations of a surface can be direct or indirect, and can be conducted with either an aided or unaided eye. The angle between the eye (or the visual aid) and the surface being examined is usually equal to or greater than 30 deg., and (according to standard specifications) the distance between the two is usually less than 24 in. The examiner needs good visual acuity to see details with or without corrective lenses. The ideal distance for the observation may change depending on the

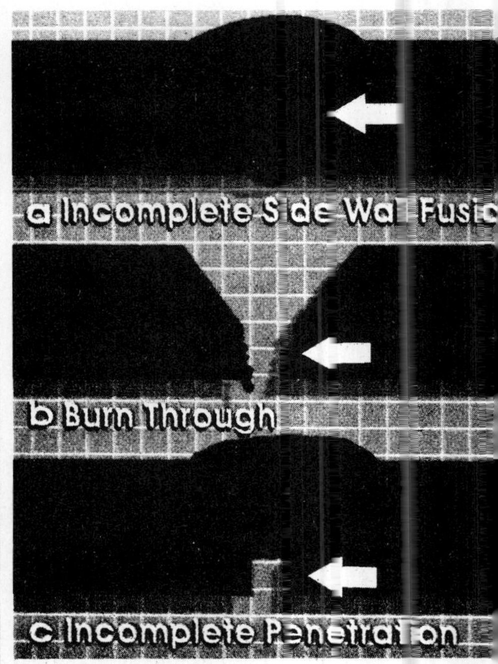

focus range of the individual examiner. Since the eye is unable to tolerate having something out of focus, a person will instinctively move to the distance that gives the sharpest image of the surface being viewed. Usually a person will have a distance range over which his or her vision meets some minimum near-surface resolution requirement. And, the distance range for most people will change with age.

Visual aids may be used to allow the tester to see inaccessible objects, to enlarge or magnify the view, or to take measurements. Visual aids include fiber-optic scopes, telescopes, microscopes, borescopes, small-diameter closed-circuit television systems, and such measuring devices as welding gages, micrometers and verniers. For example, a closed-circuit television system is often used for inspecting condensers. Such a unit is quite flexible and can be bent around tight corners if necessary.

Lighting is an important variable of visual inspections that should not be overlooked. Adequate illumination is required so the tester can visualize fine details. Small cracks or other features can be seen if the contrast from the image of the discontinuity or other feature is adequate and the detail in the image is sufficient to resolve small dimensions. Lighting with an intensity of approximately 100 foot-candles may be necessary for very small details. Contrast involves the differences in reflectivity of features, while detail involves the lateral dimension or distance over which the transitions take place. Both image features are necessary to resolve small discontinuities or other attributes.

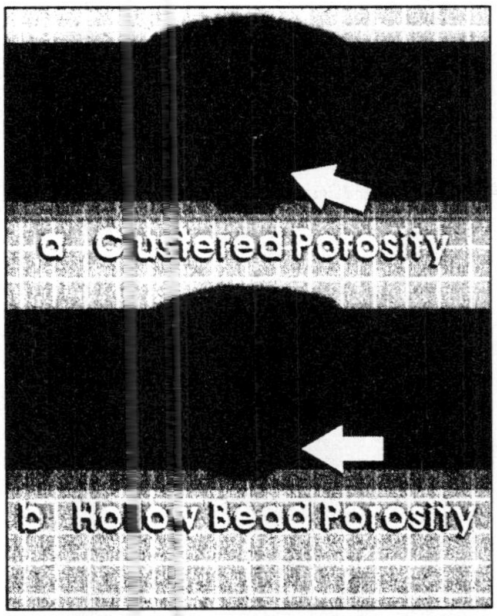

FIGURE 2 Various types of discontinuities can occur in pipe welds (left)

FIGURE 3 Porosity in welds can take one of two forms (above)

Liquid-penetrant NDT

Liquid-penetrant testing is usually thought of as simple, but it is also quite effective for finding surface discontinuities. It is very useful for nonmagnetic materials, where the magnetic-particle method (discussed later) cannot be used.

The penetrant consists of a light oil with visible or fluorescent dyes suspended in it. The penetrant flows into discontinuities by capillary action (Figure 4a), causing the liquid to move into openings in a material because of the adhesion forces between the liquid and solid. The adhesion forces are greater than the cohesive forces between the liquid molecules, causing the liquid to penetrate tiny openings in defiance of even gravity. After a period of time, the length of which depends on the material, penetrant and temperature, excess penetrant is removed from the surface, but remains in the discontinuities (Figure 4b).

Based upon the method of removing the liquid from discontinuities, penetrants are classified as water washable, post-emulsifiable, or solvent-removable. Water-washable penetrants contain an emulsifier that allows the excess penetrant to be removed by a water spray. (An emulsifier permits the mixing of two liquids that would not otherwise mix.)

Post-emulsifiable penetrants are removed with the aid of a separate emulsifier, which is sprayed on the component or is applied by immersing the component in a dip tank. From this point, the post-emulsifiable penetrant is removed in much the same way as water-washable penetrants.

Solvent-removable penetrants are generally used in field or shop applications where no water supply exists or where it is inconvenient to use water. Solvent-removable penetrants are harder to remove because the excess penetrant must be taken off by wiping the surface of the material with towels or rags several times. Solvent-dampened cloths can be used in the latter stages of removal, but much scrubbing is still needed.

TOBY ROSSER

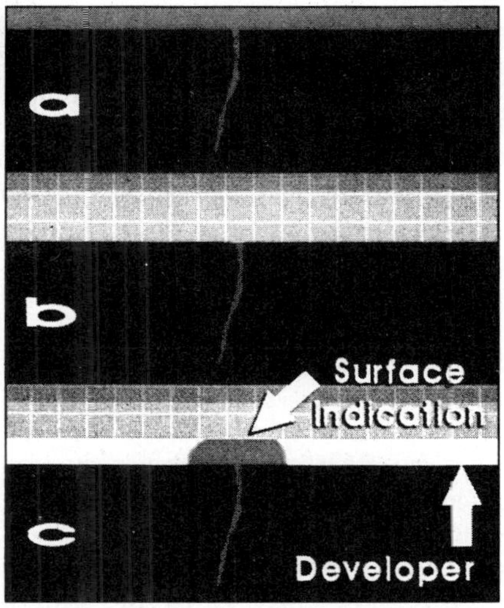

FIGURE 4. During a liquid penetrant test, the penetrant flows into a discontibuity and is trapped there (a), the excess is removed from the surface (b), then a developer is applied to blot out the penetrant to yield an indication (c)

After the penetrant is removed, a developer is applied. The developer acts as a blotter to bring the penetrant out of its entrapments (Figure 4c). The developer then spreads the penetrant over the surface near the discontinuity so that it is more visible.

Developers are chalky substances that are either water-based, dry or solvent-based. The water-based developers are either water-soluble (where particles are completely dissolved in water) or water-suspension systems (where the particles are suspended in water). The solvent-based developer has a solvent similar to a cleaner or remover solution in which the developer particles are suspended. Aqueous and dry developers have about the same sensitivity; nonaqueous developers are more sensitive because the solvent expands the liquid in the discontinuity and forces it into the developer coating on the surface.

Once the penetrant reaches the developer, a reverse capillary action occurs and the penetrant disperses in the developer's air spaces. Since the penetrant is gradually pulled from discontinuities, it takes a period of time for small reservoirs to produce a visible indication. The interpretation of the developer takes place 7–30 min after the developer is applied, at which time the size and shape of the bleedout is categorized and measured. Acceptance standards are based upon the type, size and distribution of indications.

There are various sensitivity levels available from the many different combinations of penetrants and processing conditions that one might encounter. For instance, penetrants are normally used at temperatures of 60–125°F. In this range, the viscosity and flammability of the oil carrier do not present any problems. At lower temperatures, the viscosity of oils increases, and although viscosity does not directly affect the sensitivity of the examination, it does require longer dwell times for the penetrant to enter and leave discontinuities. Thus, low-temperature applications typically require longer penetration and bleed-out times.

Higher temperatures may also affect the use of the penetrant as the carrier oil evaporates and flammability may become a problem. At temperatures below 60°F and above 125°F, one normally has to qualify the penetrant for the temperature.

Whether the dye in the penetrant is a fluorescent or a visible one also affects sensitivity. Fluorescent dyes provide contrasts of approximately 300:1 between the yellow-green fluorescence and a dark background. Visible dye indications against a white background provide about a 6:1 contrast ratio. Because the fluorescent penetrants have approximately 50 times greater contrast ratios, they can be seen much more easily, so that very small indications can be detected.

For most field applications in chemical-process-industries (CPI) plants, visible-dye, solvent-removable penetrants, along with solvent-based developers, are the most widely used. Generally, both the penetrant and developer are sprayed from a pressurized can. These types of penetrants can be used on piping and pipe welds made of austenitic stainless steels and other nonferromagnetic alloys, as well as on ferromagnetic metals where the access required for a magnetic particle examination is restricted.

The biggest advantages of using solvent-removable penetrants in CPI plants are their portability, and the need for only visible light. Fluorescent

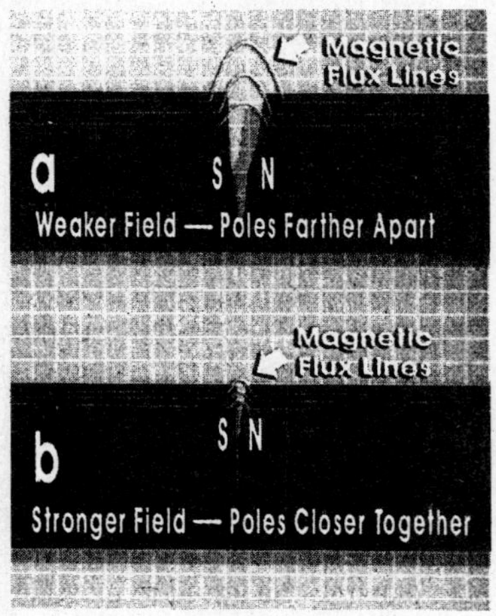

FIGURE 5. The leakage field from a wide discontinuity (a) is weaker than from a narrow one (b)

penetrants, on the other hand, must be viewed under black light with a limited white-light background, making a darkened booth or night work a necessity. In the case of stress corrosion cracks filled with corrosion products or fine narrow cracks, fluorescent penetrants may be required because of the long penetration times and poor sensitivities of the visible ones.

Performing the penetrant tests

The processing steps of a liquid-penetrant examination are fairly simple. For illustration purposes, we will assume that a solvent-based penetrant and a nonaqueous developer are being used. The procedures for other combinations differ only in the way the penetrant is removed and how the drying operation is carried out.

The first step is always the same — cleaning the part or material being tested. This is very important, since contaminants can prevent the penetrant from entering a discontinuity or they can hold the penetrant and cause a false indication. A strong solvent should be used for cleaning (most penetrant manufacturers provide one). After cleaning, a short evaporation time, usually on the order of 5 min, is needed to get rid of the solvent.

Next, the penetrant is applied. This is frequently done by using a pressurized spray can, but it may be done with a paint brush or spray gun, by pouring, or by dipping the part into an immersion bath. The penetrant must remain on the examination area for a specific period of time, usually 10–15 min, which depends primarily upon the type of materials, unless some tight cracks or stress-corrosion cracks must be found.

After the penetration time (or dwell time), the penetrant is first removed by wiping excess liquid off with a dry cloth or lint-free rag. As the penetrant is removed, the rag or cloth may be dampened with a remover or cleaner, and the surface wiped until no trace of the pen-

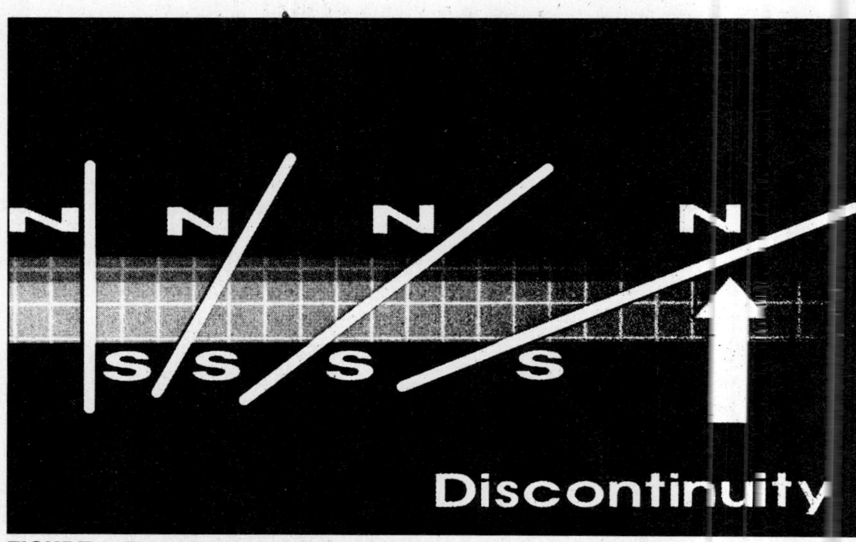

FIGURE 6. The separation of the leakage poles across a discontinuity is a function of the orientation of the magnetic field

etrant remains. This is almost impossible when there is surface roughness associated with a weld or casting. Normally, some background penetrant will remain but very small amounts are not likely to mask or interfere with true indications.

At this point, the developer is applied by spraying. A thin uniform coating is needed. Thick coats may mask fine indications, whereas a too-light coating may not provide enough blotting action to draw the penetrant from the discontinuity.

If possible, the surface should be viewed as the developer dries and the indications form. This will permit the interpreter to see whether excessive bleed-out occurs from several linear indications that may eventually converge and become rounded, appearing as a single indication. This could happen, for example, when there is a series of weld crater cracks that form a star-like pattern — after time, the bleed-out from the individual cracks merge, creating a single indication.

Indications must not be interpreted before 7 min have elapsed or beyond 30 min after the developer has been applied. When the interpreter has determined the cause and type of indication and measured its dimensions, the indications can be evaluated against the requirements of the acceptance standards.

Magnetic-particle NDT

Magnetic-particle examinations can only be performed on ferromagnetic materials, such as iron and steel. They cannot be used on such nonferromagnetic materials as aluminum, copper, titanium and lead.

Magnetic-particle testing methods are rather simple and they are fast and sensitive. They are based on the creation of a magnetic field within a ferromagnetic material, which magnetizes the material. When there is a discontinuity on or near the surface, there is a break in the continuity of the field. The magnetic flux lines are forced to the path of least resistance, which usually brings them out over the discontinuity and into the air. Where these flux lines leave and enter the material, they form magnetic poles of opposite polarity. These magnetic poles themselves form a small magnet, which attracts magnetic materials. Magnetic particles (which are small magnetizable materials) are used to detect the leakage fields, as they form indications consisting of accumulations of particles.

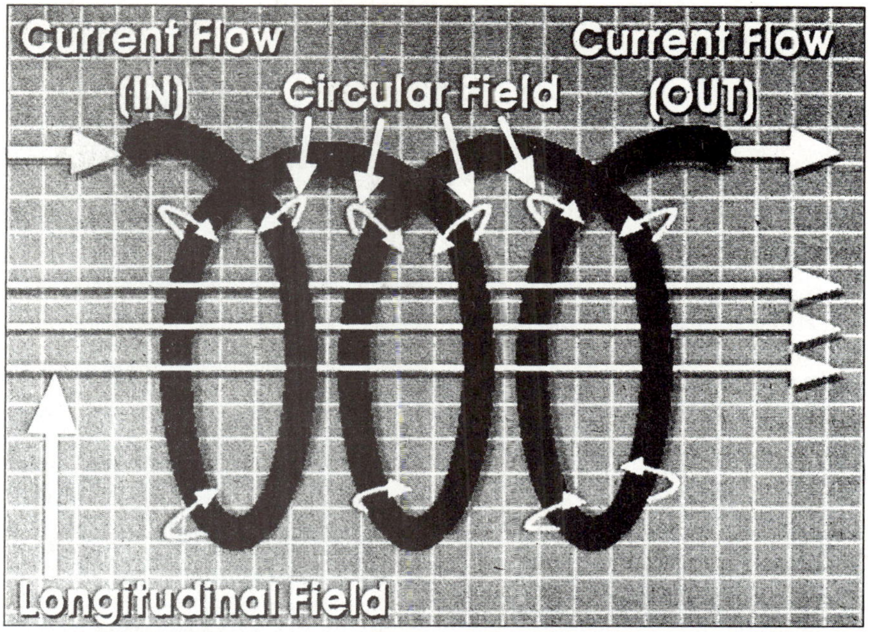

FIGURE 7. A longitudinal magnetic field is generated by the circular magnetic field that is caused by the current in the coil wire

The strength of a leakage field depends upon the number of flux lines that are forced out of the material. That number in turn depends on the strength and orientation of the magnetic field and the size, shape and location of the discontinuities. The closer to perpendicular the magnetic field is to the discontinuity and the nearer the discontinuity is to the surface, the greater will be the strength of the leakage field.

Figure 5 illustrates two surface discontinuities in a magnetic material — one wide (Figure 5a) and one narrow (Figure 5b). The narrow discontinuity will have the stronger leakage field because the leakage poles are closer together. For a given strength of magnetic field, the strength of the leakage field will be dependent upon the separation of the two leakage poles. Figure 6 shows the separation of the poles as a function of the orientation of a discontinuity. As the magnetic field is orientated away from perpendicular, the leakage poles are separated at greater distances and the leakage field is weakened. The higher the leakage field, the greater the attractive force for the magnetic particles that are sprayed or poured over the surface. The more perpendicular the magnetic field is to the discontinuity, the larger the number of particles forming the indication and the easier it will be for the inspector to detect the indication.

The magnetic particles themselves are particles of iron, certain steels, or certain iron oxides. These particles are sufficiently small to be acted upon by the leakage field — to be drawn to and held by the field. The particles may be dry and applied from salt-and-pepper-shaker-like containers or sprayed on. Wet particles may be suspended in light oil or water. The particles may be colored with a visible dye or paint for contrast with the object, or they may be coated with a fluorescent dye that will glow a bright yellow-green under black light. The contrasts are similar to those obtainable in liquid-penetrant examinations. As with the penetrants, the fluorescent particles can be seen more easily. In a visible magnetic-particle examination, the inspector needs a disciplined eye that must scan the surface to see the particles. When fluorescent magnetic particles are used, the brightness of the indications make them immediately apparent, so the deliberate scanning is not required.

Conducting the magnetic exam

Magnetic fields are generated when there is a flow of electrical current. If the current is strong enough, an adequate magnetizing force can be created within the material to be tested. When current flows, a circular magnetic field exists around and in the conductor (the piece through which the current is flowing), and is strongest at the surface of the conductor. If the conductor is ferromagnetic, the magnetic field is even stronger. A conductor can also be shaped into a coil (Figure 7) to produce longitudinal magnetism.

For magnetic-particle testing in CPI plants, two different types of equipment generate circular and longitudinal fields in the component being tested. Circular fields are produced by prods (Figure 8a). The prods, or current conductors, are placed in contact with the part of the object to be examined, and transmit current through the object. The magnetic field is perpendicular to the current flow from one prod tip to the other. The current is a direct current whose amperage is determined by the spacing between the prods. For materials less than ¾ in. thick, 90–110 amperes per inch of prod spacing is typical; for materials ¾ in. thick or more, 100–125 amperes per inch of prod spacing must be used.

Longitudinal magnetic fields are produced by a yoke (Figure 8b). The magnetic field is directed from one yoke leg to the other, which is opposite to the way a magnetic field is generated from prods. Since discontinuites can occur both along and across a weld, the use of prods or yokes requires at least two examinations, where the prods or yoke feet are aligned perpendicular and parallel to the weld. Some overlap of positions is necessary since the examination is not effective near the yoke feet or prod tips. The strength of the magnetizing force for a yoke is determined by its ability to lift a ferromagnetic weight. If the yoke uses a.c. power, it must be able to lift a 10-lb weight; if it uses d.c. power, it must be able to lift a 40-lb weight. The lift tests must be performed at the maximum yoke-leg spacing to be used in the examination.

For magnetic-particle examinations of such items as bolts, studs, shafts and nuts, a wet, horizontal, magnetic-particle examination system (Figure 9) could be employed. Such systems can produce circular and longitudinal fields from the heads and the coil that they contain. The strength of the circular

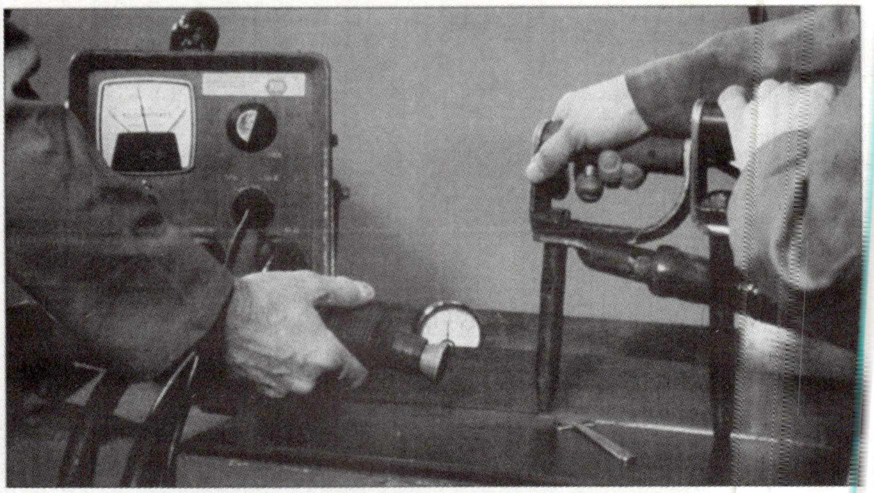

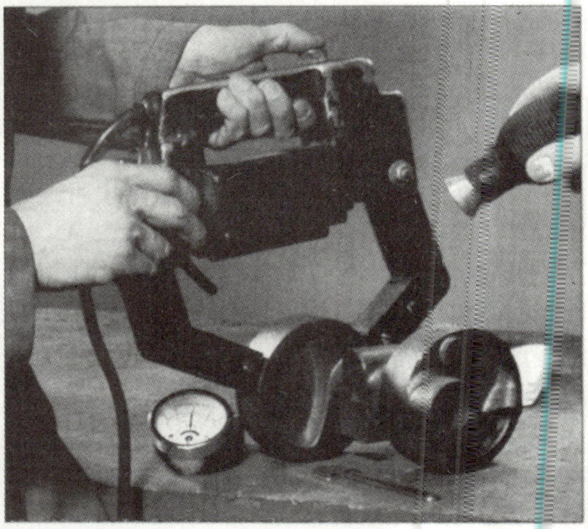

FIGURE 8. Prods (above) generate circular magnetic fields within the part, whereas a yoke (right) produces a longitudinal field

FIGURE 9. Wet, horizontal, magnetic-particle-examination equipment (below) contains heads (to the right of the inspector's hands) and a coil (to the left of his hands). In the foreground are various studs, nuts, bolts and washers that can be inspected with this unit

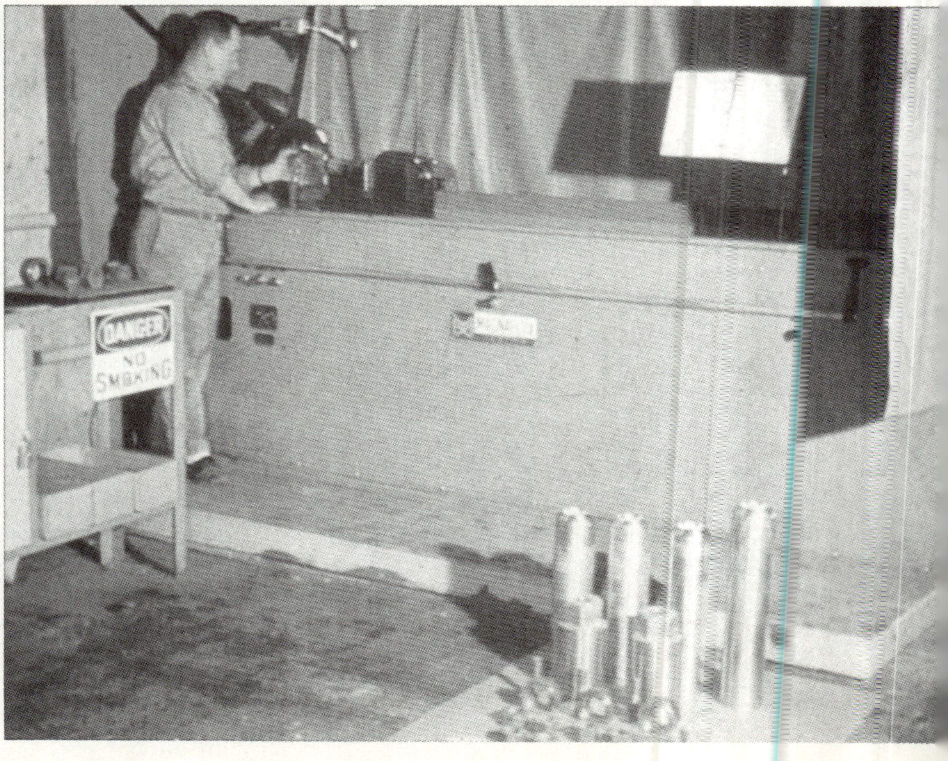

magnetization needed is a function of the part diameter; the longitudinal magnetic field is determined by taking into consideration the length-to-diameter ratio of the part and the number of turns in the coil.

A magnetic-particle examination has a few essential steps. First, as with a liquid-penetrant test, the area to be examined must be thoroughly cleaned.

The next step depends upon whether the magnetic particles are dry or wet. If wet particles are used, they are sprayed or poured on the examination surface, and then the magnetizing current is applied. If dry particles are used, the magnetizing current is applied first and then the magnetic particles are applied while the current is on.

With wet particles, the interpretation and evaluation steps take place next. With dry particles, any excess particles are blown off the surface while the magnetizing current is on (to hold any indications) and then the interpretation and evaluation can take place. Finally, the surface is demagnetized (if necessary) and any remaining particles are removed.

Proof tests

Proof tests are special types of tests that involve pressurizing piping or vessels to simulate operational stresses. As part of the proof test, a visual examination to detect any leaks may be required. The proof test can be performed with fluids, but the use of compressible gases can result in a large amount of damage should a failure occur. Usually, water is employed.

Various piping and vessel specifications and codes require proof tests. During the proof test, the component is stressed to some percentage of its design load. Pressure is monitored to detect any pressure drop, and the surface is checked for any evidence of leaks.

The test is performed before any operational stresses and conditions are present and usually serves as both a proof test and a visual examination for leakage. Because the proof-test pressure is higher than the design pressure, the test is not usually repeated.

Codes and standards applicable to CPI plants such as ANSI/ASME B31.3, Chemical Plant and Petroleum Refinery Piping, and ASME Boiler and Pressure Vessel Code Section VIII, Rules for Construction of Pressure Vessels, require proof tests. For instance, the latter (in UG-99 [*2*]) requires that every point in the vessel be able to withstand pressures at least equal to 1½ times the maximum allowable working pressure. Other means of determining a higher pressure may be required. Section VIII also requires an inspection of all joints and connections for leaks at a pressure of not less than two-thirds the test pressure.

Leak tests

Leak tests are a group of special nondestructive-testing methods that are used not only to detect and locate leaks, but to also quantify them. A leak implies a hole or passageway through which a contained fluid can escape. The identification of a leak may mean not just the identification of a single discontinuity, but rather several discontinuities that form a leakage path. The leakage rate depends upon the difference in the pressure across the pressure boundary. Different pressure differentials will produce several leak rates through the same path. Thus, both the pressure differential and the leak rate are needed to quantify the true leakage rate, and both should always be reported for the leak test.

Leak tests may be performed for a variety of reasons. One reason is to prevent or limit the loss of material, which can be hazardous and hamper system operation. Another is to maintain reliability in systems.

Leak rates may be defined in several different units. The most common units are Pascal-cubic meters per second, Pa-m^3/s, and standard cubic centimeters per second, std cm^3/s (in the latter, a pressure of 1 atm is implicit). These rates only differ by a factor of 10, with std cm^3/s being smaller.

Several different leak-test techniques are available, each with a different range of sensitivities. Only a few of the techniques will be covered here. For more information on the subject, consult Refs. [*3,4*].

Bubble emission tests

Bubble emission testing is one of the simplest leak-testing techniques. It can find small leaks in pressurized systems.

One version of bubble emission testing is an immersion bubble leak-test. The component is pressurized and then submersed in a liquid, and any stream of gas bubbles indicates the presence of a leak.

In another variation of the method, a bubble solution is applied to the surface after pressurization and the surface is viewed. The length of time for the application and viewing is important, and may have to be rather long to detect gas that has a long leak path. The leak-detector solution that is used usually consists of water with additives to give it better wetability. The solution should be neither acidic nor alkaline, and it should not react with the materials of the components being tested.

Many people are familiar with immersion bubble-emission testing, having seen it performed at the local service station on an inner tube or tire

In a third form of the bubble emission test, a vacuum box (Figures 10–12) is used. The apparatus consists of a box with a glass or plastic (e.g., Plexiglas) window and a pressure gage; in the bottom of the box there is an opening surrounded by a gasket. First, a leak-detector solution (resembling a soap solution) is applied to the surface of the object that is being tested. Then the vacuum box is placed on the object and connected to a pump, which draws a vacuum that seals the box tightly agains the surface. The vacuum pulls air from any leaks that may be present, causing a continuous stream of detector-solution bubbles to appear inside the box. If no continuous bubble forma-

FIGURE 10. Vacuum box leak test is performed on a butt joint in plate sections

tion is seen, the area being tested is acceptable.

The vacuum box can be used with or without internal pressure in the component. Sensitivity will be better with greater differential pressures. Test sensitivities of 10^{-3} Pa-m^3/s can be achieved. For pressure differentials of approximately 1 atm, sensitivities of 10^{-2} Pa-m^3/s are possible. Pressure differentials of 2 psi or more are necessary for an effective vacuum-box bubble test.

The quantity and diameters of the bubbles will be a function of the leakage rate. The bubble solution should not be allowed to dry or evaporate. In addition, if there are contaminants on the surface, false indications are possible.

FIGURE 11. This vacuum box is used to inspect straight corner joints

Pressure-change testing

Pressure-change measurements can also be used to determine leakage rates. These tests cannot be performed at atmospheric pressure, since a change in pressure is necessary for leakage to occur. Pressures either above or below atmospheric can be used.

This technique does not necessarily locate the leak path(s) or sites, but does measure total leakage. The technique is better suited for large systems, but can be modified for smaller ones.

The leakage rate of a system is equal to the total volume of the system times the pressure change observed, divided by the time interval of the test. In the Système International d'Unites (S.I.) system of units, system volume is measured in m^3, pressure change in Pa, and time in seconds.

FIGURE 12. This vacuum box checks the intersection of two corner joints

In one version of the technique, the flow-measurement technique, gas pressure is maintained constant in the system. The amount of gas added or removed from the system is used to determine the leakage rate for the specific period of time.

Measurement sensitivity and accuracy depend upon many factors, including the ability to read the minimum change in pressure that produces the maximum acceptable leakage rate. Static-pressure measurements are made at the beginning, periodically throughout, and at the conclusion of the test. If the time interval between measurements is too short, accuracy may be affected.

Temperature changes can cause pressure variations without any leakage, and must be accounted for in any calculations. Barometric pressure changes can produce pressure changes both outside and inside a system, and these must be reckoned with in any pressure-change measurement test. Finally, all absolute-pressure gages and other instrumentation must be properly calibrated.

Halogen diode detectors

Halogen gases contain fluorine, chlorine, bromine or iodine. Because of the toxicity of the elemental halogens and the low sensitivity of the diode detectors to them, these gases are not used as tracers in halogen leak detectors. Rather, the refrigerant gases R-12 (dichlorodifluoromethane) and R-22 (monochlorodifluoromethane) are used, as they are are non-toxic and inert compounds and are liquids at moderate pressures. They can be obtained from refrigeration vendors in most localities, and have been found to provide exce-

Chicago Bridge and Iron

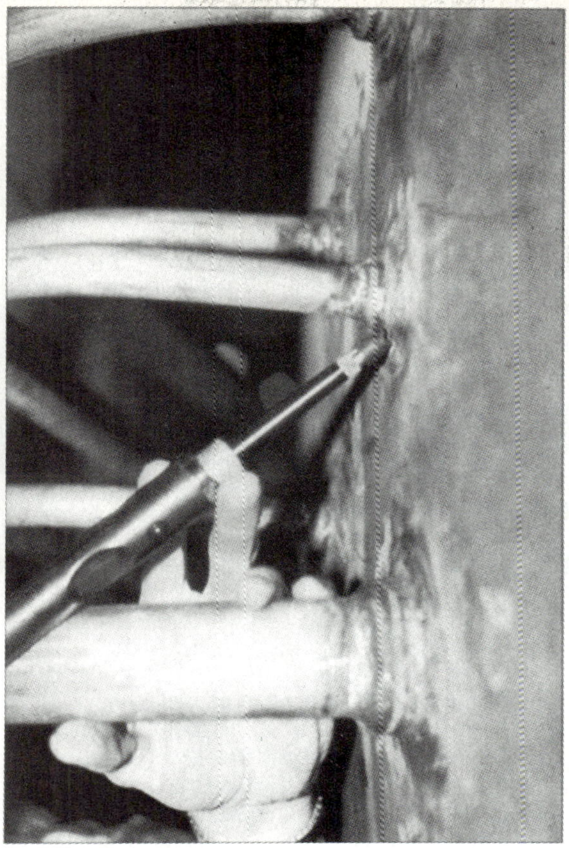

FIGURE 13. A double-walled liquid cryogenic vessel undergoes halogen leak testing (above); close-up of a sniffer probe inspecting a weld (right)

lent results in leak-test applications. However, with the increasing concern over the depletion of the earth's ozone layer by chlorofluorocarbons, such testing may become less and less common in the future.

Two techniques for applying the halogen gas to the vessel or system to be tested are available. The gas may be introduced internally and pressurized, and a sniffer probe moved over the outside surfaces (Figure 13). This is termed a static test. In a dynamic test, the component is connected to a vacuum line for evacuating the system, and a halogen sensor is placed inside this line. The halogen gas is then introduced on the outside surface, while the internal gas is removed by a pump and any leakage measured.

The detection of the halogen leaks is accomplished by using either a halide torch or a heated anode detector. The halide torch can be used where the refrigerant is mixed with air internally and pressurized. It has a burner supplied with fuel that is free of halides and provides a flame that changes color in the presence of a halogen. An air intake tube is used to search out suspected leakage areas. When some of the halogen gas burns in the torch, the normally blue flame will turn green. The sensitivity of the torch to leaks is approximately 10^{-5} Pa-m^3/s.

A heated anode detector has a hot platinum-and-ceramic heater element that emits positive ions when exposed to halogens. These ions are collected on the cathode and provide an electrical signal current to indicate a leak. Ion emission rates are a function of temperature, area, and nature and purity of the anode. The sensitivity of the heated anode detector operating in air at atmospheric pressure is approximately 10^{-10} Pa-m^3/s. Because the anode is heated, these detectors cannot be used with explosive gas mixtures.

Mass spectrometer tests

In leak detection with a mass spectrometer, a helium tracer gas is passed through a pressure boundary, and its subsequent detection indicates leaks. There are several methods of mass-spectrometer leak testing that differ in how the tracer gas is introduced and then detected in a system. Two of these methods have the most applicability to plant environments.

In the tracer probe method of testing, a piping system or vessel is evacuated and hooked up to a mass spectrometer helium detector. The helium is introduced through a small probe on the outside of the system. The probe is scanned over the various areas to locate leaks.

In the detector probe method, helium is placed on the inside of the system and the system is pressurized. The detector probe is connected to the mass spectrometer and collects gas over the surface being scanned to locate leaks.

The spectrometer is a device which allows for the detection of the helium. The collected gas is drawn into an ionization chamber and electrons are accelerated into the chamber to strike the gas molecules, ionizing them. Once ionized, the positive ions are forced out of the chamber by repeller electrodes. These positive ions are forced out through a slit and then attracted toward a ground-potential plate. The lower-mass ions are accelerated more rapidly than the higher-mass ions. A magnetic field acting on the ions forces the ions on a different path (which depends on their speed), while directing only the helium ions to the collector path. The ion current caused by only the helium ions can then be amplified and calibrated. ■

Looking ahead

The next article in this series is scheduled to appear in the September 1989 issue. It will cover the use of NDT to find discontinuities and other defects that are deeper than the surface. These include the ultrasonic, eddy-current, and radiographic methods.

References

1. ASME/ANSI B31.1, Chemical Plant and Petroleum Refinery Piping, 1987, ANSI (formerly the American National Standards Institute) and the American Soc. of Mechanical Engineers.
2. ASME Boiler and Pressure Vessel Code, Section VIII, Div. 1, Rules for Construction of Pressure Vessels, 1983 edition, American Soc. of Mechanical Engineers.
3. McMaster, R. C., Nondestructive Testing Handbook, Leak Testing, Vol. 1, American Soc. for Nondestructive Testing, Inc., and American Soc. for Metals, 1982.
4. Palluzi, R. P., Testing for Leaks in Pilot Plants, *Chem. Eng.*, Nov. 9, 1987, pp. 81–85.

ONLINE DIAGNOSTICS MAKES MANUFACTURING MORE ROBUST

PART 1

Samir Padalkar
Quant Trading Co.
Gabor Karsai, Janos Sztipanovits
Vanderbilt University
Frank DeCaria
E.I. DuPont de Nemours

Chemical process plants are operating under increasingly tight budgets and production demands. Any time and production lost when a piece of equipment fails or a process fault occurs can be a costly drain on a plant's finances.

An online system to detect, diagnose and recover from such failures can be a boon in minimizing unnecessary downtime. Here we present a practical and efficient method that has been implemented in chemical process plants and power stations to diagnose and overcome potential process upsets.

The overall methodology is part of a software program called IPCS, or Intelligent Process Control System (*1,2*). IPCS is a model-based programming system for developing complex monitoring, control, simulation, diagnostics and recovery applications. The software has an easy-to-use graphical user interface that enables engineers with minimal programming experience to create their own applications.

In contrast to the "expert system" based approaches for diagnosing faults, this real-time system uses generic models of a plant and its processes, allowing the basic IPCS system to be used and re-used in different applications (box, p. 331). Generic models also simplify maintenance tasks, such as testing, validation, upgrades and improvements of the models.

Additionally, the system guarantees real-time performance, and so it is useful in time-critical situations. Finally, it can display and update diagnostic results quickly and comprehensibly, making it suitable for plant operations.

The system's fault-detection model is based on: alarms that are triggered when preset operating parameters are exceeded; and on associations between the alarms and failure modes. An efficient fault-diagnostic algorithm determines the primary causes of the failures, by evaluating structural, probabilistic and time-based constraints that are represented in the plant's failure-propagation model. The technique works well with complex and multiple data sources, by rapidly reducing volumes of information to the most significant results.

The method also models multiple strategies for recovering from equipment faults or process failure modes. These recovery strategies range from displaying explicit messages to operators, to switching on backup equipment, to restructuring the plant's operations. The latter, for example, may include operating some sections of a plant at a lower production rate, or involve changing control and monitoring strategies, or combinations of both.

Going real time

A real-time system for fault diagnosis should meet some stringent requirements (*3*), such as providing guaranteed response times. Further, a real-time system must be able to "reason" about time, since much information can be deduced from the time and sequence of important events.

The system should also perform nonmonotonically, that is, it should be able to change a diagnosis in real time, and discard old data in response to new information. A real-time system should operate reliably and continuously when the plant is running, and, in some cases, even after a plant shuts down. Finally, a real-time system must be able to accept input from unscheduled asynchronous events, such as faults, and be able to distinguish among the various types of faults (box, p. 332).

In the IPCS system, fault detection involves developing models of normal plant operations, and establishing the criteria that signal when normal opera-

Minimize downtime by detecting, diagnosing and recovering from process faults in real time

Originally published March 1995

tions are violated. The models cover different aspects of a plant's structure and functions, such as plant behavior, process monitoring systems, equipment and process faults, and recovery strategies.

Building the models

The first step in the design of a monitoring and fault-diagnostic system is to specify the models that describe a plant. Monitoring requires knowledge of the physical processes (material and energy flows) because they represent the basic functionalities in the plant. Monitoring also requires modeling the flow of information, such as the dataflows that compute process variables and check various contraints to detect failures.

DENISE EARDMAN

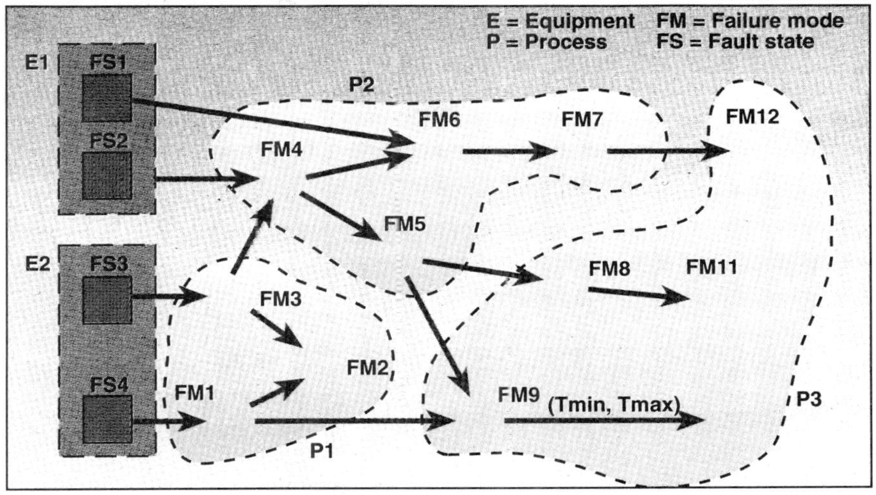

FIGURE 1. Fault states in equipment and failure modes in processes can spread from one section of a plant to another

The diagnostics system uses fault models to represent fault modes, fault propagation paths, and the physical structure of the plant (since a diagnosis has to be expressed in terms of a specific piece of hardware). These models are specified by describing various aspects of a plant and its operations, based on the following paradigms:

- Process models, which list the various functionalities in a plant
- Equipment models, which provide details of the hardware in the plant
- Activity models, which implement the desired monitoring, control, simulation, diagnostics and recovery functions, and include the required computational techniques

The process and equipment models describe what is known about the plant, while the activity models are good for specifying the software functions that need to be implemented. The models are presented in the form of various diagrams, and are stored in an Object-Oriented Database Management System (OODBMS).

To simplify the process of building complex models, IPCS employs various organizational principles, including hierarchy (to separate various levels of abstraction), multiple views, and references or links to other models. For example, process models (PM) can represent a plant's functionalities from several views:

1. Hierarchical process flowsheets
2. Mathematical models – that is, equations
3. Failure-propagation graphs
4. Discrete states and state transitions
5. Process and equipment associations

The view of the process and equipment associations has references to equipment models. A process model of a petroleum distillation operation, for example, may contain references to an equipment model describing the distillation column needed to perform the separations operation.

Equipment models (EM) have the following views:

1. Equipment structure
2. Discrete states and state transitions
3. Anticipated equipment faults

The following kinds of activity models (AM) are available:

1. Algorithms, for user-defined subroutines and libraries
2. Timers, for clock-related purposes
3. Simulations, for real-time simulation
4. External interfaces, for communicating with external data sources
5. Operator interfaces, for communicating with plant personnel
6. Compound activities, which are essentially hierarchical dataflow diagrams embedding other activities

The first five activity models are "primitives" in that they do not contain other activities. These primitives are essentially processing blocks that per-

MANY APPROACHES TO FAULT DIAGNOSIS

Traditionally, users have chosen a number of ways to implement real-time fault-diagnosis systems. The rule-based or expert-system approaches try to capture experiential knowledge about the dynamics of a plant, and try to mechanically imitate a plant operator's reactions to process upsets and equipment failures. This approach is generally applicable to small- or medium-scale plants, but poses problems for real-time diagnoses of larger units. The reason is that the size of the knowledge base needed for such plants becomes extremely large, complex and unwieldy.

Model-based systems follow a different path: They incorporate into the system in the form of mathematical or physical models information that is available or generated during the plant's design. These models can then be used for validating and interpreting sensor signals, for tracking changes in the state of a plant, and even for controlling the reconfiguration of the processing system.

One such approach uses the plant's structure and its normal functional specifications to identify fault sources. These systems can diagnose multiple faults, and avoid exponential growth in the size of the models. However, this technique, too, can pose problems for real-time diagnoses of some plants, because accurately specifying all the functional relationships existing in a plant is difficult. The technique is suited for use in modern plants, since many of them have known internal structures and well-defined operations.

Another model-based approach uses causal fault models and probabilistic inferencing for general diagnoses. Various causal models have been used in the chemical process industries: fault trees, signed digraphs, event digraphs and unsigned digraphs weighted with fault propagation probabilities and times. ❒

form a computation when new data become available on their inputs.

The power and practical applicability of the activity models is greatly enhanced by using references to process and equipment models. For example, if a process model specifies an equation that models a process, a simulation activity model can use this equation as part of a complex simulation. If the process model changes, the activity models – and the software that uses them – will also automatically change, and an error will be flagged if the user attempts to run an inconsistent application.

A key benefit of IPCS is its use of object-oriented programming techniques to design generic and highly re-usable fault models. For instance, a process model of a distillation column can be developed, refined and stored as a generic *type* that can be reused in different applications in a plant. The model types can be stored in a separate database, thus forming a library of types that can be distributed to all of a company's plants. An engineer can either use the model type as is, or can modify it.

The IPCS system consists of two main components, the Development Shell and the Target Shell. The development shell is used for building the models and, later, for editing the models to keep them updated.

The system's editing component consists of various specialized tools that provide a graphical interface – a so-called "visual programming environment" – for building the models. The use of graphics, however, is not exclusive; wherever necessary, text can be used, such as for equations.

The tools conform to the actual workstyle of design engineers who build specific monitoring and diagnostic applications. Process engineers, for example, can directly interact with the system, thus reducing the need for separate software to create the models.

The target shell defines the generic structure of the monitoring and diagnostic systems, and consists of the following primary components:

• The physical interface, which couples the IPCS to an external data-acquisition system, and receives and distributes the incoming analog and digital data among the various components of the moniotoring and diagnostic system

TYPES OF FAULTS

A typical plant may be subject to a number of faults and failures, each of which needs to be diagnosed correctly:

• Single or multiple faults, which can include failure of either a single or several components
• Propagated faults, where a fault in one component can spread to other processes, setting off many alarms
• Sensor failures, resulting from aging and degradation
• Feedback-loop failures, in which the effects of a component failure can be amplified

In addition, diagnosis of these faults is complicated by failures in specific phases of a plant, such as failure of a component used only during startup but not during normal operation. Another complication arises in time-critical applications, which must be capable of generating "rough" diagnoses that are then refined over time. ❑

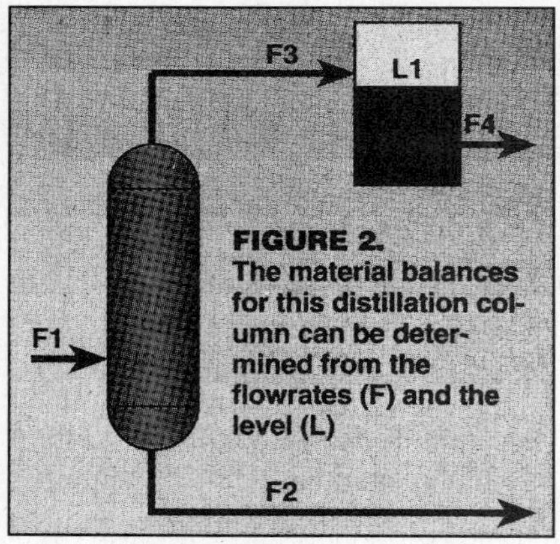

FIGURE 2. The material balances for this distillation column can be determined from the flowrates (F) and the level (L)

• The monitoring system, which includes data processing modules (such as for scaling, computing values of process variables, algorithms for fault detection, and so on), and a real-time database for storing the measured or computed process variables

• The diagnostic system, which interprets the incoming and computed alarms, and provides fault diagnostics in real time

• The operator interface, which provides different layers of access to the real-time database and to the current set of diagnostic results

• The simulation system, which includes various solvers for the numerical solution of mathematical equations

Each IPCS implementation incorporates these main components, but in each design they are configured differently, depending on the specific requirements of the plant. The IPCS program also includes a system integrator component, which builds executable applications from the models and the libraries of the target shell. This method, known as model-based program synthesis, is already being used in many areas (*4,5*).

Fault diagnostics

Fault modeling (*2*) begins by categorizing a plant's functions to generate a process hierarchy and a plant structure. The latter is then used to generate an equipment hierarchy. Each process in the process hierarchy has failure modes that represent violations of its normal behavior.

Then, by representing the causal relationship among the different process failure modes, a fault-propagation graph can be specified. In this graph, a fault-propagation link is weighted by a probability factor, and it has an associated time interval.

Fault propagations from equipment to processes can also be specified by linking equipment failed-states to process failure-modes. Similarly, alarms, which are generated from sensor values using suitable algorithms, can also be associated with process failure modes. An ON alarm triggers the diagnostic reasoning engine, which uses: the current configuration of the ON alarms; and the fault-propagation models to determine the primary cause of the fault. An example of a fault model is shown in Figure 1.

It is possible to dynamically change fault-propagation probabilities and the time intervals of any fault propagation link. This is done to handle switching from one phase of plant operation to another, for example, switching to a higher production rate. Hence, fault-propagation parameters can always be recalculated whenever the need arises, and all possible propagations with all possible parameters do not have to be entered into the fault model.

Recovering from an upset

Fault recovery is a complex problem because it requires effective coordination of several different areas, such as planning, reconfiguration, and rescheduling of plant operations. As a result, the fault-recovery method is not generic but open-ended, to enable users to add their own modeling concepts.

Some of the different kinds of responses that can be developed are:

• Switching to backup equipment

- Initiating actions based on an algorithmic activity, such as running a special computer program
- Reconfiguring monitoring or control strategies, or both

The finite-state modeling concept is the basic building block for switching to backup equipment and reconfiguring control monitoring strategies. Users can describe different states and the transitions between them, and can specify the actions to be taken in case a specific transition occurs.

For example, consider a plant that has a pump A and a backup pump B. The operating states can be Normal-Operation (pump A operational), Backup (pump A faulty, pump B operational), and Inoperational (both pumps A and B are faulty). The transition condition from Normal-Operation to Backup occurs when pump A is diagnosed as faulty, and from Backup to Inoperational when both pumps A and B are faulty.

Thus, when a transition is to be made from Normal-Operation to Backup, the action taken is to switch from pump A to the backup pump B. Similarly, control monitoring strategies can be switched from one strategy to another to reflect such changes in the state of a process.

Currently, users can take advantage of various model-building facilities in IPCS to build generic activities for switching to backup equipment and reconfiguring control monitoring strategies. Also, some fault modeling and diagnostic methods in IPCS have been designed to make it easy to specify and implement specific fault-recovery strategies. Examples include:

- Forewarning of critical process failure-modes
- Generating a specific event for each diagnosed fault source. These events often serve as inputs to fault recovery strategies
- Explicitly modeling redundant equipment, and explicitly modeling finite-state machines for activating redundant equipment
- Logging of current ON alarms and alarm-event histories

For example, if it is necessary for a fault-recovery program to make a switch from one piece of equipment to another, it would need to know on a timely basis the identities of the fault sources, a prediction of what may happen in the future, and the status of backup equipment. Thus, the diagnostics engine should supply enough information on the diagnosed faults and provide a forewarning of an incipient failure, to decide whether there is sufficient time for the chosen recovery strategy to be effective, or if there are enough non-faulty pieces of equipment to guarantee the execution of the recovery strategy.

A typical example

To illustrate the overall technique, consider a real-life application developed for a worldscale polyester-intermediates plant. The problem here is validating flow and level sensors, that is, diagnosing faulty sensors and then predicting the actual readings of the faulty sensors.

Multiple, interlocking material balances serve as the basis for fault modeling in this example. An example of the material balances on a section is shown in Figure 2.

This section consists of a distillation column with the overheads feeding a tank. Three material balances are derived from this section (based upon the law of conservation of mass, and assuming that there no leaks in the section):

Balance B1 is $F1 = F2 + F3$.

Balance B2 is $F1 = F2 + \Delta L1 + F4$.

Balance B3 is $F3 = \Delta L1 + F4$.

The flow sensors F1, F2, F3, F4, and the level sensor L1 are modeled, with each having a failed state. Violation of the material balances B1, B2, and B3 are modeled as process failure-modes (BV1, BV2, BV3).

Fault propagations are modeled from the failed state of each sensor to each violation of the material balances' failure modes. For example, fault propagations are modeled from the failed state of flow sensor F1 to the material balance failure-modes BV1 and BV2. Each fault propagation is weighted with a probability, and a time interval.

By monitoring each material balance, an alarm can be triggered if the balance is violated. Since each alarm is associated with a respective failure mode or fault propagation, users can isolate, for example, which specific sensor is faulty. Next, a prediction of the sensor's true reading is made in terms of a range determined from the material balance violations and the readings of the other non-faulty sensors.

The diagnostic application described here has 10 material balances, and monitors more than 35 sensors. Each sensor is modeled as having the failed states "read-hi," "read-lo," "stuck-hi," "stuck-lo," and "stuck." Additional failure modes and alarms are used to select between these failed states.

The results are communicated to the operator, and to the plant's process controller. Calculations based upon these sensor readings are then modified to include a correction factor that is determined from the predicted values. This IPCS application has already diagnosed at least three faulty sensors, and it and other similar applications have the potential of saving this plant more than half-a-million dollars per year. ■

Edited by Jayadev Chowdhury

Acknowledgements

The Vanderbilt University group was supported, in part, by Osaka Gas Information Research Insitute Co. Their kind support is acknowledged. The IPCS is currently being commercialized by OGIS-RI.

References

1. Padalkar, S. J., et al. Real-time fault diagnostics, *IEEE Expert*, June 1991, pp. 75–85.
2. Karsai, G., et al. Model-based intelligent process control for cogenerator plants, *Journal of Parallel and Distributed Computing*, June 1992, pp. 90–103.
3. Laffey, T.J., et al. Real-time knowledge-based systems, AI Magazine, Spring 1988, Vol.9., No.1., pp. 27–45.
4. Abbot, B.A., et al. "Model-Based Approach for Software Synthesis", *IEEE Software,* May 1993, pp. 42–52.
5. Wilkes, D.M., et al. "The Multigraph and Structural Adaptivity", *IEEE Transactions on Signal Processing*, August 1993, pp. 2,695-2,717.

Authors

Samir Padalkar is a senior software developer at Quant Trading Inc. (New York), where he develops integrated trading systems for the financial industry. He received his Ph.D. in electrical engineering from Vanderbilt University in 1990.

Gabor Karsai is assistant professor of electrical and computer engineering at Vanderbilt University (VU; Nashville, TN 37235). He has B.Sc. and M.Sc. degrees in electrical engineering from the Technical University of Budapest, Hungary, and a Ph.D. in electrical engineering from VU.

Janos Sztipanovits is professor of electrical and computer engineering at VU. He received his degrees from the Hungarian Academy of Sciences. He is a senior member of IEEE, and chairman of Intelligent Instruments and Measurement Systems of the Robotics and Automation Soc.

Frank DeCaria (B.S.Ch.E., West Virginia University), has worked for DuPont's polyester intermediates unit at Old Hickory. He works on applying fault diagnosis to online control systems.

ONLINE TESTING ASSURES INTEGRITY OF SENSORS

PART 2

J. L. Riner, H. M. Hashemian, K.M. Petersen and **D. D. Beverly**
Analysis and Measurement Services Corp.

Testing the performance and dynamic response of process instrumentation is increasingly being recognized as crucial to safety, such as in the event of a sudden change in process conditions. New regulations, such as the Process Safety Management Standard introduced in the U.S. in 1992, also stress the need for an ongoing program to assure the continued integrity of process instrumentation.

Experts estimate that 5% of the capital investment in large chemical process plants is for instrumentation, primarily because timely process data is critical to the operation and control of a plant. Much of this critical process information is provided by temperature, pressure, level and flow sensors. As a result, a plant maintenance program must identify and implement methods that can provide needed information on the health of plant sensors.

Here we present an overview of online techniques for testing the response time of resistance temperature detectors (RTDs), thermocouples and pressure transmitters. Online testing at process conditions provides information about the dynamic performance of sensors, such as the degradation that occurs over long operating periods.

TESTING RTDs

Conventionally, the response time of an RTD is measured in a laboratory using the plunge test. The test provides a baseline value of the response time that is useful for comparing a group of RTDs or RTD-thermowells, to ensure that they provide comparable and consistent response times. The plunge test is also useful for testing the response times of RTDs of various designs, or from different vendors, to determine which sensor meets a plant's needs.

In the plunge test, a sensor is exposed to a sudden change in temperature and its output is recorded until it reaches steady state. The time constant is then obtained by measuring the time that corresponds to 63.2% of the final steady-state value.

Although the definition of a time constant is analytically valid only for a first-order system, it is conventionally used for empirically establishing the response time of most temperature sensors regardless of the dynamic order. Therefore, all references to the terms "response time" or "time constant" made here correspond to this definition, regardless of the type or size of the sensor, the test conditions, or the test method.

However, since the response time of a temperature sensor is strongly affected by process conditions, plunge tests conducted in reference (laboratory) conditions do not provide accurate information about the sensor's response time after installation in a plant. To determine the response time during actual operations, an online technique, Loop Current Step Response, or LCSR, has been developed.

The LCSR test is performed by connecting the RTD to one arm of a Wheatstone bridge and changing the bridge current from a few milliamperes to a level of about 40 to 80 mA (Figure 1). The step change in current produces Joule heating in the RTD element and causes its resistance to increase in proportion to the RTD's ability to dissipate the heat to the environment.

Keeping process instruments healthy can help avoid plant failures

The transient change in RTD resistance produces a transient LCSR voltage signal at the output of the Wheatstone bridge. The LCSR transient is then fitted to a heat transfer model to yield the response time of the RTD. The technique simulates the conditions as if the RTD was exposed to a sudden change in the surrounding temperature.

The advantage of the LCSR test is that it permits online testing of installed RTDs and provides the actual "in service" response times of the RTDs. The test accounts for all installation effects on the RTD's time constant. This includes the effect of the thermowell (if one is used), the fit between the RTD and the thermowell, and all process conditions, such as temperature, pressure and flow.

AGING IN TRANSMITTERS *Degradation* • Causes	PERFORMANCE: CALIBRATION	RESPONSE TIME	TOTAL FAILURE
Partial or total loss of fill fluid • Manufacturing flaws • High pressure	Y	Y	Y
Degradation of fill fluid • Viscosity changes due to radiation and heat		Y	
Wear, friction, and sticking of mechanical linkages (especially in force balance transmitters) • Pressure fluctuations and surges • Corrosion and oxidation		Y	
Failure of seals, allowing moisture into transmitter electronics • Embrittlement and cracking of seals due to radiation and heat	Y		Y
Leakage of process fluid into cell fluid, resulting in temperature changes in sensor, viscosity & changes in fill fluid • Failure of seals • Manufacturing flaws • Rupture of sensing elements	Y	Y	Y
Changes in values of electronic components • Heat, radiation, humidity • Changes in power supply voltages • Maintenance	Y		
Changes in spring constants of bellows and diaphragms • Mechanical fatigue • Pressure cycling	Y	Y	
Blockage of holes in ceramic inserts in sensing modules or crimped capillaries • Normal aging • Manufacturing flaws • Mishandling		Y	
Drift of damping resistors • Thermal fatigue • Radiation effects • Vibration		Y	
Failure of transmitter electronics • Normal aging			Y

TABLE 1. Performance degradation of pressure transmitters due to aging can result from many factors

LCSR validation

The validity of the LCSR test depends on the following assumptions about the physical location of the sensing element in the sensing tip of the sensor:

• The heat transfer between the sensor and its surrounding fluid must be one-dimensional (radial)

• The sensing element must be located at the center of the sensor assembly, or there must be little heat capacity between the sensing element and the centerline of the sensor assembly

These assumptions must be satisfied for the heat transfer to and from the sensing element to be unidirectional and for the LCSR transient to be transformable to the plunge test transient. The only reliable and practical method to ensure that the LCSR test is valid for a specific RTD is to confirm the results experimentally.

The validation involves a plunge test followed by an LCSR test, both performed under the same test conditions. The LCSR data is then analyzed, and the response time compared with that of the corresponding plunge test result. A reasonable agreement between the plunge and the LCSR test results indicates that the sensors can be tested by the LCSR method.

DENISE EARDMAN

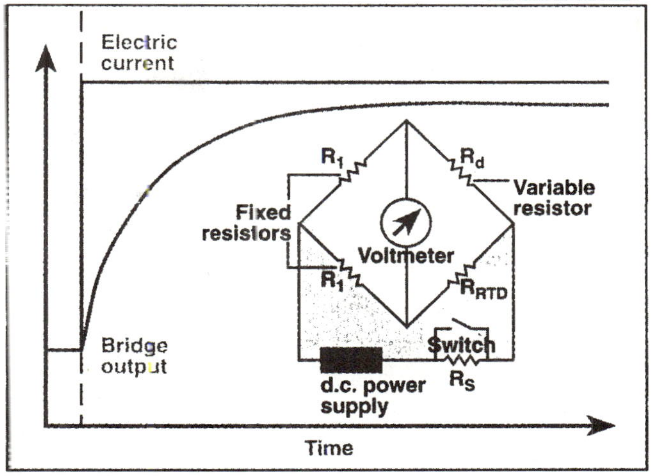

FIGURE 1. The equipment needed to test the response time of RTDs using the LCSR method

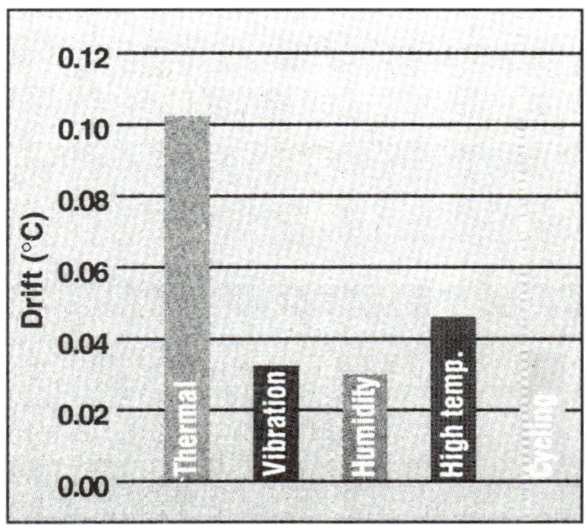

FIGURE 2. Many factors contribute to the aging of RTDs and their response times

TESTING TRANSMITTERS

Conventionally, the response time of pressure transmitters is tested with a hydraulic pressure generator to produce a test signal in the form of a step or ramp. The pressure test signal is applied both to the transmitter under test and to a high-speed reference transmitter. The outputs of the two transmitters are recorded on a dual-channel strip-chart recorder, to determine the response time of the transmitter.

Figure 3 shows how the response time (t) of a pressure transmitter is determined from the ramp test data. This response time value is sometimes referred to as the asymptotic ramp time delay.

In practice, however, the response time of a pressure transmitter is usually defined as the time delay between the response of the reference transmitter and the test transmitter as they pass through a setpoint. This response time should have nearly the same value as the response time shown in Figure 3, provided that the transmitter is linear and has a nonoscillatory response.

Noise analysis

Two online methods – noise analysis and power interrupt (PI) – have been developed and validated for response time testing of pressure transmitters. Noise analysis can be used for response time testing of most pressure transmitters, but the PI test is applicable only to force-balance pressure transmitters.

The noise analysis technique is based on analyzing the natural fluctuations that exist at the output of pres-

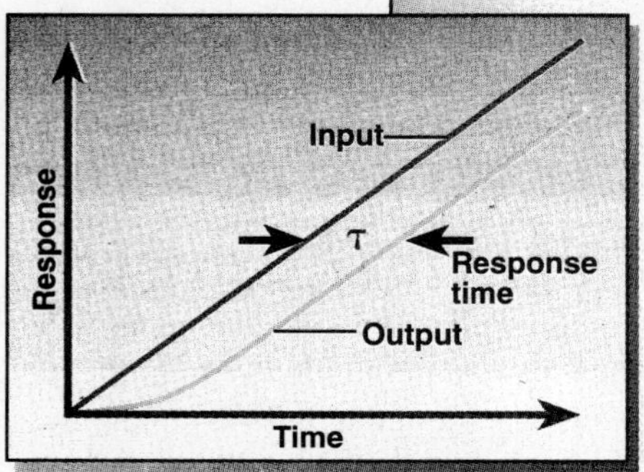

FIGURE 3. The ramp response method for calculating response time of pressure transmitters

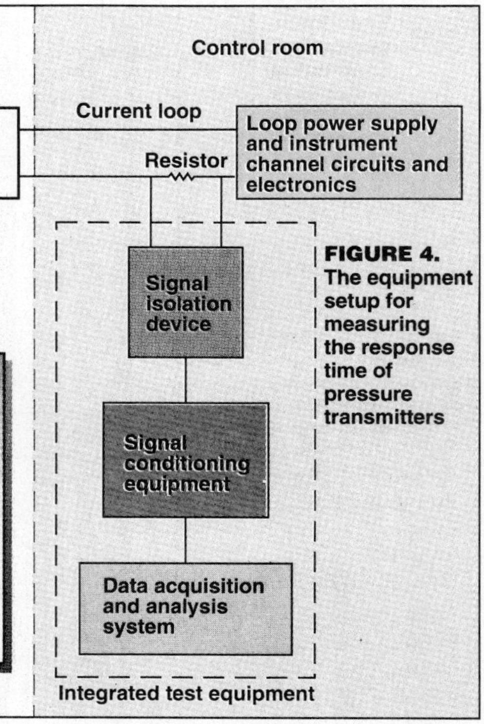

FIGURE 4. The equipment setup for measuring the response time of pressure transmitters

sure transmitters while the plant is operating. These pressure fluctuations (noise) are usually due to turbulence induced by the flow of process fluids, variations in process levels, and other naturally occurring phenomena.

The noise is superimposed on the d.c. output of the transmitter, then the d.c. component removed by passing the signal through a high-pass filter or by adding a negative bias to the output. This leaves the a.c. component, which is amplified and passed through a low-pass filter for anti-aliasing and removal of high-frequency electrical noise and interferences. The signal is then digitized by an analog-to-digital converter and analyzed.

A benefit of the noise analysis technique is that it is a passive test that can be performed remotely from the control room. The test does not interfere with plant operation, and can be performed on several transmitters at a time. The test takes about one hour for each group of transmitters, regardless of the number of transmitters tested at one time. Figure 4 shows the equipment set-up for testing each transmitter.

For each transmitter, 30 to 60 min of noise data are recorded and analyzed, to obtain the response time of the transmitter. The analysis of noise data can be performed in the frequency domain or the time domain, and is based on the assumption that the dynamic characteristics of the transmitter are linear. The frequency domain and time domain analyses are independent methods for determining the response time of pressure transmitters. It is helpful to analyze the data with both methods and average the results.

Sensor performance can degrade with age

For frequency domain analysis, the Power Spectral Density (PSD) of the noise signal is obtained through a Fast Fourier Transform (FFT) algorithm. An appropriate mathematical function is then fit to the PSD, from which the response time of the transmitter is calculated. The PSDs of pressure transmitters have various shapes depending on the plant, the transmitter installation and service, the process conditions, and other effects (Figure 5).

In the time domain analysis, the raw noise data is processed with a univariate autoregressive (AR) modeling program, to first obtain the impulse response (that is, the response to a narrow pressure pulse) and, then, the step response, from which the response time is calculated. After the data is analyzed in the frequency domain and time domain, the response time results are averaged.

The in-plant noise tests for each transmitter are often repeated for three frequency bands, from near d.c. to 10, 100 and 1,000 Hz, to cover the entire frequency spectrum over which the sensor's response time may lie. The data for each of the frequency bands are analyzed separately, the outliers rejected, and the remaining numbers averaged to provide a single frequency domain and time domain result. The two results are then averaged, to give the response time of the transmitter.

The validity of the noise analysis technique has been examined by laboratory testing of representative commercial units, and found to be within ±0.10 s of conventional ramp-and-step tests. Further, for in-plant tests, experience has shown that the noise analysis method generally provides conservative response time results, provided that the process noise that excites the transmitter is free of any interfering resonances.

Power interrupt

The second test, the power interrupt (PI), is a method for in situ response time testing of force-balance pressure transmitters. The test is based on a momentary interruption of the electrical supply power normally used to activate the transmitter.

The test is performed by turning the power to the transmitter OFF for a few seconds and then ON. As soon as the power is turned OFF, the lever system in the pressure transmitter deflects until it comes to a physical stop inside the transmitter.

When the power is turned ON, the force motor brings the lever system back to the equilibrium position. While the system is returning to the equilibrium position, the transmitter will output a transient signal that can be analyzed to give the response time of the transmitter. This response time corre-

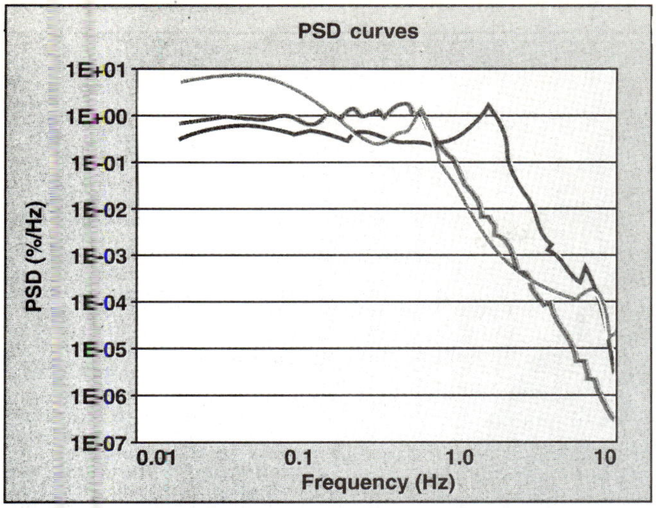

FIGURE 5. Typical Power Spectral Density (PSD) curves of pressure transmitters in nuclear plants

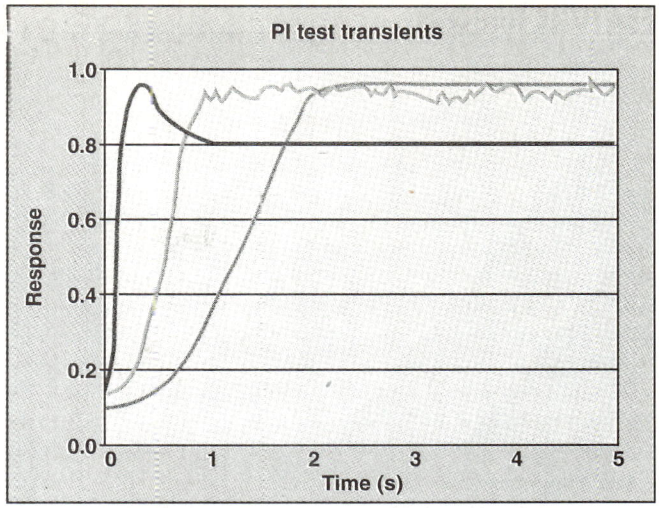

FIGURE 6. Typical transients resulting from testing the response times of force-balance pressure transmitters

sponds to the response time that would be obtained using the conventional step or ramp tests.

The PI transient typically begins with switching spikes followed by a ramp output. This lasts until the force motor has built up enough force to move the mechanical system of the transmitter away from the stop and back to the equilibrium position.

The rate of this ramp signal is referred to as the critical ramp rate. The critical ramp is followed by a transient that contains the response time of the transmitter. The transient is separated from the ramp output by a numerical algorithm. It is then analyzed by a mathematical fitting procedure to give the response time of the transmitter.

Figure 6 shows three commonly encountered PI transients observed in plants. The three PI transients include one that is clean and free of any noise or fluctuations, one that contains some low frequency oscillations, and one that has an overshoot.

The effect of the noise and fluctuations can often be reduced by repeating the PI tests several times, analyzing the individual tests, and averaging the results. In the standard procedure for the PI test, the test is usually repeated ten times for each transmitter, and the response time results of the individual tests are averaged to give the response time of the transmitter.

The PI test accounts for the response time of both the mechanical and the electronic components in the transmitter. The PI test has been found to be within better than ±0.10 s of the response time values obtained with conventional methods, provided that: the dynamic characteristics of the transmitters are predominantly linear; and the step or ramp pressure signals used are not so large as to cause the mechanical system of the transmitter to come to a physical stop inside the unit.

Aging characteristics

The performance of RTDs and transmitters are characterized by their accuracies and response times. In RTDs, accuracy is a measure of how well the RTD indicates a static temperature, and response time defines how quickly the RTD detects a temperature change. The accuracy and response time of RTDs are generally independent.

The deterioration of accuracy is called calibration drift or calibration shift, and the deterioration of response time is called response time degradation. Accuracy can be restored by recalibration (if the RTD is stable), but response time is an intrinsic characteristic that cannot be altered once the RTD is manufactured. In the case of thermowell-mounted RTDs, however, response time degradation due to movement of the RTD in its thermowell can sometimes be reversed.

Aging refers to response time degradation of RTDs with time in normal environments and under normal operating conditions. The time period varies depending on the application, but typically is the time between two maintenance overhauls, say 18 months.

Normal aging of RTDs occurs from longterm exposure to any combination of heat, humidity, vibration, temperature cycling, and mechanical shock. However, a significant calibration shift should not occur in an RTD as long as the sensing element is not stressed or contaminated after calibration and the insulation material is kept dry and in place. Any new stress, contamination, or metallurgical change in the sensing element, or moisture in the insulation material, can cause a calibration shift.

The effect of stressors, such as heat, vibration, humidity, thermal cycling and high temperatures on aging of RTDs has been tested (Figure 2). Temperature is the most important factor because sensor materials have different thermal expansion coefficients, causing the element to experience stress whenever the temperature changes. The resistance of the sensing element increases with tension stresses and decreases with compression stresses. For small temperature variations the stress reverses itself, but for large ones the effect is not reversible except by annealing.

Metallurgical changes in the sensing element, such as grain growth, can occur at temperatures above 420°C. Cold working (or work hardening) results from vibration and mechanical shock, and can be eliminated by annealing, which requires heating the RTD above 400°C.

The insulation resistance of an RTD decreases as moisture enters the sheath, and causes the RTD to indicate a lower temperature than normal. At high temperatures, moisture in the RTD is not normally a major concern because water vapor is likely to diffuse out of the RTD. However, since insulation resistance decreases significantly at high temperatures, any moisture remaining in the RTD can have a significant impact on the insulation's resistance value.

A major cause of response time degradation in RTDs is any change in the air gap of an RTD-thermowell. Changes of as little as a few hundredths of a millimeter in the size of the air gap, caused by vibration, shock and other mechanical effects during installation and operation, will change the response time significantly. If the RTD is spring-loaded into the thermowell, mechanical effects may change the insertion length or the contact pressure, increasing the size of the air gap in the thermowell and resulting in a response time increase.

Aging degradation also occurs in pressure transmitters exposed to stress for a period of time. Typical aging mechanisms that can cause a material's mechanical strength or physical properties to degrade include thermal, mechanical and electrical fatigue, wear, corrosion, erosion, embrittlement, diffusion, chemical reaction, cracking or fracture, and surface contamination. A few examples of the potential effects of some of the most dominant stressors for transmitters are listed in the Table (p. 85).

A traditional preventive-maintenance program essentially determines the probable frequency of failure of devices and systems, thus allowing them to be inspected, repaired or replaced before they fail. A mechanical-integrity program goes beyond preventive maintenance by utilizing test and inspection methods to monitor the health of plant equipment and instrumentation at routine intervals.

In fact, routine testing of temperature and pressure sensors can be performed remotely, such as from a control room, even while the plant is operating. Thus, the test methods described here are effective tools for routinely monitoring the normal aging and degradation of plant instruments, and to forestall their failure during operation. ■

Edited by Jayadev Chowdhury

Authors

J. L. Riner is an operations manager at Analysis and Measurement Services Corp. (AMS; 9111 Cross Park Drive, Knoxville, TN 37923; Tel: 615-691-1756) for the past four years. Previously, he worked for 15 years at Union Carbide in various positions, including business unit operations manager. At AMS, Mr. Riner has published eight papers on diagnostic testing and mechanical integrity. He has a B.S. in mechanical engineering from the University of Arkansas.

H. M. Hashemian, president of AMS, has been with the company for 17 years. He has published almost 100 technical papers. He has served as the principal investigator of six Phase I and three Phase II SBIR projects. He is a fellow of ISA, a senior member of IEEE, and a member of ASTM, among other societies. He has an M.S. in nuclear engineering from the University of Tennessee.

K. M. Petersen, vice president of AMS, has been with the firm for 15 years. He has published more than 25 papers. He is a senior member of ISA, and served as chairman of several ISA committees, and is the papers co-chairman for ISA's 41st International Instrument Symposium, to be held in May 1995. He has a B.S. in nuclear engineering from the University of Tennessee.

Dan D. Beverly, quality assurance manager of AMS, has been with the firm for eight years. He has co-authored several papers, and has taught several AMS seminars and short courses. He is a senior member of ISA, and a member of IEEE and ASQC. He has a B.S. in electrical engineering from the University of Tennessee.

INDEX